TEXTBOOK
OF POLYMER SCIENCE

TEXTBOOK
OF POLYMER SCIENCE

THIRD EDITION

FRED W. BILLMEYER, JR.

Professor of Analytical Chemistry
Rensselaer Polytechnic Institute, Troy, New York

A Wiley-Interscience Publication

John Wiley & Sons

New York · Chichester · Brisbane · Toronto · Singapore

Library of Congress Cataloging in Publication Data:

Billmeyer, Fred W.
    Textbook of polymer science.

    Includes bibliographies and indexes.
    1. Polymers and polymerization.   I. Title.

QD381.B52 1984   668.9   83-19870
ISBN 0-471-03196-8

Printed in the United States of America

10 9 8 7 6 5 4

PREFACE

"I am inclined to think that the development of polymerization is, perhaps, the biggest thing chemistry has done, where it has had the biggest effect on everyday life. The world would be a totally different place without artificial fibers, plastics, elastomers, etc. Even in the field of electronics, what would you do without insulation? And there you come back to polymers again." †

And indeed one does. From the lowly throwaway candy wrapper to the artificial heart, polymers touch our lives as does no other class of materials, with no end to new uses and improved products in sight. Yet, many instances of the need for better education in the polymer field, both in our universities and for the public, remain unchanged. Some of these were discussed at length in the Preface to the second edition of this book, which follows, and I shall not repeat them.

The present revision has two major directions. The first is to improve its value as a textbook. To this end I have rearranged the text to consider polymerization before describing the properties of polymers, a change that several of my colleagues feel has pedagogical advantages. I have also drawn on my files from 25 years of teaching polymer science, at the University of Delaware, the Massachusetts Institute of Technology, and Rensselaer Polytechnic Institute, to provide material for a section on Discussion Questions and Problems at the end of each chapter.

The second objective of the revision is the more common one, to bring the contents up-to-date by judicious addition, deletion, and revision, and in this I hope I have been successful. Many sections have been changed little, reflecting the maturity of certain aspects of polymer science, but the reader will find new material inserted in every chapter. A few additions of particular note are a section on polymerization reaction engineering in Chapter 6, a discussion of scaling concepts in Chapter 7, and expansion of the sections on polymer processing in Chapter 17.

†Lord Todd, president of the Royal Society of London, quoted in *Chem. Eng. News* **58**(40), 29 (1980), in answer to the question, What do you think has been chemistry's biggest contribution to science, to society?

I have tried to include brief descriptions of the new polymer materials in the marketplace in Chapters 13–16 and in a section on composite materials in Chapter 17. Unfortunately, some discussion of less timely topics had to be eliminated to prevent undue expansion of the text.

My approach to referencing the literature had to remain essentially the same as that adopted in the second edition, despite some dissenting opinions. The explosion of the literature in polymer science makes it totally impossible to provide full coverage of original articles, as was possible 20 or 25 years ago. I have therefore cited many new books, and many articles from the *Encyclopedia of Polymer Science and Technology*, the Kirk–Othmer *Encyclopedia of Chemical Technology*, third edition, and the *Modern Plastics Encyclopedia*. Each of these sources (save the last, which provides information on current commercial products and processes) was selected to provide detailed citation of the original literature, as well as more complete coverage of the topic for which it was cited.

With retirement imminent, I look back with pleasure on the preparation of this volume and its predecessors. They have brought me much pleasure, more in the friendship of many readers and colleagues than in the accomplishment. I hope that the usefulness of this last revision will surpass that of those before it.

FRED W. BILLMEYER, JR.

Troy, New York
January 1984

ACKNOWLEDGMENTS

I wish to thank many colleagues, both in the Polymer Science and Engineering Program at Rensselaer and elsewhere, for valuable suggestions that have been incorporated into this revision. Seventy-five Rensselaer students who used a first draft of about two-thirds of the text in the course Introduction to Polymer Chemistry also provided helpful ideas and corrections.

The text was capably typed and retyped by Peggy Ruggeri. To her, to my graduate students, and especially to my patient wife Annette, I owe many thanks.

F.W.B., Jr.

PREFACE TO THE SECOND EDITION

"Dear Colleague, Leave the concept of large molecules well alone . . . there can be no such thing as a macromolecule."

It is said[†] that this advice was given to Hermann Staudinger just 45 years ago, after a major lecture devoted to his evidence in favor of the macromolecular concept. Today it seems almost impossible that this violent opposition to the idea of the existence of polymer molecules could have existed in relatively recent times. Now we take for granted not only the existence of macromolecules, but their value to us in food, clothing, shelter, transportation, communication, most other aspects of modern technology, and, last but far from least, the muscles, sinews, genes, and chromosomes that constitute our bodies and intellect.

Even within the years since the first edition of *Textbook of Polymer Science* (1962) was written, the use of synthetic polymers has proliferated, as discussed in Chapter 7E. Not only has the annual production of plastics (for example) increased some 250% in the last eight years, but on a volume basis it has already exceeded that of copper and aluminum, and is expected to surpass the production of steel by the mid-1980's. One consequence of this widening use of polymeric materials is that a substantial if not major fraction of all chemists and chemical engineers, to say nothing of those in other disciplines, is employed in industry related in some way to polymers. Estimates vary, but this fraction appears to be one-third to one-half or higher.

Education in polymer science has not kept pace. By far the majority of colleges and universities in the United States have no courses in polymer science, no staff member conducting research in this area, and only cursory mention of polymers in other courses. There are, needless to say, many exceptions, ranging from the isolated effort of a single staff member to such major centers for polymer research as those

[†] Robert Olby, "The Macromolecular Concept and the Origins of Molecular Biology," *J. Chem. Educ.* **47**, 168–174 (1970).

at Case-Western Reserve, the Universities of Akron and Massachusetts, and Rens-selaer, where 10 or more staff members constitute a formal or informal polymer research center.

Fortunately, this scarcity of education in polymer science is slowly diminishing, but it is still evident in many areas. What is most unfortunate is that it appears to exist, not because of a lack of awareness but, rather, a lack of interest. For example, on several occasions a graduating Ph.D. student in my own institution has dropped by to say that he was trained in another area of chemistry and was vaguely aware of our polymer science program, but he had now accepted employment with a large company where he was told that he would be working in the polymer field, and please, was there any way he could learn all about polymers in the short time remaining before he started the job?

Perhaps we could overlook a few individual instances of this sort, but it is more disturbing to note that, while polymeric materials are widely used as previously indicated, most "materials" curricula are merely renamed metallurgy departments giving only lip service to major families of materials other than metals, such as ceramics and polymers. Again, there are many exceptions, and I do not wish to leave the impression that this is always the case. I have, however, yet to see an introductory materials course or textbook that treats polymers fairly in accord with their wide usage throughout the world.

Another dichotomy that deserves mention, and which I wish I had the ability to overcome, is exemplified by the communications gap between the polymer scientist as now trained in the university or in industry and the biologist or biomedical scientist, whose concepts of macromolecules are vastly different. Many of my colleagues seem to share my feeling that major advances can be made in the next few years by the application of polymer physics and physical chemistry to biological materials.

The foregoing commentary emphasizes my feeling that the need for education in polymer science exists more than ever today, and that it must be filled by teaching at several levels in several disciplines. Clearly, no single book can serve all needs in this field, but I hope that in its new edition *Textbook of Polymer Science* will continue to be valuable in many ways. In the revision I have attempted to keep in mind its use as supplemental reading material in such undergraduate chemistry courses as physical chemistry, organic chemistry, and instrumental analysis, where an effort is made to introduce polymer science into the chemistry curriculum at appropriate places; as supplemental reading in other curricula such as biology and its interdisciplinary offshoots, materials in the broad sense, and the environmental sciences growing in popularity today; as the textbook in polymer science and engineering courses at the undergraduate and first-year graduate levels; as supple-mentary reading to broaden the background of the student in advanced polymer courses using appropriate specialized texts; in continuing education at the post-graduate level, in universities but particularly in industry; and as a reference and guide to the literature for the practicing polymer scientist and engineer.

Just as the use of polymers has proliferated in the past decade, so has its literature. The leading journal in the field, *Journal of Polymer Science,* has subdivided into several largely independent parts. New journals have appeared, some with national society sponsorship, such as *Macromolecules,* sponsored by the American Chemical Society, others as commercial ventures. Polymer articles, and even polymer sections, appear in other journals, new and old. *POST—J,* described below, abstracts approximately 500 journals for articles on polymers.

Abstract journals devoted solely to polymers have appeared. *POST—J,* acronym for Polymer Science and Technology—Journals, and its companion *POST—P* (Patents) are published biweekly, at a rate of 500–700 abstracts per issue in mid-1970, by the Chemical Abstracts Service of the American Chemical Society.† Other such services exist, sponsored by universities and as commercial ventures.

Not only has the number of general and specialized books on polymers increased at a tremendous rate in recent years, but a number of valuable compilations has appeared. Outstanding among these are the *Encyclopedia of Polymer Science and Technology* and the *Modern Plastics Encyclopedia,* both referenced in this book.

One result of the appearance of many specialized books and encyclopedias is that it is no longer necessary for a textbook to cite the original literature in detail; indeed, it would be impossible to do so today. Therefore I have limited references to specialized books and compilations where possible, on the assumption that most of them will be as readily available as the original literature and can in their turn cover the subject more completely than is possible in the present volume. The exceptions, aside from some key references of historical value, lie in the areas where adequate specialized coverage has not yet appeared. The reader will recognize these areas by the extent and content of the bibliography at the end of each chapter.

In revising the *Textbook of Polymer Science,* I have reached the conclusion that many of the basic principles of polymer science are now well established. Examples are the kinetics of condensation and free-radical addition polymerization. In areas such as these, the reader will find only minor changes from the 1962 edition. Elsewhere the revision has been more extensive. Much new material has been added, particularly in areas still under rapid development. This has had to be counterbalanced by the omission of an equivalent amount of material that no longer represents the current state of our knowledge, or was of lesser or only historical interest, in order to keep the length (and price) of the volume in hand.

The actual arrangement of material has undergone little change. Chapters 1–4 now comprise Part I, dealing with introductory concepts and the characterization of macromolecules. Important additions in this section include discussion of solubility parameters, free-volume theories of polymer solution thermodynamics, gel permeation chromatography, vapor-phase osmometry, and scanning electron microscopy, with extensive revision of many other sections.

Part II (Chapters 5–7) deals with the structure and properties of bulk polymers and includes considerable revision of parts of Chapter 5, where a few of the concepts of crystallinity in polymers, new in 1962, have had to be modified as our knowledge in this area has grown. Chapter 7 has been revised in order of presentation, with considerable new material added.

The format and content of Part III, concerned with polymerization kinetics, have

---

†*Post—J* and *Post—P* are now combined and issued as part of the Macromolecular Sections of *Chemical Abstracts* (footnote added in 1983).

been revised primarily for the citation of recent advances and new references. The exception lies in Chapter 10, whose topic is ionic and coordination polymerization, in which field many of the concepts new in 1962 have now reached a stage of further elucidation and acceptance.

As in the earlier edition, the chapters of Part IV describe the polymerization, structure, properties, fabrication, and applications of commercially important polymers, including those used as plastics, fibers, and elastomers. The reader comparing old and new chapter titles will find that considerable rearrangement has been made, and the content is likewise extensively revised. Of particular note are new sections in Chapter 15 on aromatic heterochain, heterocyclic, ladder, and inorganic polymers.

Part V, dealing with polymer processing, has in contrast been revised primarily by the addition of new references.

It is my hope that the many readers whose kind comments on earlier editions have given me pleasure will continue to find this revision useful.

FRED W. BILLMEYER, JR.

Troy, New York
September 1970

CONTENTS

PART THREE.   CHARACTERIZATION

PART FIVE.   PROPERTIES OF COMMERCIAL POLYMERS

PART SIX. POLYMER PROCESSING

APPENDIXES

TEXTBOOK
OF POLYMER SCIENCE

PART ONE

INTRODUCTION

CHAPTER ONE

THE SCIENCE OF LARGE MOLECULES

A. BASIC CONCEPTS OF POLYMER SCIENCE

Over half a century ago, Wolfgang Ostwald (1917)† coined the term *the land of neglected dimensions* to describe the range of sizes between molecular and macroscopic within which occur most colloidal particles. The term *neglected dimensions* might have been applied equally well to the world of polymer molecules, the high-molecular-weight compounds so important to man and his modern technology. It was not until the 1930's that the science of high polymers began to emerge, and the major growth of the technology of these materials came even later. Yet today polymer dimensions are neglected no more, for industries associated with polymeric materials employ more than half of all American chemists and chemical engineers.

The science of macromolecules is divided between biological and nonbiological materials. Each is of great importance. Biological polymers form the very foundation of life and intelligence and provide much of the food on which man exists. This book, however, is concerned primarily with the chemistry, physics, and technology of nonbiological polymers. These are the synthetic materials used for plastics, fibers, and elastomers, with a few naturally occurring polymers, such as rubber, wool, and cellulose, included. Today these substances are truly indispensable to mankind, being essential to clothing, shelter, transportation, and communication, as well as to the conveniences of modern living.

A *polymer* is a large molecule built up by the repetition of small, simple chemical units. In some cases the repetition is linear, much as a chain is built up from its links. In other cases the chains are *branched* or interconnected to form three-

---

†Parenthetical years or names and years refer to items in the bibliography at the end of the chapter.

dimensional networks. The *repeat unit* of the polymer is usually equivalent or nearly equivalent to the *monomer,* or starting material from which the polymer is formed. Thus (Table 1-1) the repeat unit of poly(vinyl chloride) is —CH_2CHCl—; its monomer is vinyl chloride, CH_2=$CHCl$.

The length of the polymer chain is specified by the number of repeat units in the chain. This is called the *degree of polymerization* (DP). The molecular weight of the polymer is the product of the molecular weight of the repeat unit and the DP. Using poly(vinyl chloride) as an example, a polymer of DP 1000 has a molecular weight of 63 × 1000 = 63,000. Most high polymers useful for plastics, rubbers, or fibers have molecular weights between 10,000 and 1,000,000.

Unlike many products whose structure and reactions were well known before their industrial application, some polymers were produced on an industrial scale long before their chemistry or physics was studied. Empiricism in recipes, processes, and control tests was usual.

Gradually the study of polymer properties began. Almost all were first called anomalous because they were so different from the properties of low-molecular-weight compounds. It was soon realized, however, that polymer molecules are many times larger than those of ordinary substances. The presumably anomalous properties of polymers were shown to be normal for such materials, as the consequences of their size were included in the theoretical treatments of their properties.

Polymerization Processes. The processes of polymerization were divided by Flory (1953) and Carothers (Mark 1940) into two groups known as *condensation*

TABLE 1-1. Some Linear High Polymers, Their Monomers, and Their Repeat Units

Polymer	Monomer	Repeat Unit
Polyethylene	CH_2=CH_2	—CH_2CH_2—
Poly(vinyl chloride)	CH_2=$CHCl$	—CH_2CHCl
Polyisobutylene	CH_2=C, with CH_3 above and CH_3 below	—CH_2—C—, with CH_3 above and CH_3 below
Polystyrene[a]	CH_2=CH, with benzene ring below	—CH_2—CH—, with benzene ring below
Polycaprolactam (6-nylon)	H—$N(CH_2)_5C$—OH, with H below N and O below C	—$N(CH_2)_5C$—, with H below N and O below C
Polyisoprene (natural rubber)	CH_2=CH—C=CH_2, with CH_3 below C	—CH_2CH=C—CH_2—, with CH_3 below C

[a]By convention, the symbol ◯ is used throughout to represent the benzene ring, double bonds being omitted.

and *addition* polymerization or, in more precise terminology (Chapter 2A), *step-reaction* and *chain-reaction* polymerization.

Condensation or *step-reaction polymerization* is entirely analogous to condensation in low-molecular-weight compounds. In polymer formation the condensation takes place between two polyfunctional molecules to produce one larger polyfunctional molecule, with the possible elimination of a small molecule such as water. The reaction continues until almost all of one of the reagents is used up; an equilibrium is established that can be shifted at will at high temperatures by controlling the amounts of the reactants and products.

Addition or *chain-reaction polymerization* involves chain reactions in which the chain carrier may be an ion or a reactive substance with one unpaired electron called a *free radical*. A free radical is usually formed by the decomposition of a relatively unstable material called an *initiator*. The free radical is capable of reacting to open the double bond of a vinyl monomer and add to it, with an electron remaining unpaired. In a very short time (usually a few seconds or less) many more monomers add successively to the growing chain. Finally two free radicals react to annihilate each other's growth activity and form one or more polymer molecules.

With some exceptions, polymers made in chain reactions often contain only carbon atoms in the main chain (*homochain polymers*), whereas polymers made in step reactions may have other atoms, originating in the monomer functional groups, as part of the chain (*heterochain polymers.*)

Molecular Weight and Its Distribution. In both chain and stepwise polymerization, the length of a chain is determined by purely random events. In step reactions, the chain length is determined by the local availability of reactive groups at the ends of the growing chains. In radical polymerization, chain length is determined by the time during which the chain grows before it diffuses into the vicinity of a second free radical and the two react. In either case, the polymeric product contains molecules having many different chain lengths. Molecular weight and molecular-weight distribution in polymers is considered further in Section *D*.

Branched and Network Polymers. In contrast to the linear-chain molecules discussed so far, some polymers have branched chains, often as a result of side reactions during polymerization (Fig. 1-1*a*). The term *branching* implies that the individual molecules are still discrete; in still other cases *crosslinked* or *network structures* are formed (Fig. 1-1*b*), as in the use of monomers containing more than two reactive groups in stepwise polymerization. If, for example, glycerol is substituted for ethylene glycol in the reaction with a dibasic acid, a three-dimensional network polymer results. In recent years, a variety of branched polymer structures, some with outstanding high-temperature properties, has been synthesized (Chapter 15).

In commercial practice crosslinking reactions may take place during the fabrication of articles made with *thermosetting* resins. The crosslinked network extending throughout the final article is stable to heat and cannot be made to flow or melt. In contrast, most linear or branched polymers can be made to soften and take on

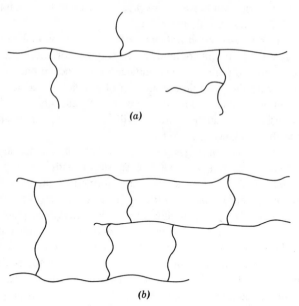

FIG. 1-1. Schematic representation of (a) branched and (b) network polymers.

new shapes by the application of heat and pressure. They are said to be *thermoplastic*.

The Texture of Polymers. The geometrical arrangement of the atoms in a polymer chain can be divided conveniently into two categories:

a. Arrangements fixed by the chemical bonding in the molecule, such as cis and trans isomers, or *d* and *l* forms. Throughout this book, such arrangements are described as *configurations*. The configuration of a polymer chain cannot be altered unless chemical bonds are broken and reformed.

b. Arrangements arising from rotation about single bonds. These arrangements, including the manifold forms that the polymer chain may have in solution, are described as *conformations*.

In dilute solution, where the polymer chain is surrounded by small molecules, or in the melt, where it is in an environment of similar chains, the polymer molecule is in continual motion because of its thermal energy, assuming many different *conformations* in rapid succession. As a polymer melt is cooled, or as this molecular motion so characteristic of polymers is restrained through the introduction of strong interchain forces, the nature of the polymer sample changes systematically in ways that are important in determining its physical properties and end uses (Fig. 1-2).

In the molten state, polymer chains move freely, though often with enormous viscosity, past one another if a force is applied. This is the principle utilized in the

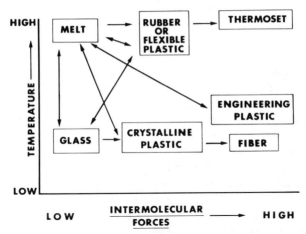

FIG. 1-2. The interrelation of the states of bulk polymers. The arrows indicate the directions in which changes from one state to another can take place (Billmeyer 1969).

fabrication of most polymeric articles and is the chief example of the plasticity from which the very name *plastics* is derived. If the irreversible flow characteristic of the molten state is inhibited by the introduction of a tenuous network of primary chemical-bond crosslinks in the process commonly called *vulcanization* (Chapter 19), but the local freedom of motion of the polymer chains is not restricted, the product shows the elastic properties we associate with typical rubbers. If, however, the interchain forces result from secondary bonds, such as the interaction of polar groups, rather than primary chemical bonds, the rubber is not one of high elasticity but has the properties of limpness and flexibility: A familiar example is the vinyl film widely used alone or in coated fabrics. Secondary-bond forces are capable of forming and breaking reversibly as the temperature is changed, as indicated by the arrows in Fig. 1-2.

Continued primary-bond crosslinking in the postpolymerization step of vulcanization converts rubber into hard rubber or ebonite, whereas crosslinking concurrent with polymerization produces a wide variety of thermosetting materials. Common examples are the phenol-formaldehyde and amine-formaldehyde families widely used as plastics.

As the temperature of a polymer melt or rubber is lowered, a point known as the *glass-transition temperature* is reached where polymeric materials undergo a marked change in properties associated with the virtual cessation of local molecular motion. Thermal energy is required for segments of a polymer chain to move with respect to one another; if the temperature is low enough, the required amounts of energy are not available. Below their glass-transition temperature, amorphous polymers have many of the properties associated with ordinary inorganic glasses, including hardness, stiffness, brittleness, and transparency.

In addition to undergoing a glass transition as the temperature is lowered, some polymers can *crystallize* at temperatures below that designated as their crystalline melting point. Not all polymers are capable of crystallizing; to oversimplify somewhat, the requirements for crystallizability in a polymer are that it have either a geometrically regular structure or that any substituent atoms or groups on the backbone chain be small enough so that, if irregularly spaced, they can still fit into an ordered structure by virtue of their small size (see Chapter 10).

The properties of crystalline polymers are highly desirable. Crystalline polymers are strong, tough, stiff, and generally more resistant to solvents and chemicals than their noncrystalline counterparts. Further improvements in these desirable properties can be brought about in at least two ways.

First, by increasing intermolecular forces through the selection of highly polar polymers, and by using inherently stiff polymer chains, crystalline melting points can be raised so that the desirable mechanical properties associated with crystallinity are retained to quite high temperatures. The resulting plastics are capable of competing with metals and ceramics in engineering applications (Chapters 12 and 15).

Second, the properties of crystalline polymers can be improved for materials in fiber form by the process of orientation or drawing. The result is the increased strength, stiffness, and dimensional stability associated with synthetic fibers (Chapter 18).

GENERAL REFERENCES

Mark 1966, 1977; Billmeyer 1972, 1982; Elias 1977, Part I; Bovey 1979.

B. THE RISE OF MACROMOLECULAR SCIENCE

Early Investigations

Natural polymers have been utilized throughout the ages. Since his beginning man has been dependent on animal and vegetable matter for sustenance, shelter, warmth, and other requirements and desires. Natural resins and gums have been used for thousands of years. Asphalt was utilized in prebiblical times; amber was known to the ancient Greeks; and gum mastic was used by the Romans.

About a century ago the unique properties of natural polymers were recognized. The term *colloid* was proposed to distinguish polymers as a class from materials that could be obtained in crystalline form. The concept was later broadened to that of the "colloidal state of matter," which was considered to be like the gaseous, liquid, and solid states. Although useful for describing many colloidal substances, such as gold sols and soap solutions, the concept of a reversibly attainable colloidal state of matter has no validity.

The hypothesis that colloidal materials are very high in molecular weight is also quite old, but before the work of Raoult and van't Hoff in the 1880's no suitable methods were available for estimating molecular weights. When experimental methods did become available, molecular weights ranging from 10,000 to 40,000 were obtained for such substances as rubber, starch, and cellulose nitrate. The existence of large molecules implied by these measurements was not accepted by the chemists of the day for two reasons.

First, true macromolecules were not distinguished from other colloidal substances that could be obtained in noncolloidal form as well. When a material of well-known structure was seen in the colloidal state, its apparent high molecular weight was considered erroneous. Thus it was assumed that Raoult's solution law did not apply to any material in the colloidal state. Second, coordination complexes and the association of molecules were often used to explain polymeric structures in terms of physical aggregates of small molecules.

In the search by the early organic chemists for pure compounds in high yields, many polymeric substances were discovered and as quickly discarded as oils, tars, or undistillable residues. A few of these materials, however, attracted interest. Poly(ethylene glycol) was prepared about 1860; the individual polymers with degrees of polymerization up to 6 were isolated and their structures correctly assigned. The concept of extending the structure to very high molecular weights by continued condensation was understood.

The Rise of Polymer Science

Acceptance of the Existence of Macromolecules.   Acceptance of the macromolecular hypothesis came about in the 1920's, largely because of the efforts of Staudinger (1920), who received the Nobel Prize in 1953 for his championship of this viewpoint. He proposed long-chain formulas for polystyrene, rubber, and polyoxymethylene. His extensive investigations of the latter polymers left no doubt as to their long-chain nature. More careful molecular-weight measurements substantiated Staudinger's conclusions, as did x-ray studies showing structures for cellulose and other polymers that were compatible with chain formulas. The outstanding series of investigations by Carothers (1929, 1931) supplied quantitative evidence substantiating the macromolecular viewpoint.

The Problem of End Groups.   One deterrent to the acceptance of the macromolecular theory was the problem of the ends of the long-chain molecules. Since the degree of polymerization of a typical polymer is at least several hundred, chemical methods for detecting end groups were at first not successful. Staudinger (1925) suggested that no end groups were needed to saturate terminal valences of the long chains; they were considered to be unreactive because of the size of the molecules. Large ring structures were also hypothesized (Staudinger 1928), and this concept was popular for many years. Not until Flory (1937) elucidated the mechanism for chain-reaction polymerization did it become clear that the ends of long-chain molecules consist of normal, satisfied valence structures. This was but one of many of his contributions to polymer science honored when Flory received the Nobel Prize in 1974. The presence and nature of end groups have since been investigated in detail by chemical and physical methods (Chapter 8A).

Molecular Weight and Its Distribution. Staudinger (1928) was among the first to recognize the large size of polymer molecules and to utilize the dependence on molecular weight of a physical property, such as dilute-solution viscosity (Staudinger 1930), for determining polymer molecular weights. He also understood clearly that synthetic polymers are polydisperse (Staudinger 1928). A few years later Lansing (1935) distinguished unmistakably among the various average molecular weights obtainable experimentally.

Configurations of Polymer Chain Atoms. Staudinger's name is also associated with the first studies (1935) of the configuration of polymer chain atoms. He showed that the phenyl groups in polystyrene are attached to alternate chain carbon atoms. This regular head-to-tail configuration has since been established for most vinyl polymers. The mechanism for producing branches in normally linear vinyl polymers was introduced by Flory (1937), but such branches were not adequately identified and characterized for another decade (see Chapter 8E). Natta (1955a,b) first recognized the presence of sterospecific regularity in vinyl polymers. He received the Nobel Prize in 1963.

Early Industrial Developments

Rubber. The modern plastics industry began with the utilization of natural rubber for erasers and in rubberized fabrics a few years before Goodyear's discovery of vulcanization in 1839. In the next decade the rubber industry arose both in England and in the United States. In 1851 hard rubber, or ebonite, was patented and commercialized.

Derivatives of Cellulose. Cellulose nitrate, or nitrocellulose, discovered in 1838, was successfully commercialized by Hyatt in 1870. His product, Celluloid, cellulose nitrate plasticized with camphor, could be formed into a wide variety of useful products by the application of heat and pressure. It has been superseded in almost all uses by more stable and more suitable polymers. Cellulose acetate was discovered in 1865, and partially acetylated products were commercialized as acetate rayon fibers and cellulose acetate plastics in the early 1900's. Cellulose itself, dissolved and reprecipitated by chemical treatment, was introduced still later as viscose rayon and cellophane.

Synthetic Polymers. The oldest of the purely synthetic plastics is the family of phenol-formaldehyde resins, of which Baekeland's Bakelite was the first commercial product. Small-scale production of phenolic resins and varnishes was begun in 1907 (Baekeland 1909).

The first commercial use of styrene was in synthetic rubbers made by copolymerization with dienes in the early 1900's. Polystyrene was produced commercially in Germany about 1930 and successfully in the United States in 1937. Large-scale production of vinyl chloride-acetate resins began in the early 1920's also. Table 1-2 shows the approximate dates of introduction of some of the synthetic plastics of greatest commercial interest.

TABLE 1-2.　Some Commercially Important Polymers and Their Dates of Introduction

Date	Polymer	Date	Polymer
1930	Styrene–butadiene rubber	1943	Silicones
1936	Poly(vinyl chloride)	1944	Poly(ethylene terephthalate)
1936	Polychloroprene (neoprene)	1947	Epoxies
1936	Poly(methyl methacrylate)	1948	ABS resins
1936	Poly(vinyl acetate)	1955	Polyethylene, linear
1937	Polystyrene	1956	Polyoxymethylene
1939	66-Nylon	1957	Polypropylene
1941	Polytetrafluoroethylene	1957	Polycarbonate
1942	Unsaturated polyesters	1964	Ionomer resins
1943	Polyethylene, branched	1965	Polyimides
1943	Butyl rubber	1970	Thermoplastic elastomers
1943	6-Nylon	1974	Aromatic polyamides

GENERAL REFERENCES

Flory 1953; Mark 1966, 1967, 1977, 1981; Staudinger 1970; Marvel 1981; Seymour 1982.

C.　MOLECULAR FORCES AND CHEMICAL BONDING IN POLYMERS

The nature of the bonds that hold atoms together in molecules is explained by quantum mechanics in terms of an atom consisting of a small nucleus, concentrating the mass and positive charge, surrounded by clouds or shells of electrons relatively far away. It is among the outermost, more loosely bound electrons, called *valence electrons*, that chemical reactions and primary-bond formation take place.

Primary Bonds

Ionic Bond.　The most stable electronic configuration for most atoms (except hydrogen) important in polymers is a complete outer shell of eight electrons, called an *octet*. In inorganic systems this structure may be obtained by the donation of an electron by one atom to another to form an *ionic* bond:

$$\mathrm{Na\cdot \ + \ \cdot \ddot{\underset{\cdot\cdot}{Cl}}: \ \longrightarrow \ Na^+ \ + : \ddot{\underset{\cdot\cdot}{Cl}}:^-}$$

These bonds are not usually found in macromolecular substances except in the use of divalent ions to provide "crosslinks" between carboxyl groups in natural resins, and in *ionomers* (Chapter 13).

Covalent Bond. These bonds are formed when one or more pairs of valence electrons are shared between two atoms, again resulting in stable electronic shells:

$$\cdot \overset{\cdot}{\underset{\cdot}{C}} \cdot + 4H \cdot \longrightarrow H : \overset{\overset{H}{\cdot\cdot}}{\underset{\underset{H}{\cdot\cdot}}{C}} : H$$

The *covalent bond* is the predominant bond in polymers.

Coordinate Bond. This bond is similar to the covalent bond in that electrons are shared to produce stable octets; but in the coordinate bond both of the shared electrons come from one atom. Addition compounds of boron trichloride are common examples:

$$\begin{matrix} : \overset{\cdot\cdot}{Cl} : \\ : \overset{\cdot\cdot}{Cl} : B \ + \ : \overset{\cdot\cdot}{O} : R \\ : \overset{\cdot\cdot}{Cl} : \quad R \end{matrix} \longrightarrow \begin{matrix} : \overset{\cdot\cdot}{Cl} : \\ : \overset{\cdot\cdot}{Cl} : B : \overset{\cdot\cdot}{O} : R \\ : \overset{\cdot\cdot}{Cl} : R \end{matrix}$$

where R is an organic group. The *coordinate* or *semipolar* bond has properties between those of the ionic and covalent bonds. No polymers containing true coordinate bonds have reached commercialization.

Metallic Bond. In the *metallic bond* the number of valence electrons is far too small to provide complete outer shells for all the atoms. The resulting bonds involve the concept of positively charged atoms embedded in a permeating "gas" of electrons free to move about at will. The metallic bond is not utilized in polymeric systems.

Typical Primary-Bond Distances and Energies. From studies of the positions of atoms in molecules and the energetics of molecular formation and dissociation, it is possible to assign typical energies and lengths to primary bonds. Table 1-3 lists some of these properties of interest in polymeric systems. The angles between successive single bonds involving the atomic arrangements usual in polymers range between 105° and 113°, not far from the tetrahedral angle of 109°28′.

Secondary-Bond Forces

Even when all the primary valences within covalent molecules are saturated, there are still forces acting between the molecules. These are generally known as *secondary valence* or *intermolecular forces*, or *van der Waals forces*. The following three types are recognized, and the first and third in particular contribute greatly to the physical properties of polymers.

Dipole Forces. When different atoms in a molecule carry equal and opposite electric charges, the molecule is said to be *polar* or to have a *dipole moment*. At

TABLE 1-3. Typical Primary-Bond Lengths and Energies

Bond	Bond Length (Å)	Dissociation Energy (kJ/mole)
C—C	1.54	347
C=C	1.34	611
C—H	1.10	414
C—N	1.47	305
C≡N	1.15	891
C—O	1.46	360
C=O	1.21	749
C—F	1.35	473
C—Cl	1.77	339
N—H	1.01	389
O—H	0.96	464
O—O	1.32	146

large distances such a molecule acts like an electrically neutral system, but at molecular distances the charge separation becomes significant and leads to a net intermolecular force of attraction. The magnitude of the interaction energy depends on the mutual alignment of the dipoles. Molecular orientation of this sort is always opposed by thermal agitation; hence the dipole force is strongly dependent upon temperature.

Induction Forces. A polar molecule also influences surrounding molecules that do not have permanent dipoles. The electric field associated with a dipole causes slight displacements of the electrons and nuclei of surrounding molecules, which lead to induced dipoles. The intermolecular force between the permanent and induced dipoles is called the *induction force*. The ease with which the electronic and nuclear displacements are made is called the *polarizability* of the molecule. The energy of the induction force is always small and independent of temperature.

Dispersion Forces. The existence of intermolecular forces in nonpolar materials, plus the small temperature dependence of intermolecular forces even where the dipole effect is known to far outweigh the induction effect, suggests the presence of a third type of intermolecular force. All molecules have time-dependent dipole moments that average out to zero and which arise from different instantaneous configurations of the electrons and nuclei. These fluctuations lead to perturbations of the electronic clouds of neighboring atoms and give rise to attractive forces called *dispersion forces*. They are present in all molecules and make up a major portion of the intermolecular forces unless very strong dipoles are present. In nonpolar materials only the dispersion forces exist. They are independent of temperature. Occasionally the term *van der Waals' forces* is applied to the dispersion forces alone.

Interrelation of Intermolecular Forces. The energy of the intermolecular attractive forces varies as the inverse sixth power of the intermolecular distance. As with primary-bond forces, repulsion arises when the atoms approach more closely than an equilibrium distance of 3–5 Å. The energy of typical secondary-bond attractive forces is 8–40 kJ/mole, divided among the three secondary-bond types according to the polarizability and dipole moment of the bonding molecules.

The Hydrogen Bond. The bond in which a hydrogen atom is associated with two other atoms is particularly important in many polymers, including proteins, and is held by many to be essential to life processes. Since the classical concepts of chemical bonding allow hydrogen to form only one covalent bond, the hydrogen bond can be considered electrostatic or ionic in character. This model does not, however, account for all the properties of the hydrogen bond; it is appealing to consider the bond covalent in some cases. The hydrogen bond occurs between two functional groups in the same or different molecules. The hydrogen is usually attached to an acidic group (a proton donor), typically a hydroxyl, carboxyl, amine, or amide group. The other group must be basic, usually oxygen, as in carbonyls, ethers, or hydroxyls; nitrogen, as in amines and amides; and occasionally halogens. The association of such polar liquid molecules as water, alcohols, and hydrofluoric acid, the formation of dimers of simple organic acids, and important structural effects in polar polymers such as nylon, cellulose, and proteins are due to hydrogen bonding.

Typically, hydrogen bonds range between 2.4 and 3.2 Å in length and between 12 and 30 kJ/mole in dissociation energy. Only fluorine, nitrogen, oxygen, and (occasionally) chlorine are electronegative enough to form hydrogen bonds.

Intermolecular Forces and Physical Properties

Secondary-bond forces are not of great importance in the formation of stable chemical compounds. They lead, rather, to the aggregation of separate molecules into solid and liquid phases. As a result, many physical properties such as volatility, viscosity, surface tension and frictional properties, miscibility, and solubility are determined largely by intermolecular forces.

The *cohesive energy* is the total energy necessary to remove a molecule from a liquid or solid to a position far from its neighbors. This is approximately equal to the heat of vaporization or sublimation at constant volume and can be estimated from thermodynamic data. The cohesive energy per unit volume, usually called the *cohesive energy density,* and its variation with molecular structure illustrate the effects of intermolecular forces on the physical properties of matter.

Volatility and Molecular Weight. The tendency of a molecule to volatilize from its liquid is a function of its total translational energy and therefore of the temperature. The boiling point depends on the relation of the translational energy to the cohesive energy and thus is a function of molecular weight in a homologous series. At high molecular weights the total cohesive energy per molecule becomes greater than primary-bond energy, and the molecules decompose before they volatilize. This point is reached at molecular weights far below those of typical polymers.

The melting point is also related to the cohesive energy, but here another important factor comes into play. This is the influence of molecular order or entropy. In thermodynamic terms, changes of state take place only when the free-energy† change in the process

$$\Delta G = \Delta H - T \Delta S$$

is favorable, and the enthalpy term ΔH may easily be outweighed by the entropy term $T \Delta S$ whenever a radical change in molecular configurations occurs in the process. Thus, in general, a high boiling point is associated with a high melting point, but the relation between melting point and molecular structure is fairly complicated. Symmetrical molecules, which have low entropies of fusion, melt at higher temperatures than do similar but less symmetrical molecules.

Effect of Polarity.   A molecule containing strongly polar groups exerts correspondingly strong attractive forces on its neighbors. This is reflected in higher boiling and melting points and other manifestations of higher cohesive energy density.

Miscibility and Solubility.   These properties are also determined by the intermolecular forces. The thermal effect on mixing or solution is the difference between the cohesive energy of the mixture and that of the individual pure components. Again entropy considerations are important, but in general a negative heat of mixing favors solubility and a positive heat of mixing favors immiscibility. The intermolecular forces therefore lead directly to the solubility law of "like dissolves like."

The role of the intermolecular forces and the cohesive energy density in determining the solubility of polymers is discussed further in Chapter 7A.

Intermolecular Forces and Polymer Types.   Table 1-4 lists the cohesive energy densities of some typical polymers. These data corroborate the conclusions of Section A regarding the texture of polymers, for, in the absence of primary-bond crosslinks, it is the intermolecular forces that provide the restraints on molecular motion, which, as illustrated in Fig. 1-2, are a major determinant of the nature of bulk polymers.

If the intermolecular forces are small and the cohesive energy is low, and the molecules have relatively flexible chains, they comply readily to applied stresses and have properties usually associated with elastomers. Somewhat higher cohesive energy densities, accompanied in some cases by bulky side groups giving stiffer chains, are characteristic of typical plastics. If the cohesive energy is higher still, the materials exhibit the high resistance to stress, high strength, and good mechanical

---

†This book follows the convention of defining the (Gibbs) *free energy* (now sometimes called the *free enthalpy*) as $G = H - TS$, and the *work content* or Helmholtz free energy as $A = E - TS$.

TABLE 1-4.   Cohesive Energy Densities of Linear Polymers[a]

Polymer	Repeat Unit	Cohesive Energy Density (J/cm^3)
Polyethylene	—CH_2CH_2—	259
Polyisobutylene	—$CH_2C(CH_3)_2$—	272
Polyisoprene	—$CH_2C(CH_3)$=$CHCH_2$—	280
Polystyrene	—$CH_2CH(C_6H_5)$—	310
Poly(methyl methacrylate)	—$CH_2C(CH_3)$ $(COOCH_3)$—	347
Poly(vinyl acetate)	—$CH_2CH(OCOCH_3)$—	368
Poly(vinyl chloride)	—CH_2CHCl—	381
Poly(ethylene terephthalate)	—$CH_2CH_2OCOC_6H_4COO$—	477
Poly(hexamethylene adipamide)	—$NH(CH_2)_6NHCO(CH_2)_4CO$—	774
Polyacrylonitrile	—CH_2CHCN—	992

[a]Walker (1952) and Small (1953).

properties typical of fibers, especially where molecular symmetry is favorable for crystallization. Chain stiffness or flexibility, referred to above, is largely determined by hindrance to free rotation about carbon–carbon single bonds in the polymer chain.

GENERAL REFERENCES

Ketelaar 1953; Cottrell 1958; Pauling 1960, 1964; Pimentel 1960; Chu 1967; Phillips 1970; Elias 1977, Part I.

D.   MOLECULAR WEIGHT AND MOLECULAR-WEIGHT DISTRIBUTION

Perhaps the most important feature distinguishing polymers from low-molecular-weight species is the existence of a distribution of chain lengths and therefore degrees of polymerization and molecular weights in all known polymers (except possibly some biological macromolecules). This distribution can be illustrated by plotting the weight of polymer of a given molecular weight against the molecular weight, as in Fig. 1-3.

Because of the existence of the distribution in any finite sample of polymer, the experimental measurement of molecular weight can give only an average value. Several different averages are important. For example, some methods of molecular-weight measurement in effect count the number of molecules in a known mass of material. Through knowledge of Avogadro's number, this information leads to the

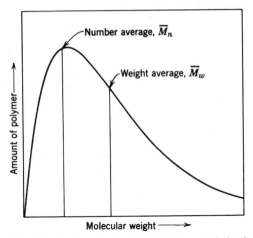

FIG. 1-3. Distribution of molecular weights in a typical polymer.

number-average molecular weight $\bar{M}_n$ of the sample. For typical polymers the number average lies near the peak of the weight-distribution curve or the most probable molecular weight.

If the sample contains N_i molecules of the ith kind, for a total number of molecules $\Sigma_{i=1}^{\infty} N_i$, and each of the ith kind of molecule has a mass m_i, then the total mass of all the molecules is $\Sigma_{i=1}^{\infty} N_i m_i$. The number-average molecular mass is

$$\bar{m}_n = \frac{\Sigma_{i=1}^{\infty} m_i N_i}{\Sigma_{i=1}^{\infty} N_i} \qquad (1\text{-}1)$$

and multiplication by Avogadro's number gives the number-average molecular weight (mole weight);

$$\bar{M}_n = \frac{\Sigma_{i=1}^{\infty} M_i N_i}{\Sigma_{i=1}^{\infty} N_i} \qquad (1\text{-}2)$$

Number-average molecular weights of commercial polymers usually lie in the range 10,000–100,000, although some materials have values of $\bar{M}_n$ 10-fold higher, and others 10-fold lower. In most cases, however, the physical properties associated with typical high polymers are not well developed if $\bar{M}_n$ is below about 10,000.

After $\bar{M}_n$, the next higher average molecular weight that can be measured by absolute methods is the weight-average molecular weight $\bar{M}_w$. This quantity is defined as

$$\bar{M}_w = \frac{\Sigma_{i=1}^{\infty} N_i M_i^2}{\Sigma_{i=1}^{\infty} N_i M_i} \qquad (1\text{-}3)$$

TABLE 1-5. Typical Ranges of $\bar{M}_w/\bar{M}_n$ in Synthetic Polymers[a]

Polymer	Range
Hypothetical monodisperse polymer	1.000
Actual "monodisperse" "living" polymers	1.01–1.05
Addition polymer, termination by coupling	1.5
Addition polymer, termination by disproportionation, or condensation polymer	2.0
High conversion vinyl polymers	2–5
Polymers made with autoacceleration	5–10
Addition polymers prepared by coordination polymerization	8–30
Branched polymers	20–50

[a]Billmeyer (1977).

It should be noted that each molecule contributes to $\bar{M}_w$ in proportion to the square of its mass: A quantity proportional to the first power of M measures only concentration, and not molecular weight. In terms of concentrations $c_i = N_i M_i$ and weight fractions $w_i = c_i/c$, where $c = \Sigma_{i=1}^{\infty} c_i$,

$$\bar{M}_w = \frac{\Sigma_{i=1}^{\infty} c_i M_i}{c} = \sum_{i=1}^{\infty} w_i M_i \tag{1-4}$$

Unfortunately, there appears to be no simple analogy for $\bar{M}_w$ akin to counting molecules to obtain $\bar{M}_n$.

Because heavier molecules contribute more to $\bar{M}_w$ than light ones, $\bar{M}_w$ is always greater than $\bar{M}_n$, except for a hypothetical monodisperse polymer. The value of $\bar{M}_w$ is greatly influenced by the presence of high-molecular-weight species, just as $\bar{M}_n$ is influenced by species at the low end of the molecular-weight distribution curve.

The quantity $\bar{M}_w/\bar{M}_n$ is a useful measure of the breadth of the molecular-weight distribution curve and is the parameter most often quoted for describing this feature. The range of values of $\bar{M}_w/\bar{M}_n$ in synthetic polymers is quite large, as illustrated in Table 1-5.

For some types of polymerization, the distribution of molecular weights (more often expressed as degrees of polymerization) can be calculated statistically; this topic is discussed in Chapter 3E. Experimental methods for measuring the molecular-weight averages defined above, among others, are the subject of Chapter 8.

GENERAL REFERENCES

Peebles 1971; Slade 1975; Billingham 1977.

DISCUSSION QUESTIONS AND PROBLEMS

1. Define the following terms: polymer, monomer, repeat unit, network, degree of polymerization, homochain polymer, heterochain polymer, thermoplastic, thermosetting, configuration, conformation.

2. Discuss some of the properties that make polymers useful materials, and show how they result from unique features of polymer structure such as high molecular weight. (This topic is amplified in later chapters.)

3. Consider three hypothetical monodisperse polymers, with $M = 10,000$, $M = 100,000$, and $M = 1,000,000$. For each, calculate $\bar{M}_w$ and $\bar{M}_n$ after adding the following to 100 parts by weight c_i of the polymer with $M = 100,000$:

   a. 20 parts by weight of the polymer with $M = 10,000$.

   b. 20 parts by weight of the polymer with $M = 1,000,000$.

   Calculate $\bar{M}_w$ and $\bar{M}_n$ after adding the following to 100 parts by number (of molecules) of the polymer with $M = 100,000$:

   c. 20 parts by number of the polymer with $M = 10,000$.

   d. 20 parts by number of the polymer with $M = 1,000,000$.

   Discuss the dependence of $\bar{M}_w$ and $\bar{M}_n$ on the presence of high- and low-molecular-weight material.

BIBLIOGRAPHY

Baekeland 1909. L. H. Baekeland, "The Synthesis, Constitution, and Uses of Bakelite," *J. Ind. Eng. Chem.* **1**, 149–161 (1909); reprinted in *Chemtech* **6**, 40–53 (1976).

Billingham 1977. N. C. Billingham, *Molar Mass Measurements in Polymer Science,* Halsted Press, John Wiley & Sons, New York, 1977.

Billmeyer 1969. Fred W. Billmeyer, Jr., "Molecular Structure and Polymer Properties," *J. Paint Technol.* **41**, 3–16; erratum, 209 (1969).

Billmeyer 1972. Fred W. Billmeyer, Jr., *Synthetic Polymers: Building the Giant Molecule,* Doubleday, Garden City, New York, 1972.

Billmeyer 1977. Fred W. Billmeyer, Jr., "The Size and Weight of Polymer Molecules," Chapter 4 in Herman S. Kaufman and Joseph J. Falcetta, eds., *Introduction to Polymer Science and Technology: An SPE Textbook,* Wiley-Interscience, New York, 1977.

Billmeyer 1982. Fred W. Billmeyer, Jr., "Polymers," pp. 745–755 in Martin Grayson, ed., *Kirk–Othmer Encyclopedia of Chemical Technology,* 3rd ed., Vol. 18, Wiley-Interscience, New York, 1982.

Bovey 1979. F. A. Bovey and F. H. Winslow, "The Nature of Macromolecules," Chapter 1 in F. A. Bovey and F. H. Winslow, eds., *Macromolecules: An Introduction to Polymer Science,* Academic Press, New York, 1979.

Carothers 1929. W. H. Carothers, "An Introduction to the General Theory of Condensation Polymers," *J. Am. Chem. Soc.* **51**, 2548–2559 (1929).

Carothers 1931. Wallace H. Carothers, "Polymerization," *Chem. Rev.* **8**, 353–426 (1931).

Chu 1967. Benjamin Chu, *Molecular Forces: Based on the Baker Lectures of Peter J. W. Debye*, Wiley-Interscience, New York, 1967.

Cottrell 1958. Tom L. Cottrell, *The Strengths of Chemical Bonds*, 2nd ed., Academic Press, New York, 1958.

Elias 1977. Hans-Georg Elias, *Macromolecules · 1 · Structure and Properties*, Plenum Press, New York, 1977.

Flory 1937. Paul J. Flory, "Mechanism of Vinyl Polymerization," *J. Am. Chem. Soc.* **59**, 241–253 (1937).

Flory 1953. Paul J. Flory, *Principles of Polymer Chemistry*, Cornell University Press, Ithaca, New York, 1953.

Ketelaar 1953. J. A. A. Ketelaar, *Chemical Constitution*, Elsevier, New York, 1953.

Lansing 1935. W. D. Lansing and E. O. Kraemer, "Molecular Weight Analysis of Mixtures by Sedimentation Equilibrium in the Svedberg Ultracentrifuge," *J. Am. Chem. Soc.* **57**, 1369–1377 (1935).

Mark 1940. H. Mark and G. Stafford Whitby, eds., *Collected Papers of Wallace Hume Carothers on High Polymeric Substances*, Interscience, New York, 1940.

Mark 1966. Herman F. Mark and the Editors of *Life*, *Giant Molecules*, Time, New York, 1966.

Mark 1967. H. F. Mark, "Polymers—Past, Present and Future," pp. 19–55 in W. O. Milligan, ed., *Proceedings of the Robert A. Welch Foundation Conferences on Chemical Research, X. Polymers*, The Robert A. Welch Foundation, Houston, 1967.

Mark 1977. Herman F. Mark and Sheldon Atlas, "Introduction to Polymer Science," Chapter 1 in Herman S. Kaufman and Joseph J. Falcetta, eds., *Introduction to Polymer Science and Technology: An SPE Textbook*, Wiley-Interscience, New York, 1977.

Mark 1981. Herman Mark, "Polymer Chemistry in Europe and America—How it all Began," *J. Chem. Educ.* **58**, 527–534 (1981).

Marvel 1981. C. S. Marvel, "The Development of Polymer Chemistry in America—The Early Days," *J. Chem. Educ.* **58**, 535–539 (1981).

Natta 1955a. G. Natta, Piero Pino, Paolo Corradini, Ferdinando Danusso, Enrico Mantica, Giorgio Mazzanti, and Giovanni Moranglio, "Crystalline High Polymers of α-Olefins," *J. Am. Chem. Soc.* **77**, 1708–1710 (1955).

Natta 1955b. G. Natta, "A New Class of α-Olefin Polymers with Exceptional Regularity of Structure" (in French), *J. Polym. Sci.* **16**, 143–154 (1955).

Ostwald 1917. Dr. Wolfgang Ostwald, *An Introduction to Theoretical and Applied Colloid Chemistry (The World of Neglected Dimensions)* (translated by Dr. Martin H. Fischer), John Wiley & Sons, New York, 1917.

Pauling 1960. Linus Pauling, *The Nature of the Chemical Bond*, 3rd ed., Cornell University Press, Ithaca, New York, 1960.

Pauling 1964. Linus Pauling and Roger Hayward, *The Architecture of Molecules*, W. H. Freeman, San Francisco, California, 1964.

Peebles 1971. Leighton H. Peebles, Jr., *Molecular Weight Distributions in Polymers*, Wiley-Interscience, New York, 1971.

Phillips 1970. James C. Phillips, *Covalent Bonding in Crystals, Molecules and Polymers*, Chicago University Press, Chicago, Illinois, 1970.

Pimentel 1960. George C. Pimentel and Aubrey L. McClellan, *The Hydrogen Bond*, W. H. Freeman, San Francisco, California, 1960.

Seymour 1982. Raymond B. Seymour, ed., *History of Polymer Science and Technology*, Marcel Dekker, New York, 1982.

Slade 1975. Philip E. Slade, Jr., *Polymer Molecular Weights*, Vol. 4 of Philip E. Slade, Jr., and Lloyd T. Jenkins, eds., *Techniques of Polymer Evaluation*, Marcel Dekker, New York, 1975.

Small 1953. P. A. Small, "Some Factors Affecting the Solubility of Polymers," *J. Appl. Chem.* **3,** 71–80 (1953).

Staudinger 1920. H. Staudinger, "Polymerization" (in German), *Ber. Dtsch. Chem. Ges. B* **53,** 1073–1085 (1920).

Staudinger 1925. H. Staudinger, "The Constitution of Polyoxymethylenes and Other High-Molecular Compounds" (in German), *Helv. Chim. Acta* **8,** 67–70 (1925).

Staudinger 1928. H. Staudinger, "The Constitution of High Polymers. XIII" (in German), *Ber. Dtsch. Chem. Ges. B* **61,** 2427–2431 (1928).

Staudinger 1930. H. Staudinger and W. Heuer, "Highly Polymerized Compounds, XXXIII. A Relation Between the Viscosity and the Molecular Weight of Polystyrenes" (in German), *Ber. Dtsch. Chem. Ges. B* **63,** 222–234 (1930).

Staudinger 1935. H. Staudinger and A. Steinhofer, "Highly Polymerized Compounds. CVII. Polystyrenes" (in German), *Justus Liebigs Ann. Chem.* **517,** 35–53 (1935).

Staudinger 1970. Herman Staudinger, *From Organic Chemistry to Macromolecules,* Wiley-Interscience, New York, 1970.

Walker 1952. E. E. Walker, "The Solvent Action of Organic Substances on Polyacrylonitrile," *J. Appl. Chem.* **2,** 470–481 (1952).

PART TWO

POLYMERIZATION

CHAPTER TWO

STEP-REACTION (CONDENSATION) POLYMERIZATION

A. CLASSIFICATION OF POLYMERS AND POLYMERIZATION MECHANISMS

In 1929 W. H. Carothers suggested a classification of polymers into two groups, *condensation* and *addition* polymers. Condensation polymers are those in which the molecular formula of the repeat unit of the polymer chain lacks certain atoms present in the monomer from which it is formed (or to which it can be degraded). For example, a polyester is formed by typical condensation reactions between bifunctional monomers, with the elimination of water:

$$x\text{HO—R—OH} + x\text{HOCO—R}'\text{—COOH} \longrightarrow$$

$$\text{HO[—R—OCO—R}'\text{—COO—]}_x\text{H} + (2x - 1)\text{H}_2\text{O}$$

Addition polymers (Chapter 3) are those in which this loss of a small molecule does not take place. The most important group of addition polymers includes those derived from unsaturated vinyl monomers:

$$\text{CH}_2{=}\text{CH} \longrightarrow \text{—CH}_2\text{—CH—CH}_2\text{—CH—}, \quad \text{etc.}$$
$$\phantom{\text{CH}_2{=}} | \qquad\qquad | \qquad\quad |$$
$$\phantom{\text{CH}_2{=}} \text{X} \qquad\qquad \text{X} \qquad\quad \text{X}$$

Carother's original distinction between addition and condensation polymers was amended by Flory, who placed emphasis on the *mechanisms* by which the two types of polymer are formed. Condensation polymers are usually formed by the stepwise intermolecular condensation of reactive groups; addition polymers ordinarily result from chain reactions involving some sort of active center.

The classification adopted in this book is based on the reaction mechanism, dividing polymerization into *step reactions,* commonly producing step-reaction or condensation polymers, and *chain reactions,* commonly producing chain-reaction or addition polymers (Mark 1950). Polymerizations are classified without regard to loss of a small molecule or type of interunit linkage. Wherever precise differentiation on the basis of mechanism is required, the terms *step reaction* and *chain reaction* are used; in deference to well-established tradition, the common terms *condensation* and *addition* are permissible where no confusion can result.

Some of the consequences of the differences between the mechanisms of chain and stepwise polymerization are shown in Table 2-1.

GENERAL REFERENCES

Mark 1940; Flory 1953; Lenz 1967; Vollmert 1973; Elias 1977.

B. MECHANISM OF STEPWISE POLYMERIZATION

Types of Condensation Polymers

Table 2-2 lists some representative products of stepwise polymerization. Most of them are, stoichiometrically, condensation polymers. Proteins and cellulose are included on the basis that they can be degraded hydrolytically to monomers differing from their repeating units by the addition of the elements of a molecule of water.

The type of products formed in a condensation reaction is determined by the *functionality* of the monomers, that is, by the average number of reactive functional groups per monomer molecule. Monofunctional monomers give only low-molecular-weight products. Bifunctional monomers give linear polymers, as illustrated

TABLE 2-1.   Distinguishing Features of Chain- and Step-Polymerization Mechanisms

Chain Polymerization	Step Polymerization
Only growth reaction adds repeating units one at a time to the chain.	Any two molecular species present can react.
Monomer concentration decreases steadily throughout reaction.	Monomer disappears early in reaction: at DP^a 10, less than 1% monomer remains.
High polymer is formed at once; polymer molecular weight changes little throughout reaction.	Polymer molecular weight rises steadily throughout reaction.
Long reaction times give high yields but affect molecular weight little.	Long reaction times are essential to obtain high molecular weights.
Reaction mixture contains only monomer, high polymer, and about 10^{-8} part of growing chains.	At any stage all molecular species are present in a calculable distribution.

[a]Degree of polymerization.

earlier in this chapter. Polyfunctional monomers, with more than two functional groups per molecule, give branched or crosslinked (three-dimensional) polymers. The properties of the linear and the three-dimensional polymers differ widely.

The mechanism of step-reaction polymerization is discussed in this section, following the classification scheme of Lenz (1967). Further information on the organic chemistry of the major types of condensation polymers having commercial utility is found in Part 4.

Carbonyl Addition–Elimination Mechanism

The most important reaction that has been used for the preparation of condensation polymers is that of addition and elimination at the carbonyl double bond of carboxylic acids and their derivatives. The generalized reaction is

$$R\overset{\overset{\displaystyle O}{\|}}{-C}-X + Y: \longrightarrow \left[R\overset{\overset{\displaystyle O:}{|}}{\underset{\underset{\displaystyle X}{|}}{-C}}-Y \right] \longrightarrow R\overset{\overset{\displaystyle O}{\|}}{-C}-Y + X:$$

where R and R' (below) may be alkyl or aryl groups, X may be OH, OR', NH$_2$, NHR', $\overset{\overset{\displaystyle O}{\|}}{OCR'}$, or Cl; and Y may be R'O$^-$, R'OH, R'NH$_2$, or R'COO$^-$. The species in the bracket is considered to be a metastable intermediate, which can either return to the original state by eliminating Y or proceed to the final state by eliminating X. The following paragraphs provide some typical examples of this reaction.

Direct Reaction. The direct reaction of a dibasic acid and a glycol to form a polyester, or a dibasic acid and a diamine to form a polyamide, works well and is widely used in practice. In esterification, a strong acid or acidic salt often serves as a catalyst. The reaction may be carried out by heating the reactants together and removing water, usually applying vacuum in the later stages.

An important modification of the direct reaction is the use of a salt, as in the preparation of poly(hexamethylene adipamide) (66-nylon) by heating the hexamethylene diamine salt of adipic acid above the melting point in an inert atmosphere. The stringent requirement of stoichiometric equivalence to obtain high molecular weight is easily met by purifying the salt by recrystallization.

Interchange. The reaction between a glycol and an ester,

$$x\text{HO—R—OH} + x\text{R}''\text{OCO—R'—COOR}'' \longrightarrow$$

$$\text{R}''\text{O(—CO—R'—COO—R—O)}_x\text{H} + (2x - 1)\text{R}''\text{—OH}$$

is often used to produce polyesters, especially where the dibasic acid has low solubility. Frequently the methyl ester is used, as in the production of poly(ethylene terephthalate) from ethylene glycol and dimethyl terephthalate. The reaction be-

TABLE 2-2. Typical Step-Reaction Polymers

Type	Interunit Linkage	Examples
Polyester	$\overset{O=}{-C-O-}$	$HO(CH_2)_xCOOH \longrightarrow HO[-(CH_2)_xCOO-]_yH + H_2O$ $HO(CH_2)_xOH + HOOC(CH_2)_xCOOH \longrightarrow HO\left[-(CH_2)_x\overset{O=}{O}C(CH_2)_x\overset{O=}{C}O-\right]_yH + H_2O$ $\begin{array}{c}CH_2OH \\ \| \\ -CHOH + HOOC(CH_2)_xCOOH \longrightarrow \text{three-dimensional network} + H_2O \\ \| \\ CH_2OH\end{array}$
Polyanhydride	$\overset{O=\quad O=}{-C-O-C-}$	$HOOC(CH_2)_xCOOH \longrightarrow HO[-CO(CH_2)_xCOO-]_yH + H_2O$
Polyacetal	$\underset{R}{\overset{H}{-O-C-O-}}$	$HO(CH_2)_xOH + CH_2(OR)_2 \longrightarrow HO[-(CH_2)_xOCH_2O-]_y(CH_2)_xOH + ROH$
Polyamide	$\overset{O=}{-C-NH-}$	$NH_2(CH_2)_xCOOH \longrightarrow H[-NH(CH_2)_xCO-]_yOH + H_2O$ $NH_2(CH_2)_xNH_2 + HOOC(CH_2)_xCOOH \longrightarrow H[-NH(CH_2)_xNHCO(CH_2)_xCO-]_yOH + H_2O$
Polyurethane	$\overset{O=}{-O-C-NH-}$	$HO(CH_2)_xOH + OCN(CH_2)_xCNO \longrightarrow [-O(CH_2)_xOCONH(CH_2)_xNHCO-]_y$
Polyurea	$\overset{O=}{-NH-C-NH-}$	$NH_2(CH_2)_xNH_2 + OCN(CH_2)_xCNO \longrightarrow [-NH(CH_2)_xNHCONH(CH_2)_xNHCO-]_y$
Silk fibroin	$\overset{O=}{-C-NH-}$	$NH_2CH_2COOH + NH_2CHRCOOH \longleftarrow H[-NHCH_2CONHCHRCO-]_yOH + H_2O$
Cellulose	$-C-O-C-$	$C_6H_{12}O_6 \longleftarrow [C_6H_{10}O_4]-O-[C_6H_{10}O_4]- + H_2O$

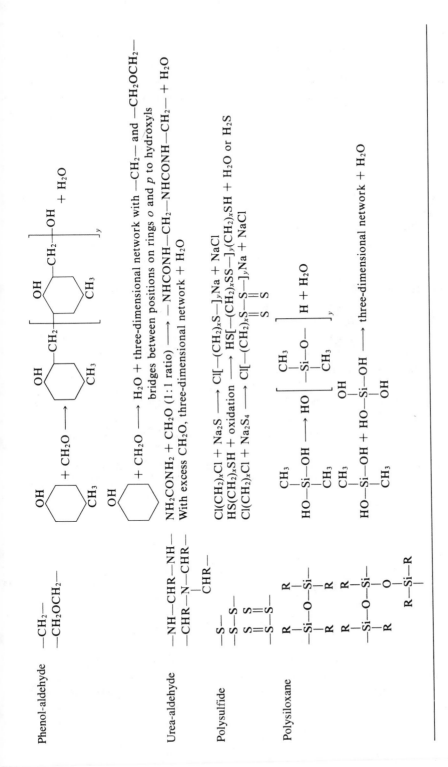

tween a carboxyl and an ester link is much slower, but other interchange reactions, such as amine–amide, amine–ester, and acetal–alcohol, are well known.

Acid Chloride or Anhydride.   Either of these species can be reacted with a glycol or an amine to give a polymer. The anhydride reaction is widely used to form an alkyd resin from phthalic anhydride and a glycol:

$$x \begin{array}{c} \\ \\ \end{array} \bigcirc \begin{array}{c} CO \\ \diagdown \\ \diagup \\ CO \end{array} O + x HOROH \longrightarrow HO \left[\begin{array}{c} -CO \diagup \bigcirc \diagdown COORO \\ \\ \end{array} \right]_x H + (x-1)H_2O$$

The condensation in bulk of an acid chloride with a glycol is not useful because of side reactions leading to low-molecular-weight products, but the reaction of an acid chloride with a diamine is a valuable means of preparing polyamides.

Interfacial Condensation.   The reaction of an acid halide with a glycol or a diamine proceeds rapidly to high-molecular-weight polymer if carried out at the interface between two liquid phases, each containing one of the reactants (Morgan 1965). Very-high-molecular-weight polymer can be formed. Typically, an aqueous phase containing the diamine or glycol and an acid acceptor is layered at room temperature over an organic phase containing the acid chloride. The polymer formed at the interface can be pulled off as a continuous film or filament. The method has been applied to the formation of polyamides, polyurethanes, polyureas, polysulfonamides, and polyphenyl esters. It is particularly useful for preparing polymers that are unstable at the higher temperatures usual in step-reaction polymerization.

Ring Versus Chain Formation.   In addition to polymer formation, bifunctional monomers may react intramolecularly to produce a cyclic product. Thus hydroxy acids may give either lactones or polymers on heating,

$$\begin{array}{ccc} & & \begin{array}{c} CO \\ \diagup \quad | \\ R \quad \quad | \\ \diagdown \quad O \end{array} \\ & \nearrow & \\ HORCOOH & & \\ & \searrow & \\ & & H[ORCO]_x OH \end{array}$$

amino acids may give lactams or linear polyamides, and so on.

The chief factor governing the type of product is the size of the ring that can be formed. If the ring contains less than five atoms or more than seven, the product is usually linear polymer. If a ring of five atoms can form, it will do so to the exclusion of linear chains. If six- or seven-membered rings can form, either type of product can result. Larger rings can sometimes be formed under special conditions.

The lack of formation of rings with less than five atoms is explained by the strain imposed by the valence angles of the ring atoms. Five-membered rings are virtually strain free, and all larger rings can be strain free if nonplanar forms are possible. As the ring size increases, the statistical probability of forming rings becomes smaller. The formation of five-membered rings to the complete exclusion of linear polymer is not thoroughly understood.

The formation of high polymer from cyclic monomers of the type just discussed is often, kinetically, an anionic chain polymerization and is discussed in Chapter 4. This class of reaction includes the commercially important production of nylon from caprolactam.

Other Mechanisms

Carbonyl Addition–Substitution Reactions.   The reaction of aldehydes with alcohols, involving first addition and then substitution at the carbonyl groups, is of great practical and historic importance in stepwise polymerization. The general reaction, leading to acetal formation, is

$$
\underset{\substack{\parallel \\ RCH}}{O} + R'OH \longrightarrow \underset{\substack{| \\ OR'}}{\overset{OH}{\underset{|}{RCH}}} \xrightarrow{\;R'OH\;} \underset{\substack{| \\ OR'}}{\overset{OR'}{\underset{|}{RCH}}}
$$

In addition to polyacetals, important polymers formed in this way include those of formaldehyde and phenol, urea, or melamine (Chapter 16).

Nucleophilic Substitution Reactions.   These reactions are important from the standpoint of commercial organic polymers, primarily because of their use in the polymerization of epoxides. The most common epoxide monomer is epichlorohydrin, which reacts with a nucleophile N: as follows:

$$
N: + \; H_2C\!\!-\!\!\underset{\diagdown\;\diagup}{\underset{O}{CHCH_2Cl}} \longrightarrow NCH_2\underset{\substack{| \\ O:^-}}{CHCH_2Cl}
$$

Typically, the nucleophile is a bifunctional hydroxy compound such as bisphenol A, and the reaction proceeds as described in Chapter 16.

A second nucleophilic substitution reaction of historic interest in polymer chemistry is that used for the production of the early polysulfide rubbers from aliphatic dichlorides and sodium sulfide:

$$
x\mathrm{ClCH_2CH_2Cl} + x\mathrm{Na_2S}_n \longrightarrow -[\mathrm{CH_2CH_2S}_n-]_x + 2x\mathrm{NaCl}
$$

This type of reaction can be used for the production of semiorganic and inorganic polymers, and is probably also the basis for the formation of natural polysaccharides and polynucleotides by polymerization in living organisms catalyzed by enzymes.

Double-Bond Addition Reactions.　Although addition reactions at double bonds are often associated with polymerization by chain mechanisms (Chapters 3 and 4), this is not necessarily the case, and many important stepwise polymerizations are based on this reaction. Of major interest among these is the ionic addition of diols to diisocyanates in the production of polyurethanes (Chapter 15):

$$x\text{HOROH} + x\text{OCNR'CNO} \longrightarrow -\left[\begin{array}{cc} \overset{\displaystyle O}{\overset{\|}{\text{OROC}}}\text{NHR'N} & \overset{\displaystyle O}{\overset{\|}{\text{HC}}}- \end{array} \right]_x$$

The free-radical addition of dithiols to unconjugated diolefins also proceeds by a stepwise mechanism.

Free-Radical Coupling.　These reactions lead to several polymerization schemes of interest, including the preparation of arylene ether polymers, polymers containing acetylene units, and arylene alkylidene polymers. If [Ox] is an oxidizing agent, the first of these can be written as

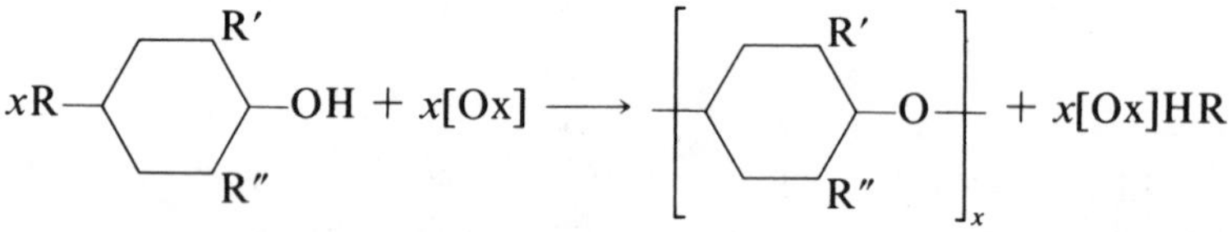

Aromatic Electrophilic–Substitution Reactions.　Reactions of this type, using standard Friedel–Crafts catalysts, produce polymers by a step-growth mechanism. A typical example is the production of poly(*p*-phenylene):

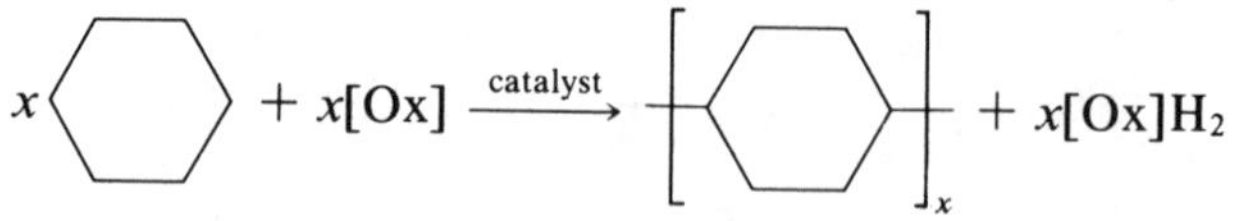

GENERAL REFERENCES

Mark 1940; Marvel 1959; Morgan 1965; Lenz 1967, Chapters 4–8; Ravve 1967, Part IV; Sorenson 1968; Solomon 1972, Chapter 1; Collins 1973, Exps. 1 and 2; Vollmert 1973, Chapter 2; Allen 1974, Chapter 5; Elias 1977, Chapter 17.1; Millich 1977; Bowden 1979, Chapter 2.4; Allcock 1981, Chapter 2; Carraher 1982.

C. KINETICS AND STATISTICS OF LINEAR STEPWISE POLYMERIZATION

Reactivity and Molecular Size

The similarity between monofunctional and polyfunctional step reactions is illustrated in the chemical equations in this chapter. Other similarities exist in the influence of temperature and catalysts on the rates of the reactions. It is tempting to substitute the simple concept of reactions between functional groups for that of the myriad separate reactions of each molecular species with all the others. With this substitution, the kinetics of stepwise polymerization is essentially identical with that of simple condensation; without it, analysis of the kinetics would be almost hopelessly difficult.

If the simplification is to be made, the rate of the reaction of a group must be independent of the size of the molecule to which it is attached. This assumption is amply justified by experimental evidence. The rate constants of condensation reactions in a homologous series reach asymptotic values independent of chain length quite rapidly and show no tendency to drop off with increases in molecular size. The diluting effect of the large chain must, of course, be taken into account by adjusting to constant molar composition of reactive groups. In addition, the rate constants for monofunctional and bifunctional reagents are identical for sufficiently long chains separating the reactive groups of the bifunctional compound.

Kinetics of Stepwise Polymerization

With the concept of functional group reactivity independent of molecular weight, the kinetics of stepwise polymerization becomes quite simple. The formation of a polyester from a glycol and a dibasic acid (Solomon 1972) may be taken as an example. It is well known that this reaction is catalyzed by acids. In the absence of added strong acid, a second molecule of the acid being esterified acts as a catalyst. The reaction is followed by measuring the rate of disappearance of carboxyl groups:†

$$-\frac{d[COOH]}{dt} = k[COOH]^2[OH] \qquad (2\text{-}1)$$

†Stepwise polymerization typically involves equilibrium reactions of the type

$$A + B \underset{k_r}{\overset{k_f}{\rightleftharpoons}} C + D$$

where the rates of the forward and reverse reactions are $k_f[A][B]$ and $k_r[C][D]$, respectively. At equilibrium these rates are equal, whence $K = k_r/k_f = [A][B]/[C][D]$. If the system is far from equilibrium, as in the initial stages of polymerization, the reverse reaction is negligibly slow, and changes in the concentrations of the reactants may be considered to result from the forward reaction alone, as in Eq. 2-1 and those following.

If the concentrations c of carboxyl and hydroxyl groups are equal,

$$\frac{-dc}{dt} = kc^3 \qquad (2\text{-}2)$$

and

$$2kt = \frac{1}{c^2} - \text{const.}$$

It is convenient to introduce the extent of reaction p, defined as the fraction of the functional groups that has reacted at time t. Then

$$c = c_0(1 - p) \qquad (2\text{-}3)$$

and

$$2c_0^2 kt = \frac{1}{(1 - p)^2} + \text{const.} \qquad (2\text{-}4)$$

or a plot of $1/(1 - p)^2$ should be linear in time. Typical experimental data (Fig. 2-1) bear this out over a wide range of reaction times, but in fact the linear portion corresponds to only a relatively narrow range of conversion, roughly 80–93%.

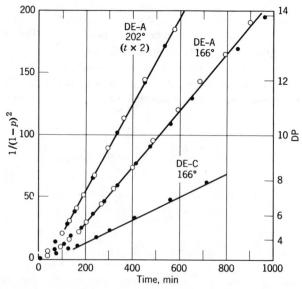

FIG. 2-1.   Reactions of (di)ethylene glycol (DE) with adipic acid (A) and caproic acid (C) (Flory 1946, 1949). Time values at 202°C have been multiplied by 2.

Deviations at lower conversions are attributed to changes in the polarity of the reaction medium as the monomers are converted to polymer (Solomon 1972), and at higher conversions to an increase in the rate of the reverse reaction (Bowden 1979, pp. 179–180).

If only bifunctional reactants are present and no side reactions occur, the number of unreacted carboxyl groups equals the total number of molecules N in the system. If acid or hydroxyl groups separately (not in pairs) are defined as structural units, the initial number of carboxyls present is equal to the total number N_0 of structural units present. The number-average degree of polymerization $\bar{x}_n$ is simply

$$\bar{x}_n = \frac{N_0}{N} = \frac{c_0}{c} = \frac{1}{1 - p} \tag{2-5}$$

The scale of DP in Fig. 2-1 shows that uncatalyzed esterifications require quite long times to reach high degrees of polymerization. Greater success is achieved by adding to the system a small amount of catalyst, whose concentration is constant throughout the reaction. In this case the concentration of the catalyst may be included in the rate constant:

$$- \frac{d[COOH]}{dt} = k'[COOH][OH]$$

$$- \frac{dc}{dt} = k'c^2 \tag{2-6}$$

$$c_0 k' t = \frac{1}{1 - p} + \text{const.}$$

Hence $\bar{x}_n$ increases linearly with reaction time (Fig. 2-2). Similar reactions have been shown to be linear up to at least $\bar{x}_n = 90$, corresponding to molecular weights of 10,000.

Statistics of Linear Step-Reaction Polymerization

Rate equations similar to Eq. 2-6 may be written for each molecular species in the reaction mixture. From these it is possible to derive the distribution of molecular weights in a step-reaction polymer. The same relation is, however, more readily derived from statistical considerations.

The statistical analysis of stepwise polymerization yields results that are equivalent to and in some cases more easily obtained than those of kinetic considerations. The method assumes independence of reaction rate and molecular size. The extent of reaction p is defined as the probability that a functional group has reacted at time t. This definition is entirely equivalent to the previous definition of p as the fraction of the functional groups that has reacted. It follows that the probability of finding a functional group unreacted is $1 - p$.

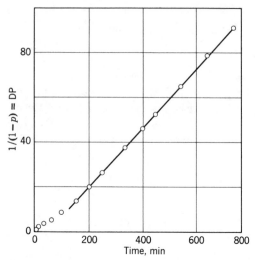

FIG. 2-2.   Reaction of ethylene glycol with adipic acid, catalyzed by 0.4 mol.% of p-toluenesulfonic acid (Flory 1946, 1949).

It is now necessary to find the probability that a given molecule selected at random is an x-mer, that is, contains x repeating units. Such a molecule contains $x - 1$ reacted functional groups (e.g., carboxyls) and, on the end, one unreacted group of this type. The probability of finding a single reacted carboxyl group in the molecule is p, and that of finding $x - 1$ of them in the same molecule is p^{x-1}. The presence of one unreacted group has a probability of $1 - p$. Hence the probability of finding the complete molecule is $p^{x-1}(1 - p)$. This is also equal to the fraction of all the molecules that are x-mers. If there are N molecules in all, the total number of x-mers is

$$N_x = Np^{x-1} (1 - p) \qquad (2\text{-}7)$$

If the total number of units present is N_0, $N = N_0(1 - p)$, and

$$N_x = N_0(1 - p)^2 p^{x-1} \qquad (2\text{-}8)$$

This is the number-distribution function for a linear stepwise polymerization at extent of reaction p.

The weight fraction w_x of x-mers is given by

$$w_x = \frac{xN_x}{N_0}$$

$$= x(1 - p)^2 p^{x-1} \qquad (2\text{-}9)$$

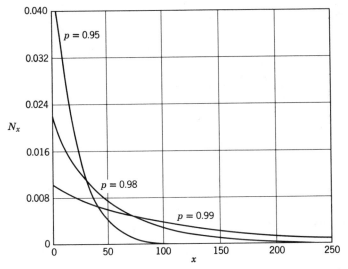

FIG. 2-3. Number- or mole-fraction distribution of chain molecules in a linear step-reaction polymer for several extents of reaction p (Flory 1946, 1949).

This is the weight-distribution function for a linear stepwise polymerization at extent of reaction p. Equations 2-8 and 2-9 are illustrated in Figs. 2-3 and 2-4, respectively. On a number basis, monomers are more plentiful than any other molecular species at all stages of the reaction. On a weight basis, however, the proportion of low-molecular-weight material is very small and decreases as the average molecular

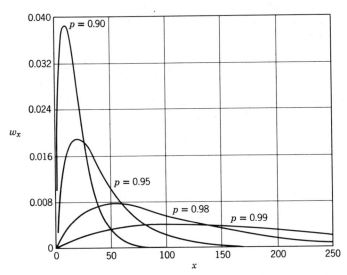

FIG. 2-4. Weight-fraction distribution of chain molecules in a linear step-reaction polymer for several extents of reaction p (Flory 1946, 1949).

weight increases. The maximum in the weight-distribution curve occurs near the number-average molecular weight.

Various average degrees of polymerization may be derived from the weight- or number-distribution functions according to the definitions of the corresponding molecular-weight averages in Chapter 1. The number-average degree of polymerization $\bar{x}_n$ calculated in this way is that given in Eq. 2-5. The weight-average degree of polymerization is

$$\bar{x}_w = \frac{1 + p}{1 - p} \tag{2-10}$$

The breadth of the molecular-weight distribution curve $\bar{x}_w/\bar{x}_n = 1 + p$; at large extents of reaction $\bar{x}_w/\bar{x}_n \to 2.0$. This has been confirmed experimentally for nylon (Howard 1961).

Molecular-Weight Control

Although initial molecular-weight control in a stepwise polymerization can be achieved by stopping the reaction (e.g., by cooling) at the desired point, the stability of the product to subsequent heating may not be adequate with respect to changes in molecular weight.

The easiest way to avoid this situation is to adjust the composition of the reaction mixture slightly away from stoichiometric equivalence, by adding either a slight excess of one bifunctional reactant or (as in molecular-weight stabilization of nylons by acetic acid) a small amount of a monofunctional reagent. Eventually the functional group deficient in amount is completely used up, and all chain ends consist of the group present in excess. If only bifunctional reactants are present, the two types of groups being designated A and B and initially present in numbers $N_A <$ N_B such that the ratio $r = N_A/N_B$, the total number of monomers present is $\frac{1}{2}$ $(N_A + N_B) = \frac{1}{2} N_A(1 + 1/r)$. At extent of reaction p (defined for A groups; for B groups, extent of reaction $= rp$), the total number of chain ends is $N_A(1 - p) + N_B(1 - rp) = N_A[1 - p + (1 - rp)/r]$. Since this is twice the number of molecules present,

$$\bar{x}_n = \frac{N_A(1 + 1/r)/2}{N_A[1 - p + (1 - rp)/r]/2} = \frac{1 + r}{1 + r - 2rp} \tag{2-11}$$

As $p \to 1$,

$$\bar{x}_n = \frac{1 + r}{1 - r} \tag{2-12}$$

Thus if 1 mol.% of stabilizing groups is added,

$$\bar{x}_n = \frac{1 + (100/101)}{1 - (100/101)} = 201$$

This illustrates the precision needed in maintaining stoichiometric balance in order to obtain high degrees of polymerization. Loss of one ingredient, side reactions, or the presence of monofunctional impurities may severely limit the degree of polymerization that can be achieved.

The analysis can be applied to the case of an added monofunctional reagent, retaining Eqs. 2-11 and 2-12 unchanged if r is appropriately defined in terms of the numbers of functional groups present.

Interchange Reactions

In interchange reactions, which may occur freely at elevated temperature, as in polymer melts, the molecular weights of the reacting molecules can change, since, for example, two molecules of average length may react to give one longer and one shorter than average. However, the total number of molecules and hence $\bar{x}_n$ do not change.

It can be shown that if free interchange takes place, the final molecular-weight distribution is always the most probable distribution defined by Eqs. 2-8 and 2-9. The change can be demonstrated by mixing two polymers of different $\bar{x}_n$ and following a weight-average property (such as melt viscosity) as a function of time at elevated temperature.

Multichain Polymer

Another type of step-reaction polymer of interest is produced by polymerizing a bifunctional monomer of type A–B with a small amount of monomer of functionality f of the type R–A$_f$. If the reaction is carried nearly to completion, the resulting polymer consists of f chains growing out from a central unit R. Network structures cannot occur, since no units of the type B–B are present. The case $f = 1$ corresponds to the most probable distribution of Eq. 2-9. The polymer formed when $f = 2$ is also linear; those with higher functionalities are branched.

Since the length of each of the branches is statistically determined independent of all the others, the probability of having a molecule with several branches all longer or shorter than average becomes smaller as the number of branches increases. The molecular-weight distribution thus becomes narrower with increasing functionality, as shown in Fig. 2-5 (Schulz 1939). The distribution breadth is given by

$$\frac{\bar{x}_w}{\bar{x}_n} = 1 + \frac{1}{f} \tag{2-13}$$

For $f = 2$, $\bar{x}_w/\bar{x}_n = 1.5$.

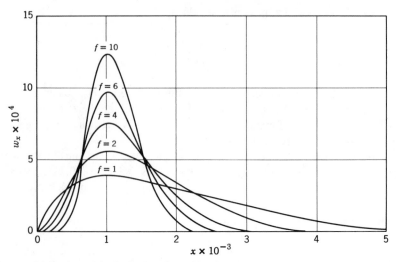

FIG. 2-5. Molecular-weight distributions for multichain polymers with functionality f as shown (Schulz 1939).

GENERAL REFERENCES

Flory 1946, 1949, 1953; Howard 1961; Lenz 1967; Solomon 1972, Chapter 1; Collins 1973, Exp. 10; Allen 1974, Chapter 5; Elias 1977, Chapter 17.2; Bowden 1979, Chapter 2.4; Odian 1981, Chapter 2.2.

D. POLYFUNCTIONAL STEP-REACTION POLYMERIZATION

Three-dimensional step-reaction polymers are produced from the polymerization of reactants with more than two functional groups per molecule. The structures of these polymers are more complex than those of linear step-reaction polymers. Three-dimensional polymerization is complicated experimentally by the occurrence of gelation, or the formation of essentially infinitely large polymer networks in the reaction mixture. The sudden onset of gelation marks the division of the mixture into two parts: the gel, which is insoluble in all nondegrading solvents, and the sol, which remains soluble and can be extracted from the gel. As the polymerization proceeds beyond the gel point, the amount of gel increases at the expense of sol, and the mixture rapidly transforms from a viscous liquid to an elastic material of infinite viscosity. An important feature of the onset of gelation is the low number-average molecular weight of the mixture at the gel point, where its weight-average molecular weight becomes infinite.

In the statistical consideration of three-dimensional step polymerization, it is assumed that all functional groups are equally reactive, independent of molecular weight or viscosity, and that all the reactions occur between functional groups on different molecules. Errors resulting from these assumptions affect only the numerical agreement between theory and experiment, and not the overall considerations of the theory.

Gelation

Prediction of the Gel Point. In order to calculate the point in the reaction at which gelation takes place, a *branching coefficient* α is defined as the probability that a given functional group on a *branch unit* (i.e., a unit of functionality greater than 2) is connected to another branch unit.

The value of α at which gelation becomes possible can be deduced from statistical considerations to be just $1/(f - 1)$; hence the critical value of α for gelation is

$$\alpha_c = \frac{1}{f - 1} \tag{2-14}$$

Here f is the functionality of the branch units; if more than one type of branch unit is present, an average f over all types of branch units may be used in Eq. 2-14.

The relation between α and the extent of reaction can be shown to be

$$\alpha = \frac{p_A p_B \rho}{p_A p_B (1 - \rho)} \tag{2-15}$$

where the extents of reaction for A and B groups are p_A and p_B, and the ratio of A groups on branch units to all A groups in the mixture is ρ. Either p_A or p_B can be eliminated from this expression by defining $r = N_A/N_B$, whence $p_B = rp_A$. Then

$$\alpha = \frac{rp_A^2 \rho}{1 - rp_A^2(1 - \rho)} = \frac{p_B^2 \rho}{r - p_B^2(1 - \rho)} \tag{2-16}$$

Simpler relations can be derived for several cases. When equal numbers of A and B groups are present, $r = 1$ and $p_A = p_B = p$:

$$\alpha = \frac{p^2 \rho}{1 - p^2(1 - \rho)} \tag{2-17}$$

When there are no A–A units, $\rho = 1$ and

$$\alpha = rp_A^2 = \frac{p_B^2}{r} \tag{2-18}$$

If both conditions apply, that is, $r = \rho = 1$,

$$\alpha = p^2 \tag{2-19}$$

Finally, with only branch units present the probability that a functional group on a branch unit leads to another branch unit is just the probability that it has reacted:

$$\text{(branch units only)} \qquad \alpha = p \qquad\qquad (2\text{-}20)$$

The equations hold for all branch-unit functionalities with the definitions given for r and ρ.

Experimental Observations of the Gel Point.   The gel point can be observed precisely as the time when the polymerizing mixture suddenly loses fluidity, for example, when bubbles no longer rise in it. If the extent of reaction has been followed as a function of time, say, by removing aliquots of the solution and titrating for the number of functional groups present, the value of p at the gel point can be determined.

In several cases investigated by Flory, observed values of α_c were always slightly higher than those theoretically required. This discrepancy is attributed to the reaction of some functional groups to form intramolecular links, which do not contribute to network structures. The reactions must therefore be carried slightly further to reach the critical point.

Molecular-Weight Distributions in Three-Dimensional Step-Reaction Polymers

The distribution functions for three-dimensional polymers are derived with somewhat more difficulty than those for the linear case, Eqs. 2-8 and 2-9. They depend upon the functionality and relative amounts of all the units involved. Only one example is discussed, for simplicity that of the reaction of equivalent quantities of two trifunctional monomers, all three of the functional groups of each monomer being equally reactive. In this case the weight-distribution function is

$$w_x = \left[\frac{(fx - x)!f}{(x - 1)!(fx - 2x + 2)!} \right] p^{x-1}(1 - p)^{fx - 2x + 2} \qquad (2\text{-}21)$$

This equation is analogous to Eq. 2-9. The first factor (in brackets) arises because of the numerous geometric isomers of a polyfunctional x-mer. (In the linear case this factor is just x.) The second term is the probability of finding $x - 1$ links in an x-mer, and the last term is the probability of finding $(f - 2)x + 2 = fx - 2x + 2$ unreacted links or ends.

Equation 2-21 is illustrated in Figs. 2-6 to 2-8. In comparison with the linear case, the weight distributions of branched step-reaction polymers of increasing functionality are progressively broader at equivalent extents of reaction (Fig. 2-6). Figure 2-7 shows that the distributions broaden out with increasing extent of reaction p (or, alternatively, α, in the trifunctional case).

In Fig. 2-8 the weight fraction of the various molecular species is plotted against

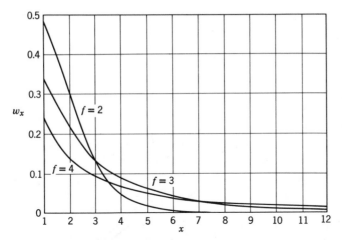

FIG. 2-6. Molecular-weight distribution in stepwise polymerization for monomers with functionality f, at $p = 0.3$ (Mark 1950).

$\alpha = p$. In contrast to the linear case (Figs. 2-7 and 2-8 versus Fig. 2-4), the weight fraction of monomer is always greater than the amount of any one of the other species, the weight fractions of higher species being successively lower. The extent of reaction at which the weight fraction of any species reaches its maximum shifts continuously to higher values for higher molecular weights. In no case, however, does the maximum occur beyond the gel point ($\alpha = \frac{1}{2}$).

Up to the gel point the sum of the weight fractions of all the species present must equal unity; beyond this point the sum of the weight fractions of all finite species drops below unity as the weight fraction of gel increases.

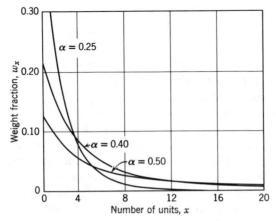

FIG. 2-7. Molecular-weight distribution in step-reaction polymer formed from trifunctional units at various stages in the reaction, denoted by $\alpha = p$ (Flory 1946, 1949).

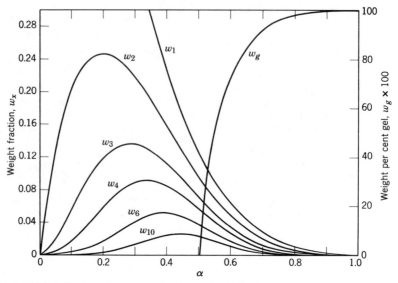

FIG. 2-8. Weight fractions of various molecular species in a trifunctional stepwise polymerization as a function of $\alpha = p$ (Flory 1946, 1949). The weight fractions of the finite species are calculated from Eq. 2-21, and that of the gel from Eq. 2-24.

The number-average degree of polymerization is given by

$$\bar{x}_n = \frac{1}{1 - fp/2} \tag{2-22}$$

and the weight-average degree by

$$\bar{x}_w = xw_x = \frac{1 + p}{1 - (f - 1)p} \tag{2-23}$$

At the gel point the weight-average degree of polymerization becomes infinite. As may be seen in Fig. 2-9, where both averages are plotted against $\alpha = p$, $\bar{x}_n$ attains only a value of 4 at this point. The very large values of $\bar{x}_w/\bar{x}_n$ near the gel point illustrate the extreme breadth of the distributions. (The portions of the curves in Fig. 2-9 above the gel point refer only to the sol fraction.)

The increase in distribution breadth with increasing degree of branching is illustrated in Fig. 2-10. These curves represent the calculated distributions for bifunctional step reactions in which the branching factor α was varied by varying the small amount of added trifunctional units. The curves were calculated for $\bar{x}_n = 50$ in each case. The curves are drawn so that the total area under each curve extended to infinite x is the same.

Equation 2-21 is valid for finite species beyond the gel point. If it is summed over all values of x and subtracted from unity, the weight fraction of gel results.

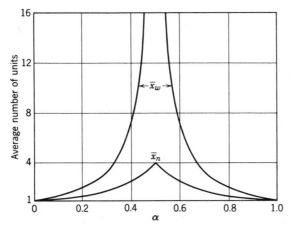

FIG. 2-9. Weight- and number-average degree of polymerization as a function of α for a trifunctional step-reaction polymer (Flory 1946, 1949).

For trifunctional units this is

$$w_g = 1 - \frac{(1 - p)^3}{p^3} \qquad (2\text{-}24)$$

The distribution functions for $\alpha > \frac{1}{2}$ are identical with those at the corresponding lower value $1 - \alpha$, except that they are reduced by a factor $w_s = 1 - w_g$, the total weight fraction of sol present at that value of α. Thus the complexity of the

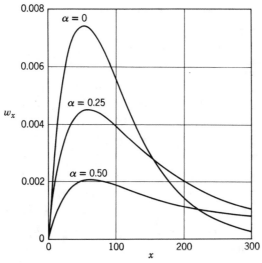

FIG. 2-10. Weight distributions for step-reaction polymers made to $\bar{x}_n = 50$ with various degrees of branching α induced by the addition of trifunctional units: $\alpha = 0$ indicates no branching, and $\alpha = 0.50$ is the critical point for gelation (Flory 1946, 1949).

distribution of the sol fraction above the gel point is presumed to become less until, at $\alpha = 1$, only monomer is left in a vanishingly small amount. The reduction of $\bar{x}_w/\bar{x}_n$ shown for values of α increasing above $\frac{1}{2}$ in Fig. 2-9 is another way of expressing the same phenomenon.

Although the discussion was confined to the case of equivalent quantities of two trifunctional monomers, and Figs. 2-6 to 2-10 refer specifically to this case, Eqs. 2-21 to 2-23 are more broadly applicable if p is replaced with α, which is then determined in the case of interest by r, p, and ρ, as indicated in Eq. 2-16.

GENERAL REFERENCES

Flory 1946, 1949, 1953; Howard 1961; Lenz 1967; Solomon 1972, Chapter 1; Allen 1974, Chapter 5; Elias 1977, Chapter 17.3; Bowden 1979, Chapter 2.4; Odian 1981.

DISCUSSION QUESTIONS AND PROBLEMS

1. Contrast step and chain polymerization with respect to types of growth reaction possible, change in monomer concentration and molecular weight of polymer with conversion, species present during the polymerization, and the effects of long reaction times on molecular weight.

2. Identify and justify the major assumptions in the kinetic treatment of stepwise polymerization.

3. Show by graphs (a) the dependence of $\bar{x}_w$ and $\bar{x}_n$ on conversion, (b) the appropriate function of p versus time, and (c) w_x and N_x versus time in stepwise polymerization.

4. Derive equations for (a) molecular-weight control by added stabilizer and (b) the weight distribution of molecular weight in linear step polymerization.

5. Discuss briefly in step polymerization (a) interchange reactions, (b) ring versus chain formation, (c) the advantages of molecular-weight control by added stabilizer, (d) how to obtain high molecular weights, (e) why an ester is used as a starting material in the preparation of poly(ethylene terephthalate), and (f) why a salt is used as a starting material in the preparation of 66-nylon.

6. Calculate $\bar{x}_n$ and $\bar{x}_w$ for an equimolar mixture of a diacid and a glycol at the following extents of reaction: 0.500, 0.750, 0.900, 0.950, 0.980, 0.990, 0.995.

7. a. Explain the significance of each term in the equation for the weight distribution of molecular weight in linear step polymerization.

   b. Calculate the weight fraction of trimer in such a polymerization carried to 99% conversion.

   c. Calculate $\bar{x}_w$ and $\bar{x}_n$ for the above case.

8. A laboratory preparation of 610-nylon was made with 0.2 mol.% acetic acid (based on carboxyl groups) present as a viscosity stabilizer. The reaction was carried to completion.

a. Calculate $\bar{M}_w$ and $\bar{M}_n$.

b. In a subsequent preparation, by error 2 mol.% acetic acid was used. This reaction was also carried to completion. The two batches of polymer, of equal weight, were mixed and melted prior to spinning. The error was noted when the melt viscosity was found to change with time. Describe qualitatively the dependence of the viscosity on time and discuss its source. Calculate the final values of $\bar{M}_w$ and $\bar{M}_n$ for the mixture.

c. In an effort to salvage some high-molecular-weight polymer, the mixture was extracted with a solvent in which only low-molecular-weight species were soluble. Analysis of the extract showed that it contained species up through the tetramer. Assuming all such species were removed, calculate the weight fraction removed in the extract and the values of $\bar{M}_w$ and $\bar{M}_n$ for the remaining extracted polymer.

9. A polyester, made with equivalent quantities of a dibasic acid and a glycol, is to be stabilized in molecular weight at $\bar{x}_n = 100$ by adding methanol.

a. How much methanol is required?

b. Calculate $\bar{x}_w$ and the weight and number fractions of monomer in the resulting polymer.

c. By error, the same number of moles of hydroxyl was added as glycerol instead of methanol. Will the mixture gel if the reaction is carried to completion?

10. When caprolactam is polymerized into 6-nylon, 10 wt.% of the monomer is left unchanged, and the remaining 90% is converted into polymer having the Flory distribution of molecular weight and $\bar{x}_w = 600$. Later the mixture was extracted by a process that removed all species up through that with $x = 5$. Calculate $\bar{x}_n$ for the following:

a. The polymer with $\bar{x}_w = 600$.

b. The mixture of 10% monomer and 90% polymer.

c. The extract, excluding the cyclic monomer.

d. The extracted polymer.

BIBLIOGRAPHY

Allcock 1981. Harry R. Allcock and Frederick W. Lampe, *Contemporary Polymer Chemistry*, Prentice-Hall, Englewood Cliffs, New Jersey, 1981.

Allen 1974. P. E. M. Allen and C. R. Patrick, *Kinetics and Mechanisms of Polymerization Reactions*, John Wiley & Sons, New York, 1974.

Bowden 1979. M. J. Bowden, "Formation of Macromolecules," Chapter 2 in F. A. Bovey and F. H. Winslow, eds., *Macromolecules: An Introduction to Polymer Science,* Academic Press, New York, 1979.

Carothers 1929. W. H. Carothers, "An Introduction to the General Theory of Condensation Polymers," *J. Am. Chem. Soc.* **51,** 2548–2559 (1929).

Carraher 1982. Charles E. Carraher and Jack Preston, eds., *Interfacial Synthesis, Recent Advances,* Vol. 3, Marcel Dekker, New York, 1982.

Collins 1973. Edward A. Collins, Jan Bareš, and Fred W. Billmeyer, Jr., *Experiments in Polymer Science,* Wiley-Interscience, New York, 1973.

Elias 1977. Hans-Georg Elias, *Macromolecules · 2 · Synthesis and Materials* (translated by John W. Stafford), Plenum Press, New York, 1977.

Flory 1946. Paul J. Flory, "Fundamental Principles of Condensation Polymerization," *Chem. Rev.* **39,** 137–197 (1946).

Flory 1949. Paul J. Flory, "Condensation Polymerization and Constitution of Condensation Polymers," pp. 211–283 in R. E. Burk and Oliver Grummitt, eds., *High Molecular Weight Organic Compounds* (*Frontiers in Chemistry*), Vol. 6, Interscience, New York, 1949.

Flory 1953. Paul J. Flory, *Principles of Polymer Chemistry,* Cornell University Press, Ithaca, New York, 1953.

Howard 1961. G. J. Howard, "The Molecular Weight Distribution of Condensation Polymers," pp. 185–231 in J. C. Robb and F. W. Peaker, eds., *Progress in High Polymers,* Vol. 1, Academic Press, New York, 1961.

Lenz 1967. Robert W. Lenz, *Organic Chemistry of Synthetic High Polymers,* Wiley-Interscience, New York, 1967.

Mark 1940. H. Mark and G. Stafford Whitby, eds., *Collected Papers of Wallace Hume Carothers on High Polymeric Substances,* Interscience, New York, 1940.

Mark 1950. H. Mark and A. V. Tobolsky, *Physical Chemistry of High Polymeric Systems,* Interscience, New York, 1950.

Marvel 1959. C. S. Marvel, *An Introduction to the Organic Chemistry of High Polymers,* John Wiley & Sons, New York, 1959.

Millich 1977. Frank Millich and Charles E. Carraher, Jr., eds., *Interfacial Synthesis,* Vol. 1; *Fundamentals,* Vol. 2, *Polymer Applications and Technology,* Marcel Dekker, New York, 1970.

Morgan 1965. Paul W. Morgan, *Condensation Polymers: By Interfacial and Solution Methods,* Wiley-Interscience, New York, 1965.

Odian 1981. George Odian, *Principles of Polymerization,* 2nd ed., John Wiley & Sons, New York, 1981.

Ravve 1967. A. Ravve, *Organic Chemistry of Macromolecules,* Marcel Dekker, New York, 1967.

Schulz 1939. G. V. Schulz, "The Kinetics of Chain Polymerization. V. The Effect of Various Reaction Species on the Polymolecularity" (in German), *Z. Phys. Chem.* **B43,** 25–46 (1939).

Solomon 1972. D. H. Solomon, ed., *Step-Growth Polymerizations,* Vol. 3 in George E. Ham, ed., *Kinetics and Mechanisms of Polymerization,* Marcel Dekker, New York, 1972.

Sorenson 1968. Wayne R. Sorenson and Tod W. Campbell, *Preparative Methods of Polymer Chemistry,* 2nd ed., Wiley-Interscience, New York, 1968.

Vollmert 1973. Bruno Vollmert, *Polymer Chemistry* (translated by Edmund H. Immergut), Springer-Verlag, New York, 1973.

CHAPTER THREE

RADICAL CHAIN (ADDITION) POLYMERIZATION

A. MECHANISM OF VINYL POLYMERIZATION

The polymerization of unsaturated monomers typically involves a chain reaction. It can be initiated by methods typical for simple gas-phase chain reactions, including the action of ultraviolet light. It is susceptible to retardation and inhibition. In a typical chain polymerization, one act of initiation may lead to the polymerization of thousands of monomer molecules.

The characteristics of chain polymerization listed in Table 2-1 suggest that the active center responsible for the growth of the chain is associated with a single polymer molecule through the addition of many monomer units. Thus polymer molecules are formed from the beginning, and almost no species intermediate between monomer and high-molecular-weight polymer are found. Of several postulated types of active center, three have been found experimentally: cation, anion, and free radical. Free-radical polymerization is discussed in this chapter, and the related cases of ionic and coordination polymerization are described in Chapter 4.

The concept of vinyl polymerization as a chain mechanism is not new, dating back to Staudinger's work in 1920. In 1937 Flory showed conclusively that radical polymerization proceeds by and requires the steps of initiation, propagation, and termination typical of chain reactions in low-molecular-weight species.

The carbon–carbon double bond is, because of its relatively low stability, particularly susceptible to attack by a free radical. The reaction of the double bond with a radical proceeds well for compounds of the type CH_2=CHX and CH_2=CXY, called vinyl monomers. (Monomers in which fluorine is substituted for hydrogen may be included in this class.) The polymerization of monomers with more than one double bond is considered in Chapter 6.

Not all vinyl monomers yield high polymer as a result of radical polymerization. Aliphatic hydrocarbons other than ethylene polymerize only to oils, and 1,2-di-substituted ethylenes not at all. Among compounds of the type $CH_2=CXY$, those in which both groups are larger than CH_3 polymerize slowly, if at all, by radical mechanisms.

Generation of Free Radicals.   Many organic reactions take place through inter-mediates having an odd number of electrons and, consequently, an unpaired elec-tron. Such intermediates are known as free radicals. They can be generated in a number of ways, including thermal decomposition of organic peroxides or hydro-peroxides (Mageli 1968) or azo or diazo compounds (Zand 1965).

Two reactions commonly used to produce radicals for polymerization are the thermal or photochemical decomposition of benzoyl peroxide,

$$(C_6H_5COO)_2 \longrightarrow 2C_6H_5COO\cdot \longrightarrow 2C_6H_5\cdot + 2CO_2$$

and of azobisisobutyronitrile,

$$(CH_3)_2CN{=}NC(CH_3)_2 \longrightarrow 2(CH_3)_2C\cdot + N_2$$
$$\quad\; | \qquad\quad | \qquad\qquad\qquad\quad |$$
$$\;\; CN \quad\;\; CN \qquad\qquad\qquad\;\; CN$$

The stability of radicals varies widely. Primary radicals are less stable and more reactive than secondary radicals, which are in turn less stable than tertiary ones. (Some tertiary radicals, such as the triphenylmethyl, can be isolated in the solid state without decomposition.) The phenyl radical is more reactive than the benzyl radical, the allyl radical is quite unreactive, and so on.

To be useful in initiating polymerization, a compound undergoing thermal de-composition to free radicals should have a first-order decomposition rate constant of 10^{-5} to 10^{-6} sec^{-1} at the desired polymerization temperature, usually 50–150°C.

Although thermal decomposition is a common means of generating radicals, it has a disadvantage in that the rate of generation of free radicals cannot be controlled rapidly because of the heat capacity of the system. *Photoinitiated polymerization,* on the other hand, can be controlled with high precision, since the generation of radicals can be made to vary instantaneously by controlling the intensity of the initiating light (Oster 1969). Light of short enough wavelength (i.e., high enough energy per quantum) can initiate polymerization directly. It is customary, however, to use a *photochemical initiator* such as benzoin or azobisisobutyronitrile, which is decomposed into free radicals by ultraviolet light in the 3600-Å region, where direct initiation through decomposition of monomer does not occur. In photo-polymerization these initiators are used at temperatures low enough so that they do not undergo appreciable thermal decomposition. Convenient sources of 3600-Å light are available in the form of mercury-arc lamps or fluorescent lamps with special phosphors, and lasers.

High-energy radiation from a wide variety of sources, including electrons, gamma rays, x-rays, and slow neutrons, is effective in producing free radicals that can initiate polymerization (Dole 1972, Wilson 1974). The reactions involved are varied and nonspecific, resulting from the gross damage to molecular structures resulting from the transfer of large amounts of energy. Monomers can be polymerized in the solid as well as the liquid and gaseous states (Eastmond 1970). The reactions of high-energy radiation with polymers are discussed in Chapter 6D.

Radicals are also generated as a result of oxidation–reduction reactions (*redox initiation*). Appropriate to an aqueous medium, redox initiation is often used with emulsion polymerization, and is discussed with that topic in Chapter 6B. *Electrochemical initiation* (Friedlander 1966, Funt 1967) results from the generation of free radicals by electrode reactions.

In all of these methods it is only the means of generating the free radicals that varies; the remaining steps in the polymerization process are unchanged.

The generation of radicals for the initiation of polymerization has been reviewed (Bevington 1961, North 1966, O'Driscoll 1969).

Initiation. When free radicals are generated in the presence of a vinyl monomer, the radical adds to the double bond with the regeneration of another radical. If the radical formed by decomposition of the initiator I is designated by $R\cdot$,

$$I \longrightarrow 2R\cdot$$

$$R\cdot + CH_2{=}CHX \longrightarrow RCH_2\overset{\displaystyle H}{\underset{\displaystyle X}{\overset{|}{\underset{|}{C}}}}\cdot$$

The regeneration of the radical is characteristic of chain reactions.

Evidence for the radical mechanism of addition polymerization comes not only from the capability of radicals to accelerate vinyl polymerization, but also from the demonstration that the polymers so formed contain fragments of the radicals. The presence of heavy atoms, such as bromine or iodine, or radioactive atoms in the initiator has been shown many times to lead to polymers from which these atoms cannot be removed.

The efficiency with which radicals initiate chains can be estimated by comparing the amount of initiator decomposed with the number of polymer chains formed. The decomposition of the initiator can usually be followed by analytic methods. The most direct method of finding the initiator efficiency then depends upon analyzing the polymer for initiator fragments. Most initiators in typical vinyl polymerizations have efficiencies between 0.6 and 1.0; that is, between 60 and 100% of all the radicals formed ultimately initiate polymer chains. The major cause of low efficiency is recombination of the radical pairs before they move apart (cage effect) (Noyes 1965).

Propagation. The chain radical formed in the initiation step is capable of adding successive monomers to propagate the chain:

$$
R—(CH_2CHX—)_xCH_2\overset{H}{\underset{X}{C}}\cdot + CH_2{=}CHX \longrightarrow R—(CH_2CHX—)_{x+1}CH_2\overset{H}{\underset{X}{C}}\cdot
$$

Termination. Propagation would continue until the supply of monomer was exhausted were it not for the strong tendency of radicals to react in pairs to form a paired-electron covalent bond with loss of radical activity. This tendency is compensated for in radical polymerization by the small concentration of radical species compared to monomers.

The termination step can take place in two ways: *combination* or *coupling*;

$$
—CH_2\overset{H}{\underset{X}{C}}\cdot + \cdot\overset{H}{\underset{X}{C}}CH_2— \longrightarrow —CH_2\overset{H}{\underset{X}{C}}{-}\overset{H}{\underset{X}{C}}CH_2—
$$

or *disproportionation*;

$$
—CH_2\overset{H}{\underset{X}{C}}\cdot + \cdot\overset{H}{\underset{X}{C}}CH_2— \longrightarrow —CH_2\overset{H}{\underset{X}{C}}{-}H + \overset{H}{\underset{X}{C}}{=}CH—
$$

in which hydrogen transfer results in the formation of two molecules with one saturated and one unsaturated end group. Each type of termination is known. For example, studies (Bevington 1954*a,b*) of the number of initiator fragments per molecule showed that polystyrene terminates predominantly by combination, whereas poly(methyl methacrylate) terminates entirely by disproportionation at polymerization temperatures above 60°C, and partly by each mechanism at lower temperatures.

The termination reaction, controlled by diffusion in many cases, has been reviewed (North 1974).

Radical–Molecule Reactions

Although the three steps of initiation, propagation, and termination are both necessary and sufficient for chain polymerization, other steps can take place during polymerization. As these often involve the reaction between a radical and a molecule, they are conveniently so classified.

Chain Transfer. It was recognized by Flory (1937) that the reactivity of a radical could be transferred to another species, which would usually be capable of continuing the chain reaction. The reaction involves the transfer of an atom between the radical and the molecule. If the molecule is saturated, like a solvent or other additive, the atom must be transferred to the radical:

$$
\begin{array}{c}
\quad\ \text{H} \qquad\qquad\qquad\qquad\qquad \text{H} \\
\quad\ | \qquad\qquad\qquad\qquad\qquad\ | \\
-\text{CH}_2\text{C}\cdot + \text{CCl}_4 \longrightarrow -\text{CH}_2\text{CCl} + \cdot\text{CCl}_3 \\
\quad\ | \qquad\qquad\qquad\qquad\qquad\ | \\
\quad\ \text{X} \qquad\qquad\qquad\qquad\qquad \text{X}
\end{array}
$$

If the molecule is unsaturated, like a monomer, the atom transferred (usually hydrogen) can go in either direction:

$$
\begin{array}{c}
\qquad\qquad\qquad\qquad\qquad\qquad -\text{CH}_2\text{CH}_2\text{X} + \text{CH}_2{=}\text{C}\cdot \\
\qquad\qquad\qquad\qquad\qquad\qquad\qquad\qquad\qquad\qquad | \\
\quad\ \text{H} \qquad\qquad\qquad\quad \nearrow \qquad\qquad\qquad\qquad\ \text{X} \\
\quad\ | \\
-\text{CH}_2\text{C}\cdot + \text{CH}_2{=}\text{CHX} \\
\quad\ | \qquad\qquad\qquad\quad \searrow \qquad\qquad\qquad\qquad\ \text{H} \\
\quad\ \text{X} \qquad\qquad\qquad\qquad\qquad\qquad\qquad\qquad | \\
\qquad\qquad\qquad\qquad\qquad\qquad -\text{CH}{=}\text{CHX} + \text{CH}_3\text{C}\cdot \\
\qquad\qquad\qquad\qquad\qquad\qquad\qquad\qquad\qquad\qquad | \\
\qquad\qquad\qquad\qquad\qquad\qquad\qquad\qquad\qquad\ \text{X}
\end{array}
$$

The major effect of chain transfer to a saturated small molecule (solvent, initiator, or deliberately added *chain-transfer agent*) is the formation of additional polymer molecules for each radical chain initiated. Transfer to polymer and transfer to monomer with subsequent polymerization of the double bond lead to the formation of branched molecules. The latter reaction has a pronounced effect on molecular-weight distribution (Section *C*) and is important in the production of graft copolymers (Chapter 5*D*).

The efficiency of compounds as chain-transfer agents varies widely with molecular structure. Aromatic hydrocarbons are rather unreactive unless they have benzylic hydrogens. Aliphatic hydrocarbons become more reactive when substituted with halogens. Carbon tetrachloride and carbon tetrabromide are quite reactive. The reactivity of various polymer radicals to transfer varies widely. Transfer reactions offer a valuable means for comparing radical reactivities.

Inhibition and Retardation. A retarder is defined as a substance that can react with a radical to form products incapable of adding monomer. If the retarder is very effective, no polymer may be formed; this condition is sometimes called *inhibition* and the substance an *inhibitor*. The distinction is merely one of degree. The two phenomena are illustrated in Fig. 3-1, where the rate of polymerization is displayed in idealized fashion. The action of a retarder is twofold: It both reduces the concentration of radicals and shortens their average lifetime and thus the length of the polymer chains.

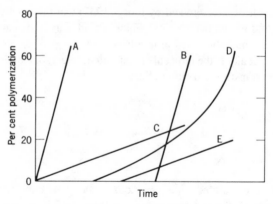

FIG. 3-1. Idealized time-conversion curves for typical inhibition and retardation effects (Goldfinger 1967): curve A, normal polymerization; curve B, simple inhibition; curve C, simple retardation; curve D, nonideal inhibition; curve E, inhibition followed by retardation.

In the simplest case, the retarder may be a free radical, such as triphenylmethyl or diphenylpicrylhydrazyl, which is too unreactive to initiate a polymer chain. The mechanism of retardation is simply the combination or disproportionation of radicals. If the retarder is a molecule, the chemistry of retardation is more complex, but the product of the reaction must be a radical too unreactive to initiate a chain.

Inhibitors are useful in determining initiation rates, since their reaction with radicals is so rapid that the decomposition of inhibitor is independent of its concentration but gives directly the rate of generation of radicals. As a result, the length of the *induction period* before polymerization starts is directly proportional to the number of inhibitor molecules initially present. This number then represents the number of radicals produced during the time of the induction period. For this analysis to be valid it must be known that one molecule of inhibitor is exactly equivalent to one radical. There are a number of reasons why this may not be so, and each case must be investigated separately.

Configuration of Monomer Units in Vinyl Polymer Chains

The addition of a vinyl monomer to a free radical can take place in either of two ways:

$$R\cdot + CH_2{=}CHX \nearrow \quad R{-}CH_2{-}\underset{\underset{X}{|}}{CH}\cdot \quad (I)$$

$$\searrow \quad R{-}\underset{\underset{X}{|}}{CH}{-}CH_2\cdot \quad (II)$$

the reaction leading to the more stable product being favored. Since the unpaired electron can participate in resonance with the substituent X in structure I but not in II, reaction I is favored. Steric factors also favor reaction I.

The occurrence of reaction I exclusively would lead to a *head-to-tail* configuration in which the substituents occur on alternate carbon atoms:

$$-CH_2-\underset{\underset{X}{|}}{CH}-CH_2-\underset{\underset{X}{|}}{CH}-CH_2-\underset{\underset{X}{|}}{CH}-, \quad \text{etc.}$$

Alternative possibilities are a *head-to-head, tail-to-tail* configuration:

$$-CH_2-\underset{\underset{X}{|}}{CH}-\underset{\underset{X}{|}}{CH}-CH_2-CH_2-\underset{\underset{X}{|}}{CH}-\underset{\underset{X}{|}}{CH}-CH_2, \quad \text{etc.}$$

or a random structure containing both arrangements. The possibility of obtaining a regular head-to-head, tail-to-tail configuration exclusively by chain polymerization is remote, and it appears that only an occasional monomer unit enters the chain in a reverse manner to provide an isolated head-to-head, tail-to-tail linkage.

Laboratory Methods in Vinyl Polymerization

Small-scale laboratory polymerizations, carried to low conversion with pure monomer and minor amounts of initiator and other agents, are of great importance in obtaining kinetic data for substantiating the kinetics and mechanisms of polymerizations. Laboratory manuals are available describing the methods and experiments that are most useful (Sorenson 1968, McCaffery 1970, Braun 1971, Collins 1973). The experimental data of interest include the overall rate of polymerization, the rate of initiation (and transfer where present), and the molecular weight of the polymer. The determination of molecular weight is discussed in Chapter 8. Industrial-scale polymerization is discussed in Chapter 6 and Part 4.

Rate of Polymerization. A chemical method or isolation and weighing of polymers can be used to measure the rate of polymerization, but it is customary to work in an evacuated and sealed system to avoid extraneous effects due to the presence of oxygen. Polymerization rate is then followed through changes in a physical property of the system, such as density, refractive index, or ultraviolet or infrared absorption. Measurement of density is usually the most sensitive and convenient technique. The reaction is carried out in a *dilatometer*, a vessel equipped with a capillary tube in which the liquid level can be measured precisely. The decrease in volume on polymerization is relatively large, being 21% for methyl methacrylate and 27% for vinyl acetate. In practice, a few hundredths percent polymerization can be detected. The experiments are usually carried out in the region below 5% polymer so as to avoid deviations from constant polymerization rate due to depletion of reactants and other factors. The dilatometer is placed in a constant-temperature bath to ensure isothermal conditions throughout the experiment.

Rate of Initiation. The number of polymer chains initiated per unit time can be calculated from the number of free radicals produced in the system, if the initiator efficiency is known. In photopolymerization the number of free radicals produced is twice the number of quanta absorbed, since one quantum of light decomposes a single initiator molecule into two free radicals. The number of quanta absorbed in the reaction vessel can be measured in a separate experiment.

The initiation rate obtained in this way can be checked by computing the number of polymer molecules formed per unit time from the overall rate of polymerization and $\bar{x}_n$, provided that chain transfer is absent and that the mechanism of termination is known. Since this is usually not the case, an independent check on the rate of initiation is desirable. This may be obtained by the use of an inhibitor or retarder for the polymerization.

GENERAL REFERENCES

Marvel 1959; Bevington 1961; Ham 1967; Lenz 1967; Collins 1973; Allen 1974; Tedder 1974; Bamford 1976; Allcock 1981, Chapters 3 and 5.

B. KINETICS OF VINYL RADICAL POLYMERIZATION

Conversion of Monomer to Polymer

The chemical equations of Section A for initiation, propagation, and termination of vinyl radical polymerization can be generalized in the following mathematical scheme.

Initiation in the presence of an initiator I may be considered in two steps: first, the rate-determining decomposition of the initiator into free radicals $R \cdot$,

$$\text{I} \xrightarrow{k_d} 2R \cdot \tag{3-1}$$

and, second, the addition of a monomer unit to form a chain radical $M_1 \cdot$,

$$R \cdot + M \xrightarrow{k_a} M_1 \cdot \tag{3-2}$$

where the k's in these and subsequent equations are *rate constants*, with subscripts designating the reactions to which they refer.

The successive steps in propagation,

$$M_1\cdot + M \xrightarrow{k_p} M_2\cdot$$
$$M_2\cdot + M \xrightarrow{k_p} M_3\cdot$$
$$\vdots$$
$$M_x\cdot + M \xrightarrow{k_p} M_{x+1}\cdot$$

$\qquad$ (3-3)

are assumed all to have the same rate constant k_p, since radical reactivity is presumed to be independent of chain length.

The termination step involves combination

$$M_x\cdot + M_y\cdot \xrightarrow{k_{tc}} M_{x+y}$$

$\qquad$ (3-4)

or disproportionation

$$M_x\cdot + M_y\cdot \xrightarrow{k_{td}} M_x + M_y$$

$\qquad$ (3-5)

Except where it is necessary to distinguish between the two mechanisms, the termination rate constant is denoted k_t.

The rates of the three steps may be written in terms of the concentrations (in brackets) of the species involved and the rate constants. The rate of initiation is

$$v_i = \left(\frac{d[M\cdot]}{dt}\right)_i = 2fk_d[I]$$

$\qquad$ (3-6)

where the factor f represents the fraction of the radicals formed by Eq. 3-1 that is successful in initiating chains by Eq. 3-2. The rate of termination is

$$v_t = -\left(\frac{d[M\cdot]}{dt}\right)_t = 2k_t[M\cdot]^2$$

$\qquad$ (3-7)

For many cases of interest the concentration of free radicals $[M\cdot]$ becomes essentially constant very early in the reaction, as radicals are formed and destroyed at identical rates. In this *steady-state* condition $v_i = v_t$ and Eqs. 3-6 and 3-7 may be equated to solve for $[M\cdot]$:

$$[M\cdot] = \left(\frac{fk_d[I]}{k_t}\right)^{1/2}$$

$\qquad$ (3-8)

The rate of propagation is essentially the same as the overall rate of disappearance of monomer, since the number of monomers used in Eq. 3-2 must be small compared to that used in Eq. 3-3 if polymer is obtained. Then

$$v_p = -\frac{d[M]}{dt} = k_p[M][M\cdot]$$
(3-9)

or, with substitution from Eq. 3-8,

$$v_p = k_p \left(\frac{f k_d[I]}{k_t}\right)^{1/2} [M]$$
(3-10)

Thus the overall rate of polymerization should, in the early stages of the reaction, be proportional to the square root of the initiator concentration and, if f is independent of [M], to the first power of the monomer concentration. This is true if the initiator efficiency is high. With very low efficiency, f may be proportional to [M], making v_p proportional to $[M]^{3/2}$.

The proportionality of the overall rate to the square root of initiator concentration has been confirmed experimentally in a large number of cases. Typical experimental data are shown in Fig. 3-2. The straight lines are drawn with the theoretical slope of $\frac{1}{2}$.

The deviation for polystyrene at low initiator concentration (Fig. 3-2c) is due to the presence of some thermal initiation. This phenomenon, peculiar to very pure

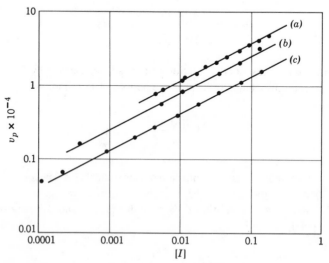

FIG. 3-2. Dependence of rate of polymerization v_p on concentration of initiator [I] for (a) methyl methacrylate with azobisisobutyronitrile, (b) methyl methacrylate with benzoyl peroxide, and (c) styrene with benzoyl peroxide (Mayo 1959).

styrene and a few of its derivatives, without any initiator present, involves the thermal dimerization of styrene followed by a reaction with still another styrene to produce a radical capable of initiating polymerization (Ebdon 1971).

Evaluation of Rate Constants. If the rate constant for initiator decomposition and the initiator efficiency are known, the ratio of rate constants k_p^2/k_t can be evaluated from the overall polymerization rate: From Eqs. 3-6 and 3-10 we have

$$\frac{k_p^2}{k_t} = \frac{2v_p^2}{v_i[M]^2} \tag{3-11}$$

These rate constants can be separated only by non-steady-state methods, described below. Measurements of k_p^2/k_t as a function of temperature lead to useful information about the thermochemistry of polymerization (Section C).

Overall Rate as a Function of Conversion. If the initiator concentration does not vary much during the course of polymerization and the initiator efficiency is independent of monomer concentration, polymerization proceeds by first-order kinetics; that is, the polymerization rate is proportional to monomer concentration. In some systems, such as the benzyol peroxide-initiated polymerization of styrene, the reaction is accurately first order up to quite high conversions (Fig. 3-3).

The polymerization of certain monomers undiluted or in concentrated solution is accompanied by a marked deviation from first-order kinetics in the direction of an increase in reaction rate and molecular weight termed *autoacceleration* or the

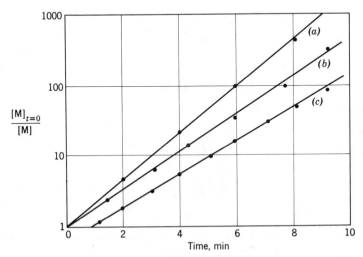

FIG. 3-3. Polymerization of vinyl phenylbutyrate in dioxane with benzoyl peroxide initiator at 60°C plotted as a first-order reaction (Marvel 1940). At $t = 0$, monomer concentration was (a) 2.4 g/100 ml, (b) 6.0 g/100 ml, and (c) 7.3 g/100 ml.

gel effect (Norrish 1939, Schulz 1947, Trommsdorff 1948). The effect is particularly pronounced with methyl methacrylate, methyl acrylate, or acrylic acid (Fig. 3-4). It is independent of initiator and is due to a decrease in the rate at which the polymer molecules diffuse through the viscous medium, thus lowering the ability of two long-chain radicals to come together and terminate. Because termination is in fact diffusion controlled for most liquid-phase polymerizations, even at low conversion, the dependence of diffusion rate on the viscosity of the medium leads to the gel effect at high polymer concentrations, with high-molecular-weight polymer, or in the presence of inert solutes increasing the viscosity of the medium (North 1974).

The decrease in termination rate leads to an increase in overall polymerization rate and in molecular weight, since the lifetime of the growing chains is increased. At quite high conversions (70–90%) the rate of polymerization drops to a very low value as the system becomes glassy and monomer can no longer be supplied to the growing chain ends.

Degree of Polymerization and Chain Transfer

Kinetic Chain Length. The *kinetic chain length* v is defined as the number of monomer units consumed per active center. It is therefore given by $v_p/v_i = v_p/v_t$:

$$v = \frac{k_p}{2k_t} \frac{[M]}{[M \cdot]}$$

(3-12)

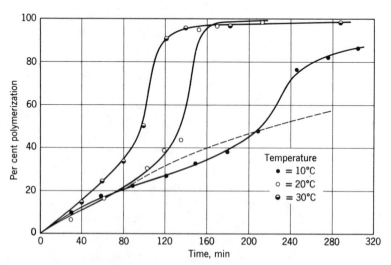

FIG. 3-4. Autoacceleration of polymerization rate in poly(methyl methacrylate) (Naylor 1953).

Eliminating the radical concentration by means of Eq. 3-9,

$$v = \frac{k_p^2 \, [\text{M}]^2}{2k_t \, v_p} \tag{3-13}$$

For initiated polymerization, use of Eq. 3-8 leads to

$$v = \frac{k_p}{2(f k_d k_t)^{1/2}} \frac{[\text{M}]}{[\text{I}]^{1/2}} \tag{3-14}$$

If no reactions take place other than those already discussed, the kinetic chain length should be related to $\bar{x}_n$, the degree of polymerization: For termination by combination $\bar{x}_n = 2v$, and for disproportionation $\bar{x}_n = v$. This is found to be precisely true for some systems, but for others wide deviations are noted in the direction of more polymer molecules than active centers. These deviations are the result of chain-transfer reactions:

$$\text{M}_x\cdot + \text{XP} \longrightarrow \text{M}_x\text{M} + \text{P}\cdot$$

$$\text{P}\cdot + \text{M} \longrightarrow \text{PM}\cdot, \quad \text{etc.}$$

where P may be monomer, initiator, solvent, or other added chain-transfer agent. (Transfer to polymer is omitted, since no new polymer molecule is produced.)
The degree of polymerization is therefore

$$\bar{x}_n = \frac{\text{rate of growth}}{\Sigma \text{ rates of all reactions leading to dead polymer}}$$

$$= \frac{v_p}{f k_d[\text{I}] + k_{tr,\text{M}}[\text{M}][\text{M}\cdot] + k_{tr,\text{S}}[\text{S}][\text{M}\cdot] + k_{tr,\text{I}}[\text{I}][\text{M}\cdot]} \tag{3-15}$$

where the terms in the denominator represent termination by combination and transfer to monomer, solvent, and initiator, respectively. If termination is by disproportionation, the first term becomes $2f k_d[\text{I}]$. If transfer constants are defined as

$$C_\text{M} = \frac{k_{tr,\text{M}}}{k_p}, \qquad C_\text{S} = \frac{k_{tr,\text{S}}}{k_p}, \qquad C_\text{I} = \frac{k_{tr,\text{I}}}{k_p} \tag{3-16}$$

then (assuming termination by combination)

$$\frac{1}{\bar{x}_n} = \frac{k_t}{k_p^2} \frac{v_p}{[\text{M}]^2} + C_\text{M} + C_\text{S} \frac{[\text{S}]}{[\text{M}]} + C_\text{I} \frac{k_t}{k_p^2 f k_d} \frac{v_p^2}{[\text{M}]^3} \tag{3-17}$$

The above analysis assumes that the radical formed in the transfer process is approximately as reactive as the original chain radical, otherwise retardation or inhibition results.

Transfer to Solvent. In the presence of a solvent, and by properly choosing conditions to keep other types of chain transfer to a minimum, Eq. 3-17 reduces to

$$\frac{1}{\bar{x}_n} = \left(\frac{1}{\bar{x}_n}\right)_0 + C_S \frac{[S]}{[M]} \tag{3-18}$$

where $(1/\bar{x}_n)_0$ combines the polymerization and transfer to monomer terms. The linear dependence of $1/\bar{x}_n$ on $[S]/[M]$ is illustrated in Fig. 3-5. In experiments of this type the ratio $v_p/[M]^2$ must be held constant throughout if $(\bar{x}_n)_0$ is to be identified with the observed degree of polymerization in the absence of solvent.

Transfer to Monomer and Initiator. If only the terms corresponding to polymerization, transfer to monomer, and transfer to initiator are kept in Eq. 3-17, the resulting equation for $\bar{x}_n$ is quadratic in v_p. This behavior has been observed, for

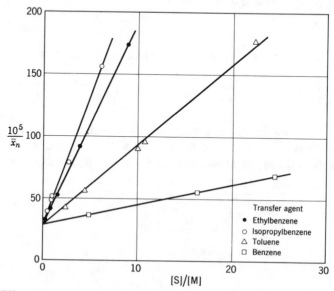

FIG. 3-5. Effect of chain transfer to solvent on the degree of polymerization of polystyrene (Gregg 1947).

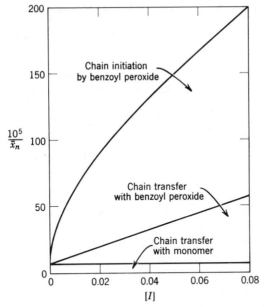

FIG. 3-6. Contribution of various sources of termination in the polymerization of styrene with benzoyl peroxide at 60°C (Mayo 1951).

example, in the benzoyl peroxide-initiated polymerization of styrene. The contribution of transfer to monomer is constant and independent of polymerization rate, and that of transfer to initiator increases rapidly with increasing rate, since high rate requires high initiator concentration (Fig. 3-6).

Control of Molecular Weight by Transfer. Chain-transfer agents with transfer constants near unity are quite useful in depressing molecular weight in polymerization reactions. This is often of great commercial importance, for example, in the polymerization of dienes to synthetic rubbers, where chain length must be controlled for ease of processing.

The choice of transfer constant near unity ensures that the transfer agent, or *regulator*, is consumed at the same rate as the monomer so that [S]/[M] remains constant throughout the reaction. Too large quantities are needed of chain-transfer agents with constants much lower than unity, and agents with transfer constants greater than about 5 are used up too early in the polymerization. Aliphatic mercaptans are suitable transfer agents for several common monomers.

Table 3-1 lists transfer constants for several solvents and radicals. The effect of chain-transfer reactions on chain length and structure is summarized in Table 3-2. The chemistry and kinetics of chain transfer have been reviewed (Palit 1965, Jenkins 1974).

TABLE 3-1.   Chain Transfer Constants for Various Solvents and Radicals at 60°C Except as Noted[a]

Solvent	$C_s \times 10^4$		
	Styrene	Methyl Methacrylate	Vinyl Acetate
Benzene	0.023	0.040	1.2
Cyclohexane	0.031	0.10 (80°C)	7.0
Toluene	0.125	0.20	22
Chloroform	0.5	1.77	150
Ethylbenzene	0.67	1.35 (80°C)	5.5
Triethylamine	7.1	8.3	370
Tetrachloroethane	18 (80°C)	0.155	107
Carbon tetrachloride	90	2.40	9600
Carbon tetrabromide	22,000	2700	28,700 (70°C)

[a] Young (1975).

Inhibition and Retardation

The kinetics of chain polymerization in the presence of added inhibitor or retarder can be described (Goldfinger 1967; Odian 1981) by adding to the scheme of Eqs. 3-1 to 3-5 the reaction

$$M_x\cdot + Z \xrightarrow{k_z} M_x + Z\cdot \tag{3-19}$$

where Z is the inhibitor. It is assumed that the radical Z· does not initiate polymerization, and is terminated without regenerating Z. Application of the steady-

TABLE 3-2.   Effect of Chain-Transfer Reactions on Chain Length and Structure[a]

Reaction	Effect on			
	v_p	$\bar{M}_n$	$\bar{M}_w$	Structure
To small molecule, giving active radical	None	Decreases	Decreases	None
To small molecule, retardation or inhibition	Decreases	May decrease or increase	May decrease or increase	None
To polymer, intermolecular	None	None	Increases	Produces long branches
To polymer, intramolecular	None	None	Increases	Produces short branches

[a] Collins (1973).

TABLE 3-3. Inhibitor Constants for Various Inhibitors and Monomers at 50°C Except as Noted[a]

Inhibitor	C_Z			
	Styrene	Methyl Acrylate	Methyl Methacrylate	Vinyl Acetate
Nitrobenzene	0.326	0.00464	—	11.2
Trinitrobenzene	64.2	0.204	—	404
p-Benzoquinone	518	—	5.5 (44°C)	—
DPPH	—	—	2000 (44°C)	—
Oxygen	14,600	—	33,000	—

[a]Ulbricht (1975).

state assumption leads, in direct analogy with Eq. 3-17, to the relation (again for termination by combination)

$$\frac{1}{\bar{x}_n} = \frac{k_t}{k_p^2} \frac{v_p}{[M]^2} + C_Z \frac{[Z]}{[M]} \tag{3-20}$$

where $C_Z = k_z/k_p$.

Table 3-3 lists inhibition constants for some common monomers and inhibitors.

Determination of Individual Rate Constants

It was shown above that the rate constants for propagation and termination occur together as the ratio k_p^2/k_t in the equations for both overall polymerization rate and kinetic chain length. Thus these rate constants cannot be evaluated separately through steady-state measurements. Recourse must be had to the study of transient phenomena, such as the rate of polymerization before the free-radical concentration has reached its steady-state value. The duration of this transient region depends on the length of time a free radical exists from its formation in the initiation step to its demise in termination.

The Sector Method. If a photopolymerization is begun by turning on light of intensity I, the concentration of free radicals grows gradually to its steady-state value over a time interval that is a function of the lifetime of the radicals. When the light is turned off, the radical concentration decays in a similar manner. At the steady state the rate of polymerization is proportional to the square root of the light intensity: $v_p \simeq I^{1/2}$. On the other hand, if the flashes are so rapid that the radical concentration changes very little during a single flash, the effect is the same as that produced by reducing the intensity by a constant factor but leaving it on all the time: If the light and dark periods are equal, $v_p \simeq (\frac{1}{2}I)^{1/2}$. The rates for long and short flashes differ, in this case, by $\sqrt{2}$ (Fig. 3-7).

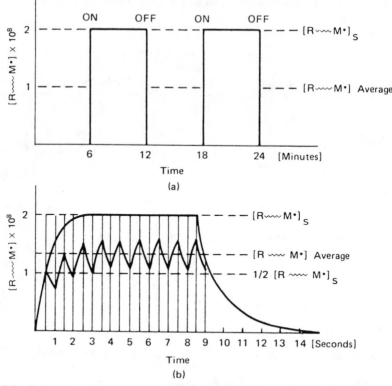

FIG. 3-7. Radical concentration versus time in the sector method: (a) long flashes compared to τ, for which the maximum radical concentration is, for example, 2×10^{-8} mole/liter and its average is 1×10^{-8}, and (b) short flashes, for which the average radical concentration is approximately $\sqrt{2} \times 10^{-8}$ (Vollmert 1973, courtesy Springer-Verlag, New York).

The rate of flashing at which the transition occurs can be related to the mean lifetime τ_s of the free radicals. This is defined as

$$\tau_s = \frac{\text{number of radicals}}{\text{number disappearing per unit time}} = \frac{[M\cdot]}{2k_t[M\cdot]^2} = \frac{1}{2k_t[M\cdot]} \quad (3\text{-}21)$$

or, if the radical concentration is substituted from Eq. 3-9,

$$\tau_s = \frac{k_p}{2k_t} \frac{[M]}{v_p} \quad (3\text{-}22)$$

Thus τ_s, [M], and the overall polymerization rate v_p suffice to calculate k_p/k_t. With the ratio k_p^2/k_t obtained from other data, the individual rate constants may be evaluated.

The relation between τ_s and the flashing rate is complex; in practice, the ex-

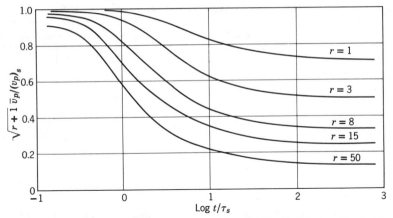

FIG. 3-8. Ratio of polymerization rates for flashing and steady illumination plotted against $\log(t/\tau_s)$ for several values of the ratio r of dark to light periods (Burnett 1947).

perimental curves of polymerization rate (or radical concentration) against flashing rate are compared to the theoretical relation (Fig. 3-8). In Figure 3-8 the ratio t/τ_s, where t is the time light is on, is plotted against $\bar{v}_p/(v_p)_s$, where $\bar{v}_p$ is the rate with flashing illumination and $(v_p)_s$ is the rate with steady illumination, for several values of the ratio r of dark to light intervals.

In practice, the light is usually flashed by placing a motor-driven *rotating sector* between the light source and the reaction vessel.

In the case of vinyl acetate, the radical lifetime is of the order of 4 sec, depending, of course, upon polymerization rate, according to Eq. 3-22. The rate constants show that the growing chain adds about 10^4 monomers per second; the kinetic chain length is the product of this figure and τ_s, or about 4×10^4. The value of $\bar{x}_n$ is about 10 times less than this because of chain transfer to monomer; thus about 10 transfers must occur during the life of a growing chain. This represents a relatively large amount of transfer to monomer.

Other Methods. The kinetic scheme for *emulsion polymerization*, described in Chapter 6, allows separation of the individual rate constants, since the growth of a single chain effectively takes place in an isolated system.

In a *viscosity method* (Bamford 1948), now little used, the polymerization is carried out in a viscometer, measurement of molecular weight by viscosity being substituted for measurement of rate. The change in viscosity after photoinitiation is stopped provides data obtained under transient conditions.

Dead-end polymerization (Tobolsky 1960, Böhme 1966), which refers to polymerization with a limited amount of initiator, allows the determination of the initiation rate constant k_d and initiator efficiency f. As the initiator is depleted, the reaction stops short of completion at the point where the half-life of the propagating chain is comparable to that of the initiator.

TABLE 3-4. Rate Constants for Propagation and Termination at 50°C[a]

Monomer	k_p (liter/mole-sec)	k_t (liter/mole-sec $\times$ 10^{-7})
Styrene	209	115
Methyl methacrylate	410	24
Methyl acrylate	1000	3.5
Vinyl acetate	2640	117
Vinyl chloride	11000	2100

[a]Korus (1975).

Values of the propagation and termination constants for several monomers are tabulated in Table 3-4. The rate constants vary over wide ranges. Since they depend upon the reactivity of both the radical and the monomer, it is difficult to draw conclusions about the effect of structure on reactivity from these data alone. Co-polymerization experiments (Chapter 5) are more useful in this respect.

GENERAL REFERENCES

Flory 1953; Walling 1957; Lenz 1967; Collins 1973, Exp. 3; Vollmert 1973, Chapter 211; Allen 1974, Chapter 7; Elias 1977, Chapter 20; Bamford 1976; Bowden 1979, Chapter 2.2; Allcock 1981, Chapter 12; Odian 1981, Chapter 3.

C. MOLECULAR WEIGHT AND ITS DISTRIBUTION

The distribution of molecular weights in vinyl polymers can readily be calculated if the reaction is restricted to low conversion, where polymer made throughout the reaction has the same average molecular weight. If termination is by disproportionation (or transfer), let p be defined as the probability that a growing chain radical will propagate rather than terminate: $p = v_p/(v_p + v_t)$.

The probability of formation of an x-mer, as a result of $x - 1$ propagations and a termination, is $p^{x-1}(1 - p)$. The situation is in complete analogy with that for linear stepwise polymerization, and the expressions for $\bar{x}_n$, N_x, w_x, and $\bar{x}_w$ given in Eqs. 2-5 and 2-8 to 2-10 apply. If high polymer is formed, that is, $p \rightarrow 1$, $\bar{x}_w/\bar{x}_n = 2$.

Termination by combination is somewhat more complex, but it can be shown that, in analogy to the case of multichain polymer (Eq. 2-13) with $f = 2$, $\bar{x}_w/\bar{x}_n = 1.5$ if transfer is absent.

The measurement of $\bar{x}_w/\bar{x}_n$ would distinguish between the possibilities of termination by combination and by disproportionation. However, this measurement is not easily carried out with the desired precision; a preferable alternative procedure involves measuring the number of initiator fragments per molecule of polymer.

For values of p near 1, the most probable distribution of Eq. 2-9 reduces to

$$w_x = xp^x(\ln p)^2$$

$$\bar{x}_n = \frac{-1}{\ln p} \qquad (3\text{-}23)$$

$$\bar{x}_w = \frac{-2}{\ln p}$$

It can be shown that both this distribution, which describes low-conversion chain polymers terminating by disproportionation, and that describing termination by combination are special cases of a more general distribution function (Zimm 1948):

$$w_x = \left(\frac{y^{z+1}}{z!}\right)x^z e^{-xy}$$

$$\bar{x}_n = \frac{z}{y} \qquad (3\text{-}24)$$

$$\bar{x}_w = \frac{z+1}{y}$$

For $z = 1$, Eq. 3-24 reduces to Eq. 3-23 with $y = -\ln p$. For $z = 2$ it describes the case of termination by combination. Higher values of z correspond to narrower distributions such as might be found in fractionated samples. In contrast to the most probable distribution, the number distribution corresponding to Eq. 3-24 does not indicate a large number of very small molecules; hence it is preferred for describing polymers from which the small molecules have been removed, for example, by precipitation of the polymer from solution.

As polymerization proceeds, the degree of polymerization of the polymer being formed changes according to Eq. 3-14. If the rate of initiation is constant as is usual, $\bar{x}_n$ decreases throughout the reaction as [M] decreases, approaching zero at complete conversion. At the same time $\bar{x}_w/\bar{x}_n$ increases with increasing conversion, approaching infinity as a limit. Typical values of $\bar{x}_w/\bar{x}_n$ for high-conversion vinyl polymers seldom exceed 5, however, except in special cases, such as autoacceleration in the polymerization of methyl methacrylate, where values of $\bar{x}_w/\bar{x}_n$ as high as 10 have been found. The above distribution functions are not well suited to the description of these broader distributions. Among empirical functions suggested for these cases is the following form, which has a high-molecular-weight "tail" when the exponent ρ is positive and less than 1 (here Γ is the gamma function):

$$w(x) = \left[\rho \Big/ \Gamma\left(\frac{z+1}{\rho}\right)\right] y^{(z+1)/\rho} x^z e^{-yx^\rho}$$

$$\bar{x}_n = \Gamma\left(\frac{z+1}{\rho}\right) \Big/ y^{1/\rho} \tag{3-25}$$

$$\bar{x}_w = \Gamma\left(\frac{z+2}{\rho}\right) \Big/ y^{1/\rho}\Gamma\left(\frac{z+1}{\rho}\right)$$

Where chain transfer to polymer is involved, extremely broad molecular weight distributions can result. The abnormal breadth arises from the fact that each transfer reaction adds a branch point to the molecule; after the branch has grown, the molecule has increased in molecular weight. But the probability of branching is proportional (approximately) to weight-average molecular weight; hence branched molecules tend to become more highly branched and still larger. The result is that the distribution of molecular weights has a long, high-molecular-weight "tail."

Among the empirical distribution functions suitable for describing fractions from which low-molecular-weight species have been removed is the so-called logarithmic normal distribution (Lansing 1935):

$$w(x) = \frac{e^{-\sigma^2/2}}{\sigma\sqrt{\pi}} \frac{1}{x_0} e^{-(\ln x - \ln x_0)^2/2\sigma^2}$$

$$\bar{x}_n = x_0 e^{\sigma^2/2} \tag{3-26}$$

$$\bar{x}_w = x_0 e^{3\sigma^2/2}$$

In this distribution the logarithm of the molecular weight follows a normal distribution function. The ratio $\bar{x}_w/\bar{x}_n$ is identical to $\bar{x}_z/\bar{x}_w$.

Other empirical equations have been proposed for cases of unusually broad distributions by Wesslau (1956), Tung (1956), Gordon (1961) and Roe (1961).

Polymers with extremely narrow molecular-weight distribution are produced by certain types of anionic polymerization ("living" polymers, Chapter 4C). Their molecular-weight distribution is described by the Poisson distribution (Flory 1940):

$$w(x) = \frac{\mu}{\mu+1} \frac{xe^{-\mu}\mu^{x-2}}{(x-1)!}$$

$$\bar{x}_n = \mu + 1 \tag{3-27}$$

$$\frac{\bar{x}_w}{\bar{x}_n} = 1 + \frac{\mu}{(\mu+1)^2}$$

GENERAL REFERENCES

Flory 1953; Walling 1957; Eastmond 1967; Peebles 1971; Bamford 1976; Elias 1977, Chapter 20; Allcock 1981, Chapter 12; Odian 1981, Chapter 3.

D. EFFECTS OF TEMPERATURE AND PRESSURE ON CHAIN POLYMERIZATION

Thermochemical and thermodynamic data give valuable insight into the mechanisms of vinyl polymerization reactions. By the usual Arrhenius treatment, the temperature dependence of the rates of the various steps in radical chain polymerization can be separated into an energy of activation representing the amount of energy that the reactant molecules must have to be able to react on collision, and a frequency factor which allows estimation of the fraction of the collisions that lead to reaction. Enthalpies of polymerization are readily measured and may also be calculated from bond dissociation energies. Differences between the two values give an indication of the amount of steric hindrance in the polymer. Free energies of polymerization reactions can be calculated for some simple cases and give information on the thermodynamic equilibria involved in polymerization.

Enthalpies and Free Energies of Polymerization

Enthalpies of Polymerization. These may be obtained from the heats of combustion of monomer and polymer or directly by a calorimetric experiment; they may also be calculated from bond dissociation energies. When the two results are compared, it is seen that the observed enthalpy of polymerization is equal to or somewhat lower than the calculated heat. For vinyl compounds with two substituents on the same carbon atom the difference is 3–8 kJ/mole. This is attributed to steric repulsion between the substituent groups in the polymer. For monosubstituted vinyl derivatives the difference is much lower; only for styrene is it definitely greater than 4 kJ/mole.

Enthalpies of polymerization (Table 3-5) vary over a wide range, reflecting differences in resonance stabilization and conjugation, steric strain in monomer and polymer, and (if present) dipole interactions and hydrogen bonding.

Entropies and Free Energies of Polymerization. The entropy of polymerization is largely determined by the loss of the translational entropy of the monomer on polymerization. Values of ΔS are therefore relatively insensitive to monomer type (Table 3-5). Both ΔH and ΔS are negative, but the exothermic nature of the polymerization reaction outweighs the influence of the entropy term in

$$\Delta G = \Delta H - T\Delta S$$

TABLE 3-5.   Enthalpies and Entropies of Polymerization at 25°C[a]

Monomer	$-\Delta H$ (kJ/mole)	$-\Delta S$ (J/mol-deg)
Tetrafluoroethylene	156	112
Vinyl chloride	96	—
Ethylene	95	100
Vinyl acetate	88	110
Propylene	86	116
Acrylonitrile	77	109
Isoprene	74	101
Butadiene	73	86
Styrene	70	105
Methyl methacrylate	56	117
Isobutylene	52	120
α-Methylstyrene	35	104

[a]Joshi (1967).

so that the free energies are negative at usual temperatures. Experimental methods for obtaining these quantities have been reviewed (Joshi 1967).

At elevated temperatures the entropy term increases, and when $\Delta H = T\Delta S$, $\Delta G = 0$ and the polymerization reaction is in equilibrium with the reverse reaction of depolymerization. This temperature is known as the *ceiling temperature* for the polymer in question. Some ceiling temperatures for pure monomers are 61°C for α-methylstyrene, 220°C for methyl methacrylate, and 310°C for styrene.

The calculation of the enthalpies and entropies of reaction of free radicals has been carried out for hydrocarbons and fluorocarbons and applied to the synthesis equations for polyethylene and polytetrafluoroethylene (Bryant 1951, 1962). Values of enthalpy and free energy for the addition of monomer to the radical become less negative as the length of the chain radical increases in the first few propagation steps. After the growing free radical reaches a length of about eight carbon atoms, all successive steps have the same enthalpy and free energy of reaction, the same as those for the overall polymerization step.

Activation Energies and Collision Factors

Overall Polymerization.   The ratio of rate constants k_p^2/k_t or $k_p/\sqrt{k_t}$ is obtained from the overall polymerization rate and the rate of initiation (Eq. 3-11). Expressing a rate constant by an Arrhenius-type equation gives, for example,

$$k_p = A_p e^{-E_p/RT} \tag{3-28}$$

where A_p is the collision frequency factor for propagation and E_p the energy of activation for propagation. Application to the rate-constant ratio gives

$$\frac{k_p}{\sqrt{k_t}} = \left(\frac{A_p}{\sqrt{A_t}}\right) e^{-(E_p - E_t/2)/RT} \tag{3-29}$$

By plotting $\log(k_p/\sqrt{k_t})$ against $1/T$, $E_p - \frac{1}{2}E_t$ and $A_p/\sqrt{A_t}$ can be evaluated. For many polymerization reactions, $E_p - \frac{1}{2}E_t$ is about 20–25 kJ/mole. This is equivalent to an increase in $k_p/\sqrt{k_t}$ of about 30–35% for every 10°C near room temperature. Equation 3-29 refers only to the ratio of rate constants $k_p/\sqrt{k_t}$ and does not include the effect of temperaure on the rate of initiation. In photopolymerization, where the rate of initiation is independent of temperature, Eq. 3-29 represents the complete activation energy.

In an initiated polymerization, v_p is proportional to $k_p\sqrt{k_d/k_t}$, and the observed activation energy is $\frac{1}{2}E_d + (E_p - \frac{1}{2}E_t)$. The activation energy E_d for spontaneous decomposition of the initiator is about 125 kJ/mole for benzoyl peroxide or azobisisobutyronitrile. The activation energy for initiated polymerization is therefore slightly greater than 80 kJ/mole. This corresponds to a two- to threefold increase in rate for a 10°C temperature change.

The activation energy for thermal polymerization is of the same order of magnitude, yet this reaction is much slower than normal because of the extremely low collision factor for thermal initiation. The normal value of A for a bimolecular reaction is $10^{11}-10^{13}$; for the thermal initiation of polystyrene it is about 10^4-10^6.

Degree of Polymerization.   If transfer reactions are negligible, the temperature coefficient of molecular weight depends on the initiation process. If an initiator is used, it follows from Eq. 3-14, by virtue of the differential form $d \ln k/dT = E/RT^2$ of the Arrhenius equation, that

$$\frac{d \ln \bar{x}_n}{dT} = \frac{E_p - E_t/2 - E_d/2}{RT^2} \tag{3-30}$$

Since $E_p - \frac{1}{2}E_t$ is 20–25 kJ and E_d is about 125 kJ, this quantity is negative and the molecular weight decreases with increasing temperature. The same is true in thermal polymerization.

In photopolymerization in the absence of transfer, $d \ln \bar{x}_n/dT = (E_p - \frac{1}{2}E_t)/RT^2$, which is positive. This is the only case in which molecular weight increases with temperature.

If transfer reactions are controlling, it is found that molecular weight decreases with increasing temperature. By plotting $\log C_s$ or $-\log \bar{x}_n$ against $1/T$, $E_{tr,s} - E_p$ and $A_{tr,s}/A_p$ may be evaluated. The quantity $E_{tr,s}$ usually exceeds E_p by 20–60 kJ/mole, higher activation energies going with poorer transfer agents. Since the frequency factor for transfer usually exceeds that for propagation, the higher activation energy is responsible for the slower rate of transfer compared to monomer addition.

Absolute Rates. The activation energies and frequency factors for propagation and termination can be separated if the individual rate constants are known. For most monomers E_p is near 30 kJ/mole, and E_t varies from 12 to 20 kJ/mole. The frequency factors vary rather widely, suggesting that steric effects are more important than activation energies in setting the values of the rate constants. For example, A_p is much lower for methyl methacrylate with two substituents on the same ethylenic carbon atom that it is for less hindered monomers. All the values of A_p are considerably lower than normal, because it is likely of low probability of the transition leading to unpairing of the double-bond electrons in the monomer.

Polymerization–Depolymerization Equilibrium. Some polymers decompose by stepwise loss of monomer units in a reaction that is essentially the reverse of polymerization (see Chapter 6). The depolymerization reaction can be incorporated into the kinetic scheme for chain polymerization. At ordinary temperatures the scheme is not changed because the rate constant for depolymerization is small. The activation energy of depolymerization is quite high (40–110 kJ/mole) compared to that of propagation, however, and at some elevated temperature the rates of polymerization and depolymerization become equal and the ceiling temperature is reached, above which polymer free radicals, in the presence of monomer at 1-atm pressure, depolymerize rather than grow.

Effects of Pressure

The effects of high pressures (up to, say, several thousand atmospheres) on polymerization go far beyond the direct effect of pressure on the volume of the system. Both equilibrium constants (recall that $\Delta G = -RT \ln K$) and rate constants are affected, the former through the thermodynamic relation

$$\left(\frac{\partial RT \ln K}{\partial P}\right)_T = -\Delta V \tag{3-31}$$

and the latter through the Eyring rate theory (Glasstone 1941) by replacing K with k and ΔV with the volume change between the reactants and the transition state.

Since the volume change on polymerization is usually negative, an increase in pressure favors the formation of polymer and is accompanied by an increase in the ceiling temperature, often 15–20°C per 1000 atm.

The polymerization of styrene at high pressures has been studied in more detail than that of other monomers (Merrett 1951, Nicholson 1956a,b). The rate of dissociation of initiator is reduced, but the rate of polymerization is sufficiently increased to outweigh this, the overall polymerization rate being seven or eight times higher at 3000 atm than at ordinary pressures. The rate of termination, being diffusion controlled, is decreased. All these factors are in the direction of higher molecular weights, but the full extent of the predicted increase is not observed, since the rate of transfer to monomer also increases with pressure and controls the molecular weight at 3000 atm and above.

GENERAL REFERENCES

Flory 1953; Walling 1957; Joshi 1967; Sawada 1969; Allen 1974, Chapter 4; Ivin
   1974; Weale 1974; Allcock 1981, Chapter 10; Odian 1981, Chapter 3.

DISCUSSION QUESTIONS AND PROBLEMS

1. Write chemical equations for the following reactions in the benzoyl peroxide-initiated polymerization of vinyl chloride: initiation, propagation, termination by combination and by disproportionation, transfer to monomer and to polymer.

2. State and justify the steady-state assumptions that are used in deriving the kinetic equations for radical chain polymerization.

3. Show by graphs the relations between the following quantities in radical chain polymerization:

   a. Conversion versus time, for the simple reaction and with (1) autoacceleration, (2) inhibition, and (3) retardation.

   b. Degree of polymerization and concentration of chain-transfer agent.

4. Write kinetic equations for initiated radical chain polymerization showing (a) how rate is related to concentrations of initiator, radicals, and monomer and (b) how degree of polymerization is related to the same quantities.

5. Discuss briefly the following phenomena in radical chain polymerization: (a) methods of generating free radicals; (b) inhibition and retardation; (c) the temperature dependence of rate and degree of polymerization; (d) absolute rate constants and radical lifetimes; (e) how to distinguish between termination by combination and by disproportionation; and (f) how to distinguish between transfer to monomer and to initiator.

6. Rate the following steps in the polymerization of ethylene in order of free energy, and indicate which are thermodynamically favorable: propagation, termination, transfer to monomer, transfer to polymer.

7. a. Which of the following monomers would you expect to polymerize readily by a free-radical mechanism? Why?

   $$CH_2{=}C(CH_3)_2, \qquad CH_2{=}CHCH_3, \qquad CH_2{=}CHCH{=}CH_2$$

   b. Which of the above compounds would you expect to be most susceptible to attack by free radicals? Why?

   c. What relation do you see between the answers to (a) and (b)?

8. Each of the following equations contains errors. Correct each equation and tell what it describes.

   a. $v_p = k_p^{1/2} (fk_d/k_t) [I]^{1/2} [M]$.

   b. $v = (k_p/2k_t) [M \cdot] [M]$.

   c. $\bar{x}_n = (\bar{x}_n)_0 + C_S [S]/[M]$.

   d. $\tau_s = (k_p^2/2k_t) [M]/v_t$.

9. Show how degree of polymerization depends on temperature in (a) polymerization initiated by thermal decomposition of an initiator, (b) photopolymerization, and (c) when transfer reactions are controlling.

10. If the overall activation energy of an initiated polymerization is 90 kJ/mole and the activation energy of decomposition of the initiator is 130 kJ/mole, how does the degree of polymerization change with temperature between 50 and 100°C? Assume no transfer reactions.

11. Given that $[M] = 10$ mole/liter, $[M \cdot] = 1 \times 10^{-8}$ mole/liter, $k_p = 150$ liter/mole-sec, and $v_i = 3 \times 10^{-9}$ mole/liter-sec, calculate (a) v_p in moles per liter-second, (b) the rate of polymerization in weight percent per hour, (c) k_t, (d) the radical lifetime, and (e) v. What can you say about the reaction if the value of $\bar{x}_n$ is (a) 10,000, (b) 5000, or (c) 1000?

12. A monomer with molecular weight 100 and density 1 photopolymerizes in bulk at a rate of 3.6 wt.% per hour when the rate of initiation is 1×10^{-9} mole/liter-sec. The radical lifetime is 10 sec. Calculate $[M \cdot]$, k_p, k_t, $\bar{M}_n$, and $\bar{M}_w$, assuming termination by combination and no transfer.

13. Mayo (1951) gave the following data for the benzoyl peroxide-initiated polymerization of styrene at 60°C. Here $f = 0.60$, $k_d = 3.2 \times 10^{-6}$ sec^{-1}, and the molar volume of styrene is 120 cm^3. Calculate the transfer constant to initiator.

$v_p \times 10^4$ mole/liter-sec	$\bar{x}_n$	$v_p \times 10^4$	$\bar{x}_n$
0.05	8300	0.40	1550
0.07	6700	0.60	1170
0.09	5900	0.80	770
0.13	4500	1.25	510
0.20	3300	1.60	340
0.26	2200		

14. Canale (1960) reported studies on the radical polymerization of methyl α-cyanoacrylate, $CH_2{=}C(CN)COOCH_3$, at 60°C using azobisisobutyronitrile as initiator. Here $fk_d = 1.1 \times 10^{-5}$ sec^{-1}. Monomer density at 60°C is 1.067 g/ml. Polymer density at 60°C is 1.289 g/ml. The fractional rate of monomer conversion per hour determined at the beginning of each run was 0.035 when

$[I] = 4 \times 10^{-4}$, 0.125 when $[I] = 5.04 \times 10^{-3}$, and 0.21 when $[I] = 1.52 \times 10^{-2}$ mole/liter. Determine k_p^2/k_t, indicating any assumptions made.

15. Derive kinetic equations for radical chain polymerization in the presence of an inhibitor such that essentially all the chains are terminated by reaction with the inhibitor. Show how the rate of polymerization depends on the concentrations of monomer, initiator, and inhibitor at the steady state.

16. Polymerization by diradicals is initiated by the reaction $M + \cdot P \cdot$, where $\cdot P \cdot$ is a diradical that can grow from both ends.

    a. Write all possible reactions for propagation (2), termination by combination (3), termination by disproportionation (3), and transfer to monomer (2). Note that reactions converting $\cdot P \cdot$ to monoradicals $Q \cdot$ must be included.

    b. Write kinetic equations for each step, expressing rate constants as multiples of k_p, k_{tc}, k_{td}, and k_{tr}. Remember that "cross-product" terms must be counted twice.

    c. Using this kinetic scheme, show the following.

       1. The rate of initiation is

          $$v_i = (k_{tc} + k_{td})\,[T]^2$$

          where $[T] = [Q] + 2[P]$ is the total concentration of radical ends.

       2. The overall polymerization rate is

          $$v_p = -\frac{d[M]}{dt} = \frac{k_p\,[M]\,v_i^{1/2}}{(k_{tc} + k_{td})^{1/2}}$$

       3. The rate of production of polymer is

          $$\frac{d[\Sigma M_x]}{dt} = k_{tr}\,[T]\,[M] + k_{td}\,[T]^2$$

    (Note: The steady-state condition says that $d[P]/dt = 0$ and $d[Q]/dt = 0$. The consequences of the latter are important.)

    d. Derive an expression for $1/\bar{x}_n$ and simplify it for the cases of termination by disproportion only and by combination only.

BIBLIOGRAPHY

Allcock 1981. Harry R. Allcock and Frederick W. Lampe, *Contemporary Polymer Chemistry*, Prentice-Hall, Englewood Cliffs, New Jersey, 1981.

Allen 1974. P. E. M. Allen and C. R. Patrick, *Kinetics and Mechanisms of Polymerization Reactions*, John Wiley & Sons, New York, 1974.

Bamford 1948. C. H. Bamford and M. J. S. Dewar, "Studies in Polymerization. I. A Method for Determining the Velocity Constants in Polymerization Reactions and its Application to Styrene," *Proc. R. Soc. London* **A192,** 309–328 (1948).

Bamford 1976. C. H. Bamford and C. F. H. Tippers, eds., "Free-Radical Polymerization," Vol. 14A in *Comprehensive Chemical Kinetics,* Elsevier, Amsterdam, 1976.

Bevington 1954a. J. C. Bevington, H. W. Melville, and R. P. Taylor, "The Termination Reaction in Radical Polymerization. Polymerizations of Methyl Methacrylate and Styrene at 25°," *J. Polym. Sci.* **12,** 449–459 (1954).

Bevington 1954b. J. C. Bevington, H. W. Melville, and R. P. Taylor, "The Termination Reaction in Radical Polymerizations. II. Polymerization of Styrene at 60° and of Methyl Methacrylate at 0 and 60°, and the Copolymerization of These Monomers at 60°," *J. Polym. Sci.* **14,** 463–476 (1954).

Bevington 1961. J. C. Bevington, *Radical Polymerization,* Academic Press, New York, 1961.

Böhme 1966. R. D. Böhme and A. V. Tobolsky, "Dead-End Polymerization," pp. 594–605 in Herman F. Mark, Norman G. Gaylord, and Norbert M. Bikales, eds., *Encyclopedia of Polymer Science and Technology,* Vol. 4, Wiley-Interscience, New York, 1966.

Bowden 1979. M. J. Bowden, "Formation of Macromolecules," Chapter 2 in F. A. Bovey and F. H. Winslow, eds., *Macromolecules: An Introduction to Polymer Science,* Academic Press, New York, 1979.

Braun 1971. Dietrich Braun, Harald Cherdon, and Werner Kern, *Techniques of Polymer Synthesis and Characterization,* Wiley-Interscience, New York, 1971.

Bryant 1951. W. M. D. Bryant, "Free Energies of Formation of Hydrocarbon Free Radicals. I. Application to the Mechanism of Polythene Synthesis," *J. Polym. Sci.* **6,** 359–370 (1951).

Bryant 1962. W. M. D. Bryant, "Free Energies of Formation of Fluorocarbons and their Radicals. Thermodynamics of Formation and Depolymerization of Polytetrafluoroethylene," *J. Polym. Sci.* **56,** 277–296 (1962).

Burnett 1947. G. M. Burnett and H. W. Melville, "Determination of the Velocity Coefficients for Polymerization Processes. I. The Direct Photopolymerization of Vinyl Acetate," *Proc. R. Soc. London* **A189,** 456–480 (1947).

Canale 1960. A. J. Canale, W. E. Goode, J. B. Kinsinger, J. R. Panchak, R. L. Kelso, and R. K. Graham, "Methyl α-Cyanoacrylate. I. Free-Radical Homopolymerization," *J. Appl. Polym. Sci.* **4,** 231–236 (1960).

Collins 1973. Edward A. Collins, Jan Bareš, and Fred W. Billmeyer, Jr., *Experiments in Polymer Science,* Wiley-Interscience, New York, 1973.

Dole 1972. Malcolm Dole, ed., *The Radiation Chemistry of Macromolecules,* Academic Press, New York, 1972.

Eastmond 1967. G. C. Eastmond, "Free-Radical Polymerization," pp. 361–431 in Herman F. Mark, Norman G. Gaylord, and Norbert M. Bikales, eds., *Encyclopedia of Polymer Science and Technology,* Vol. 7, Wiley-Interscience, New York, 1967.

Eastmond 1970. G. C. Eastmond, "Solid-State Polymerization," Chapter 1 in A. D. Jenkins, ed., *Progress in Polymer Science,* Vol. 2, Pergamon Press, New York, 1970.

Ebdon 1971. J. R. Ebdon, "Thermal Polymerization of Styrene—A Critical Review," *Br. Polym. J.* **3,** 9–12 (1971).

Elias 1977, Hans-Georg Elias, *Macromolecules ·2· Synthesis and Materials* (translated by John W. Stafford), Plenum Press, New York, 1977.

Flory 1937. Paul J. Flory, "Mechanism of Vinyl Polymerizations," *J. Am. Chem. Soc.* **59,** 241–253 (1937).

Flory 1940. Paul J. Flory, "Molecular Size Distribution in Ethylene Oxide Polymers," *J. Am. Chem. Soc.* **62,** 1561–1565 (1940).

Flory 1953. Paul J. Flory, *Principles of Polymer Chemistry,* Cornell University Press, Ithaca, New York, 1953.

Friedlander 1966. Henry Z. Friedlander, "Electrochemical Initiation," pp. 629–641 in Herman F. Mark, Norman G. Gaylord, and Norbert M. Bikales, eds., *Encyclopedia of Polymer Science and Technology,* Vol. 5, Wiley-Interscience, New York, 1966.

Funt 1967. B. L. Funt, "Electrolytically Controlled Polymerizations," *Macromol. Rev.* **1,** 35–56 (1967).

Glasstone 1941. Samuel Glasstone, Keith J. Laidler, and Henry Eyring, *The Theory of Rate Processes,* McGraw-Hill, New York, 1941.

Goldfinger 1967. George Goldfinger, William Yee, and Richard D. Gilbert, "Inhibition and Retardation," pp. 644–664 in Herman F. Mark, Norman G. Gaylord, and Norbert M. Bikales, eds., *Encyclopedia of Polymer Science and Technology,* Vol. 7, Wiley-Interscience, New York, 1967.

Gordon 1961. Manfred Gordon and Ryong-Joon Roe, "Surface-Chemical Mechanism of Heterogeneous Polymerization and Derivation of Tung's and Wesslau's Molecular Weight Distribution," *Polymer* **2,** 41–59 (1961).

Gregg 1947. R. A. Gregg and F. R. Mayo, "Chain Transfer in the Polymerization of Styrene. III. The Reactivities of Hydrocarbons Toward the Styrene Radical," *Disc. Faraday Soc.* **2,** 328–337 (1947).

Ham 1967. George E. Ham, "General Aspects of Free-Radical Polymerization," Chapter 1 in George E. Ham, ed., *Vinyl Polymerization,* Vol. 1, Part I, Marcel Dekker, New York, 1967.

Ivin 1974. K. J. Ivin, "Thermodynamics of Addition Polymerization Processes," Chapter 16 in A. D. Jenkins and A. Ledwith, eds., *Reactivity, Mechanism and Structure in Polymer Chemistry,* John Wiley & Sons, New York, 1974.

Jenkins 1974. A. D. Jenkins, "The Reactivity of Polymer Radicals in Propagation and Transfer Reactions," Chapter 4 in A. D. Jenkins and A. Ledwith, eds., *Reactivity, Mechanism and Structure in Polymer Chemistry,* John Wiley & Sons, New York, 1974.

Joshi 1967. R. M. Joshi and B. J. Zwolinski, "Heats of Polymerization and their Structural and Mechanistic Implications," Chapter 8 in George E. Ham, ed., *Vinyl Polymerization,* Vol. 1, Part I, Marcel Dekker, New York, 1967.

Korus 1975. R. Korus and K. F. O'Driscoll, "Propagation and Termination Constants in Free Radical Polymerization," pp. II-45–II-52 in J. Brandrup and E. H. Immergut, eds., with the collaboration of W. McDowell, *Polymer Handbook,* 2nd ed., Wiley-Interscience, New York, 1975.

Lansing 1935. W. D. Lansing and E. O. Kraemer, "Molecular Weight Analysis of Mixtures by Sedimentation Equilibrium in the Svedberg Ultracentrifuge," *J. Am. Chem. Soc.* **57,** 1369–1377 (1935).

Lenz 1967. Robert W. Lenz, *Organic Chemistry of Synthetic High Polymers,* Wiley-Interscience, New York, 1967.

McCaffery 1970. Edward M. McCaffery, *Laboratory Preparation for Macromolecular Chemistry,* McGraw-Hill, New York, 1970.

Mageli 1968. O. L. Mageli and J. R. Kolcynski, "Peroxy Compounds," pp. 814–841 in Herman F. Mark, Norman G. Gaylord, and Norbert M. Bikales, eds., *Encyclopedia of Polymer Science and Technology,* Vol. 9, Wiley-Interscience, New York, 1968.

Marvel 1940. C. S. Marvel, Joseph Dec, and Harold G. Cooke, Jr., "Optically Active Polymers from Active Vinyl Esters. A Convenient Method of Studying the Kinetics of Polymerization," *J. Am. Chem. Soc.* **63,** 3499–3504 (1940).

Marvel 1959. Carl S. Marvel, *An Introduction to the Organic Chemistry of High Polymers,* John Wiley & Sons, New York, 1959.

Mayo 1951. Frank R. Mayo, R. A. Gregg, and Max S. Matheson, "Chain Transfer in the Polymerization of Styrene. VI. Chain Transfer with Styrene and Benzoyl Peroxide; the Efficiency of Initiation and the Mechanism of Chain Termination," *J. Am. Chem. Soc.* **73,** 1691–1700 (1951).

Mayo 1959. Frank R. Mayo, "Contributions of Vinyl Polymerization to Organic Chemistry," *J. Chem. Educ.* **36,** 157–160 (1959).

Merrett 1951. F. M. Merrett, Ph.D., and R. G. W. Norrish, Ph.D., "Aspects of Polymerization at High Pressures," *Proc. R. Soc. London* **A206,** 309–334 (1951).

Naylor 1953. M. A. Naylor and F. W. Billmeyer, Jr., "A New Apparatus for Rate Studies Applied to the Photopolymerization of Methyl Methacrylate," *J. Am. Chem. Soc.* **75,** 2181–2185 (1953).

Nicholson 1956a. A. E. Nicholson and R. G. W. Norrish, "The Decomposition of Benzoyl Peroxide in Solution at High Pressures," *Disc. Faraday Soc.* **22,** 97–103 (1956).

Nicholson 1956b. A. E. Nicholson and R. G. W. Norrish, "Polymerization of Styrene at High Pressures using the Sector Technique," *Disc. Faraday Soc.* **22,** 104–113 (1956).

Norrish 1939. R. G. W. Norrish and E. F. Brookman, "The Mechanism of Polymerization Reactions. I. The Polymerization of Styrene and Methyl Methacrylate," *Proc. R. Soc. London* **A171,** 147–171 (1939).

North 1966. Alastair M. North, *The Kinetics of Free Radical Polymerization,* Pergamon Press, New York, 1966.

North 1974. A. M. North, "The Influence of Chain Structure on the Free-Radical Termination Reaction," Chapter 5 in A. D. Jenkins and A. Ledwith, eds., *Reactivity, Mechanism and Structure in Polymer Chemistry,* John Wiley & Sons, New York, 1974.

Noyes 1965. Richard M. Noyes, "Cage Effect," pp. 796–801 in Herman F. Mark, Norman G. Gaylord, and Norbert M. Bikales, eds., *Encyclopedia of Polymer Science and Technology,* Vol. 2, Wiley-Interscience, New York, 1965.

Odian 1981. George Odian, *Principles of Polymerization,* 2nd ed., John Wiley & Sons, New York, 1981.

O'Driscoll 1969. Kenneth F. O'Driscoll and Premamoy Ghosh, "Initiation in Free Radical Polymerization," Chapter 3 in Teiji Tsuruta and Kenneth F. O'Driscoll, eds., *Structure and Mechanism in Vinyl Polymerization,* Marcel Dekker, New York, 1969.

Oster 1969. Gerald Oster, "Photopolymerization and Photocrosslinking," pp. 145–156 in Herman F. Mark, Norman G. Gaylord, and Norbert M. Bikales, eds., *Encyclopedia of Polymer Science and Technology,* Vol. 10, Wiley-Interscience, New York, 1969.

Palit 1965. Santi R. Palit, Satya R. Chatterjee, and Asish R. Mukherjee, "Chain Transfer," pp. 575–610 in Herman F. Mark, Norman G. Gaylord, and Norbert M. Bikales, eds., *Encyclopedia of Polymer Science and Technology,* Vol. 3, Wiley-Interscience, New York, 1965.

Peebles 1971. Leighton H. Peebles, Jr., *Molecular Weight Distributions in Polymers,* Wiley-Interscience, New York, 1971.

Roe 1961. Ryong-Joon Roe, "A Test Between Rival Theories for Molecular Weight Distributions in Heterogeneous Polymerization: The Effect of Added Terminating Agent," *Polymer* **2,** 60–73 (1961).

Sawada 1969. Hideo Sawada, "Thermodynamics of Polymerization," *J. Macromol. Sci. Rev. Macromol. Chem.* **3,** 313–395 (1969).

Schulz 1947. G. V. Schulz and G. Harborth, "On the Mechanism of the Explosive Course of the Polymerization of Methyl Methacrylate" (in German), *Makromol. Chem.* **1,** 106–139 (1947).

Sorenson 1968. Wayne R. Sorenson and Tod W. Campbell, *Preparative Methods of Polymer Chemistry,* 2nd ed., Wiley-Interscience, New York, 1968.

Staudinger 1920. H. Staudinger, "Polymerization" (in German), *Ber. Dtsch. Chem. Ges. B.* **53,** 1073–1085 (1920).

Tedder 1974. J. M. Tedder, "The Reactivity of Free-Radicals," Chapter 2 in A. D. Jenkins and A. Ledwith, eds., *Reactivity, Mechanism and Structure in Polymer Chemistry,* John Wiley & Sons, New York, 1974.

Tobolsky 1960. A. V. Tobolsky, C. E. Rodgers, and R. D. Brickman, "Dead-End Radical Polymerization. II," *J. Am. Chem. Soc.* **82,** 1277–1280 (1960).

Trommsdorff 1948. Ernst Trommsdorff, Herbert Köhle, and Paul Lagally, "On the Polymerization of Methyl Methacrylate" (in German), *Makromol. Chem.* **1,** 169–198 (1948).

Tung 1956. L. H. Tung, "Fractionation of Polyethylene," *J. Polym. Sci.* **20,** 495–506 (1956).

Ulbricht 1975. J. Ulbricht, "Inhibitors and Inhibition Constants in Free Radical Polymerization," pp. II-53–II-55 in J. Brandrup and E. H. Immergut, eds., with the collaboration of W. McDowell, *Polymer Handbook,* 2nd ed., Wiley-Interscience, New York, 1975.

Vollmert 1973. Bruno Vollmert, *Polymer Chemistry* (translated by Edmund H. Immergut), Springer-Verlag, New York, 1973.

Walling 1957. Cheves Walling, *Free Radicals in Solution,* John Wiley & Sons, New York, 1957.

Weale 1974. K. E. Weale, "The Influence of Pressure on Polymerization Reactions," Chapter 6 in A. D. Jenkins and A. Ledwith, eds., *Reactivity, Mechanism and Structure in Polymer Chemistry,* John Wiley & Sons, New York, 1974.

Wesslau 1956. Hermann Wesslau, "The Molecular Weight Distributions of Some Low Pressure Polyethylenes" (in German), *Makromol. Chem.* **20,** 111–142 (1956).

Wilson 1974. Joseph E. Wilson, *Radiation Chemistry of Monomers, Polymers & Plastics,* Marcel Dekker, New York, 1974.

Young 1975. Lewis J. Young, "Transfer Constants to Monomer, Polymer, Catalyst and Solvent in Free Radical Polymerization," pp. II-57–II-104 in J. Brandrup and E. H. Immergut, eds., with the collaboration of W. McDowell, *Polymer Handbook,* 2nd ed., Wiley-Interscience, New York, 1975.

Zand 1965. Robert Zand, "Azo Catalyst," pp. 278–295 in Herman F. Mark, Norman G. Gaylord, and Norbert M. Bikales, eds., *Encyclopedia of Polymer Science and Technology,* Vol. 2, Wiley-Interscience, New York, 1965.

Zimm 1948. Bruno H. Zimm, "Apparatus and Methods for Measurement and Interpretation of the Angular Variation of Light Scattering: Preliminary Results on Polystyrene Solutions," *J. Chem. Phys.* **16,** 1099–1116 (1948).

CHAPTER FOUR

IONIC AND COORDINATION CHAIN (ADDITION) POLYMERIZATION

A. SIMILARITIES AND CONTRASTS IN IONIC POLYMERIZATION

Chain-reaction polymerization is known to occur by several mechanisms other than those involving free radicals discussed in Chapter 3. Prominent among these are reactions in which the chain carriers are *carbenium ions*† (*cationic polymerization,* Section *B*) or *carbanions* (*anionic polymerization,* Section *C*). In addition, polymerization may take place by mechanisms involving coordination compounds among the monomer, the growing chain, and a catalyst, usually a solid (*coordination or insertion polymerization,* Section *D*). Still other chain polymerization reactions include ring opening and are discussed in Section *E*.

The mechanisms of these polymerizations are less thoroughly understood than that of radical polymerization, for several reasons. Reaction systems are often heterogeneous, involving inorganic catalysts‡ and organic monomers. Usually large effects may be produced by a third component (*cocatalyst*) present in very low concentration. Polymerization often leads to very-high-molecular-weight polymer at an extremely high rate, increasing enormously the difficulty in obtaining kinetic data or even reproducible results.

---

†The term *carbenium ion* replaces *carbonium ion* for species like CH_3^+ at the suggestion of Olah (1972).
‡The word *catalyst* should be used when the agent exerts its influence throughout the growth of the chain and is not consumed in the reaction, and *initiator* when it is used up at the instant of initiation. Because of uncertainty over the mechanisms of ionic polymerization, both terms are found. In addition, Kennedy (1977) suggests that the terms *catalyst* and *cocatalyst* be replaced by *coinitiator* and *initiator*, respectively, in a complete reversal of terminology from that used traditionally and in this book.

As indicated in Table 4-1, the type of monomer polymerizing best by each mechanism is related to the polarity of the monomer and the acid–base strength of the ion formed. Monomers with electron-donating groups attached to the double-bonded carbons form stable carbenium ions and polymerize best with cationic catalysts. Conversely, monomers with electron-withdrawing substituents form stable anions and require anionic catalysts. Free-radical polymerization may be considered an intermediate case that is favored by moderate electron withdrawal from the double bond plus conjugation in the monomer. Many monomers can polymerize by more than one mechanism, including coordination polymerization.

Initiation of ionic polymerization usually involves the transfer of an ion or an electron to or from the monomer, with the formation of an ion pair. The counterion of this pair stays close to the growing chain end throughout its lifetime, particularly in media of low dielectric constant. Several possible structures, including bound or contact ion pairs, solvated or solvent-separated ion pairs, and free ions can sometimes be distinguished experimentally or may be in equilibrium.

In contrast to radical polymerization, termination in ionic polymerization never involves the reaction between two growing chains, but usually involves the uni-molecular reaction of a chain with its counterion or a transfer reaction leaving a species too weak to propagate. An agent that retards radical polymerization, such as oxygen, often has little influence on ionic polymerization; but impurities that neutralize the catalysts prevent the reaction, and many aromatic, heterocyclic, olefinic, and acetylenic compounds retard or terminate ionic polymerization. In addition to these differences, ionic polymerization can usually be distinguished from free-radical polymerization by its very high rate and low (often below 0°C) reaction temperature.

TABLE 4-1. Types of Chain Polymerization Suitable for Common Monomers[a]

Monomer Type	Polymerization Mechanism[b]			
	Radical	Cationic	Anionic	Coordination
Ethylene	+	+	−	+
Propylene and α-olefins	−	−	−	+
Isobutylene	−	+	−	−
Dienes	+	−	+	+
Styrene	+	+	+	+
Vinyl chloride	+	−	−	+
Vinylidene chloride	+	−	+	−
Vinyl fluoride	+	−	−	−
Tetrafluoroethylene	+	−	−	+
Vinyl ethers	−	+	−	+
Vinyl esters	+	−	−	−
Acrylic and methacrylic esters	+	−	+	+
Acrylonitrile	+	−	+	+

[a]Lenz (1967).
[b] + = high polymer formed; − = no reaction or oligomers only.

Kinetics

It is convenient to consider some common features of cationic and anionic polymerization.

Formation of Initiator.   In many instances the species that initiates ionic polymerization is produced by ionization of the added initiator, for example, H^+ from $BF_3 \cdot H_2O$ or $C_5H_{11}^-$ from potassium amyl. Sometimes the added substance reacts with the monomer or with the solvent to produce the initiating species, as in the reaction of a strong base with dimethyl sulfoxide to form the initiating anion $CH_3SOCH_2^-$.

Initiation.   In the step of initiation the initiating species can react with the monomer in one of two ways: transfer of one electron without the formation of a bond between the two and transfer of two electrons with bond formation.

The transfer of a single electron from initiator to monomer always produces a radical ion, two of which generally dimerize to produce a diion. Three cases can be differentiated. The first is electron transfer from an electron donor to the monomer, for example, from the naphthalide ion (formed by the ionization of sodium naphthalide) to styrene:

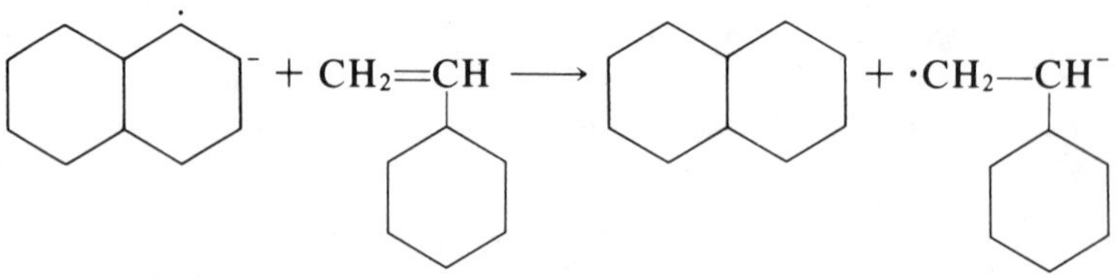

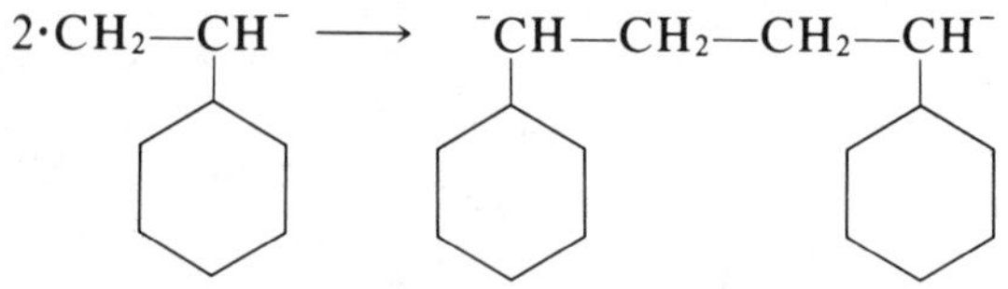

The second and third cases of single-electron transfer are, respectively, electron transfer from a monomer to an electron acceptor and the formation of a charge-transfer complex between a monomer with electron-donor or electron-acceptor properties and another molecule.

When two electrons are transferred, the monomer and the initiating species are bonded together. Again three cases can be distinguished:

a.   A cation can add to a monomer in a nucleophilic reaction to form a monomer cation, for example, in the initiation of isobutylene polymerization by boron trifluoride.

$$BF_2OH^-H^+ \; + \; CH_2 \; = \; C(CH_3)_2 \longrightarrow BF_3OH^-C^+(CH_3)_3$$

b. An anion can add to a monomer in an electrophilic reaction to form a monomer anion, as in the initiation of styrene polymerization by potassium amyl:

$$K^+C_5H_{11}^- + CH_2{=}CH \longrightarrow K^+C_5H_{11}CH_2CH^-$$

c. A neutral molecule and a monomer can add to produce a zwitterion, as in the case of initiation of polymerization of a lactone with a tertiary amine:

$$R_3N + \overline{OCH_2CH_2CO} \longrightarrow R_3N^+CH_2CH_2COO^-$$

Propagation and Termination. Whether the initiation step has led to a monoion or a diion, propagation takes place by the further addition of monomers to the growing chain, accompanied by its counterion. Termination involves the destruction of the growing anion or cation, possibly by recombination with the counterion or reaction with another species, but in many instances such a reaction does not take place and there is no termination. These cases, known as "living polymers," are described in Section *C*.

GENERAL REFERENCES

Elias 1977, Chapter 18.1; Odian 1981, Chapters 5-1 and 5-5.

B. CATIONIC POLYMERIZATION

Typical catalysts for cationic polymerization include, in order of importance, aprotonic acids (Lewis acids and Friedel–Crafts halides), protonic (Brönsted) acids, and stable carbenium-ion salts. All these are strong electron acceptors. Many of them, particularly the Lewis acids, require a cocatalyst, usually a Lewis base or other proton donor, to initiate polymerization. Those monomers with electron-donating 1,1-substituents that can form stable carbenium ions are polymerized by cationic mechanisms.

High rate of polymerization at low temperatures is a characteristic of ionic polymerizations. It is often difficult to establish uniform reaction conditions before the reactants are consumed. The polymerization of isobutylene by $AlCl_3$ or BF_3 takes place within a few seconds at $-100°C$, producing polymer of molecular weight up to several million. Both rate and molecular weight are much lower at room temperature. These considerations limit the industrial application of cationic polymerization to the above-mentioned example.

Mechanism

As discussed in Section A, the most satisfactory theory of cationic polymerization involves the carbenium ion as the chain carrier. For example, in the polymerization of isobutylene with boron trifluoride catalyst, the first step is the reaction of the catalyst and a cocatalyst, for example, water, to form a *catalyst–cocatalyst complex* that donates a proton to an isobutylene molecule to give a carbenium ion, $(CH_3)_3C^+$. This ion then reacts with monomer with the reformation of a carbenium ion at the end of each step. The "head-to-tail" addition of monomer to ion is the only one possible for energetic reasons. Since the reaction is in general carried out in a hydrocarbon medium of low dielectric constant, the anion and the growing cationic end form an ion pair.

The termination reaction can take place by the rearrangement of the ion pair to yield a polymer molecule with terminal unsaturation, plus the original complex,

$$CH_3[(CH_3)_2CCH_2]_xC(CH_3)_2^+(BF_3OH)^-$$

$$\downarrow$$

$$H^+(BF_3OH)^- + CH_3[(CH_3)_2CCH_2]_xC\underset{CH_3}{\overset{CH_2}{<}}$$

Here the catalyst–cocatalyst complex is regenerated, and many kinetic chains can be produced from each catalyst–cocatalyst species. An alternative termination mechanism, in which the catalyst or cocatalyst combines with the growing chain, is unlikely in this example, but can take place when a covalent bond is formed, as in the polymerization of styrene with trifluoroacetic acid:

$$H[CH_2CH]_xCH_2CH^{+-}OCOCF_3 \longrightarrow H[CH_2CH]_xCH_2CHOCOCF_3$$

Chain transfer to monomer can also take place, and chain transfer to polymer is known and leads to branched polymers.

The efficiency of the catalyst is dependent on the acid strength of the catalyst–cocatalyst complex. Also, the efficiency of the catalyst as a terminator should be related to the base strength of the complex anion. Thus the more active a molecule is as a catalyst, the less active it is as a terminator. Although the requirements of the isobutylene–BF_3 system are fully satisfied by the mechanism just given, generalization of the scheme should be applied with caution. The experimental data on other systems, for example, do not conclusively prove the necessity of a cocatalyst.

Kinetics

Although many cationic polymerizations proceed so rapidly that it is difficult to establish the steady state, the following kinetic scheme appears to be valid. Let the catalyst be designated by A and the cocatalyst by RH. Initiation, propagation, termination, and transfer may be represented as follows:

$$
\begin{aligned}
A + RH &\overset{K}{\rightleftharpoons} H^{+\,-}AR \\
H^{+\,-}AR + M &\xrightarrow{k_i} HM^{+\,-}AR \\
HM_x^{+\,-}AR + M &\xrightarrow{k_p} HM_{x+1}^{+\,-}AR \\
HM_x^{+\,-}AR &\xrightarrow{k_t} M_x + H^{+\,-}AR \\
HM_x^{+\,-}AR + M &\xrightarrow{k_{tr}} M_x + HM^{+\,-}AR
\end{aligned}
\tag{4-1}
$$

The rate of initiation is

$$
v_i = Kk_i[A][RH][M]
\tag{4-2}
$$

where [A] is the catalyst concentration. (If the formation of $H^{+\,-}AR$ is the rate-determining step, v_i may be independent of [M] and the kinetic scheme should be appropriately modified.)

Since, in strong contrast to radical polymerization, termination is first order,

$$
v_t = k_t[M^+]
\tag{4-3}
$$

where $[M^+]$ is written as an abbreviation for $[HM^{+\,-}AR]$. By the steady-state assumption (which appears to hold in most cases, despite uncertainty as to whether such a state is actually achieved),

$$
[M^+] = \frac{Kk_i}{k_t}[A][RH][M]
\tag{4-4}
$$

The overall polymerization rate is

$$
v_p = k_p[M][M^+] = K\frac{k_i k_p}{k_t}[A][RH][M]^2
\tag{4-5}
$$

If termination predominates over transfer,

$$
\bar{x}_n = \frac{v_p}{v_t} = \frac{k_p[M^+][M]}{k_t[M^+]} = \frac{k_p}{k_t}[M]
\tag{4-6}
$$

whereas if transfer predominates,

$$\bar{x}_n = \frac{v_p}{v_{tr}} = \frac{k_p[M^+][M]}{k_{tr}[M^+][M]} = \frac{k_p}{k_{tr}} \tag{4-7}$$

Since v_p is proportional to $k_i k_p / k_t$, and, if termination predominates, $\bar{x}_n$ is proportional to k_p / k_t, the overall rate of polymerization increases with decreasing temperature if $E_t > E_i + E_p$, and the molecular weight increases as the temperature of reaction decreases if $E_t > E_p$. Since the propagation step involves the approach of an ion to a neutral molecule in a medium of low dielectric constant, no activation energy is necessary. The termination step, on the other hand, requires the rearrangement of an ion pair and thus involves an appreciable activation energy. These conditions are exactly the reverse of those found in free-radical polymerization.

The dielectric constant of the medium has a significant effect on the nature of the ion pair, which can range between the extremes of complete covalency and completely free ions. It is often convenient to consider two types of propagating species, an ion pair and a free ion, in equilibrium. The nature of the solvent and the counterion can alter the course of the polymerization greatly by shifting this equilibrium. Large increases in v_p and $\bar{x}_n$ are usually observed as the solvating power of the medium is increased: The concentration of the free ion, which propagates more rapidly, is increased at the same time. Similarly, the larger and the less tightly bound the counterion, the more rapid is the propagation.

GENERAL REFERENCES

Plesch 1963, 1968; Eastham 1965; Lenz 1967; Collins 1973, Exp. 6; Baker 1974; Ledwith 1974a; Elias 1977, Chapters 18-2 and 18-4; Russell 1977; Bowden 1979, Chapter 2.2.3; Odian 1981, Chapter 5-2; Kennedy 1982a,b.

C. ANIONIC POLYMERIZATION

Anionic polymerization was carried out on a commercial scale for many years before the nature of the polymerization was recognized, in the production of the buna-type synthetic rubbers in Germany and Russia by the polymerization of butadiene with sodium or potassium as the catalyst. The first anionic chain reaction to be so identified was the polymerization of methacrylonitrile by sodium in liquid ammonia at $-75°C$ (Beaman 1948). In modern times, the growth of commercial products of anionic polymerization has been phenomenal (Hsieh 1981). Most of these polymerizations are based on the use of organometal catalyst systems, which allow unprecedented control over polymer structure (Chapters 5D, 12E, and 13). In addition to the monomer types listed in Table 4-1, monomers that polymerize by ring scission (Section E) may do so by anionic mechanisms.

Mechanism

The conventional method of initiation of ionic chains involves the addition of a negative ion to the monomer, with the opening of a bond or ring and growth at one end:

$$NH_2^- + CH_2{=}CHX \longrightarrow H_2NCH_2CHX^-$$

$$CH_3O^- + \underset{\underset{O}{\diagdown\diagup}}{CH_2CH_2} \longrightarrow CH_3OCH_2CH_2O^-$$

(In the equations of this section the counterions are omitted; however, their role is essentially the same as in cationic polymerization.) Simultaneous growth from more than one center can be obtained from polyvalent ions such as those derived from

$$C\left[\left\langle\bigcirc\right\rangle Li\right]_4 \quad \text{or} \quad N(CH_2CH_2O^-Na^+)_3.$$

The more basic the ion, the better it serves to initiate chains. Thus OH^- will not initiate the anionic polymerization of styrene, NH_2^- initiates fairly well, but $\overset{\ominus}{CH_2^-}$ is powerful. Similarly, more acid monomers require less basic ions, with the acidity of the monomer depending on the strength of the $X{-}M^-$ bond formed in initiation and the stability of the resulting ion.

Initiation can also occur by the transfer of an electron to a monomer of high electron affinity. If D or D^- is an electron donor,

or

$$D + M \longrightarrow D^+ + M^-$$

$$D^- + M \longrightarrow D + M^-$$

Presumably, M^- is less reactive than a true carbanion or a free radical, but the addition of a monomer to M^- gives a species that contains one radical end and one anion end:

$$CH_2{=}\bar{C}HX + CH_2{=}CHX \longrightarrow \underset{\text{anion end}}{:\bar{C}H_2CHXCH_2}\underset{\text{radical end}}{\overset{\bullet}{C}HX}$$

This species can add monomer from the two ends by different mechanisms. Two radical ends may dimerize, however, leaving a divalent anion to propagate.

Propagation in anionic polymerization may be conventional or may be more complex, as in the elimination of CO_2 from N-carboxyanhydrides. In contrast to radical polymerization, the β (unsubstituted) carbon atom at the end of the growing chain is the site of addition of the next monomer.

As in cationic polymerization, termination is always unimolecular, usually by transfer. The recombination of a chain with its counterion or the transfer of a hydrogen to give terminal unsaturation, frequent in cationic polymerization, is unlikely in anionic mechanisms, as may be recognized by considering the small likelihood of transferring H^- when the counterion is Na^+. Thus in anionic polymerization termination usually involves transfer, and the kinetic chain is broken only if the new species is too weak to propagate.

This leads to the unique situation in which, by careful purification to eliminate all species to which transfer can occur, the termination step is effectively eliminated and the growing chains remain active indefinitely. This case is described below under "Living" Polymers.

Kinetics

The kinetics of anionic polymerization may be illustrated by the polymerization of styrene with potassium amide in liquid ammonia (Higginson 1952, Wooding 1952):

$$KNH_2 \xrightarrow{K} K^+ + NH_2^-$$

$$NH_2^- + M \xrightarrow{k_i} NH_2M^-$$

$$NH_2M_x^- + M \xrightarrow{k_p} NH_2M_{x+1}^- \quad (4\text{-}8)$$

$$NH_2M_x^- + NH_3 \xrightarrow{k_{tr}} NH_2M_xH + NH_2^-$$

The usual kinetic analysis leads for high degree of polymerization to

$$v_p = \frac{Kk_pk_i}{k_{tr}} \frac{[NH_2^-]}{[NH_3]} [M]^2 \quad (4\text{-}9)$$

and

$$\bar{x}_n = \frac{k_p[M]}{k_{tr}[NH_3]} \quad (4\text{-}10)$$

"Living" Polymers

Since the termination step usually involves transfer to some species not essential to the reaction, anionic polymerization with carefully purified reagents may lead to systems in which termination is lacking. The resulting species, called "living" polymers (Henderson 1968; Szwarc 1968b), can be prepared, for example, by polymerizing styrene with sodium naphthalene. Kinetic analysis shows that the polymer can have an extremely narrow distribution of molecular weight and for all practical purposes be essentially monodisperse; if initiation is rapid compared to propagation, the molecular-weight distribution is the Poisson function (Chapter 3C) for which $\bar{x}_w/\bar{x}_n \simeq 1 + 1/\bar{x}_n$. The polymer can be "killed" by addition of a terminating agent, for example, water, at the end of the reaction.

The living polymer technique provides an unique opportunity for the preparation of block copolymers (Chapter 5D) of precisely defined composition (Henderson 1968; Fetters 1969; Hsieh 1981).

GENERAL REFERENCES

Overberger 1965; Lenz 1967; Szwarc 1968a; Morton 1969, 1977, 1983; Collins
    1973, Exp. 9; Dart 1974; Parry 1974; Bywater 1975; Elias 1977, Chapters
    18.2 and 18.4; Bowden 1979, Chapter 2.2.4; Odian 1981, Chapter 5-3.

D. COORDINATION POLYMERIZATION

In the early 1950's a major new polymerization technique was discovered that led
to the production of polymers with unusual stereospecific structures (Chapters 10A
and 13). Although there were earlier indications of this type of polymerization
(Schildknecht 1947), the field actually came into existence with the work of Ziegler
(1955) and Natta (1955), who developed new polymerization catalysts with unique
stereoregulating powers. Their Nobel Prize addresses (Ziegler 1964; Natta 1965)
form excellent introductions to the field.

Since the earliest days, a number of names have been applied to the polymerization technique using what have become known as Ziegler or Ziegler–Natta catalysts. Some have emphasized the stereoregulating features of certain of these
polymerizations, but that is not an essential feature. The term *polyinsertion* is still
used, suggestive of chain propagation by the insertion of a monomer between the
catalyst and the polymer chain to which it is attached, but the term *coordination
polymerization*, used early to suggest the essential feature of a coordination complex
with the catalyst, has remained and has been justified by mechanistic studies.

Coordination polymerization is far from simple in terms of mechanism, kinetics,
or application. The most important catalysts are solids, leading to heterogeneous
polymerization systems, but soluble catalyst systems are known. Additional complications to the study of coordination systems arise because the same catalysts,
under other conditions, can initiate polymerization by cationic, anionic, or even
free-radical mechanisms. Quite often coordination polymerization is carried out
using a catalyst in the form of a slurry of small solid particles in an inert medium
(a *fluid-bed* process) or a supported solid catalyst. Heterogeneous polymerization
is not essential for coordination polymerization and the development of stereospe-
cificity, however, for sufficiently polar monomers.

Ziegler–Natta Catalysts

The Ziegler–Natta or, often, Ziegler catalysts have the remarkable property of
polymerizing a wide variety of monomers to linear and stereoregular polymers.
Ethylene is polymerized to a highly linear chain, in contrast to the products of
radical polymerization (Chapter 13). Polypropylene may be made in either the
isotactic or syndiotactic form, but higher α-olefins yield only isotactic polymers.
Dienes such as butadiene and isoprene may be polymerized to products that are
almost exclusively cis-1,4, trans-1,4, isotactic-1,2, or syndiotactic-1,2, depending
on choice of catalyst and conditions. Many other examples exist.

TABLE 4-2.   Components of Typical Ziegler–Natta Catalysts[a]

Organometallic Compound	Transition-Metal Salt
Triethyl aluminum	Titanium tetrachloride
Diethyl aluminum chloride	Vanadium trichloride
Diethyl aluminum chloride	Triacetyl acetone vanadium
Diethyl aluminum chloride	Triacetyl acetone chromium
Diethyl aluminum chloride	Cobalt chloride–pyridine complex
Butyl lithium	Titanium tetrachloride
Butyl magnesium iodide	Titanium trichloride
Ethyl aluminum dichloride	Dichlorodicyclopentadienyl titanium

[a]Henrici-Olivé (1969).

Ziegler catalysts are complexes formed by the interaction of alkyls of metals of groups I–III in the periodic table with halides and other derivatives of transition metals of groups IV–VIII. Some typical compositions are listed in Table 4-2. There are literally hundreds of combinations that are active in polymerization. Current thinking is that the transition metal, with its d orbitals available for both σ and π bonding, is the catalyst center (Cossee 1964). A coordination complex is formed with the organometallic part of the system supplying some of the ligands. At least one empty coordination site accommodates the growing polymer chain; this is indicated here by the square box:

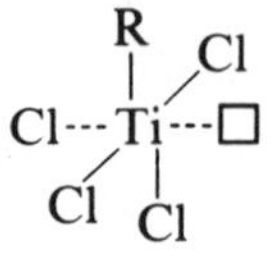

Here a monometallic catalyst is shown; where an organometallic compound plays a role, the metal atoms may be bonded by chlorine or alkyl bridges (Fig. 4-1). The titanium has been reduced from Ti(IV) to Ti(III) and the chlorine bridge is a so-called three-center bond.

Mechanisms.   Because mechanistic studies are so much easier when carried out in a homogeneous system, much information has been obtained from experiments with soluble Ziegler–Natta catalysts. Henrici-Olivé (1981) studied the system *bis*-cyclopentadienyl titanium(IV) dichloride (Cp_2TiCl_2),

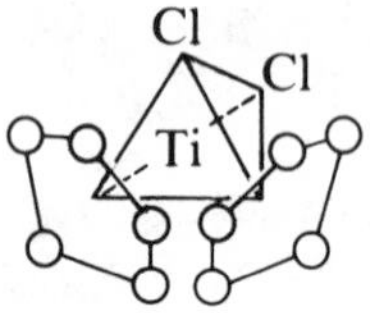

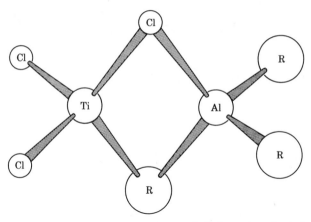

FIG. 4-1. Structure postulated for the metal halide–metal alkyl complex of a typical coordination catalyst.

and an aluminum alkyl, usually dichloroethyl aluminum. The active catalyst species was identified as

$$\begin{array}{c} Cp \overset{Et}{\underset{Cp}{\overset{|}{\diagdown}}} \overset{Cl}{\underset{Ti}{\diagup}} \overset{Et}{\underset{Al}{\diagup}} \\ Cp \diagup \quad \diagdown Cl \diagup \quad \diagdown Cl \end{array}$$

The transition metal was identified as the active site. The growth step in the polymerization of ethylene by catalysts of this general type is presumed by most investigators to be that shown in Fig. 4-2. The ethylene is coordinated to the free site at the transition metal and inserted between the metal and the alkyl group R or, later, the growing chain. Both the course and the rate of the reaction can be influenced by the nature of the other ligands present.

A chain-transfer reaction exists in which a β-hydrogen is transferred from the growing chain to the transition metal, leaving the chain with a vinyl end group. This is a molecular-weight-determining step, though in the commercial production of Ziegler polyethylene it is unimportant, as the molecular weights are usually quite high.

Only heterogeneous Ziegler–Natta catalysts yield isotactic polymers from non-polar monomers such as α-olefins. The nature of the catalyst surface appears to be important in imparting stereoregularity in these cases. Propagation probably takes place at active sites on the crystal surface of the transition-metal compound, $TiCl_3$, for example. The sites are formed by the reaction of the group I–III soluble metalorganic compound, for example, $Al(C_2H_5)_3$ or $Al(C_2H_5)_2Cl$, with the crystal surface, yielding species of the type shown in Fig. 4-1. The chemical and crystal structures of the catalysts determine the orientation of the monomer as it is added to the growing chain. The driving force for isotactic propagation thus results from

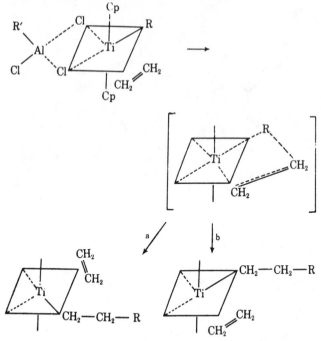

FIG. 4-2. Growth step postulated for the polymerization of ethylene by a soluble Ziegler–Natta catalyst. (Reprinted with permission from Henrici-Olivé 1981. Copyright 1981 by the American Chemical Society.)

interaction between the monomer and the ligands of the transition metal at the active site (Boor 1967). Several other models have been proposed, but none appear to be as widely applicable.

Kinetics. Considering here the important but complex case of heterogeneous polymerization, we denote the surface of the transition-metal compound by a heavy bar, ■■, the metal alkyl by AR, and the monomer by M. The following reactions are pertinent:

Adsorption of AR from solution to form the active site, and of M from solution,

$$\blacksquare\blacksquare + AR \underset{}{\overset{K_1}{\rightleftharpoons}} \blacksquare\blacksquare—AR$$

$$\blacksquare\blacksquare + M \underset{}{\overset{K_2}{\rightleftharpoons}} M—\blacksquare\blacksquare \tag{4-11}$$

Initiation,

$$M—\blacksquare\blacksquare—AR \overset{k_i}{\longrightarrow} \blacksquare\blacksquare—A—M—R \tag{4-12}$$

Propagation,

$$\text{M}\underline{\quad\blacksquare\quad}\text{—A—M}_x\text{—R} \xrightarrow{k_p} \underline{\quad\blacksquare\quad}\text{—A—M}_{x+1}\text{—R} \qquad (4\text{-}13)$$

Transfer with monomer,

$$\text{M}\underline{\quad\blacksquare\quad}\text{—A—M}_x\text{—R} \xrightarrow{k_{tr}} \underline{\quad\blacksquare\quad}\text{—A—M—R} + \text{M}_x \qquad (4\text{-}14)$$

where M_x is polymer.

Other transfer reactions are possible. Termination of the kinetic chain can occur by several reactions, an example being that with monomer to form an inactive site:

$$\text{M}\underline{\quad\blacksquare\quad}\text{—A—M}_x\text{—R} \xrightarrow{k_t} \underline{\quad\blacksquare\quad}\text{—A—M—} + \text{M}_x\text{R} \qquad (4\text{-}15)$$

Rate equations can be derived if the adsorption law is known. If it is the usual Langmuir isotherm, in which the equilibria of Eq. 4-11 are attained and maintained rapidly and in competition, the fractions of the available surface covered by AR and M, θ_A and θ_M, are given by

$$\theta_A = \frac{K_1[\text{AR}]}{1 + K_1[\text{AR}] + K_2[\text{M}]} \qquad (4\text{-}16)$$

$$\theta_M = \frac{K_2[\text{M}]}{1 + K_1[\text{AR}] + K_2[\text{M}]} \qquad (4\text{-}17)$$

The overall rate is given by

$$\frac{-d[\text{M}]}{dt} = (k_p + k_{tr})\theta_M[\text{C*}] \qquad (4\text{-}18)$$

where [C*] is the concentration of growing chains. Assuming that [C*] reaches the steady state, we can equate the equations for initiation and termination:

$$k_i\theta_A\theta_M = k_t[\text{C*}]\theta_M \qquad (4\text{-}19)$$

from which, by elimination of [C*], we obtain

$$-\frac{d[\text{M}]}{dt} = \frac{(k_p + k_{tr})k_i}{k_t}\frac{K_1K_2[\text{AR}][\text{M}]}{(1 + K_1[\text{AR}] + K_2[\text{M}])^2} \qquad (4\text{-}20)$$

Similarly, an expression for $\bar{x}_n$ can be written. These equations are oversimplified for clarity by assuming only one transfer and one termination reaction. Tait (1975) has extended and confirmed these kinetics for the polymerization of 4-methylpentane-1 by vanadium trichloride and triisobutyl aluminum.

Other Coordination Polymerizations

Polymerization With Supported Metal Oxide Catalysts.   Catalysts consisting of metals or metal oxides adsorbed on or complexed with the surface of a solid carrier are of commercial interest for the low-pressure polymerization of olefins. Although often used in fixed-bed processes, the catalysts can be slurried and employed in fluid-bed operation like the Ziegler–Natta catalysts.

Common catalyst compositions include oxides of chromium or molybdenum, or cobalt and nickel metals, supported on silica, alumina, titania, zirconia, or activated carbon. The mechanisms and kinetics of the reactions are still largely unknown, although many proposals like those made for Ziegler–Natta polymerization have been presented.

"Alfin" Polymerization.   The "Alfin" catalyst (Morton 1964; Reich 1966) consists of a suspension, in an inert solvent like pentane, of a mixture of an alkylenyl sodium compound (such as allyl sodium), an alkoxide of a secondary alcohol (such as sodium isopropoxide), and an alkali halide (such as sodium chloride). The catalyst is highly specific for the polymerization of dienes into the 1,4-forms.

GENERAL REFERENCES

Gaylord 1959; Natta 1966; Ketley 1967; Lenz 1967; Ledwith 1974*b*; Witt 1974; Elias 1977, Chapter 19; Vandenberg 1977; Boor 1979; Bowden 1979, Chapter 2.2.5; Odian 1981, Chapter 8-4.

E.   RING-OPENING POLYMERIZATION

Polymers may be prepared by routes involving ring opening, which must be classed stoichiometrically as addition, since no small molecule is split off in the reaction. Some typical examples are given in Table 4-3. The polymerization of these compounds has some aspects of both chain and step polymerization as far as kinetics and mechanism are concerned. It resembles chain polymerization in that it proceeds by the addition of monomer (but never of larger units) to growing chain molecules. However, the chain-initiating and subsequent addition steps may be similar and proceed at similar rates; if so, these are not chain reactions in the kinetic sense. As in stepwise polymerization, the polymer molecules continue to increase in molecular weight throughout the reaction.

Most of these cyclic compounds polymerize by ionic chain mechanisms in the presence of strong acids or bases when water and alcohols are excluded. These

TABLE 4-3. **Typical Ring-scission Polymers**

Type	Example
Lactone	$O(CH_2)_xCO \longrightarrow -[O(CH_2)_xCO-]_y$
Lactam	$HN(CH_2)_xCO \longrightarrow -[HN(CH_2)_xCO-]_y$
Cyclic ether	$(CH_2)_xO \longrightarrow -[(CH_2)_xO-]_y$
Cyclic anhydride	$CO(CH_2)_xCO \longrightarrow -[CO(CH_2)_xCOO-]_y$ O
N-Carboxyanhydride	$COCHRNHCO \longrightarrow -[COCHRNH-]_y + CO_2$ O

polymerizations are often very rapid. They are of commercial interest in the polymerization of caprolactam, ethylene oxide, and 3,3-*bis*-(chloromethyl)oxetane. These are true chain reactions.

GENERAL REFERENCES

Furukawa 1963; Lenz 1967; Bowden 1979, Chapter 2.3; Odian 1981, Chapter 7.

DISCUSSION QUESTIONS AND PROBLEMS

1. Unlike free-radical polymerization, both cationic and anionic polymerizations show a marked dependence on the type of solvent used. Discuss the causes and nature of these effects.

2. Write chemical equations for the major steps in the polymerization of the following:
   a. Isobutylene by stannic chloride.
   b. Styrene by sodium naphthalene.
   c. Ethylene by titanium tetrachloride and diethyl aluminum chloride.

3. Write kinetic equations, as far as they are known, for each of the polymerizations of Question 2.

4. Discuss the influence of ion pairs on polymerization by ionic mechanisms.

5. The Nobel Prize was awarded to Karl Ziegler and Giulio Natta for the discovery and development of what are now called Ziegler–Natta catalysts. Discuss the structure of these catalysts and the mechanisms of the coordination polymerization in which they are used.

6. Anionic polymerizations often do not have facile termination steps. Discuss the consequences of this fact and the resulting "living" polymers.

7. Contrast termination mechanisms in cationic and free-radical polymerization, and indicate an easy way, based on kinetics, to distinguish between the two.

BIBLIOGRAPHY

Baker 1974. R. Baker, "Carbonium Ions," Chapter 8 in A. D. Jenkins and A. Ledwith, eds., *Reactivity, Mechanism and Structure in Polymer Chemistry,* John Wiley & Sons, New York, 1974.

Beaman 1948. Ralph G. Beaman, "Anionic Chain Polymerization," *J. Am. Chem. Soc.* **70,** 3115–3118 (1948).

Boor 1967. J. Boor, Jr., "The Nature of the Active Site in the Ziegler-Type Catalyst," *Macromol. Revs.* **2,** 115–268 (1967).

Boor 1979. John Boor, Jr., *Ziegler–Natta Catalysts and Polymerization,* Academic Press, New York, 1979.

Bowden 1979. M. J. Bowden, "Formation of Macromolecules," Chapter 2 in F. A. Bovey and F. H. Winslow, eds., *Macromolecules—An Introduction to Polymer Science,* Academic Press, New York, 1979.

Bywater 1975. S. Bywater, "Anionic Polymerization," Chapter 2 in A. D. Jenkins, ed., *Progress in Polymer Science,* Vol. 4, Pergamon Press, New York, 1975.

Collins 1973. Edward A. Collins, Jan Bareš, and Fred W. Billmeyer, Jr., *Experiments in Polymer Science,* Wiley-Interscience, New York, 1973.

Cossee 1964. P. Cossee, "Ziegler–Natta Catalysis. I. Mechanism of Polymerization of α-Olefins with Ziegler–Natta Catalysts," *J. Catal.* **3,** 80–88 (1964).

Dart 1974. E. C. Dart, "Carbanions," Chapter 10 in A. D. Jenkins and A. Ledwith, eds., *Reactivity, Mechanism and Structure in Polymer Chemistry,* John Wiley & Sons, New York, 1974.

Eastham 1965. A. M. Eastham, "Cationic Polymerization," pp. 35–59 in Herman F. Mark, Norman G. Gaylord, and Norbert M. Bikales, eds., *Encyclopedia of Polymer Science and Technology,* Vol. 3, Wiley-Interscience, New York, 1965.

Elias 1977. Hans-Georg Elias, *Macromolecules · 2 · Synthesis and Materials* (translated by John W. Stafford), Plenum Press, New York, 1977.

Fetters 1969. L. J. Fetters, "Synthesis of Block Polymers by Homogeneous Anionic Polymerization," *J. Polym. Sci.* **C26,** 1–35 (1969).

Furukawa 1963. Junji Furukawa and Takeo Saegusa, *Polymerization of Aldehydes and Oxides,* Wiley-Interscience, New York, 1963.

Gaylord 1959. Norman G. Gaylord and Herman F. Mark, *Linear and Stereoregular Addition Polymers,* Interscience, New York, 1959.

Henderson 1968. J. F. Henderson and M. Szwarc, "The Use of Living Polymers in the Preparation of Polymer Structures of Controlled Architecture," *Macromol. Rev.* **3,** 317–401 (1968).

Henrici-Olivé 1969. Gisela Henrici-Olivé and Salvador Olivé, *Polymerisation—Katalyse, Kinetik, Mechanismen* (in German), Verlag Chemie, Weinheim, West Germany, 1969.

Henrici-Olivé 1981. G. Henrici-Olivé and S. Olivé, "Mechanism for Ziegler–Natta Catalysis," *Chemtech* **11,** 746–752 (1981).

Higginson 1952. W. C. E. Higginson and N. S. Wooding, "Anionic Polymerisation. Part I. The Polymerisation of Styrene in Liquid Ammonia Solution Catalysed by Potassium Amide," *J. Chem. Soc.* **1952,** 760–774 (1952).

Hsieh 1981. Henry L. Hsieh, Ralph C. Farrar, and Kishore Udipi, "Anionic Polymerization: Some Commercial Applications," *Chemtech* **11,** 626–633 (1981).

Kennedy 1977. J. P. Kennedy, S. Y. Huang, and S. C. Feinberg, "Cationic Polymerization with Boron Halides. III. BCl₃ Coinitiator for Olefin Polymerization," *J. Polym. Sci. Polym. Chem. Ed.* **15,** 2801–2819 (1977).

Kennedy 1982a. Joseph P. Kennedy and Ernest Marechal, *Carbocationic Polymerization,* John Wiley & Sons, New York. 1982.

Kennedy 1982b. Joseph P. Kennedy, Tibor Kelen, and Ferenc Tudos, "Quasiliving Carbocationic Polymerization. I: Classification of Living Polymerizations in Carbocationic Systems," *J. Macromol. Sci. Chem.* **A18,** 1189–1207 (1982); R. Faust, A. Fehervari, and Joseph P. Kennedy, "II. The Discovery: The α-Methylstyrene System," *J. Macromol. Sci. Chem.* **A18,** 1209–1228 (1982); and following papers.

Ketley 1967. A. D. Ketley, ed., *The Stereochemistry of Macromolecules,* Vol. 1, Marcel Dekker, New York, 1967.

Ledwith 1974a. A. Ledwith and D. C. Sherrington, "Reactivity and Mechanism in Cationic Polymerization," Chapter 9 in A. D. Jenkins and A. Ledwith, eds., *Reactivity, Mechanism and Structure in Polymer Chemistry,* John Wiley & Sons, New York, 1974.

Ledwith 1974b. A. Ledwith and D. C. Sherrington, "Reactivity and Mechanism in Polymerization by Complex Organometallic Derivatives," Chapter 12 in A. D. Jenkins and A. Ledwith, eds., *Reactivity, Mechanism and Structure in Polymer Chemistry,* John Wiley & Sons, New York, 1974.

Lenz 1967. Robert W. Lenz, *Organic Chemistry of Synthetic High Polymers,* Wiley-Interscience, New York, 1967.

Morton 1964. Avery A. Morton, "Alfin Catalysts," pp. 629–638 in Herman F. Mark, Norman G. Gaylord, and Norbert M. Bikales, eds., *Encyclopedia of Polymer Science and Technology,* Vol. 1, Wiley-Interscience, New York, 1964.

Morton 1969. Maurice Morton, "Anionic Polymerization," Chapter 5 in George E. Ham, ed., *Vinyl Polymerization,* Part I, Vol. 2, Marcel Dekker, New York, 1969.

Morton 1977. Maurice Morton and Lewis J. Fetters, "Anionic Polymerizations and Block Copolymers," Chapter 9 in Calvin E. Schildknecht, ed., with Irving Skeist, *Polymerization Processes,* Wiley-Interscience, New York, 1977.

Morton 1983. Maurice Morton, *Anionic Polymerization: Principles and Practice,* Academic Press, New York, 1983.

Natta 1955. G. Natta, Piero Pino, Paolo Corradini, Ferdinando Danusso, Enrico Mantica, Giorgio Mazzanti, and Giovanni Moranglio, "Crystalline High Polymers of α-Olefins," *J. Am. Chem. Soc.* **77,** 1708–1710 (1955).

Natta 1965. Giulio Natta, "Macromolecular Chemistry: From the Stereospecific Polymerization to the Symmetric Autocatalytic Synthesis of Molecules," *Science* **147,** 261–272 (1965).

Natta 1966. Giulio Natta and Umberto Giannini, "Coordinate Polymerization," pp. 137–150 in Herman F. Mark, Norman G. Gaylord, and Norbert M. Bikales, eds., *Encyclopedia of Polymer Science and Technology,* Vol. 4, Wiley-Interscience, New York, 1966.

Odian 1981. George Odian, *Principles of Polymerization,* 2nd ed., John Wiley & Sons, New York, 1981.

Olah 1972. George A. Olah, "The General Concept and Structure of Carbocations Based on Differentiation of Trivalent ('Classical') Carbenium Ions from Three-Center Bound Penta- or Tetracoordinated ('Nonclassical') Carbonium Ions. The Role of Carbocations in Electrophilic Reactions," *J. Am. Chem. Soc.* **94,** 808–820 (1972).

Overberger 1965. C. G. Overberger, J. E. Mulvaney, and Arthur M. Schiller, "Anionic Polymerization," pp. 95–137 in Herman F. Mark, Norman G. Gaylord, and Norbert M. Bikales, eds., *Encyclopedia of Polymer Science and Technology,* Vol. 2, Wiley-Interscience, New York, 1965.

Parry 1974. A. Parry, "Anionic Polymerization," Chapter 11 in A. D. Jenkins and A. Ledwith, eds., *Reactivity, Mechanism and Structure in Polymer Chemistry,* John Wiley & Sons, New York, 1974.

Plesch 1963. P. H. Plesch, ed., *The Chemistry of Cationic Polymerization,* Macmillan, New York, 1963.

Plesch 1968. P. H. Plesch, "Cationic Polymerization," pp. 137–188 in J. C. Robb and F. W. Peaker, eds., *Progress in High Polymers,* Vol. 2, CRC Press, Cleveland, Ohio, 1968.

Reich 1966. Leo Reich and A. Schindler, *Polymerization by Organometallic Compounds,* Wiley-Interscience, New York, 1966.

Russell 1977. Kenneth E. Russell and Geoffrey J. Wilson, "Cationic Polymerizations," Chapter 10 in Calvin E. Schildknecht, ed., with Irving Skeist, *Polymerization Processes,* Wiley-Interscience, New York, 1977.

Schildknecht 1947. C. E. Schildknecht, A. O. Zoss, and Clyde McKinley, "Vinyl Alkyl Ethers," *Ind. Eng. Chem.* **39,** 180–186 (1947).

Szwarc 1968a. Michael Szwarc, *Carbanions, Living Polymers and Electron-Transfer Processes,* John Wiley & Sons, New York, 1968.

Szwarc 1968b. M. Szwarc, " 'Living' Polymers," pp. 303–325 in Herman F. Mark, Norman G. Gaylord, and Norbert M. Bikales, eds., *Encyclopedia of Polymer Science and Technology,* Vol. 8, Wiley-Interscience, New York, 1968.

Tait 1975. Peter J. T. Tait, "Ziegler–Natta Polymerization—Model, Mechanism, and Kinetics," *Chemtech* **5,** 688–692 (1975).

Vandenberg 1977. Edwin J. Vandenberg and Ben C. Repka, "Ziegler-Type Polymerizations," Chapter 11 in Calvin E. Schildknecht, ed., with Irving Skeist, *Polymerization Processes,* Wiley-Interscience, New York, 1977.

Witt 1974. D. R. Witt, "Reactivity and Mechanism with Chromium Oxide Polymerization Catalysts," Chapter 13 in A. D. Jenkins and A. Ledwith, eds., *Reactivity, Mechanism and Structure in Polymer Chemistry,* John Wiley & Sons, New York, 1974.

Wooding 1952. N. S. Wooding and W. C. E. Higginson, "Anionic Polymerisation. Part III. The Polymerisation of Styrene in Liquid Ammonia Catalysed by Potassium," *J. Chem. Soc.* **1952,** 1178–1180.

Ziegler 1955. Karl Ziegler, E. Holzkamp. H. Breil, and H. Martin, "Polymerization of Ethylene and Other Olefins" (in German), *Angew. Chem.* **67,** 426 (1955).

Ziegler 1964. Karl Ziegler, "Consequences and Development of an Invention" (in German), *Angew. Chem.* **76,** 545–553 (1964).

CHAPTER FIVE

COPOLYMERIZATION

A. KINETICS OF COPOLYMERIZATION

Early Experiments in Copolymerization

Although the polymerization of organic compounds has been known for over 100 years, the simultaneous polymerization (*copolymerization*) of two or more monomers was not investigated until about 1911, when *copolymers* of olefins and diolefins were found to have rubbery properties and were more useful than *homopolymers* made from single monomers.

In the 1930's it was found that monomers differed markedly in their tendencies to enter into copolymers. Staudinger (1939) fractionated a vinyl chloride–vinyl acetate copolymer made from a mixture of equimolar quantities of the two monomers. He found no polymer containing equal amounts of each monomer but, instead, found vinyl chloride : vinyl acetate ratios of 9:3, 7:3, 5:3, and 5:7 among the fractions.

At about the same time, acrylic esters were found to enter copolymers with vinyl chloride faster than did the second monomer. The first polymer formed was rich in the acrylate; later, as the amount of acrylate in the monomer system was depleted, the polymer became richer in vinyl chloride. Maleic anhydride and other monomers that homopolymerize with great difficulty were found to copolymerize readily with such monomers as styrene and vinyl chloride. Pairs of monomers, such as stilbene–maleic anhydride and isobutylene–fumaric ester, in which neither monomer polymerizes alone, readily gave high-molecular-weight copolymers in which the monomers appeared in a 1:1 ratio, no matter which was in excess in the monomer feed.

More recently, the ability to produce polymers containing long sequences of two or more different monomers (*block* and *graft copolymers*) has led to new products with unique and valuable properties.

The kinetic scheme for chain-reaction copolymerization is developed for radical reactions; extension to ionic systems is straightforward.

The Copolymer Equation

In 1936 Dostal made the first attack on the mechanism of copolymerization by assuming that the rate of addition of monomer to a growing free radical depends only on the nature of the end group on the radical chain. Thus monomers M_1 and M_2 lead to radicals of types $M_1\cdot$ and $M_2\cdot$. There are four possible ways in which monomer can add:

Reaction	Rate	
$M_1\cdot + M_1 \longrightarrow M_1\cdot$	$k_{11}[M_1\cdot][M_1]$	
$M_1\cdot + M_2 \longrightarrow M_2\cdot$	$k_{12}[M_1\cdot][M_2]$	
$M_2\cdot + M_1 \longrightarrow M_1\cdot$	$k_{21}[M_2\cdot][M_1]$	(5-1)
$M_2\cdot + M_2 \longrightarrow M_2\cdot$	$k_{22}[M_2\cdot][M_2]$	

The kinetics of copolymerization was more fully elucidated in 1944 by Alfrey, Mayo, Simha, and Wall. To Dostal's reaction scheme they added the assumption of the steady state applied to each radical type separately, that is, the concentrations of $M_1\cdot$ and $M_2\cdot$ must each remain constant. It follows that the rate of conversion of $M_1\cdot$ to $M_2\cdot$ must equal that of conversion of $M_2\cdot$ to $M_1\cdot$:

$$k_{21}[M_2\cdot][M_1] = k_{12}[M_1\cdot][M_2] \tag{5-2}$$

The rates of disappearance of the two types of monomer are given by

$$-\frac{d[M_1]}{dt} = k_{11}[M_1\cdot][M_1] + k_{21}[M_2\cdot][M_1]$$

$$-\frac{d[M_2]}{dt} = k_{12}[M_1\cdot][M_2] + k_{22}[M_2\cdot][M_2] \tag{5-3}$$

By defining $r_1 = k_{11}/k_{12}$ and $r_2 = k_{22}/k_{21}$ and combining Eqs. 5-2 and 5-3, it can be shown that the composition of copolymer being formed at any instant is given by

$$\frac{d[M_1]}{d[M_2]} = \frac{[M_1]}{[M_2]} \frac{r_1[M_1] + [M_2]}{[M_1] + r_2[M_2]} \tag{5-4}$$

This is known as the *copolymer equation;* it has been verified by many experimental investigations of copolymer composition.

Monomer Reactivity Ratios

The *monomer reactivity ratios* r_1 and r_2 are the ratios of the rate constant for a given radical adding its own monomer to the rate constant for its adding the other monomer. Thus $r_1 > 1$ means that the radical $M_1 \cdot$ prefers to add M_1; $r_1 < 1$ means that it prefers to add M_2. In the system styrene (M_1)–methyl methacrylate (M_2), for example, $r_1 = 0.52$ and $r_2 = 0.46$; each radical adds the other monomer about twice as fast as its own.

Since the rate constants for initiation and termination do not appear in Eq. 5-4, the composition of the copolymer is independent of overall reaction rate and initiator concentration. The reactivity ratios are unaffected in most cases by the presence of inhibitors, chain transfer agents, or solvents. Even in heterogeneous systems they remain unchanged, unless the availability of the monomers is altered by their distribution between phases. A change from a free radical to an ionic mechanism, however, changes r_1 and r_2 markedly.

A few typical values of monomer reactivity ratios are given in Table 5-1. An extensive compilation is given by Young (1975a).

Types of Copolymerization

Ideal. A copolymer system is said to be *ideal* when the two radicals show the same preference for adding one of the monomers over the other: $k_{11}/k_{12} = k_{21}/k_{22}$, or $r_1 = 1/r_2$, or $r_1 r_2 = 1$. In this case the end group on the growing chain has no influence on the rate of addition, and the two types of units are arranged at random along the chain in relative amounts determined by the composition of the feed and

TABLE 5-1. Typical Monomer Reactivity Ratios[a]

Monomer 1	Monomer 2	r_1	r_2	$T(°C)$
Acrylonitrile	1,3-Butadiene	0.02	0.3	40
	Methyl methacrylate	0.15	1.22	80
	Styrene	0.04	0.40	60
	Vinyl acetate	4.2	0.05	50
	Vinyl chloride	2.7	0.04	60
1,3-Butadiene	Methyl methacrylate	0.75	0.25	90
	Styrene	1.35	0.58	50
	Vinyl chloride	8.8	0.035	50
Methyl methacrylate	Styrene	0.46	0.52	60
	Vinyl acetate	20	0.015	60
	Vinyl chloride	10	0.1	68
Styrene	Vinyl acetate	55	0.01	60
	Vinyl chloride	17	0.02	60
Vinyl acetate	Vinyl chloride	0.23	1.68	60

[a]Young (1975a).

the relative reactivities of the two monomers. The copolymer equation reduces to $d[M_1]/d[M_2] = r_1[M_1]/[M_2]$.

Alternating. Here each radical prefers to react exclusively with the other monomer: $r_1 = r_2 = 0$. The monomers alternate regularly along the chain, regardless of the composition of the monomer feed. The copolymer equation simplifies to $d[M_1]/d[M_2] = 1$.

Most actual cases lie between the ideal and the alternating systems: $0 < r_1 r_2 < 1$. A third possibility, with both r_1 and r_2 greater than unity, corresponds to the tendency to form block copolymers (Section D).

Instantaneous Composition of Feed and Polymer

Let F_1 and F_2 be the mole fractions of monomers 1 and 2 in the polymer being formed at any instant:

$$F_1 = 1 - F_2 = \frac{d[M_1]}{d([M_1] + [M_2])} \qquad (5\text{-}5)$$

If f_1 and f_2 similarly represent mole fractions in the monomer feed,

$$f_1 = 1 - f_2 = \frac{[M_1]}{[M_1] + [M_2]} \qquad (5\text{-}6)$$

the copolymer equation can be written as

$$F_1 = \frac{r_1 f_1^2 + f_1 f_2}{r_1 f_1^2 + 2 f_1 f_2 + r_2 f_2^2} \qquad (5\text{-}7)$$

It is apparent that in general F_1 does not equal f_1, and both f_1 and F_1 change as the polymerization proceeds. The polymer obtained over a finite range of conversion consists of a continuous distribution of varying composition, described in Section B.

Equation 5-7 may be used to calculate curves of feed versus instantaneous polymer composition for various monomer reactivity ratios. Such curves for a series of ideal copolymerizations ($r_1 r_2 = 1$) are shown in Fig. 5-1. Except for pairs of monomers having very similar reactivities, only a small range of feeds gives copolymers containing appreciable amounts of both components.

Several curves for nonideal cases are shown in Fig. 5-2. These curves show the effect of increasing tendency toward alternation. As alternation increases, more and more feeds yield a copolymer containing a good deal of each component. This tendency makes practical the preparation of many important copolymers.

If one of the monomers is very much more reactive than the other, the first polymer formed contains mostly the more reactive monomer. Later in the poly-

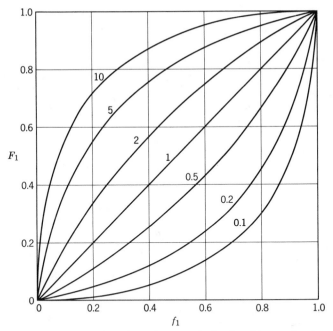

FIG. 5-1. Instantaneous composition of copolymer (mole fraction F_1) as a function of monomer composition (mole fraction f_1) for ideal copolymers with the values of $r_1 = 1/r_2$ indicated.

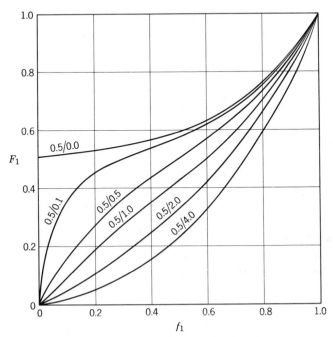

FIG. 5-2. Instantaneous composition of copolymer F_1 as a function of monomer composition f_1 for the values of the reactivity ratios r_1/r_2 indicated.

merization this monomer is largely used up, and the last polymer formed consists mainly of the less reactive monomer. Styrene and vinyl acetate form such a system.

For cases in which both r_1 and r_2 are less (or greater) than unity, the curves of Fig. 5-2 cross the line representing $F_1 = f_1$. At the point of intersection polymerization proceeds without a change in the composition of feed or polymer. This is known as *azeotropic copolymerization*. Solution of Eq. 5-4 with $d[M_1]/d[M_2] = [M_1]/[M_2]$ gives the critical composition for the azeotrope:

$$\frac{[M_1]}{[M_2]} = \frac{1 - r_2}{1 - r_1} \tag{5-8a}$$

or, from Eq. 5-7,

$$(f_1)_c = \frac{1 - r_2}{2 - r_1 - r_2} \tag{5-8b}$$

Evaluation of Monomer Reactivity Ratios

The usual experimental determination of r_1 and r_2 involves polymerizing to low conversion for a variety of feed compositions. The polymers are isolated and their compositions measured. One of several methods of analysis of the data may be used:

a. Direct curve fitting on polymer–monomer composition plots. This is a poor method, since the composition curve is rather insensitive to small changes in r_1 and r_2.
b. Plots of r_1 versus r_2. The copolymer equation may be solved for one of the reactivity ratios:

$$r_2 = \frac{[M_1]}{[M_2]} \left[\frac{d[M_2]}{d[M_1]} \left(1 + \frac{[M_1]}{[M_2]} r_1 \right) - 1 \right] \tag{5-9}$$

Each experiment with a given feed gives a straight line; the intersection of several of these allows the evaluation of r_1 and r_2 (Fig. 5-3). If the experimental errors are high, the lines may not intersect in a single point; the region within which the intersections occur gives information about the precision of the experimental results.
c. Plots of F_1 versus f_1. Equation 5-7 can be rearranged to

$$\frac{f_1(1 - 2F_1)}{(1 - f_1)F_1} = r_2 + \frac{f_1^2(F_1 - 1)}{(1 - f_1)^2 F_1} r_1 \tag{5-10}$$

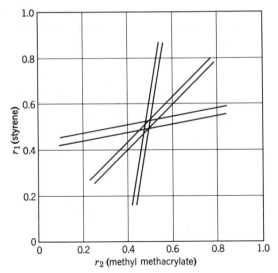

FIG. 5-3. Evaluation of monomer reactivity ratios by graphic solution of the copolymer equation (Burnett 1954, data of Mayo 1944).

This is the equation of a straight line with slope r_1 and intercept r_2. Each experiment gives one point on the line. The least-squares treatment of a series of such points gives the best values of r_1 and r_2 in a straightforward way.

d. Analysis of data giving the copolymer composition as a function of conversion, using the integrated form of the copolymer equation given in Section B (Meyer 1966).

Rate of Copolymerization

The discussion so far has considered only the relative rates of the four possible propagation steps in a binary system. The overall rate of copolymerization depends also on the rates of initiation and termination.

If the initiator releases radicals that have a high efficiency of combination with each monomer, the two types of initiation step need not be considered separately. Steady-state conditions apply both to the total radical concentration and to the separate concentrations of the two radicals. Three types of termination must be considered, involving all possible pairs of radical types. The steady-state condition for all the radicals requires that the initiation and termination rates be equal:

$$v_i = 2k_{t_{11}}[M_1\cdot]^2 + 2k_{t_{12}}[M_1\cdot][M_2\cdot] + 2k_{t_{22}}[M_2\cdot]^2 \qquad (5\text{-}11)$$

Here $k_{t_{11}}$ and $k_{t_{22}}$ are the termination rate constants for pairs of like radicals, and $k_{t_{12}}$ is the constant for the cross-reaction. If this condition is applied to the equation for the overall polymerization rate,

$$-\frac{d([M_1] + [M_2])}{dt} = v_p$$

$$= \frac{(r_1[M_1]^2 + 2[M_1][M_2] + r_2[M_2]^2)(v_i^{1/2}/\delta_1)}{\{r_1^2[M_1]^2 + 2(\phi r_1 r_2 \delta_2/\delta_1)[M_1][M_2] + (r_2\delta_2/\delta_1)^2[M_2]^2\}^{1/2}} \quad (5\text{-}12)$$

where

$$\delta_1 = \left(\frac{2k_{t_{11}}}{k_{11}^2}\right)^{1/2}, \qquad \delta_2 = \left(\frac{2k_{t_{22}}}{k_{22}^2}\right)^{1/2} \quad (5\text{-}13)$$

and

$$\phi = \frac{k_{t_{12}}}{2(k_{t_{11}})^{1/2}(k_{t_{22}})^{1/2}} \quad (5\text{-}14)$$

The δ values are simply related to quantities of the type $k_p^2/2k_t$, as seen in Chapter 3; ϕ compares the cross-termination rate constant to the geometric mean of the termination rate constants for like pairs of radicals. A value of ϕ greater than 1 means that cross termination is favored, and conversely.

If r_1 and r_2 are known, measurement of the rate of copolymerization allows ϕ to be determined. A typical result is shown in Fig. 5-4. The value of ϕ greater

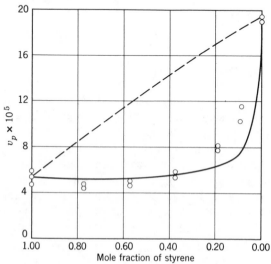

FIG. 5-4. Copolymerization rate data for the system styrene–methyl methacrylate at 60°C (Walling 1949). The broken line assumes $\phi = 1$; the solid line is calculated for $\phi = 13$.

than unity is typical of most copolymerizations; indeed, cross termination is markedly favored, except for nearly ideal copolymerizations. It is likely that differences in polarity are responsible for the preference for cross termination, as well as for deviations from ideal copolymerization. As a result, the copolymerization of pairs of monomers of quite different polarities is likely to be slower than the mean of the rates for the separate homopolymerizations.

The above treatment assumes that the termination process is controlled by the chemical nature of the growing-chain radical ends. It is known, however, that termination in radical polymerization is generally diffusion controlled. It is often useful to express the rate equation in terms of a single termination constant k_t (Atherton 1962, North 1963):

$$v_p = \frac{(r_1[M_1]^2 + 2[M_1][M_2] + r_2[M_2]^2)v_i^{1/2}}{k_t^{1/2}(r_1[M_1]/k_{11} + r_2[M_2]/k_{22})} \tag{5-15}$$

Ideally, k_t might be expected to be the mole-fraction-weighted average of the termination rate constants for the two homopolymerizations, but this appears to be an oversimplification, possibly due to penultimate effects (see below).

Extensions of Copolymerization Kinetics

Multicomponent Copolymerization. The mathematics of the copolymer equation has been extended to the cases of three or more monomers (Ham 1964), but the quantitative treatment is quite complex; for example, three monomers require the specification of nine propagation and six termination reactions, and six reactivity ratios. Some simplification results from rather liberal assumptions about the steady state and the probabilities of finding certain sequences of monomers in the resulting polymer. The predicted results are in good agreement with experiment.

Penultimate and Remote-Unit Effects. Ham (1964) and co-workers have developed the early suggestion of Merz (1946), that units before the last may affect the reactivity of a chain radical. This is often the case, the effect being particularly pronounced with monomers of type CHX=CHX, where X is a polar substituent. It has been detected in copolymers of styrene with maleic anhydride and fumaronitrile.

When the penultimate monomer unit does have some influence on the subsequent addition of monomer, four reactivity ratios, corresponding to the four possible types of radicals, must be considered. Effects of units still farther back along the chain have been detected experimentally in some cases.

Depropagation During Copolymerization. If one or both of the monomers has a tendency to depropagate during a copolymerization, an additional complication is introduced (Lowry 1960). This deviation from the simple treatment can be detected by the dependence of the copolymer composition on the absolute concentration of the monomers, even though their relative concentration is invariant.

Complex Participation. The presence of complexes between comonomers, for example, charge-transfer complexes (Section *B*) also can cause deviations from normal copolymerization kinetics, often accompanied by an enhanced tendency toward alternation. Four additional propagation steps and six reactivity ratios are required in the analysis. As in the case of depropagation, copolymer composition is predicted to vary with temperature and monomer concentration. Without the availability of complete and accurate experimental data it is often difficult to assign the cause of deviations from simple copolymer kinetics to one or another of the above models.

GENERAL REFERENCES

Ham 1964, 1966; Bamford 1976; Collins 1973, Exp. 5; Elias 1977, Chapter 22; Odian 1981, Chapter 6.

B. COMPOSITION OF COPOLYMERS

The copolymer equation (Eq. 5-4) predicts the average composition of the polymer formed at any instant in the polymerization. Not only may statistical fluctuations of composition about this average occur, but also polymer formed during a finite interval may contain a range of compositions resulting from variation of feed composition. In this case, the overall polymer composition can be calculated by integrating the copolymer equation.

Statistical Fluctuation of Instantaneous Copolymer Composition

Stockmayer (1945) showed that the spread in copolymer composition due to statistical fluctuations is small in practical cases. Alternating copolymers give the minimum spread in composition; even in ideal copolymers, where the spread is greater and is a function of the degree of polymerization, the effect is not large. For a copolymer containing on the average 50% of each monomer, only 12% of the polymer molecules contain less than 43% of either monomer at $\bar{x}_n = 100$, while only 12% of the molecules contain less than 49.3% of either monomer at $\bar{x}_n = 10,000$. The distribution of compositions follows a Gaussian curve. This small spread in composition is neglected in the discussion to follow.

   The probability of occurrence of sequences of n similar monomer units can be calculated statistically for an ideal copolymer (Alfrey 1952): The number $N(n)$ of such sequences is

$$N(n) = p^{n-1} (1 - p) \qquad (5\text{-}16)$$

where p is the probability of addition of that monomer,

$$p = \frac{r_1[M_1]}{r_1[M_1] + [M_2]} \tag{5-17}$$

Except for a monomer present in large excess, the sequences are always short. Any tendency toward alternation clearly leads to shorter sequences than those in an ideal copolymer.

Integration of the Copolymer Equation

The direct integration of the copolymer equation gives a result that is not convenient for calculations. The most convenient method (Skeist 1946) for calculating copolymer composition and distribution involves the use of Eq. 5-7. For a system in which $F_1 > f_1$, when dM moles of monomer have polymerized, the polymer contains $F_1 \, dM$ moles of monomer 1, and the feed contains $(f_1 - df_1)(M - dM)$ moles of monomer 1. For a material balance,

$$Mf_1 - (M - dM)(f_1 - df_1) = F_1 \, dM \tag{5-18}$$

Combining this with Eq. 5-7 gives

$$\frac{dM}{M} = \frac{df_1}{F_1 - f_1} \tag{5-19}$$

which can easily be integrated by numerical or graphic means to give the desired composition of polymer as a function of conversion. One convenient closed-form result of this integration is (Meyer 1965, 1967)

$$\log\left(\frac{M}{M_0}\right) = \frac{r_2}{1 - r_2} \log\left(\frac{f_1}{(f_1)_0}\right) + \frac{r_1}{1 - r_1} \log\left(\frac{f_2}{(f_2)_0}\right)$$

$$+ \frac{1 - r_1 r_2}{(1 - r_1)(1 - r_2)} \log\left(\frac{(f_1)_0 - \epsilon}{f_1 - \epsilon}\right) \tag{5-20}$$

where $\epsilon = (1 - r_2)/(2 - r_1 + r_2)$.

Variation of Copolymer Composition With Conversion

Ideal System. The system styrene (M_1)–2-vinylthiophene (M_2) is nearly ideal, with $r_1 = 0.35$ and $r_2 = 3.10$. The variation of the composition of polymer formed at any instant in this system with conversion and initial feed composition is shown in Fig. 5-5. The greater reactivity of the vinylthiophene causes it to be depleted until at the last only styrene remains, and pure polystyrene is the final product

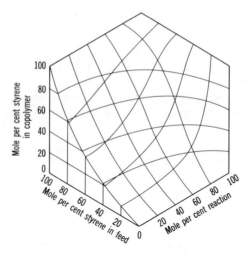

FIG. 5-5. Variation in instantaneous composition of copolymer with initial feed and conversion for the system styrene–2-vinylthiophene (Mayo 1950).

formed. The overall polymer composition varies less with conversion and ultimately approaches the initial feed composition as required for a material balance.

The polymer composition distribution for 100% conversion for several initial feeds is shown in the bar graphs in Fig. 5-6. Each block represents a 5% interval in copolymer composition. The U-shaped distribution curves arise whenever the monomer reactivity ratios are less than 0.5 and greater than 2, respectively. They indicate that appreciable quantities of two distinct compositions predominate, with little material of intermediate composition.

Alternating System.   The system styrene (M_1)–diethyl fumarate (M_2) with $r_1 = 0.30$ and $r_2 = 0.07$ has a strong tendency to alternate and an azeotropic composition at 57 mol.% styrene. The diagrams for this copolymer are more complex than those for the previous system. Feeds near the azeotropic composition remain almost unchanged up to rather high conversion, but eventually all feeds containing more than 57 mol.% styrene drift toward pure styrene, and those containing less styrene than the azeotropic amount yield pure diethyl fumarate. Thus (Fig. 5-7) pure polystyrene is formed at the last from all initial feeds higher in styrene than the azeotropic, and pure poly(diethyl fumarate) from all initial feeds containing less than 57 mol% styrene. The overall polymer composition shows markedly less change with conversion for all feeds and is of course invariant at the azeotropic composition.

Attainment of Homogeneity in Copolymers.   It is often desirable to produce a polymer that is homogeneous in composition rather than one having a broad distribution of compositions. The foregoing discussion suggests two ways of doing this: (a) The reaction can be stopped short of complete conversion. For example, the styrene–diethyl fumarate copolymer from feed containing 40% styrene, when

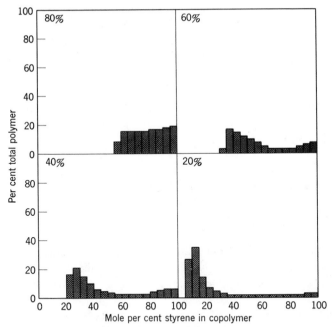

FIG. 5-6. Distribution of copolymer composition at 100% conversion, for the indicated values of initial mole percent styrene in the feed, for the system styrene–2-vinylthiophene (Burnett 1954, after Mayo 1950).

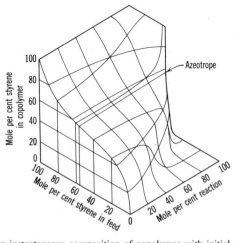

FIG. 5-7. Variation in instantaneous composition of copolymer with initial feed and conversion for the system styrene–diethyl fumarate (Mayo 1950).

stopped at 75% conversion, contains only polymer having between 44 and 52% styrene. (b) The polymerization can be started with a feed giving the desired polymer and the feed composition maintained by adding an increment of more reactive monomer as the polymerization proceeds. Both methods find commercial use.

GENERAL REFERENCES

Alfrey 1952; Odian 1981, Chapter 6.

C. MECHANISMS OF COPOLYMERIZATION

Free-Radical Copolymerization

The order of reactivity of monomers toward free radicals not only is a function of the reactivity of the monomers, but also depends on the nature of the attacking radical. This is illustrated by the tendency of many monomers to alternate in a copolymer chain. The two factors, general reactivity and alternating tendency, are predominant in determining the behavior of monomers in copolymerization.

The reactivity of monomers and radicals in copolymerization is determined by the nature of the substituents on the double bond of the monomer. These substituents influence reactivity in three ways: They may activate the double bond, making the monomer more reactive; they may stabilize the resulting radical by resonance; or they may provide steric hindrance at the reaction site.

Reactivity of Monomers.   The relative reactivities of monomers to a reference radical can be derived from the monomer reactivity ratio. The inverse of this ratio is the rate of reaction of the reference radical with another monomer, relative to that with its own monomer. If the latter rate is taken as unity, relative monomer reactivities can be examined. A few such rates are listed in Table 5-2; more extensive tabulations are given in the General References. As a different reference point is

TABLE 5-2.   Relative Reactivities of Monomers to Reference Radicals at 60°C

Monomer	Styrene	Methyl Methacrylate	Acrylo-nitrile	Vinyl Chloride	Vinyl Acetate
		Reference Radical			
Styrene	(1.0)	2.2	25	50	100.
Methyl methacrylate	1.9	(1.0)	6.7	10	67
Acrylonitrile	2.5	0.82	(1.0)	25	20
Vinylidene chloride	5.4	0.39	1.1	5	10
Vinyl chloride	0.059	0.10	0.37	(1.0)	4.4
Vinyl acetate	0.019	0.05	0.24	0.59	(1.0)

taken for each radical, values in different columns are not comparable, although ratios of values can be compared from column to column.

Although there are some exceptions, the effectiveness of substituents in enhancing the reactivity of the monomer lies in about the order

$$-C_6H_5 > -CH{=}CH_2 > -COCH_3 > -CN > -COOR > -Cl$$
$$> CH_2Y > -OCOCH_3 > -OR$$

The effect of a second substituent on the same carbon atom is usually additive.

This order of reactivity corresponds to the resonance stabilization of the radical formed after the addition. In the case of styrene the radical can resonate among forms of the type

The radical is stabilized with a resonance energy of about 80 kJ/mole. At the other extreme, substituents that have no double bonds conjugated with the ethylenic double bond give radicals having very low resonance energies (5–15 kJ/mole), since only polar or unbonded forms can contribute to the resonance.

Substituents that stabilize the product radical tend also to stabilize the monomer, but the amount of stabilization is much smaller than for the radical; in styrene the monomer is stabilized to the extent of about 12 kJ/mole. Thus the stabilizing effect of a substituent on the product radical is compensated for only to a limited extent by stabilization of the monomer.

Resonance Stabilization. Resonance stabilization of monomers and radicals can be considered in terms of a potential energy diagram of the type shown in Fig. 5-8. The energy changes shown represent situations before and after each of four possible reactions in which a resonance-stabilized (M_s) or unstabilized (M) monomer adds to a stabilized ($R_s\cdot$) or unstabilized ($R\cdot$) radical. The set of four curves labeled with the four reaction possibilities represents the combined energy of the growing-chain radical and the monomer before reaction, while the two curves labeled "new radical" represent the potential energy of the new bond formed by their combination. Activation energies are represented by solid arrows, and heats of reaction by broken arrows. If the entropies of activation are all about the same (a good assumption in the absence of steric hindrance), the order of reaction rates is predicted to be the opposite of that of the activation energies:

$$R\cdot + M_s > R\cdot + M > R_s\cdot + M_s > R_s\cdot + M$$

This series summarizes much of the rate information available for both copolymerizations and homopolymerizations.

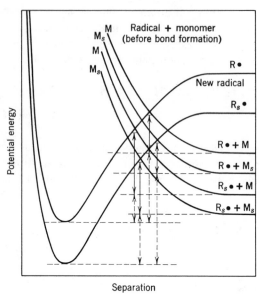

FIG. 5-8. Potential energy of a radical–monomer pair as a function of reaction coordinate; see text for further explanation (Walling 1957).

Reactivity of Radicals. The reactivity of radicals to a reference monomer can be derived by combining the absolute propagation rate constants k_{11} with reactivity ratios to obtain the rate constants k_{12} for radical 1 adding monomer 2. Typical data are listed in Table 5-3.

As expected, resonance stabilization depresses the reactivity of radicals so that their order of reactivity is the reverse of that of their monomers; styrene is one of the least reactive radicals and vinyl acetate one of the most reactive. The effect of a substituent in depressing the activity of a radical is much greater than its effect in enhancing the activity of the monomer. For example, the styrene radical is about 1000 times less reactive than the vinyl acetate radical to a given monomer if alternation effects are disregarded, but styrene monomer is only about 50 times more reactive than vinyl acetate to a given radical.

Steric Effects. The effect of steric hindrance in reducing reactivity may be demonstrated by comparing the reactivities of 1,1- and 1,2-disubstituted olefins to reference radicals. As indicated in the preceding discussion, the addition of a second 1-substituent usually increases reactivity 3- to 10-fold; however, the same substituent in the 2-position usually decreases reactivity 2- to 20-fold. The extent of reduction in reactivity depends on energy differences between cis and trans forms and on the possibilities for resonance.

Charge-Transfer Complexes. Two monomers with widely different polarities, for example, an electron donor and an electron acceptor, can form a charge-transfer complex that may act as a single species in polymerization, leading to 1:1 alternating

TABLE 5-3.   Radical–Monomer Propagation Rate Constants at 60°C (Liter/Mole-sec)

| | Radical | | | | | |
Monomer	Butadiene	Styrene	Methyl Methacrylate	Methyl Acrylate	Vinyl Chloride	Vinyl Acetate
Butadiene	100	250	2,820	42,000	350,000	—
Styrene	74	145	1,520	14,000	600,000	230,000
Methyl methacrylate	134	278	705	4,100	123,000	150,000
Methyl acrylate	132	206	370	2,090	200,000	23,000
Vinyl chloride	11	8	70	230	12,300	10,000
Vinyl acetate	—	2.6	35	520	7,300	2,300

polymers. The formation of such complexes is confirmed by spectroscopic evidence. Typical electron acceptors are methyl methacrylate, methyl acrylate, and acrylonitrile; typical donors for these complexes are vinyl chloride, vinyl acetate, propylene, and isobutylene. In many instances the presence of a Lewis acid enhances the tendency for complex formation. It is in this way that monomers such as isobutylene and propylene, which do not homopolymerize because of degradative chain transfer, can participate in free-radical polymerization.

Polarity and Alternation.   By examining the tendency for alternation as given by the product r_1r_2, it is possible to tabulate monomers in a series arranged so that two monomers farther apart in the series have a greater tendency to alternate. Such an arrangement is shown in Table 5-4. Values of r_1r_2 generally decrease as monomers farther apart in the table are compared. There are some exceptions, suggesting specific interactions, probably steric in nature. For example, vinyl acetate alternates more than styrene with vinylidene chloride, and less than styrene with acrylonitrile.

There is little doubt that alternation is primarily due to the polarity of the double bond in the monomer: The order of monomers in Table 5-4 closely parallels the order of the tendency of substituents around the double bond to donate electrons to the bond (hydrocarbon, acetoxy) or to withdraw them (carbonyl, cyano).

As might be expected, the tendency toward alternation (if not too great) parallels the tendency toward decreased reactivity induced by substitution of electron-withdrawing groups at the double bond.

The Alfrey–Price Treatment   (Alfrey 1964). Alfrey and Price (Alfrey 1947) attempted to express the factors of general reactivity and polarity quantitatively. They wrote the rate constants for copolymerization in the form

$$k_{12} = P_1 Q_2 \exp(-e_1 e_2) \tag{5-21}$$

where exp denotes the exponential, P_1 is the general reactivity of the radical $M_1 \cdot$, Q_2 is the reactivity of the monomer M_2, and e_1 and e_2 are proportional to the electrostatic interaction of the permanent charges on the substituents in polarizing

TABLE 5-4. Product of Reactivity Ratios With Monomers Arranged in Order of Alternating Tendency

Vinyl acetate							
—	Butadiene						
0.55	0.98	Styrene					
0.39	0.31	0.34	Vinyl chloride				
0.30	0.19	0.24	1.0	Methyl methacrylate			
0.6	<0.1	0.16	0.96	0.61	Vinylidene chloride		
0.21	0.0006	0.016	0.11	0.18	0.35	Acrylonitrile	
0.0049	—	0.021	0.056	—	0.55	—	Diethyl fumarate

the double bond. This form is analogous to Hammett's (1940) equation for the effects of nuclear substituents on the reactivity of aromatic compounds. From the definitions of the monomer reactivity ratios it follows that

$$r_1 = \frac{Q_1}{Q_2} \exp[-e_1(e_1 - e_2)]$$

$$r_2 = \frac{Q_2}{Q_1} \exp[-e_2(e_2 - e_1)]$$

(5-22)

and

$$r_1 r_2 = \exp[-(e_1 - e_2)^2]$$

(5-23)

The Alfrey–Price equation assumes that the same value of e applies to both monomer and radical. This assumption is justified only by the success of the Q–e scheme in interpreting relative reactivities.

It should be possible to assign Q and e values to a series of monomers from the data on a few copolymerization experiments and compute values of r_1 and r_2 for all the possible arrangements of the monomers in pairs (Price 1948). Unfortunately, this practical application of the scheme is hampered by large uncertainties in the values of Q and e that can be assigned with the present experimental error. The scheme is not entirely sound from the theoretical standpoint and is best regarded as semiempirical. In a sense, it represents the transcription of the reactivity and polarity series of Tables 5-2 and 5-4 to equation form. Young (1975b) provides an extensive tabulation of values of Q and e.

Ionic Copolymerization

The order of monomer reactivity in *cationic copolymerization* is quite different from that in free-radical polymerization and is distinguished by the high reactivities of vinyl ethers and isobutylene. However, the reactivities correspond to the anticipated effect of substituents upon the reactivity of double bonds toward electrophilic reagents. The differences of reactivity arise from the change in electron availability in the double bond and the resonance stabilization of the resulting carbenium ion.

Compared with radical-induced copolymerization, the carbenium-ion reactions show a much wider range of reactivity among the common monomers, and no tendency toward alternation in the copolymer. These characteristics mean that relatively few monomer pairs easily yield copolymers containing large proportions of both monomers.

The data on *anionic copolymerization* suggest that the order of monomer reactivity is much different from both free-radical and carbenium-ion polymerization. As anticipated, the order of reactivity in the anionic process is determined by the ability of substituents to withdraw electrons from the double bond and to stabilize the carbanion formed. Anionic and coordination polymerizations involving monomers of like electronegativity are usually ideal, since the unsubstituted carbon atom at the end of the growing chain has little influence on the course of the reaction. When the electronegativities of the monomers are widely different, however, the rates of the cross-propagation reactions may be quite different also. In extreme cases, only one monomer can add at all.

Both cationic and anionic reactivity ratios can be profoundly affected by solvent and counterion effects. These are both complicated and variable, but it appears that if the catalyst and its counterion form a tightly bound pair, solvent effects can be large, and conversely.

The contrast among the three methods of polymerization is brought out very clearly by the results of copolymerization studies using the three modes of initiation. Figure 5-9 shows the composition of the initial polymer formed from a feed of styrene and methyl methacrylate using different initiators. These results verify the different polymerization mechanisms and illustrate the extreme variations of reactivities that monomers exhibit with different type initiators. Again, the results reflect the particular selection of catalyst, solvent, and counterion almost as much as the difference between mechanisms.

Step-Reaction Copolymerization

Since the reactivities of all functional groups in simple bifunctional stepwise polymerization are essentially identical, irrespective of the length of the molecule to which they are attached, comonomers in such systems are randomly distributed along the chain in amounts proportional to their concentrations in the feed. Thus the "reactivity ratios" for stepwise polymerization are unity. Departures from this generalization are occasionally encountered, however; for example, the esterification rate of succinic acid $HOOC(CH_2)_2COOH$, differs from that of adipic acid,

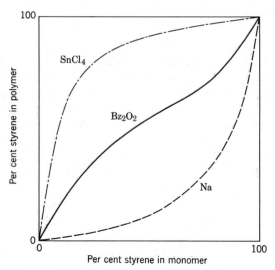

FIG. 5-9. Instantaneous composition of copolymer as a function of monomer composition for the system styrene–methyl methacrylate polymerized by cationic (SnCl$_4$), free-radical (benzoyl peroxide, Bz$_2$O$_2$), and anionic (Na) mechanisms (Landler 1950, Pepper 1954).

HOOC(CH$_2$)$_4$COOH, because of the proximity of the carboxyl groups in the former compound. For similar reasons, ethylene glycol and 1,4-butanediol have different esterification rates. Also, 1,3-butanediol differs in reactivity from 1,4-butanediol because one of the hydroxyl groups of the former is secondary, whereas both of the latter are primary. Where such reactivity differences exist, the tendency is for the more reactive species to enter into the polymer first.

GENERAL REFERENCES

Ham 1964; Lenz 1967; Bamford 1976; Elias 1977, Chapter 22; Schulz 1979; Odian 1981, Chapter 6.

D. BLOCK AND GRAFT COPOLYMERS

It is shown in Section *B* that the probability of finding long sequences of one monomer in an ordinary random copolymer is very small, except in the trivial case where one monomer is present in large excess. Methods of synthesis of polymers containing such long sequences are of interest, however, since they may lead to polymers with properties widely different from those of either homopolymers or random copolymers (as discussed in Chapter 12). Polymers with long sequences of two monomers can have two arrangements of chains: In *block copolymers* the sequences follow one another along the main polymer chain,

$$-AABB-BBAA-AABB-, \quad \text{etc.}$$

whereas in *graft copolymers* sequences of one monomer are "grafted" onto a "backbone" of the second monomer type,

$$
\begin{array}{ccc}
-AAA-AAA-AAA-, & \text{etc.} \\
B \quad\quad B \quad\quad B \\
B \quad\quad B \quad\quad B \\
| \quad\quad\; | \quad\quad\; |
\end{array}
$$

The ultimate aim in preparing either type is to obtain the desired copolymer free from homopolymer or other unwanted species, on a scale that allows evaluation of physical properties. Few of the techniques described in the literature satisfy these requirements.

The industrial importance of both block and graft copolymers has increased markedly in recent years. These commerical polymers are discussed in Part 4.

Block Copolymers

Polymers With Labile End Groups.   The usual method of preparing block copolymers utilizes a polymer with end groups that can be made to react under different conditions. In stepwise polymerization the method is trivially simple: Two polyesters, for example, of different type are separately prepared, mixed, and allowed to polymerize further. Ring-scission polymerization can be used effectively.

By means of free-radical techniques, block copolymers can be made in several ways. Labile end groups (as, for example, bromine incorporated through termination by transfer to CBr_4) can be activated by thermal or ultraviolet-light scission of the C—Br bond. Difunctional initiators can be used to produce block copolymers if the two initiator groups can be decomposed independently, or polymer radicals can be formed by milling or masticating the polymer. In each case, of course, the polymer radical must be produced in the presence of a second monomer to yield the copolymer.

By far the most important synthesis of block copolymers is that of Szwarc (1968*a,b*) for producing "living" polymers by unterminated anionic polymerization (Chapter 4*C*). Addition of a second momomer to the still active polymer leads to block copolymer uncontaminated with homopolymer and with blocks of accurately known and controlled length. The process can be repeated, with care, to produce multiblock polymers. Several commercial block copolymers are produced in this way and are discussed in Part 4; see also Fetters (1969).

Graft Copolymers

Graft copolymerization results from the formation of an active site at a point on a polymer molecule other than its end, and exposure to a second monomer. Most graft copolymers are formed by radical polymerization. The major activation reaction is chain transfer to polymer (Chapter 3). In many instances the transfer reaction involves abstraction of a hydrogen atom. An important commercial application is the grafting of polystyrene or styrene–acrylonitrile copolymer onto butadiene or acrylonitrile–butadiene copolymer rubber in the production of ABS resins (Chapter 14A). Ultraviolet or ionizing radiation, or redox initiation, among other methods, can also be used to produce the polymer radicals leading to graft copolymers.

GENERAL REFERENCES

Burlant 1960; Hoffman 1964; Ceresa 1965, 1973, 1976; Battaerd 1967; Aggarwal 1970; Allport 1973; Noshay 1977; Morton 1977; Schildknecht 1977.

DISCUSSION QUESTIONS AND PROBLEMS

1. Define the reactivity ratios r_1 and r_2, and indicate their values for (a) ideal, (b) alternating, and (c) azeotropic copolymerization.

2. Derive the copolymer equation, stating the assumptions used.

3. Discuss the chain structures obtained in alternating, random, block, and graft copolymers.

4. Consider the copolymerization of methyl acrylate (1) and vinyl chloride (2). The following compositions were found:

Mole Percent Methyl Acrylate in Feed	Mole Percent Methyl Acrylate in Polymer
7.5	44.1
15.4	69.9
23.7	75.3
32.6	82.8
42.1	86.4
52.1	90.0
74.4	96.8
86.7	98.3

a. Calculate the reactivity ratios. Explore the three methods described in the text and comment on their relative merits.

b. For a 50:50 mole percent feed, and using for uniformity the literature values $r_1 = 9.0$ and $r_2 = 0.083$, calculate and plot curves of the following:

   1. Remaining feed composition versus conversion.

   2. Instantaneous copolymer composition versus conversion.

   3. Overall copolymer composition versus conversion.

   4. Distribution of copolymer compositions.

5. Using the reactivity ratios given in Table 5-1, check the findings of Staudinger (1939) quoted in the first section of this chapter for the compositions of fractions isolated from a copolymer of vinyl acetate and vinyl chloride.

6. Given the following values of Q and e, calculate monomer reactivity ratios for the copolymerization of the monomers in pairs. Compare your results with those given in Table 5-1 and discuss the merits of the Q–e scheme for these monomers.

Monomer	Q	e
1,3-Butadiene	2.39	-1.05
Methyl methacrylate	0.74	0.40
Styrene	1.00	-0.80
Vinyl chloride	0.044	0.20

BIBLIOGRAPHY

Aggarwal 1970. S. L. Aggarwal, ed., *Block Polymers,* Plenum Press, New York, 1970.

Alfrey 1944. Turner Alfrey, Jr., and G. Goldfinger, "The Mechanism of Copolymerization," *J. Chem. Phys.* **12,** 205–209 (1944).

Alfrey 1947. Turner Alfrey, Jr., and Charles C. Price, "Relative Reactivities in Vinyl Copolymerization," *J. Polym. Sci.* **2,** 101–106 (1947).

Alfrey 1952. Turner Alfrey, Jr., John J. Bohrer, and H. Mark, *Copolymerization,* Interscience, New York, 1952.

Alfrey 1964. T. Alfrey, Jr., and L. J. Young, "The Q–e Scheme," Chapter 2 in George E. Ham, ed., *Copolymerization,* Wiley-Interscience, New York, 1964.

Allport 1973. D. C. Allport and W. H. Janes, eds., *Block Polymers,* Halsted Press, John Wiley & Sons, New York, 1973.

Atherton 1962. J. N. Atherton and A. M. North, "Diffusion-Controlled Termination in Free Radical Copolymerization," *Trans. Faraday Soc.* **58,** 2049–2057 (1962).

Bamford 1976. C. H. Bamford and C. F. H. Tippers, "Free-Radical Polymerisation," Vol. 14A in *Comprehensive Chemical Kinetics,* Elsevier, Amsterdam, 1976.

Battaerd 1967. H. A. J. Battaerd and G. W. Tregear, *Graft Copolymers,* John Wiley & Sons, New York, 1967.

Burlant 1960. William J. Burlant and Allan S. Hoffman, *Block and Graft Polymers,* Reinhold, New York, 1960.

Burnett 1954. G. M. Burnett, *Mechanism of Polymer Reactions,* Interscience, New York, 1954.

Ceresa 1965. R. J. Ceresa, "Block and Graft Copolymers," pp. 485–528 in Herman F. Mark, Norman G. Gaylord, and Norbert M. Bikales, eds., *Encyclopedia of Polymer Science and Technology,* Vol. 2, Wiley-Interscience, New York, 1965.

Ceresa 1973. R. J. Ceresa, ed., *Block and Graft Polymerization,* Vol. 1, Wiley-Interscience, New York, 1973.

Ceresa 1976. R. J. Ceresa, ed., *Block and Graft Polymerization,* Vol. 2, Wiley-Interscience, New York, 1976.

Collins 1973. Edward A. Collins, Jan Bareš, and Fred W. Billmeyer, Jr., *Experiments in Polymer Science,* Wiley-Interscience, New York, 1973.

Dostal 1936. H. Dostal, "A Basis for the Reaction Kinetics of Mixed Polymerization" (in German), *Monatsh. Chem.* **69,** 424–426 (1936).

Elias 1977. Hans-Georg Elias, *Macromolecules · 2 · Synthesis and Materials* (translated by John W. Stafford), Plenum Press, New York, 1977.

Fetters 1969. L. J. Fetters, "Synthesis of Block Polymers by Homogeneous Anionic Polymerization," *J. Polym. Sci.* **C26,** 1–35 (1969).

Ham 1964. George E. Ham, ed., *Copolymerization,* Wiley-Interscience, New York, 1964.

Ham 1966. George E. Ham, "Copolymerization," pp. 156–244 in Herman F. Mark, Norman G. Gaylord, and Norbert M. Bikales, eds., *Encyclopedia of Polymer Science and Technology,* Vol. 4, Wiley-Interscience, New York, 1966.

Hammett 1940. Louis P. Hammett, *Physical Organic Chemistry,* McGraw-Hill, New York, 1940.

Hoffman 1964. Allan S. Hoffman and Robert Bacskai, "Block and Graft Copolymerizations," Chapter 4 in George E. Ham, ed., *Copolymerization,* Wiley-Interscience, New York, 1964.

Landler 1950. Yvan Landler, "On Ionic Copolymerization" (in French), *C. R. Acad. Sci.* **230,** 539–541 (1950).

Lenz 1967. Robert W. Lenz, *Organic Chemistry of Synthetic High Polymers,* Wiley-Interscience, New York, 1967.

Lowry 1960. George G. Lowry, "The Effect of Depropagation on Copolymer Composition. I. General Theory for One Depropagating Monomer," *J. Polym. Sci.* **42,** 463–477 (1960).

Mayo 1944. Frank R. Mayo and Frederick M. Lewis, "Copolymerization. I. A Basis for Comparing the Behavior of Monomers in Copolymerization; The Copolymerization of Styrene and Methyl Methacrylate," *J. Am. Chem. Soc.* **66,** 1594–1601 (1944).

Mayo 1950. Frank R. Mayo and Cheves Walling, "Copolymerization," *Chem. Rev.* **46,** 191–287 (1950.)

Merz 1946. E. Merz, T. Alfrey, and G. Goldfinger, "Intramolecular Reactions in Vinyl Polymers as a Means of Investigation of the Propagation Step," *J. Polym. Sci.* **1,** 75–82 (1946).

Meyer 1965. Victor E. Meyer and George G. Lowry, "Integral and Differential Binary Copolymerization Equations," *J. Polym. Sci.* **A3,** 2843–2851 (1965).

Meyer 1966. Victor E. Meyer, "Copolymerization of Styrene and Methyl Methacrylate. Reactivity Ratios from Conversion-Composition Data," *J. Polym. Sci.* **A-1 4,** 2819–2830 (1966).

Meyer 1967. Victor E. Meyer and Richard K. S. Chan, "Computer Calculations of Batch-Type Copolymerization Behavior," *Am. Chem. Soc. Div. Polym. Chem. Prepr.* **8** (1), 209–215 (1967).

Morton 1977. Maurice Morton and Lewis J. Fetters, "Anionic Polymerizations and Block Copolymers," Chapter 9 in Calvin E. Schildknecht, ed., with Irving Skeist, *Polymerization Processes,* Wiley-Interscience, New York, 1977.

North 1963. A. M. North, "The ϕ-Factor in Free-Radical Copolymerization," *Polymer* **4,** 134–135 (1963).

Noshay 1977. Allen Noshay and James E. McGrath, *Block Copolymers—Overview and Critical Survey,* Academic Press, New York, 1977.

Odian 1981. George Odian, *Principles of Polymerization,* 2nd ed., John Wiley & Sons, New York, 1981.

Pepper 1954. D. C. Pepper, "Ionic Polymerisation," *Chem. Soc. Q. Rev.* **8,** 88–121 (1954).

Price 1948. Charles C. Price, "Some Relative Monomer Reactivity Factors," *J. Polym. Sci.* **3,** 772–775 (1948).

Schildknecht 1977. Calvin E. Schildknecht, "Industrial Graft Copolymerizations," Chapter 8 in C. E. Schildknecht, ed., with Irving Skeist, *Polymerization Processes,* Wiley-Interscience, New York, 1977.

Schulz 1979. D. N. Schulz and D. P. Tate, "Copolymers," pp. 798–818 in Martin Grayson, ed., *Kirk–Othmer Encyclopedia of Chemical Technology,* 3rd ed., Vol. 6, Wiley-Interscience, New York, 1979.

Simha 1944. Robert Simha and Herman Branson, "Theory of Chain Copolymerization Reactions," *J. Chem. Phys.* **12,** 253–267 (1944).

Skeist 1946. Irving Skeist, "Copolymerization: The Composition Distribution Curve," *J. Am. Chem. Soc.* **68,** 1781–1784 (1946).

Staudinger 1939. H. Staudinger and J. Schneiders, "Macromolecular Compounds. CCXXXI. Polyvinylchlorides" (in German), *Justus Liebigs Ann. Chem.* **541,** 151–195 (1939).

Stockmayer 1945. W. H. Stockmayer, "Distribution of Chain Lengths and Composition in Copolymers," *J. Chem. Phys.* **13,** 199–207 (1945).

Szwarc 1968a. Michael Szwarc, *Carbanions, Living Polymers and Electron-Transfer Processes,* John Wiley & Sons, New York, 1968.

Szwarc 1968b. M. Szwarc, " 'Living' Polymers," pp. 303–325 in Herman F. Mark, Norman G. Gaylord, and Norbert M. Bikales, eds., *Encyclopedia of Polymer Science and Technology,* Vol. 8, Wiley-Interscience, New York, 1968.

Wall 1944. Frederick T. Wall, "The Structure of Copolymers. II," *J. Am. Chem. Soc.* **66,** 2050–2057 (1944).

Walling 1949. Cheves Walling, "Copolymerization. XIII. Over-all Rates in Copolymerization. Polar Effects in Chain Initiation and Termination," *J. Am. Chem. Soc.* **71,** 1930–1935 (1949).

Walling 1957. Cheves Walling, *Free Radicals in Solution,* John Wiley & Sons, New York, 1957.

Young 1975a. Lewis J. Young, "Copolymerization Reactivity Ratios," pp. II-105-II-386 in J. Brandrup and E. H. Immergut, eds., with the collaboration of W. McDowell, *Polymer Handbook,* 2nd ed., Wiley-Interscience, New York, 1975.

Young 1975b. Lewis J. Young, "Copolymerization Reactivity Ratios," pp. II-387-II-404 in J. Brandrup and E. H. Immergut, eds., with the collaboration of W. McDowell, *Polymer Handbook,* 2nd ed., Wiley-Interscience, New York, 1975.

CHAPTER SIX

POLYMERIZATION CONDITIONS AND POLYMER REACTIONS

The discussion of polymerization kinetics in Chapters 2–5 has dealt, with the exception of the systems discussed in Chapter 4D, with polymerization of pure monomer or of homogeneous solutions of monomer and polymer in a solvent. Certain other types of polymerizing systems are of great interest because they offer practical advantages in industrial applications. Among these are polymerizations in which the monomer is carried in an *emulsion* or in *suspension* in an aqueous phase. Some of the advantages and disadvantages of these systems are itemized in Table 6-1. For purposes of discussion, it is convenient to classify polymerization systems as homogeneous or heterogeneous. These two categories of polymerization processes are discussed in Sections *A* and *B*, respectively.

As the industrial production of polymers has increased over the years, there has developed an increasing interest in the application of chemical engineering principles to polymerization. The first books and review articles on this subject are just beginning to appear. Polymerization reaction engineering is discussed in Section *C*.

Applications based on chemical reactions of polymers, during or after polymerization, are also increasing. Section *D* includes discussion of several of these, including crosslinking, both during and after polymerization; two other important postpolymerization phenomena, degradation (which frequently involves just the reverse reactions of polymerization) and reactions induced by high-energy radiation (which often involve radicals and lead to crosslinking or degradation); and the use of polymers as reagents or substrates for chemical reactions.

TABLE 6-1. Comparison of Polymerization Systems

Type	Advantages	Disadvantages
Homogeneous		
Bulk (batch type)	Minimum contamination. Simple equipment for making castings.	Strongly exothermic. Broadened molecular-weight distribution at high conversion. Complex if small particles required.
Bulk (continuous)	Lower conversion per pass leads to better heat control and narrower molecular-weight distribution.	Requires agitation, material transfer, separation, and recycling.
Solution	Ready control of heat of polymerization. Solution may be directly usable.	Not useful for dry polymer because of difficulty of complete solvent removal.
Heterogeneous		
Suspension	Ready control of heat of polymerization. Suspension or resulting granular polymer may be directly usable.	Continuous agitation required. Contamination by stabilizer possible. Washing, drying, possibly compacting required.
Emulsion	Rapid polymerization to high molecular weight and narrow distribution, with ready heat control. Emulsion may be directly usable.	Contamination with emulsifier, etc., almost inevitable, leading to poor color and color stability. Washing, drying, and compacting may be required.

A. POLYMERIZATION IN HOMOGENEOUS SYSTEMS

Bulk Polymerization. Polymerization in bulk, perhaps the most obvious method of synthesis of polymers, is widely practiced in the manufacture of condensation polymers, where the reactions are only mildly exothermic, and most of the reaction occurs when the viscosity of the mixture is still low enough to allow ready mixing, heat transfer, and bubble elimination. Control of such polymerizations is relatively easy.

Bulk polymerization of vinyl monomers is more difficult, since the reactions are highly exothermic and, with the usual thermally decomposed initiators, proceed at a rate that is strongly dependent on temperature. This, coupled with the problem in heat transfer incurred because viscosity increases early in the reaction, leads to difficulty in control and a tendency to the development of localized "hot spots" and "runaways." Except in the preparation of castings, for example, of poly(methyl methacrylate), bulk polymerization is seldom used commercially for the manufacture of vinyl polymers.

Solution Polymerization.   Polymerization of vinyl monomers in solution is advantageous from the standpoint of heat removal (e.g., by allowing the solvent to reflux) and control, but has two potential disadvantages. First, the solvent must be selected with care to avoid chain transfer and, second, the polymer should preferably be utilized in solution, as in the case of poly(vinyl acetate) to be converted to poly(vinyl alcohol) and some acrylic ester finishes, since the complete removal of solvent from a polymer is often difficult to the point of impracticality.

GENERAL REFERENCES

Ringsdorf 1965; Lenz 1967; Matsumoto 1977; Schildknecht 1977*a,b*.

B. POLYMERIZATION IN HETEROGENEOUS SYSTEMS

Polymerization From Gaseous Monomers

The polymerization of gaseous monomers can take place with the formation of a liquid phase (polymer melt) or a solid polymer. In each case the polymerization of ethylene provides the most important industrial example. Both polymerizations are discussed in Chapter 13*A*.

The high-pressure polymerization of ethylene to branched polyethylene takes place by a free-radical mechanism in the presence of a liquid phase, at temperatures above the melting point of the polymer. The polymerization is carried only to low conversion, with the remaining ethylene recovered when the pressure is lowered and recycled.

The low-pressure polymerization of ethylene to linear polyethylene takes place by coordination polymerization using a catalyst suspended in gaseous ethylene in a fluid-bed reactor. The ethylene polymerizes to the solid phase, with the small amount of the catalyst required remaining in the solid polymer.

Emulsion Polymerization

In emulsion polymerization two immiscible liquid phases are present, an aqueous continuous phase and a nonaqueous discontinuous phase consisting of monomer and polymer, as described below. The initiator is located in the aqueous phase, and the monomer–polymer particles are quite small, of the order of 0.1 μm in diameter. The kinetics of emulsion polymerization differs importantly from that of bulk polymerization, for which a limitation is set by Eq. 3-13—for combination,

$$\bar{x}_n = \frac{k_p^2}{k_t} \frac{[M]^2}{v_p} \tag{6-1}$$

TABLE 6-2. "Mutual" Recipe for the
Emulsion Copolymerization of Styrene and
Butadiene at 50°C

Ingredient	Parts by Weight
Butadiene	75
Styrene	25
Water	180
Soap	5.0
Lorol mercaptan[a]	0.50
Potassium persulfate	0.30

[a]Crude dodecyl mercaptan.

—on the molecular weight obtainable at a given rate of polymerization. With this limitation removed, emulsion systems allow higher-molecular-weight polymer to be produced at higher rates than do bulk or suspension systems.

A typical recipe for an emulsion polymerization is given in Table 6-2. This is the famous "Mutual" recipe used in the polymerization of styrene–butadiene synthetic rubber during World War II. The ingredients, in addition to monomers and water, are a fatty-acid soap, a mercaptan-type chain-transfer agent, and the water-soluble persulfate initiator.

The soap plays an important role in emulsion polymerization. At the beginning of the reaction it exists in the form of *micelles,* aggregates of 50–100 soap molecules probably having a layered structure like that shown in Fig. 6-1. Part of the monomer enters the micelles, but most of it exists as droplets a micrometer or more in diameter.

In the ideal case no polymer is formed in the monomer droplets. Polymerization can take place (at a very low rate) in the homogeneous phase in the absence of soap, but this cannot account for the bulk of the polymer formed. At the beginning of the reaction, polymer is formed in the soap micelles; these represent a favorable environment for the free radicals generated in the aqueous phase, because of the relative abundance of monomer and the high surface/volume ratio of the micelles compared to the monomer droplets. As polymer is formed, the micelles grow by the addition of monomer from the aqueous phase (and ultimately from the monomer droplets).

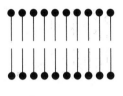

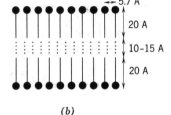

(a)   (b)

FIG. 6-1. Idealized structure of a soap micelle (a) without and (b) with solubilized monomer.

Soon (2–3% polymerization) the polymer particles grow much larger than the original micelles and absorb almost all the soap from the aqueous phase. Any micelles not already activated disappear; further polymerization takes place within the polymer particles already formed. The monomer droplets are unstable at this stage; if agitation is stopped, they coalesce into a continuous oil phase containing no polymer. The droplets act as reservoirs of monomer, which is fed to the growing polymer particles by diffusion through the aqueous phase. The polymer particles may contain about 50% monomer up to the point at which the monomer droplets disappear, at 60–80% polymerization. The rate of polymerization is constant over most of the reaction up to this point, but then falls off as monomer is depleted in the polymer particles. Rate increases with increasing soap (and initial micelle) concentration.

Smith–Ewart Kinetics   (Smith 1948a,b). In an ideal emulsion system, free radicals are generated in the aqueous phase at a rate of about 10^{13} per cubic centimeter per second. There are about 10^{11} polymer particles per cubic centimeter. Simple calculations show that termination of the free radicals in the aqueous phase is negligible and that diffusion currents are adequate for the rapid diffusion of free radicals into the polymer particles—on the average, about one per particle every 10 sec.

It can also be calculated from the known termination rate constants that two free radicals within the same polymer particle would mutually terminate within a few thousandths of a second. Therefore each polymer particle must contain most of the time either one or no free radicals. At any time half of the particles (on the average) contain one free radical, the other half none. The rate of polymerization per cubic centimeter of emulsion is

$$v_p = k_p[M] \frac{N}{2} \tag{6-2}$$

where N is the number of polymer particles per cubic centimeter. Since the monomer concentration is approximately constant, the rate depends principally on the number of particles present and not on the rate of generation of radicals.

The degree of polymerization also depends upon the number of particles:

$$\bar{x}_n = \frac{k_p N[M]}{\rho} \tag{6-3}$$

where ρ is the rate of generation of radicals. Unlike v_p, $\bar{x}_n$ is a function of the rate of free-radical formation. In bulk polymerization rate can be increased only by increasing the rate of initiation; this, however, causes a decrease in the degree of polymerization. In emulsion polymerization the rate may be increased by increasing the number of polymer particles. If the rate of initiation is kept constant, the degree of polymerization increases rather than decreases as the rate rises. Since the number of polymer particles is determined by the number of soap micelles initially present, both rate and molecular weight increase with increasing soap concentration. The Smith–Ewart kinetics require that

$$v_p \propto N, \; [I]^{0.4}, \; [E]^{0.6}$$
$$N \propto [I]^{0.4}, \; [E]^{0.6} \tag{6-4}$$
$$\bar{x}_n \propto N, \; [E]^{0.6}, \; [I]^{-0.6}$$

where [E] is the soap or emulsifier concentration.

Deviations From Smith–Ewart Kinetics.   The above kinetic scheme is highly idealized, though valuable for its simplicity. It explains adequately only a small portion of the vast literature on emulsion polymerization, though it works well for monomers such as styrene, butadiene, and isoprene, whose water solubility is very low, less than 0.1%. Among the circumstances under which the Smith–Ewart kinetics fails to apply quantitatively are the following:

a.   Larger particles ($>0.1 - 0.15$ μm in diameter), which can accommodate more than one growing chain simultaneously.

b.   Monomers with higher water solubility (1–10%), such as vinyl chloride, methyl methacrylate, vinyl acetate, and methyl acrylate. Here initiation in the aqueous phase, followed by precipitation of polymer, becomes important. These particles may absorb emulsifier, decreasing [E] and thus N, or they may serve as sites for polymerization, increasing N.

c.   Chain transfer to emulsifier. This often takes place in large enough amount to suggest that growing chains are localized near the surfaces of the particles, where the soap exists.

Inverse Emulsion Systems.   Emulsion polymerization can also be carried out in systems using an aqueous solution of a hydrophilic monomer, such as acrylic acid or acrylamide, emulsified in a continuous oil phase using an appropriate water-in-oil emulsifier. Either oil- or water-soluble initiators can be used. The mechanism seems to be that of normal emulsion polymerization, but the emulsions are often less stable.

"Redox" Initiation.   The decomposition of peroxide-type initiators in aqueous systems is greatly accelerated by the presence of a reducing agent. This acceleration allows the attainment of high rates of radical formation at low temperatures in emulsion systems.

A typical redox system is that of ferrous iron and hydrogen peroxide. In the absence of a polymerizable monomer the peroxide decomposes to free radicals as follows:

$$H_2O_2 + Fe^{2+} \longrightarrow HO\cdot + OH^- + Fe^{3+}$$

Another widely used peroxide-type initiator is the persulfate ion. With a reducing agent R the reaction is

$$S_2O_8^{2-} + R \longrightarrow SO_4^{2-} + SO_4^{-} \cdot + R^+$$

The reducing agent is often the thiosulfate ion,

$$S_2O_8^{2-} + S_2O_3^{2-} \longrightarrow SO_4^{2-} + SO_4^{-} \cdot + S_2O_3^{-} \cdot$$

or the bisulfite ion,

$$S_2O_8^{2-} + HSO_3^{-} \longrightarrow SO_4^{2-} + SO_4^{-} \cdot + HSO_3 \cdot$$

Many other redox initiator systems have been used.

Other Heterogeneous Liquid Systems

Suspension Polymerization. The term *suspension polymerization* refers to polymerization in an aqueous system with monomer as a dispersed phase, resulting in polymer as a dispersed solid phase. The process is distinguished from superficially similar emulsion polymerization by the location of the initiator and the kinetics obeyed: In typical suspension polymerization the initiator is dissolved in the monomer phase, and the kinetics is the same as that of bulk polymerization.

The dispersion of monomer into droplets much larger than those found in emulsion polymerization, typically 0.01–0.5 cm in diameter, is maintained by a combination of agitation and the use of water-soluble stabilizers. These may include finely divided insoluble organic or inorganic materials that interfere with agglomeration mechanically, electrolytes to increase the interfacial tension between the phases, and water-soluble polymers to increase the viscosity of the aqueous phase. The tendency to agglomerate may become critical when the polymerization has advanced to the point where the polymer beads become sticky. At the completion of the reaction the polymer is freed of stabilizer by washing and is dried. For some applications (Chapter 17), the polymer beads can be used directly, whereas for others compaction is required. A typical recipe for suspension polymerization is given in Table 6-3. The method is used commercially to prepare hard, glassy vinyl polymers such as polystyrene, poly(methyl methacrylate), poly(vinyl chloride), poly(vinylidene chloride), and polyacrylonitrile.

Precipitation Polymerization. In the preparation of a polymer insoluble in its monomer, or in polymerization in the presence of a nonsolvent for the polymer, marked deviations from the kinetics of homogeneous radical polymerization may occur. The normal bimolecular termination reaction is not effective, as the result of trapping or occlusion of radicals in the unswollen, tightly coiled precipitating polymer. The theory has been confirmed by the demonstration of the presence of radicals in the polymer both by chemical methods and by electron paramagnetic

TABLE 6-3. Typical Recipes for Suspension Polymerization

Component	Parts per 100 Parts of Monomer	
	Methyl Methacrylate	Vinyl Chloride
Peroxide initiator	~0.5	0.1–0.5
Water	~350	150–350
Stabilizers[a]	0.01–1	0.01–1

[a]Typical stabilizers include methyl cellulose, gelatin, poly(vinyl alcohol), and sodium polyacrylate. Minor amounts of emulsifiers and buffers are also used.

resonance (Chapter 9B). The lifetime of the trapped radicals is many hours at room temperature, and if polymer containing such radicals is heated in the presence of monomer to a temperature where the mobility of the radicals is increased, extremely rapid polymerization takes place.

The polymerization of vinyl chloride in bulk or in the presence of a nonsolvent (Mickley 1962) follows a rate equation of the form

$$v_p = k_p \left(\frac{f k_d [I]}{k_t} \right)^{1/2} \{[M] + f(P)\} \tag{6-5}$$

where the first term inside the braces arises from the normal rate equation for polymerization in the homogeneous liquid phase, and the second term represents the increment in rate due to polymerization in the precipitated polymer particles. The function $f(P)$ is proportional to the polymer concentration [P] at low conversion and to $[P]^{2/3}$ at later times. Radical occlusion occurs, but the depth of radical penetration into the polymer particles is small, limiting radical activity to thin surface layers in larger particles present at higher conversions. Termination occurs primarily in the liquid phase, probably as a result of transfer to monomer within the polymer particles, and subsequent diffusion of the monomer radicals to the liquid phase. In contrast, transfer to monomer in the polymerization of acrylonitrile is so slow that permanent radical occlusion occurs.

Solid-Phase Polymerization

A large number of olefin and cyclic monomers can be polymerized from the crystalline solid state. Among those that react rapidly under these conditions are styrene, acrylonitrile, methacrylonitrile, formaldehyde, trioxane, β-propiolactone, diketene, vinyl stearate, vinyl carbazole, and vinyl pyrrolidone.

Polymerization always appears to be associated with defects in the monomer crystals, most likely line defects (Chapter 10D); otherwise, it would be required that the monomer and the polymer be isomorphous, that is, that they have the same crystal structure, lattice parameters, and so on. This seems extremely unlikely and has not been observed.

GENERAL REFERENCES

Lenz 1967; Schildknecht 1977c; Odian 1981; and specifically the following:

Emulsion Polymerization.   Duck 1966; Vanderhoff 1969; Collins 1973, Exps. 4 and 8; Cooper 1974; Blackley 1975; Gardon 1977; Piirma 1979; Eliseeva 1981; Odian 1981, Chapter 4.

Suspension Polymerization.   Farber 1970; Munzer 1977.

Precipitation Polymerization.   Jenkins 1967.

Solid-Phase Polymerization.   Lenz 1967; Bamford 1969.

C. POLYMERIZATION REACTION ENGINEERING

While polymerization and the reactions of polymers are in many respects similar to ordinary chemical reactions, there are some significant differences that make the former unique in the sense of reactor and reaction engineering. These are high viscosities and low diffusion rates associated with concentrated polymer solutions and polymer melts. This section describes some engineering aspects of polymerization related to these unique polymer properties; brief descriptions of the production of important polymers are given in Chapters 13–16.

Reactor Design Theory

Although the high viscosity and low diffusivity associated with polymer melts and concentrated solutions lead to complications in designing polymerization reactors, they actually simplify the associated theoretical analysis. The reason for this is that there is no turbulence in most polymer systems, and the consequent steady laminar flow can be analyzed more easily.

The goal of such an analysis is to describe the flow rates, temperatures, and compositions at any point within the reactor. This is done by solving the partial differential equations that govern momentum, heat, and mass transfer. Three simultaneous nonlinear partial differential equations of motion, the Navier–Stokes equations, describe fluid velocities, but often the geometry is such that one or at most two of these suffice. Temperatures are found by solving one partial differential equation involving the viscosity, and there is a single partial differential equation for mass transfer for each component leading to information on concentrations. Usually it suffices to consider only two components, monomer and polymer. The equations are fully described in textbooks on transport phenomena (e.g., Bird 1960), and for polymer systems they can often be solved exactly. Important results of the solutions, depending strongly on the type of mixing in the reactor, are distributions of residence times and thermal times.

Types of Reactors

Batch Reactors. Some of the polymerizations carried out in batches are those performed in emulsion or suspension, most condensation reactions, and the early stages of a few bulk or solution vinyl polymerizations. The reactants are usually well mixed by agitation, and the reaction initiated by heating the vessel externally; later, the reaction exotherm supplies the needed heat, and heat control may be required, for example, by refluxing solvent. Bulk systems are usually terminated, quenched, or transferred to other equipment before the viscosity becomes too high. Runaway reactions must be guarded against, and the safest practice is to design the reactor to contain the maximum temperature and pressure that could result.

Tubular-Flow Reactors. The design objective of a tubular-flow reactor is to provide progressive flow with little or no internal recycle, achieving a uniform distribution of thermal and residence times. To do this it is necessary to promote radial mixing or flow redistribution, and this is not easy to do without turbulence. A number of stationary techniques can be used, such as coiling the tube into a helix, introducing flow redistribution devices (so-called motionless mixers), or utilizing differences of temperature or pressure along the tube. The polymerization of ethylene at high pressures is an example of the use of a stationary-wall tubular reactor.

Radial mixing can also be achieved mechanically, as with the use of extruders or similar devices as reactors. Especially in cases where polymer must be melted or volatiles removed, extruders provide a very favorable reaction environment.

Stirred-Tank Reactors. A major design consideration for stirred-tank reactors (sometimes, autoclaves) is the removal of the heat of polymerization. This is usually achieved by running the reaction with a small positive exotherm, and maintaining the desired operating temperature by control of the amount of cooling. The usual design objective of these reactors is to promote mixing throughout the volume of the reactor in order to minimize temperature and composition variations. An almost inevitable result of this recirculation is an exponential distribution of reaction times and a high recycle rate.

The kinetics of polymerizations are sufficiently complex to be affected by mixing at the molecular scale. Yields can be affected, but only when there is a gel effect and termination is diffusion controlled. The molecular-weight distribution is more readily affected, but the magnitude of the effect depends upon the type of mixing achieved in the reactor.

With a reasonably designed agitation system, perfect mixing can be achieved on the macro scale in a stirred-tank reactor, while low diffusivities lead to poor mixing on the micro scale. A reactor operating in this way is termed segregated. The perfect mixer, segregated stirred-tank reactor, and batch or tubular reactor represent three extremes of macro and micro mixing behavior (Fig. 6-2).

The effect of micromixing on the molecular-weight distribution has been

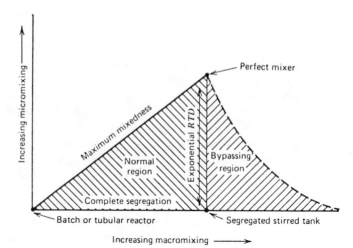

FIG. 6-2. Mixing space at the micro and macro levels, showing the locations of typical polymerization reactors. (Reprinted with permission from Nauman 1982. Copyright 1982 by John Wiley & Sons.)

studied by Tadmor (1966) and Biesenberger (1966). They showed that when the lifetime of a growing chain is long compared to the residence time, as in condensation polymerization, the molecular-weight distributions obtained with perfect mixing or in segregated stirred-tank reactors are extremely broad compared to those obtained in batch reactors (Fig. 6-3). On the other hand, in free-radical polymerization where the chain lifetime is short, the distribution is narrowest with perfect mixing (Fig. 6-4).

GENERAL REFERENCES

Nauman 1982; Biesenberger 1983.

D. CHEMICAL REACTIONS OF POLYMERS

The concept of the reactivity of functional groups being independent of molecular weight, used in developing the kinetics and statistics of stepwise polymerization (Chapter 2), applies to all functional groups, regardless of their location on the polymer chain. As a result, polymers undergo chemical reactions and serve as reagents much as do low-molecular-weight compounds, provided that reactants can be supplied to the sites of reaction. To accomplish this, most polymer reactions of importance are carried out in solution. Many of these reactions are discussed elsewhere in this book, as indicated by cross-references.

A well-known sequence of polymer reactions is the conversion of poly(vinyl acetate) through poly(vinyl alcohol) to a poly(vinyl acetal) (Chapter 14C). In the

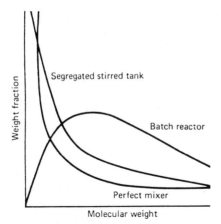

FIG. 6-3. Molecular-weight distributions obtained with various types of mixing in condensation polymerization. (Reprinted with permission from Nauman 1982. Copyright 1982 by John Wiley & Sons.)

latter reaction, as in the reaction of metals to remove chlorine from poly(vinyl chloride), functional groups react in pairs along the chain, with occasional groups isolated and incapable of reaction.

Polyesters are readily hydrolyzed unless low solubility or steric hindrance interferes. Thus, linear condensation polyesters and polyacrylates hydrolyze readily, whereas poly(ethylene terephthalate) and crosslinked alkyd and "polyester" resins are insoluble. Polymethacrylates are inert because of steric hindrance.

The nitration, sulfonation, and reduction of styrene polymers are widely used to produce ion-exchange resins (Chapter 14A). The acetylation, nitration, and xan-

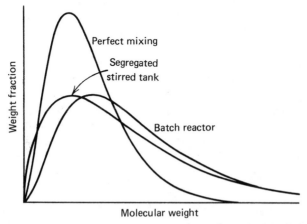

FIG. 6-4. Molecular-weight distributions obtained with various types of mixing in free-radical polymerization. (Reprinted with permission from Nauman 1982. Copyright 1982 by John Wiley & Sons.)

thation of cellulose (Chapter 15C) are important reactions, as are vulcanization and other reactions of natural and synthetic rubbers (Chapters 13D–F and 19).

In addition, unsaturated polymers can undergo such reactions as isomerization, cyclization, addition, epoxidation, and hydrogenation, while saturated polymeric hydrocarbons can be substituted on the main chain, the side chain, or the nucleus if aromatic. Terminally reactive polymers can produce block copolymers, whereas branching reactions can lead to graft copolymers (Chapter 5D). Coupling reactions can promote increases in molecular weight, as in the polyurethanes (Chapter 15B), or crosslinking, as in a variety of thermosetting resins (Chapter 16). The class of oxidation–reduction or redox polymers is discussed by Cassidy (1965).

The advantages of polymers as reagents have been reviewed (Kraus 1979). They include the ease of separation of the (polymeric) reagent from the reaction mixture; the mutual inaccessibility of reagents attached to crosslinked polymers, so that otherwise incompatible reagents can be used simultaneously ("wolf-and-lamb chemistry"); the ability of polymers to immobilize attached species, providing high local concentrations at high gross dilutions; and the possibility of tailoring the polymer to provide a special microenvironment, for example, steric or electric in nature, at the site of reaction.

Other major classes of polymer reactions are considered in the sections that follow.

Crosslinking Reactions

Monomers with two double bonds can be arbitrarily divided into 1,3-*diene* and *divinyl* compounds. With the exception of cases in which the entrance of one double bond in the monomer into a polymer chain leaves the second double bond so unreactive that it will not polymerize, these monomers can produce branched or crosslinked polymers when present during polymerization. It is usual to reduce the effect of the crosslinking reaction by utilizing only small quantities of the difunctional monomer in a copolymerization.

By the proper choice of relative reactivity of the two double bonds, it is possible to reduce the reactivity of one just enough so that it will not enter polymerization under the same conditions as the other, but can be made to react under more drastic conditions. This leads to postpolymerization crosslinking reactions of which *vulcanization* reactions are an example.

If the two double bonds are well separated, the reactivity of one is not affected by the polymerization of the other: An example is 2,11-dodecadiene. If the two double bonds are close enough together that the polymerization of one can shield the other sterically, or if they are conjugated, as in the dienes, a difference in reactivity can be expected. Some examples are given in Table 6-4.

Crosslinking During Polymerization. Consider the copolymerization of vinyl monomer A with divinyl monomer B—B, where all groups are equally reactive, as in the copolymerization of methyl methacrylate and ethylene glycol dimethacrylate. Let the concentrations of A, B—B, and B groups be [A], [BB], and [B],

TABLE 6-4. Examples of Monomers With Two Double Bonds

Dependency of Reactivity	Symmetrical	Unsymmetrical
Independent	Ethylene glycol dimethacrylate	Allyl acrylate
Intermediate	Divinylbenzene	p-Isopropenylstyrene
Interdependent	Butadiene	Chloroprene

respectively, where $[B] = 2[BB]$. Since the A and B groups are equally reactive, their ratio b/a in the copolymer is

$$\frac{b}{a} = \frac{[B]}{[A]} \qquad (6\text{-}6)$$

a simplification of the copolymer equation, Eq. 5-4. At extent of reaction p, here the fraction of the double bonds reacted, there are present $p[A]$ reacted A groups, $(1 - p)[A]$ unreacted A groups, $p^2[BB]$ doubly reacted BB molecules, $2p(1 - p)[BB]$ singly reacted BB molecules, and $(1 - p)^2[BB]$ unreacted BB molecules. In terms of B groups, the latter quantities become $(1 - p)^2[B]$ (unreacted) B groups on unreacted BB molecules, $p(1 - p)[B]$ unreacted or reacted B groups on singly reacted BB molecules, and $p^2[B]$ (reacted) B groups on doubly reacted BB molecules.

In this mixture the number of crosslinks is $p^2[BB]$, and the number of chains is derived in terms of the degree of polymerization x: number of chains = (total number of reacted units)$/x = ([A] + [B])p/x$. At the critical extent of reaction p_c for the onset of gelation, the number of crosslinks per chain is $\frac{1}{2}$, whence

$$p_c = \frac{[A] + [B]}{[B]\bar{x}_w} \qquad (6\text{-}7)$$

since Flory has shown that the appropriate average of x is $\bar{x}_w$. Equation 6-7 holds except where gelation occurs at very low extent of reaction; here the distribution of crosslinks is not random. Similar considerations apply to unsymmetrical divinyl monomers and dienes. In either case the important practical result of delaying the onset of gelation until high conversions can be achieved in the same ways:

a. By reducing the mole fraction of the divinyl or diene monomer.
b. By reducing the weight-average chain length, possibly by the use of chain-transfer agents.
c. By reducing the relative reactivity of the second double bond.

Nonconjugated dienes that have the double bonds separated by three atoms can be polymerized to give soluble, noncrosslinked polymers by an alternating intra–

intermolecular *cyclopolymerization* mechanism (Butler 1966, Gibbs 1967). Successive propagation steps involve, alternately, the addition of a monomer and the formation of a six-membered ring:

$$
\begin{array}{cc}
& H_2 \\
& C \\
-CH_2-HC\cdot & \diagdown CH \\
| & | \\
H_2C & CH_2 \\
& \diagdown X \diagup
\end{array}
\qquad
\begin{array}{cc}
& H_2 \\
& C \\
-CH_2-HC & \diagdown CH\cdot \\
| & | \\
H_2C & CH_2 \\
& \diagdown X \diagup
\end{array}
$$

where X is one of a variety of groups. Five- and seven-membered rings form from the appropriate polymers with more difficulty, and some open-chain unsaturated groups occur in the polymer.

Crosslinking After Polymerization: Vulcanization. This section is concerned only with the statistics of vulcanization and the distribution of molecular weights and crosslinks in the resulting polymer. Chemical aspects of vulcanization are discussed in Chapter 19.

The random crosslinking of double bonds bears formal resemblance to the formation of three-dimensional networks from polyfunctional monomers by stepwise polymerization. In keeping with previous nomenclature, the fraction of the monomer units on a chain that can be crosslinked is defined as q and the degree of polymerization of the chain as x. The "functionality" of the chain is its total number of vulcanizable groups qx. The extent of reaction p is the fraction of the total available crosslinks that have been formed.

Gelation occurs at a critical value of p, p_c, where there is on the average one crosslink for every two chains:

$$p_c q x = 1 \qquad (6\text{-}8)$$

As in the case of three-dimensional step-reaction polymers, after gelation the finite species constitute only part of the material, the rest being in the form of gel networks. As p increases from p_c to 1, the weight fraction of the finite species drops from unity to zero.

The case above does not correspond to actual vulcanization because it neglects the molecular-weight distribution of the chains initially present. If this is considered, the detailed analysis becomes complicated, but some important features remain about the same as for the monodisperse case. The gel point is given by the equation

$$p_c q \bar{x}_w = 1 \qquad (6\text{-}9)$$

where $\bar{x}_w$ is the weight-average degree of polymerization of the starting polymer.

Differences in chemical composition introduced by copolymerization or mixing may have more effect on the type of network structure than does molecular weight distribution. If, for example, a small amount of high-functionality polymer is mixed with a polymer of low functionality, the resulting vulcanizate will have regions of very tight crosslinking as inclusions in a looser network, leading to pronounced nonuniformity of physical properties. The foregoing treatment also assumes that the points of crosslinking are distributed at random. This is not always true; for example, if vulcanization is carried out in dilute solution, intramolecular crosslinking may occur before two molecules come near enough to form intermolecular links. Thus gelation is delayed considerably beyond the point predicted by random statistics.

Degradation

In this section the term *degradation* is taken to mean reduction of molecular weight. There are two general types of polymer degradation processes, corresponding roughly to the two types of polymerization, step reaction and chain reaction.

Random degradation is analogous to stepwise polymerization. Here chain rupture or scission occurs at random points along the chain, leaving fragments that are usually large compared to a monomer unit.

Chain depolymerization involves the successive release of monomer units from a chain end in a *depropagation* or *unzippering* reaction that is essentially the reverse of chain polymerization.

These two types may occur separately or in combination, may be initiated thermally or by ultraviolet light, oxygen, ozone, or other foreign agent, and may occur entirely at random or preferentially at chain ends or at other weak links in the chain.

It is possible to differentiate between the two processes in some cases by following the molecular weight of the residue as a function of the extent of reaction. Molecular weight drops rapidly as random degradation proceeds, but may remain constant in chain depolymerization, as whole molecules are reduced to monomer that escapes from the residual sample as a gas. Examination of the degradation products also differentiates between the two processes: The ultimate product of random degradation is likely to be a disperse mixture of fragments of molecular weight up to several hundred, whereas chain depolymerization yields large quantities of monomer.

Random Degradation. The kinetics and statistics of random degradation can be treated in exact analogy with the kinetics of linear stepwise polymerization. If p is defined, as before, as the extent of reaction, then $1 - p =$ (number of broken bonds)/(total number of bonds) is the extent of degradation. If the degradation is random, the number of bonds broken per unit time is constant as long as the total number of bonds present is large compared to the number broken. It follows that the number of chain ends increases linearly with time; hence $1/\bar{x}_n$ increases linearly with time. The acid-catalyzed homogeneous degradation of cellulose is a random degradation of this type.

An example of random degradation initiated by attack at a "weak link" in a chain is the ozonolysis of the isobutylene–isoprene copolymer butyl rubber. Here the initial attack is at the double bonds of the isoprene residues.

Chain Depolymerization. Chain depolymerization is a free-radical process that is essentially the reverse of chain polymerization. The point of initial attack may be at the chain end or at a "weak link" that may arise from a chain defect, such as an initiator fragment or a peroxide or ether link arising from polymerization in the presence of oxygen. The slightly higher activity of a tertiary hydrogen atom may be enough to provide a site for the initiation of the degradation process.

Poly(methyl methacrylate) degrades thermally by this process. The yield of monomer is 100% of the weight of polymer lost over a large fraction of the reaction. Polystyrene shows an intermediate behavior: The depropagation reaction ceases before the chain is completely destroyed. In some cases, such as olefin–SO₂ copolymers, an equilibrium can easily be reached between propagation and depropagation.

Kinetics of Degradation. A general kinetic scheme has been formulated that appears to cover all types of depolymerization. It is based on the concept of inverse chain polymerization and includes the steps of initiation, depropagation, termination, and chain transfer. The important feature of this scheme is the inclusion of the chain-transfer step, for it can be shown that the kinetics of random degradation result if the kinetic chain length before transfer is just the breaking of one bond.

The transfer reaction probably occurs rapidly by the abstraction of a hydrogen atom from a polymer. The chain that was attacked is likely to split into a radical and one or more inactive fragments at the elevated temperatures where degradation is rapid. Evidence for the transfer reaction includes the observations that tertiary hydrogen atoms at branch points in polyethylene are preferentially attacked and that the degradation of poly(methyl methacrylate) that has been copolymerized with a little acrylonitrile is quite different from that of pure poly(methyl methacrylate) because of the activity of the α-hydrogens on the acrylonitrile units.

The two factors that appear to be important in determining the course of degradation are the reactivity of the depropagating radical and the availability of reactive hydrogen atoms for transfer. With the possible exception of styrene, where the radical is stabilized by resonance, all polymers containing α-hydrogens, such as polyacrylates and polyacrylonitrile, give poor yields of monomer.

Conversely, the methacrylates give high yields of monomer because of the active radical and the α-methyl group that blocks the possibility of chain transfer. Polytetrafluoroethylene gives high yields of monomer because the strong C—F bond is not easily broken to allow transfer.

The scheme does not apply to polymers such as poly(vinyl acetate) and poly(vinyl chloride), where degradation results from the removal of side groups rather than from chain scission.

Degradation Products. Study of the products of thermal depolymerization in vacuum has shown that the chemical nature and relative amounts of these products are remarkably independent of the temperature and extent of the degradation reaction. Polystyrene, for example, degraded to 40% styrene, 2.4% toluene, and other products having an average molecular weight of 264, at temperatures between 360 and 420°C, and extents of degradation from 4 to 100%. The amounts of monomer produced by various polymers are shown in Table 6-5.

TABLE 6-5. Percent Monomer Resulting From Thermal Degradation

Polymer	Percent Monomer	
	Weight	Mole
Poly(methyl methacrylate)	100	100
Poly(α-methylstyrene)	100	100
Polyisobutylene	32	78
Polystyrene	42	65
Polybutadiene	14	57
SBR (butadiene–styrene copolymer)	12	52
Polyisoprene	11	44
Polyethylene	3	21

Environmentally degradable polymers have received much publicity in recen years and have been reviewed (Taylor 1979).

Radiation Chemistry of Polymers

The interaction of high-energy radiation with molecular substances involves the following sequence of events, regardless of the source of energy (photons, protons, electrons, neutrons, etc.). The molecules are first excited and ionized. Secondary electrons are emitted with relatively low speeds and produce many more ions along their tracks. Within 10^{-12} sec or so, molecular rearrangements take place in the ions and excited molecules, accompanied by thermal deactivation or the dissociation of valence bonds. As far as subsequent reactions are concerned, bond dissociation is the more important. It leads to the production of ions or radicals whose lifetimes depend on diffusion rates and may be weeks or months in solids at low temperatures.

The major effects in polymers arise from the dissociation of primary valence bonds into radicals, whose existence can be demonstrated by chemical methods or by EPR spectroscopy (Chapter 9B). The dissociations of C—C and C—H bonds lead to different results, degradation and crosslinking, which may occur simultaneously.

Degradation. The major result of radiation is degradation by chain scission in 1,1-disubstituted polymers, such as poly(methyl methacrylate) and its derivatives, polyisobutylene, and poly(α-methylstyrene), and in polymers containing halogen, such as poly(vinyl chloride), poly(vinylidene chloride), and polytetrafluoroethylene. The tendency toward degradation is related to the absence of tertiary hydrogen atoms, a weaker than average C—C bond (low heat of polymerization), or unusually strong bonds (such as C—F) elsewhere in the molecule.

Degradation is, of course, evidenced by decrease in molecular weight, the weight-average molecular weight being inversely proportional to the amount of radiation received. In polymers with bulky side chains, such as poly(methyl methacrylate), extensive degradation of the side chains to gaseous products also occurs.

Crosslinking.   Crosslinking is the predominant reaction on the irradiation of polystyrene, polyethylene and other olefin polymers, polyacrylates and their derivatives, and natural and synthetic rubbers. It is accompanied by the formation of gel and ultimately by the insolubilization of the entire specimen. Radiation crosslinking has a beneficial effect on the mechanical properties of some polymers and is carried out commercially to produce polyethylene with enhanced form stability and resistance to flow at high temperatures (Chapter 13*A*).

Other Reactions.   Radiation crosslinking is often accompanied by the formation of *trans*-vinylene unsaturation, both reactions resulting in the formation of hydrogen gas. If the irradiation is carried out in the presence of air, surface oxidation results. The resulting peroxides may be decomposed later in graft copolymerization (Chapter 5*D*).

GENERAL REFERENCES

Lenz 1967; Elias 1977; Loan 1979; Allcock 1981, Chapter 9; Odian 1981, Chapter 9; and specifically the following:

> *Crosslinking.*   Alfrey 1952; Temin 1966; Elias 1977, Chapter 23.5; Odian 1981, Chapters 6-6 and 9-2.
>
> *Degradation.*   Simha 1952; Madorsky 1964; Grassie 1966; Jellinek 1966, 1977; Pinner 1967; Hawkins 1972; Elias 1977, Chapters 23.6 and 24; Kelen 1982; Schnabel 1982.
>
> *Radiation Chemistry.*   Bovey 1958; Charlesby 1960; Chapiro 1962, 1969.

DISCUSSION QUESTIONS AND PROBLEMS

1. Discuss methods of controlling the heat evolved in vinyl polymerization, comparing the merits of bulk, solution, suspension, and emulsion polymerization and of the use of batch, tubular, and stirred-tank reactors.

2. Discuss the merits of the above methods and equipment with respect to the purification of the polymer produced.

3. Define and describe micelles and discuss their role in emulsion polymerization.

4. List all the necessary ingredients, and describe their functions, for the recipe for emulsion polymerization used in Collins (1973, Exp. 8).

5. Contrast the kinetics of suspension and emulsion polymerization and discuss the relative merits of the two methods.

6. A 20 wt.% aqueous solution of acrylamide is polymerized adiabatically with a redox initiator, using a starting temperature of 30°C. The heat of polymer-

ization is -74 J/mole-deg. Assume that the heat capacity of the reactor and its contents is 4 J/deg. What is the final temperature? What is the maximum solution concentration that could be used without danger of a runaway reaction?

7. An emulsion polymerization of styrene is run at 30°C with 10^{14} particles (0.2 μm in diameter) per milliliter and 10^{13} radicals produced per milliliter per second.

    a. Starting with Fick's law for the diffusion flux I across a surface of area A,

    $$I = -DA \frac{dc}{dx}$$

    where c is concentration and x is distance, derive an equation for the diffusion flux of small species into the monomer–polymer particles. State your assumptions.

    b. Calculate whether or not monomer can be adequately supplied to the monomer–polymer particles.

    c. Justify the assumption that appreciable termination does not take place in the aqueous phase. Styrene has a density of 0.90 g/ml and its solubility in water is 0.034 wt.%. Assume that for small molecules $D = 1 \times 10^{-5}$ cm²/sec. At 30°C, $k_p = 2.5 \times 10^7$ liter/mole-sec and $k_t = 55$ liter/mole-sec.

8. Discuss batch, stirred-tank, and tubular reactors with respect to mixing, residence time, thermal time, and the molecular-weight distribution produced for condensation and for free-radical polymerization.

9. If a vinyl monomer is copolymerized with 2 mol.% (based on vinyl groups) of a symmetrical independent divinyl monomer to $\bar{x}_w = 1000$, at what conversion will gelation take place? Comment on the accuracy of your result.

10. Discuss causes for the initiation of, and methods for preventing, random degradation and chain depolymerization.

11. Discuss the factors favoring crosslinking and degradation in the effect of ionizing radiation on polymers.

BIBLIOGRAPHY

Alfrey 1952. Turner Alfrey, Jr., John J. Bohrer, and H. Mark, *Copolymerization,* Interscience, New York, 1952.

Allcock 1981. Harry R. Allcock and Frederick W. Lampe, *Contemporary Polymer Chemistry,* Prentice-Hall, Englewood Cliffs, New Jersey, 1981.

Bamford 1969. C. H. Bamford and G. C. Eastmond, "Solid-Phase Addition Polymerization," *Q. Rev. Chem. Soc.* **23,** 271–299 (1969).

Biesenberger 1966. J. A. Biesenberger and Z. Tadmor, "Residence Time Dependence of Molecular Weight Distributions in Continuous Polymerizations," *Polym. Eng. Sci.* **5,** 299–305 (1966).

Biesenberger 1983. Joseph A. Biesenberger and Donald H. Sebastian, *Principles of Polymerization Engineering,* Wiley-Interscience, New York, 1983.

Bird 1960. R. Byron Bird, Warren E. Stewart, and Edwin N. Lightfoot, *Transport Phenomena,* John Wiley & Sons, New York, 1960.

Blackley 1975. D. C. Blackley, *Emulsion Polymerisation: Theory and Practice,* Halsted Press, John Wiley & Sons, New York, 1975.

Bovey 1958. Frank A. Bovey, *The Effects of Ionizing Radiation on Natural and Synthetic High Polymers,* Interscience, New York, 1958.

Butler 1966. George B. Butler, "Cyclopolymerization," pp. 568–599 in Herman F. Mark, Norman G. Gaylord, and Norbert M. Bikales, eds., *Encyclopedia of Polymer Science and Technology,* Vol. 4, Wiley-Interscience, New York, 1966.

Cassidy 1965. Harold G. Cassidy and Kenneth A. Kun, *Oxidation–Reduction Polymers (Redox Polymers),* Wiley-Interscience, New York, 1965.

Chapiro 1962. Adolphe Chapiro, *Radiation Chemistry of Polymeric Systems,* Wiley-Interscience, New York, 1962.

Chapiro 1969. A Chapiro, Robert B. Fox, Robert F. Cozzens, W. Brenner, and W. Kupfer, "Radiation-Induced Reactions," pp. 702–783 in Herman F. Mark, Norman G. Gaylord, and Norbert M. Bikales, eds., *Encyclopedia of Polymer Science and Technology,* Vol. 11, Wiley-Interscience, New York, 1969.

Charlesby 1960. Arthur Charlesby, *Atomic Radiation and Polymers,* Pergamon Press, New York, 1960.

Collins 1973. Edward A. Collins, Jan Bareš, and Fred W. Billmeyer, Jr., *Experiments in Polymer Science,* Wiley-Interscience, New York, 1973.

Cooper 1974. W. Cooper, "Emulsion Polymerization," Chapter 7 in A. D. Jenkins and A. Ledwith, eds., *Reactivity, Mechanism and Structure in Polymer Chemistry,* John Wiley & Sons, New York, 1974.

Duck 1966. Edward W. Duck, "Emulsion Polymerization," pp. 801–859 in Herman F. Mark, Norman G. Gaylord, and Norbert M. Bikales, eds., *Encyclopedia of Polymer Science and Technology,* Vol. 5, Wiley-Interscience, New York, 1966.

Elias 1977. Hans-Georg Elias, *Macromolecules · 2 · Synthesis and Materials* (translated by John W. Stafford), Plenum Press, New York, 1977.

Eliseeva 1981. V. I. Eliseeva, S. S. Ivanchev, A. I. Kuchanov, and A. V. Lebedev, *Emulsion Polymerization and its Applications in Industry* (translated by Sylvia J. Teague), Plenum Press, New York, 1981.

Farber 1970. Elliott Farber, "Suspension Polymerization," pp. 552–571 in Herman F. Mark, Norman G. Gaylord, and Norbert M. Bikales, eds., *Encyclopedia of Polymer Science and Technology,* Vol. 13, Wiley-Interscience, New York, 1970.

Gardon 1977. J. L. Gardon, "Emulsion Polymerization," Chapter 6 in Calvin E. Schildknecht, ed., with Irving Skeist, *Polymerization Processes,* Wiley-Interscience, New York, 1977.

Gibbs 1967. William E. Gibbs and John M. Barton, "The Mechanism of Copolymerization of Non-conjugated Diolefins," Chapter 2 in George E. Ham, ed., *Vinyl Polymerization,* Part I, Marcel Dekker, New York, 1967.

Grassie 1966. N. Grassie, "Degradation," pp. 647–716 in Herman F. Mark, Norman G. Gaylord, and Norbert M. Bikales, eds., *Encyclopedia of Polymer Science and Technology,* Vol. 4, Wiley-Interscience, New York, 1966.

Hawkins 1972. W. Lincoln Hawkins, ed., *Polymer Stabilization,* Wiley-Interscience, New York, 1972.

Jellinek 1966. H. H. G. Jellinek, "Depolymerization," pp. 740–793 in Herman F. Mark, Norman G. Gaylord, and Norbert M. Bikales, eds., *Encyclopedia of Polymer Science and Technology,* Vol. 4, Wiley-Interscience, New York, 1966.

Jellinek 1977. H. H. G. Jellinek, ed., *Aspects of Degradation and Stabilization of Polymers,* Elsevier, New York, 1977.

Jenkins 1967. A. D. Jenkins, "Occlusion Phenomena in the Polymerization of Acrylonitrile and Other Monomers," Chapter 6 in George E. Ham, ed., *Vinyl Polymerization,* Part I, Marcel Dekker, New York, 1967.

Kelen 1982. Tibor Kelen, *Polymer Degradation,* Van Nostrand Reinhold, New York, 1982.

Kraus 1979. M. Kraus and A. Patchornik, "Polymeric Reagents," *Chemtech* **9,** 118–128 (1979).

Lenz 1967. Robert W. Lenz, *Organic Chemistry of Synthetic High Polymers,* Wiley-Interscience, New York, 1967.

Loan 1979. L. D. Loan and F. H. Winslow, "Reactions of Macromolecules," Chapter 7 in F. A. Bovey and F. H. Winslow, eds., *Macromolecules: An Introduction to Polymer Science,* Academic Press, New York, 1979.

Madorsky 1964. Samuel L. Madorsky, *Thermal Degradation of Organic Polymers,* Wiley-Interscience, New York, 1964.

Matsumoto 1977. M. Matsumoto, K. Takakura, and T. Okaya, "Radical Polymerizations in Solution," Chapter 7 in Calvin E. Schildknecht, ed., with Irving Skeist, *Polymerization Processes,* Wiley-Interscience, New York, 1977.

Mickley 1962. Harold S. Mickley, Alan S. Michaels, and Albert L. Moore, "Kinetics of Precipitation Polymerization of Vinyl Chloride," *J. Polym. Sci.* **60,** 121–140 (1962).

Munzer 1977. M. Munzer and E. Trommsdorf, "Polymerizations in Suspension," Chapter 5 in Calvin E. Schildknecht, ed., with Irving Skeist, *Polymerization Processes,* Wiley-Interscience, New York, 1977.

Nauman 1982. E. B. Nauman, "Synthesis and Reactor Design," Chapter 10 in Mahendra D. Baijal, ed., *Plastics Polymer Science and Technology,* John Wiley & Sons, New York, 1982.

Odian 1981. George Odian, *Principles of Polymerization,* 2nd ed., John Wiley & Sons, New York, 1981.

Piirma 1979. Irja Piirma, ed., *Emulsion Polymerization,* Academic Press, New York, 1979.

Pinner 1967. S. H. Pinner, ed., *Weathering and Degradation of Plastics,* Gordon and Breach, New York, 1967.

Ringsdorf 1965. H. Ringsdorf, "Bulk Polymerization," pp. 642–666 in Herman F. Mark, Norman G. Gaylord, and Norbert M. Bikales, eds., *Encyclopedia of Polymer Science and Technology,* Vol. 2, Wiley-Interscience, New York, 1965.

Schildknecht 1977a. C. E. Schildknecht, "Cast Polymerizations," Chapter 2 in Calvin E. Schildknecht, ed., with Irving Skeist, *Polymerization Processes,* Wiley-Interscience, New York, 1977.

Schildknecht 1977b. C. E. Schildknecht, "Other Bulk Polymerizations," Chapter 4 in Calvin E. Schildknecht, ed., with Irving Skeist, *Polymerization Processes,* Wiley-Interscience, New York, 1977.

Schildknecht 1977c. Calvin E. Schildknecht, ed., with Irving Skeist, *Polymerization Processes,* Wiley-Interscience, New York, 1977.

Schnabel 1982. Wolfram Schnabel, *Polymer Degradation: Principles and Practical Applications,* Macmillan, New York, 1982.

Simha 1952. Robert Simha and L. A. Wall, "Kinetics of Chain Depolymerization," *J. Phys. Chem.* **56,** 707–715 (1952).

Smith 1948a. Wendell V. Smith, "The Kinetics of Styrene Emulsion Polymerization," *J. Am. Chem. Soc.* **70,** 3695–3702 (1948).

Smith 1948b. Wendell V. Smith and Roswell H. Ewart, "Kinetics of Emulsion Polymerization," *J. Chem. Phys.* **16,** 592–599 (1948).

Tadmor 1966. Zehev Tadmor and Joseph A. Biesenberger, "Influence of Segregation on Molecular Weight Distribution in Continuous Linear Polymerizations," *Ind. Eng. Chem. Fund.* **5,** 336–343 (1966).

Taylor 1979. Lynn Taylor, "Degradable Plastics: Solution or Illusion?," *Chemtech* **9,** 542–548 (1979).

Temin 1966. Samuel C. Temin and Allan R. Shultz, "*Crosslinking,*" pp. 331–414 in Herman F. Mark, Norman G. Gaylord, and Norbert M. Bikales, eds., *Encyclopedia of Polymer Science and Technology,* Vol. 4, Wiley-Interscience, New York, 1966.

Vanderhoff 1969. John W. Vanderhoff, "Mechanism of Emulsion Polymerization," Chapter 1 in George E. Ham, ed., *Vinyl Polymerization,* Vol. I, Part II, Marcel Dekker, New York, 1969.

PART THREE

CHARACTERIZATION

CHAPTER SEVEN

POLYMER SOLUTIONS

A. CRITERIA FOR POLYMER SOLUBILITY

The Solution Process.   Dissolving a polymer is a slow process that occurs in two stages. First, solvent molecules slowly diffuse into the polymer to produce a swollen gel. This may be all that happens—if, for example, the polymer–polymer inter-molecular forces are high because of crosslinking, crystallinity, or strong hydrogen bonding. But if these forces can be overcome by the introduction of strong polymer–solvent interactions, the second stage of solution can take place. Here the gel gradually disintegrates into a true solution. Only this stage can be materially speeded by agitation. Even so, the solution process can be quite slow (days or weeks) for materials of very high molecular weight.

Polymer Texture and Solubility.   Solubility relations in polymer systems are more complex than those among low-molecular-weight compounds, because of the size differences between polymer and solvent molecules, the viscosity of the system, and the effects of the texture and molecular weight of the polymer. In turn, the presence or absence of solubility as conditions (such as the nature of the solvent, or the temperature) are varied can give much information about the polymer; this is in fact the topic of most of this chapter.

From what has already been said, it is clear that the topology of the polymer is highly important in determining its solubility. Crosslinked polymers do not dissolve, but only swell if indeed they interact with the solvent at all. In part, at least, the degree of this interaction is determined by the extent of crosslinking: Lightly cross-linked rubbers swell extensively in solvents in which the unvulcanized material would dissolve, but hard rubbers, like many thermosetting resins, may not swell appreciably in contact with any solvent.

The absence of solubility does not imply crosslinking, however. Other features may give rise to sufficiently high intermolecular forces to prevent solubility. The presence of crystallinity is the common example. Many crystalline polymers, particularly nonpolar ones, do not dissolve except at temperatures near their crystalline melting points. Because crystallinity decreases as the melting point is approached (Chapter 10) and the melting point is itself depressed by the presence of the solvent, solubility can often be achieved at temperatures significantly below the melting point. Thus linear polyethylene, with crystalline melting point $T_m = 135°C$, is soluble in many liquids at temperatures above 100°C, while even polytetrafluoroethylene, $T_m = 325°C$, is soluble in some of the few liquids that exist above 300°C. More polar crystalline polymers, such as 66-nylon, $T_m = 265°C$, can dissolve at room temperature in solvents that interact strongly with them (for example, to form hydrogen bonds).

There is little quantitative information about the influence of branching on solubility; in general, branched species appear to be more readily soluble than their linear counterparts of the same chemical type and molecular weight.

Of all these systems, the theory of solubility, based on the thermodynamics of polymer solutions, is highly developed only for linear polymers in the absence of crystallinity. This theory is described in Sections C and D. Here the chemical nature of the polymer is by far the most important determinant of solubility, as is elucidated in the remainder of this section. The influence of molecular weight (within the polymer range) is far less, but it is of great importance to fractionation processes (Sections D and E), which yield information about the distribution of molecular weights in polymer samples.

Solubility Parameters.   Solubility occurs when the free energy of mixing

$$\Delta G = \Delta H - T\Delta S$$

is negative. It was long thought that the entropy of mixing ΔS was always positive, and therefore the sign of ΔG was determined by the sign and magnitude of the heat of mixing ΔH. For reasonably nonpolar molecules and in the absence of hydrogen bonding, ΔH is positive and was assumed to be the same as that derived for the mixing of small molecules. For this case, the heat of mixing per unit volume can be approximated (Hildebrand 1950) as

$$\Delta H = v_1 v_2 (\delta_1 - \delta_2)^2$$

where v is volume fraction and subscripts 1 and 2 refer to solvent and polymer, respectively. The quantity δ^2 is the cohesive energy density or, for small molecules, the energy of vaporization per unit volume. The quantity δ is known as the *solubility parameter*. (This expression for the heat of mixing is one of several alternatives used in theories of the thermodynamics of polymer solutions; in Section C, ΔH is written in a different but equivalent way.)

The value of the solubility-parameter approach is that δ can be calculated for both polymer and solvent. As a first approximation, and in the absence of strong interactions such as hydrogen bonding, solubility can be expected if $\delta_1 - \delta_2$ is less than 3.5–4.0, but not if it is appreciably larger.

This approach to polymer solubility, pioneered by Burrell (1955), has been extensively used, particularly in the paint industry. A few typical values of δ_1 and δ_2 are given in Table 7-1; for polymers, they are the square roots of the cohesive-energy densities of Table 1-4. Extensive tabulations have been published (Burrell 1975, Hoy 1970). Perhaps the easiest way to determine δ_2 for a polymer of known structure is by the use of the molar-attraction constants E of Table 7-2,

$$\delta_2 = \frac{\rho \Sigma E}{M}$$

where values of E are summed over the structural configuration of the repeating unit in the polymer chain, with repeat molecular weight M and density ρ.

The original solubility-parameter approach was developed for nonpolar systems. Modifications to include polarity and hydrogen bonding have led to three-dimensional solubility-parameter schemes, which lack the simplicity of the single-parameter method but are more widely applicable. Despite its shortcomings, the concept is nevertheless still extremely useful and should not be abandoned without test.

In contrast to the above considerations of the thermodynamics of dissolution of polymers, the rate of this step depends primarily on how rapidly the polymer and the solvent diffuse into one another (Ueberreiter 1962, Asmussen 1962). Solvents that promote rapid solubility are usually small, compact molecules, but these kinetically good solvents need not be thermodynamically good as well. Mixtures of a kinetically good and a thermodynamically good liquid are often very powerful and rapid polymer solvents.

TABLE 7-1. Typical Values of the Solubility Parameter δ for Some Common Polymers and Solvents[a]

Solvent	$\delta_1 [(J/cm^3)^{1/2}]$	Polymer	$\delta_2 [(J/cm^3)^{1/2}]$
n-Hexane	14.8	Polytetrafluoroethylene	12.7
Carbon tetrachloride	17.6	Poly(dimethyl siloxane)	14.9
		Polyethylene	16.2
Toluene	18.3	Polypropylene	16.6
2-Butanone	18.5	Polybutadiene	17.6
Benzene	18.7	Polystyrene	17.6
Cyclohexanone	19.0	Poly(methyl methacrylate)	18.6
Styrene	19.0	Poly(vinyl chloride)	19.4
Chlorobenzene	19.4	Poly(vinyl acetate)	21.7
Acetone	19.9	Poly(ethylene terephthalate)	21.9
Tetrahydrofuran	20.3		
Methanol	29.7	66-Nylon	27.8
Water	47.9	Polyacrylonitrile	31.5

[a]Collins (1973).

TABLE 7-2. Molar Attraction Constants E^a

Group	E [(J-cm^3)$^{1/2}$/mole]	Group	E [(J-cm^3)$^{1/2}$/mole]
—CH$_3$	303	NH$_2$	463
—CH$_2$—	269	—NH—	368
>CH—	176	—N—	125
>C<	65	C≡N	725
CH$_2$=	259	NCO	733
—CH=	249	—S—	429
>C=	173	Cl$_2$	701
—CH=aromatic	239	Cl primary	419
>C=aromatic	200	Cl secondary	425
—O—ether, acetal	235	Cl aromatic	329
—O—epoxide	360	F	84
—COO—	668	Conjugation	47
>C=O	538	cis	−14
—CHO	599	trans	−28
(CO)$_2$O	1159	Six-membered ring	−48
—OH→	462	ortho	−19
OH aromatic	350	meta	−13
—H acidic dimer	−103	para	−82

^{a}Hoy (1970).

GENERAL REFERENCES

Hildebrand 1950; Gardon 1965; Hoy 1970; ASTM D3132; Burrell 1975; Morawetz 1975, Chapter 2; Olabisi 1979, Chapter 2.3; Snyder 1980.

B. CONFORMATIONS OF DISSOLVED POLYMER CHAINS

As specified in Chapter 1, those arrangements of the polymer chain differing by reason of rotations about single bonds are termed *conformations*.† In solution, a polymer molecule is a randomly coiling mass most of whose conformations occupy many times the volume of its segments alone. The average density of segments within a dissolved polymer molecule is of the order of 10^{-4}–10^{-5} g/cm^3. The size of the molecular coil is very much influenced by the polymer–solvent interaction forces. In a thermodynamically "good" solvent, where polymer–solvent contacts

---

†Here we differ from some authorities, notably Flory (1969), who prefers the use, well established in statistical mechanics, of the term *configuration* here. We follow the convention of organic chemistry in which configuration designates stereochemical arrangement (see Chapter 10A).

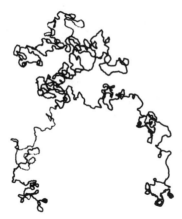

FIG. 7-1. Model of one of the many conformations of a random-coil chain of 1000 links (Treloar 1958).

are highly favored, the coils are relatively extended. In a "poor" solvent they are relatively contracted. It is the purpose of this section to describe the conformational properties of both ideal and real polymer chains. The conformations of polymer chains in the crystalline state are considered in Chapter 10*B*.

The importance of the random-coil nature of the dissolved, molten, amorphous, and glassy states of high polymers cannot be overemphasized. As the following chapters show, many important physical as well as thermodynamic properties of high polymers result from this characteristic structural feature. The random coil (Fig. 7-1) arises from the relative freedom of rotation associated with the chain bonds of most polymers and the formidably large number of conformations accessible to the molecule.

One of these conformations, the fully extended chain (often an all-trans planar zigzag carbon chain; Fig. 7-2), has special interest because its length, the *contour length* of the chain, can be calculated in a straightforward way. In all other cases the size of the random coil must be expressed in terms of statistical parameters such as the root-mean-square distance between its ends, $(\overline{r^2})^{1/2}$, or its *radius of gyration,* the root-mean-square distance of the elements of the chain from its center of gravity, $(\overline{s^2})^{1/2}$. For linear polymers that are not appreciably extended beyond their most probable shape, the mean-square end-to-end distance and the square of the radius of gyration are simply related: $\overline{r^2} = \overline{6s^2}$. For extended chains $\overline{r^2} > \overline{6s^2}$. The use of the radius of gyration is sometimes preferred because it can be determined experimentally, as described in Chapter 8*E*.

The Freely Jointed Chain. A simple model of a polymer chain consists of a series of x links of length l joined in a linear sequence with no restrictions on the angles between successive bonds. The probability W that such an array has a given end-to-end distance r can be calculated by the classical random-flight method (Ray-

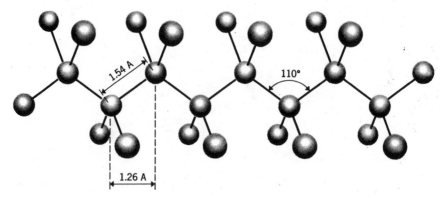

FIG. 7-2. The fully extended all-trans conformation of the carbon–carbon chain (Billmeyer 1969).

leigh 1919). The most important result of the calculation is that the end-to-end distance is proportional to the square root of the number of links:

$$(\overline{r_f^2})^{1/2} = l x^{1/2}$$

(The subscript f indicates the random-flight end-to-end distance.) Thus $(\overline{r_f^2})^{1/2}$ is proportional to $M^{1/2}$, or $\overline{r_f^2}/M$ is a characteristic property of the polymer chain structure, independent of molecular weight or length.

The distribution of end-to-end distances over the space coordinates $W(x,y,z)$ is given by the Gaussian distribution function shown graphically in Fig. 7-3. This is the density distribution of end points and shows that if one end of the chain is taken at the origin, the probability is highest of finding the other end in a unit volume

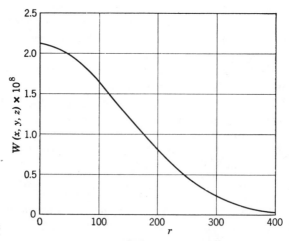

FIG. 7-3. Gaussian distribution of end-to-end distances in a random coil of 10^4 links each 2.5 Å long; $W(x,y,z)$ is expressed in angstroms to the inverse third power and r in angstroms.

near the origin. This probability decreases continuously with increasing distance from the origin. On the other hand, the probability of finding a chain end within the volume of a spherical shell between distances r and $r + dr$ from the origin has a maximum, as shown in Fig. 7-4. It should be noted that the Gaussian distribution does not fall to zero at large extensions and so must fail to describe the actual conformations of the chain where r nears the contour length. Over most of the range of interest for the dilute-solution properties of polymers, this is of little consequence; better approximations are available where needed, as in the treatment of rubber elasticity (Chapter 11*B*).

The freely jointed or random-flight model seriously underestimates the true dimensions of real polymer molecules for two reasons: First, restrictions to completely free rotation, such as fixed valence angles, the correct weights for trans and gauche conformations, and statistical deviations from ideal trans and gauche states, and other short-range interactions described below, lead to larger dimensions than calculated above; and second, long-range interactions resulting from the inability of chain atoms far removed from one another to occupy the same space at the same time result in a similar effect.

Short-Range Interactions. One of the triumphs of modern polymer chemistry is the extent to which Flory and his collaborators (Flory 1969), among others, have calculated completely and accurately the effect of short-range interactions on the dimensions of random-coil polymers. Several effects are involved. The restriction to fixed bond angle θ expands the chain by a factor of $[(1 - \cos \theta)/(1 + \cos \theta)]^{1/2}$, equal to $\sqrt{2.0}$ for carbon–carbon bonds. Restricted rotation, whether resulting from steric hindrances and resulting potential-energy barriers or from resonance leading to rigid planar conformations, increases dimensions still more. Finally, conformations that would place two atoms close together along the chain too close to one another are not allowed, leading to further expansion. The most important of these is the so-called *pentane interference* (Taylor 1948) between the first and fifth chain atoms in a sequence. Bovey (1979) summarizes these conformational features.

The net results of these short-range interactions can be expressed as a *charac-*

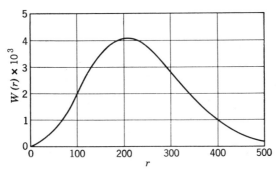

FIG. 7-4. Radial distribution of end-to-end distances for the coil of Fig. 7-3. Here $W(r)$ is expressed in angstrom^{-1} units.

teristic ratio of the square of the actual chain dimensions in the absence of long-range interactions (called the *unperturbed dimensions* and given the symbol $\overline{r_0^2}$) and the square of the random-flight end-to-end distance, l^2x. Typical values of the characteristic ratio for some common polymers are given in Table 7-3. It should be pointed out that despite short-range interactions, the chains are still random in overall conformation if long enough.

Long-Range Interactions and the Excluded Volume. Correction of polymer chain dimensions for short-range interactions still fails to eliminate conformations in which two widely separated chain segments occupy the same space. Each segment of a real chain exists within a volume from which all other segments are excluded. The theoretical calculation of the *excluded volume* and its effect on the dimensions of the polymer chain has remained a major unsolved problem in polymer science for many years. A statistical approach has been more fruitful and has allowed the simulation of chain conformations by digital computer up to chain lengths of several hundred segments. The results of these calculations of "self avoiding walks" and experimental studies (light scattering, dilute-solution viscosity, sedimentation; Chapter 8C–E) lead to the result that the mean-square end-to-end distance is proportional to $x^{0.6}$. This exponent was first calculated many years ago by Flory (1953). It is also predicted by the application of scaling concepts (de Gennes 1979, Sections C and D).

Practically, the effect of the long-range interactions is to cause a further expansion of the chain over its unperturbed dimensions, since more of the compact conformations, with small values of $\overline{r^2}$, must be excluded. The actual dimensions of the real chain can be expected to exceed its unperturbed dimensions by an expansion factor α; thus $(\overline{r^2})^{1/2} = \alpha(\overline{r_0^2})^{1/2}$. The value of α depends on the nature of the

TABLE 7-3. The Characteristic Ratio $\overline{r_0^2}/l^2x$ for Some Common Polymers, Evaluated at the Limit of High Chain Length[a]

Polymer	$\dfrac{\overline{r_0^2}}{l^2x}$
Polyethylene	6.7
Polystyrene (atactic)	10.0
Polypropylene (isotactic)	5.7
Poly(methyl methacrylate)	
Atactic	6.9
Isotactic	9.3
Syndiotactic	~7
Poly(ethylene oxide)	4.0
66-Nylon	5.9

[a]Flory (1969).

solvent: A solvent for which α is large is said to be a thermodynamically "good" solvent for that polymer, and vice versa.

In a sufficiently poor solvent, or at a sufficiently low temperature (since solvent power and α vary with the temperature), it is possible to achieve the condition $\alpha = 1$, where the chain attains its unperturbed dimensions. This special point is called the *Flory temperature* Θ; a solvent used at $T = \Theta$ is called a Θ *solvent*. The calculation of α and Θ and their relation to thermodynamic quantities is treated in Section C.

Neutron-scattering experiments (Chapter 8C) show that the above considerations apply as well to chain conformations in the melt and the glassy state.

Nonlinear Chains.   A branched molecule occupies a smaller volume than a linear one with the same number of segments, that is, the same molecular weight. It is convenient to express this diminution of size as a factor $g = \overline{s^2}$ (branched)$/\overline{s^2}$ (linear), which can be calculated statistically for various degrees and types of branching (Zimm 1949). For random branching, five trifunctional branch points per molecule reduce g to about 0.70. The change in size is the basis of a method to measure branching (Chapter 8E).

Billmeyer (1972, 1977) has used a "pop-it" bead model to illustrate the dimensions of polymer chains.

GENERAL REFERENCES

Flory 1953, 1969; Volkenstein 1963; Birshtein 1966; Bovey 1969, 1979, Chapter 3.9; Lowry 1970; Hopfinger 1973; Vollmert 1973, Chapter 32; Morawetz 1975, Chapter 3; Elias 1977, Chapter 4; de Gennes 1979, Chapter 1.

C. THERMODYNAMICS OF POLYMER SOLUTIONS†

The behavior of polymers toward solvents is characteristic and different from that of low-molecular-weight substances. The size and conformations of dissolved polymer molecules require special theoretical treatment to explain their solution properties. Conversely, it is possible to obtain information about the size and shape of polymer molecules from studies of their solution properties.

The first group of properties of interest includes those depending upon equilibrium between two phases, one or both of which is a solution of the polymer. In this section we discuss situations in which one phase is pure solvent, and in Section D solubility phenomena in which both phases contain polymer.

---

†For reviews of thermodynamics as background to this subject see Hildebrand (1950) and Lewis (1961).

Thermodynamics of Simple Liquid Mixtures

From the condition for equilibrium between two phases may be derived relations such as that for the free energy of dilution of a solution:

$$\Delta G_A = kT \ln \left(\frac{p_A}{p_A{}^0}\right) \tag{7-1}$$

where ΔG is the free energy of dilution resulting from the transfer of one molecule of liquid A from the pure liquid state with vapor pressure $p_A{}^0$ to a large amount of solution with vapor pressure p_A. This expression is written in terms of one of the *colligative properties* of solutions: vapor-pressure lowering, freezing-point depression, boiling-point elevation, and osmotic pressure. The values of these properties cannot be related to the composition of the system by pure thermodynamic reasoning. It is necessary to know the type of variation of one property with concentration; the variation of the others may then be deduced.

Ideal Solutions. In the simplest type of mixing, the molecules of components A and B have roughly the same size and shape and similar force fields. They may then form an *ideal solution*, defined as one in which Raoult's law is obeyed. This law states that the partial vapor pressure of each component in the mixture is proportional to its mole fraction. Therefore

$$p_A = p_A{}^0 \frac{N_A}{N_A + N_B} = p_A{}^0 n_A \tag{7-2}$$

where n denotes mole fraction, and Eq. 7-1, for example, becomes

$$\Delta G_A = kT \ln n_A \tag{7-3}$$

The total free energy of mixing is

$$\Delta G = N_A \, \Delta G_A + N_B \, \Delta G_B$$
$$= kT(N_A \ln n_A + N_B \ln n_B) \tag{7-4}$$

The conditions for ideal mixing imply that the heat of mixing $\Delta H = 0$; that is, the components mix without change in enthalpy. Since $\Delta G = \Delta H - T \, \Delta S$, the entropy of mixing is given by

$$\Delta S = -k(N_A \ln n_A + N_B \ln n_B) \tag{7-5}$$

which is positive for all compositions, so that, by the second law, spontaneous mixing occurs in all proportions.

Other Types of Mixing. In practice, few liquid mixtures obey Raoult's law. Three types of deviations are distinguished:

a. "Athermal" solutions, in which $\Delta H = 0$ but ΔS is no longer given by Eq. 7-5.
b. "Regular" solutions, in which ΔS has the ideal value but ΔH is finite.
c. "Irregular" solutions, in which both ΔH and ΔS deviate from the ideal values.

It is usually found in systems of similar-sized molecules that ΔS is nearly ideal when $\Delta H = 0$; therefore athermal solutions are nearly ideal. Many mixtures are found for which ΔH is finite, however. Such cases arise when the intermolecular force fields around the two types of molecule are different. Expressions for the heat of mixing may be derived in terms of the cohesive-energy density of these force fields (Section A).

Entropy and Heat of Mixing of Polymer Solutions

Deviations From Ideal Behavior. Polymer solutions invariably exhibit large deviations from Raoult's law, except at extreme dilutions, where ideal behavior is approached as an asymptotic limit. At concentrations above a few percent, deviations from ideality are so great that the ideal law is of little value for predicting or correlating the thermodynamic properties of polymer solutions. Even if the mole fraction is replaced with the volume fraction, in view of the different sizes of the polymer and solvent molecules, there is not a good correlation with the experimental results.

Entropy of Mixing. Deviations from ideality in polymer solutions arise largely from small entropies of mixing. These are not abnormal, but are the natural result of the large difference in molecular size between the two components. They can be interpreted in terms of a simple molecular model. The molecules in the pure liquids and the mixture are assumed to be representable without serious error by a lattice. A two-dimensional representation of such an arrangement for nonpolymer liquids is shown in Fig. 7-5a. Whereas the molecules of a pure component can be arranged in only one way on such a lattice, assuming that they cannot be distinguished from one another, the molecules of a mixture of two components can be arranged on a lattice in a large but calculable number of ways, W. By the Boltzmann relation, the entropy of mixing $\Delta S = k \ln W$. Equation 7-5 results for the case of molecules that can replace one another indiscriminately on the lattice.

It is assumed that the polymer molecules consist of a large number x of chain segments of equal length, flexibly joined together. Each link occupies one lattice site, giving the arrangement of Fig. 7-5b. The solution is assumed to be concentrated enough that the occupied lattice sites are distributed at random rather than lying in well-separated regions of x occupied sites each. It can now be seen qualitatively

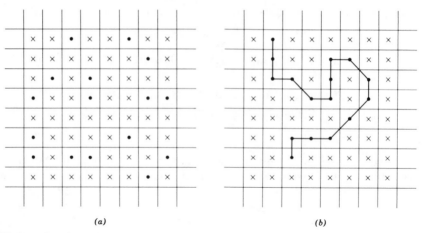

FIG. 7-5.   Two-dimensional representation of (a) nonpolymer liquids and (b) a polymer molecule located in the liquid lattice.

why the entropy of mixing of polymer solutions is small compared to that with normal solutes. There are fewer ways in which the same number of lattice sites can be occupied by polymer segments: Fixing one segment at a lattice point severely limits the number of sites available for the adjacent segment. The approximate calculation of W for such a model is due separately to Flory (1942) and Huggins (1942a,b,c); their results (which differ only in minor detail) are known as the *Flory–Huggins theory* of polymer solutions. The entropy of mixing is analogous to that given in Eq. 7-5 for simple liquids: For polymer solutions

$$\Delta S = -k(N_1 \ln v_1 + N_2 \ln v_2) \tag{7-6}$$

where subscript 1 denotes the solvent and 2 the polymer, and v_1 and v_2 are *volume fractions* defined as

$$v_1 = \frac{N_1}{N_1 + xN_2}$$

$$v_2 = \frac{xN_2}{N_1 + xN_2} \tag{7-7}$$

Heat and Free Energy of Mixing.   The heat of mixing of polymer solutions is analogous to that of ordinary solutions:[†]

$$\Delta H = \chi_1 kTN_1 v_2 \tag{7-8}$$

[†]The nomenclature of Flory is used throughout. Huggins writes $\mu_h{}^0$ instead of χ_1, and Hildebrand expresses ΔH in terms of the solubility parameter, as in Section A.

where χ_1 characterizes the interaction energy per solvent molecule divided by kT. Combining Eqs. 7-6 and 7-8 gives the Flory–Huggins expression for the free energy of mixing of a polymer solution with normal heat of mixing:

$$\Delta G = kT(N_1 \ln v_1 + N_2 \ln v_2 + \chi_1 N_1 v_2) \tag{7-9}$$

From this expression may be derived many useful relations involving experimentally obtainable quantities. For example, the partial molar free energy of mixing is

$$\Delta \bar{G}_1 = kT \left[\ln (1 - v_2) + \left(1 - \frac{1}{x} \right) v_2 + \chi_1 v_2^2 \right] \tag{7-10}$$

and from it is obtained the osmotic pressure

$$\pi = -\frac{kT}{V_1} \left[\ln (1 - v_2) + \left(1 - \frac{1}{x} \right) v_2 + \chi_1 v_2^2 \right] \tag{7-11}$$

where V_1 is the molecular volume of the solvent. If the logarithmic term in Eq. 7-11 is expanded and only low powers of v_2 are retained, the following equation for the osmotic pressure is obtained:

$$\pi = \frac{kT}{V_1} \left(\frac{v_2}{x} + (\tfrac{1}{2} - \chi_1)v_2^2 + \cdots \right) \tag{7-12}$$

In expansions of this type the coefficient of v_2^2 is known as the *second virial coefficient* A_2 (see also Chapter 8B). Here it is defined as

$$A_2 = \frac{\bar{v}_2^2}{N_0 V_1} (\tfrac{1}{2} - \chi_1) \tag{7-13}$$

where $\bar{v}_2$ is the specific volume of the polymer. This simple treatment does not describe the temperature and molecular-weight dependence of A_2 in good agreement with experiment.

Experimental Results With Polymer Solutions

The first system for which accurate experimental results were compared with the Flory–Huggins theories was that of rubber in benzene (Gee 1946, 1947). Free energies for the system were calculated from vapor-pressure measurements by Eq. 7-1, and heats of solution were measured in a calorimeter. The predicted concentration dependence of the partial molar heat of dilution

$$\Delta \bar{H}_1 = kT\chi_1 v_2^2 \tag{7-14}$$

was not observed. Despite this fact, the entropy of mixing, calculated from the heat and free energy, was in fair agreement with theory except for the dilute-solution region.

The properties of other systems do not in general conform as well to the predictions of the theory. For example, the anticipated independence of χ_1 of v_2 is usually not observed (Fig. 7-6) (but see the corresponding-state theories described below).

Dilute Solutions

Flory–Krigbaum Theory. The lattice model used in the Flory–Huggins treatment neglects the fact that a very dilute polymer solution must be discontinuous in structure, consisting of domains or clusters of polymer chain segments separated on the average by regions of polymer-free solvent. Flory and Krigbaum (1950) assume such a model in which each cloud of segments is approximately spherical, with a density that is a maximum at the center and decreases in an approximately Gaussian function with distance from the center. Within the volume occupied by

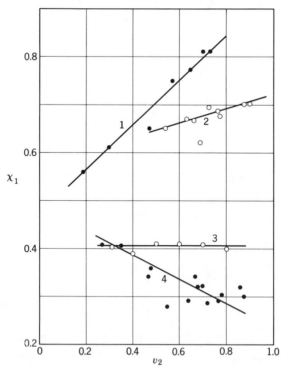

FIG. 7-6. Experimentally observed variation of χ_1 with concentration: curve 1, poly(dimethyl siloxane) in benzene (Newing 1950); curve 2, polystyrene in benzene (Bawn 1950); curve 3, rubber in benzene (Gee 1946, 1947); curve 4, polystyrene in toluene (Bawn 1950).

the segments of one molecule, all other molecules tend to be excluded. Within such an excluded volume (Section B) occur long-range intramolecular interactions whose thermodynamic functions can be derived. The partial molar heat content, entropy, and free energy of these interactions are, respectively,

$$\Delta \bar{H}_1 = kT\kappa_1 v_2^2$$
$$\Delta \bar{S}_1 = k\psi_1 v_2^2 \qquad (7\text{-}15)$$
$$\Delta \bar{G}_1 = kT(\kappa_1 - \psi_1)v_2^2$$

where, by comparison with Eq. 7-12,

$$\kappa_1 - \psi_1 = \chi_1 - \tfrac{1}{2} \qquad (7\text{-}16)$$

It is convenient to define as a parameter the *Flory temperature* Θ such that $\Theta = \kappa_1 T / \psi_1$. It follows that

$$\psi_1 - \kappa_1 = \psi_1\left(1 - \frac{\Theta}{T}\right) = \tfrac{1}{2} - \chi_1 \qquad (7\text{-}17)$$

and that at the temperature $T = \Theta$ the partial molar free energy due to polymer–solvent interactions is zero and deviations from ideal solution behavior vanish. The excluded volume becomes smaller as the solvent becomes poorer and vanishes at $T = \Theta$, where the molecules interpenetrate one another freely with no net interactions. At temperatures below Θ they attract one another and the excluded volume is negative. If the temperature is much below Θ, precipitation occurs (Section D).

The parameter α can be evaluated in terms of thermodynamic quantities:

$$\alpha^5 - \alpha^3 = 2C_m\psi_1\left(1 - \frac{\Theta}{T}\right)M^{1/2} \qquad (7\text{-}18)$$

where C_m lumps together numerical and molecular constants. This equation, which involves the assumption of validity of theories involving intermolecular interactions as well as intramolecular interactions, must be accepted with some reservations. It predicts that α increases without limit with increasing molecular weight; since $\overline{r_0^2}$ is proportional to M, $\overline{r^2}/M$ should increase with M. It also leads to the fact that, at $T = \Theta$, $\alpha = 1$ and molecular dimensions are unperturbed by intramolecular interactions. Since α depends on the entropy parameter ψ, it is larger in better solvents.

Corresponding-State Theories

From what has been said, it can be seen that both the Flory–Huggins and Flory–Krigbaum theories have serious shortcomings. Both are based on and conserve the important features of the theories of "regular" solutions of small molecules. Only the appropriate combinatorial entropy of mixing has been modified to fit the polymer case. The most important assumption that is retained is that there is no volume change on mixing.

The deviations between theory and experiment for the interaction parameter χ_1 led to its reinterpretation as a combined entropy and enthalpy parameter, as in Eqs. 7-11 and 7-12. The entropy term was considered to be a small negative correction, corresponding to the plausible idea of a small increase in order as polymer–polymer and solvent–solvent interactions were replaced with polymer–solvent interactions. Typical values of ψ_1 and κ_1 show that this is not the case, for the entropic contribution to χ_1 is positive and much larger than the enthalpic term. Still other failures of the traditional theories of the thermodynamics of polymer solutions are discussed in Section D.

These difficulties have been overcome in corresponding-state theories first developed around 1952 by Prigogine (1957) and his co-workers, and extended and put into practice by Flory (Flory 1965, Eichinger 1968a–d), and others. The major new factor in the theories is the recognition of the dissimilarity in the free volumes of the polymer and the solvent as a result of their great difference in size, the usual solvent being much more expanded than the polymer. Mixing is not unlike the condensation of a gas (the solvent) into a dense medium (the polymer). The total volume change on mixing is usually negative and is accompanied with a negative (exothermic) ΔH and a negative contribution to ΔS. Thus the total ΔH on mixing consists of the usual positive interaction term (Eq. 7-8) and the new negative term; both contribute to a new interaction parameter like χ_1. Similarly, the entropy of mixing consists of the Flory–Huggins combinatorial term (Eq. 7-6) plus the new negative term; only the latter contributes to the new χ. The relative magnitude of these contributions is discussed further in Section D. Their exact mathematical formulation is beyond the scope of this section.

It can be said, however, that a basic assumption of the free-volume theories is that all liquids, polymers when noncrystalline, and mixtures of these follow the same reduced equation of state. In Flory's formulation (see, for example, Carpenter 1970) this can be written as

$$\frac{\tilde{p}\tilde{V}}{\tilde{T}} = \frac{\tilde{V}^{1/3}}{\tilde{V}^{1/3} - 1} - \frac{1}{\tilde{V}\tilde{T}} \tag{7-19}$$

which simplifies further at low pressures, including atmospheric, where $\tilde{p} \to 0$. Here $\tilde{p}$, $\tilde{V}$, $\tilde{T}$ are the ratios of the real pressure, volume, and temperature to reduction parameters that can be evaluated from the thermal coefficient of expansion and the isothermal compressibility. It is possible to calculate reduced parameters for mixtures as well as for the pure components; from the resulting equations of state we can evaluate the volume change on mixing, the heat and entropy of mixing, and such details as the dependence of the new χ on concentration. It has been found that volume changes on mixing can be predicted to within 10–15% of the experimental values, and the correct variation of χ with concentration is predicted, in contrast to the result of the Flory–Huggins theory discussed above.

Scaling Concepts

From the previous discussion we know the properties of polymer chains in both dilute and concentrated solutions. Dilute nonoverlapping coils in a good solvent are swollen, with a size proportional to $x^{0.60}$. On the other hand, in very concentrated solutions or melts the chains are essentially ideal, with a size proportional to $x^{0.50}$, and they interpenetrate one another very strongly. What has happened in between? The only available theory for a long time was that of Flory and Huggins, and it is not adequate at low and intermediate concentrations. The solution has now been elucidated both experimentally, through neutron-scattering experiments (Chapter 8C), and theoretically, through the application of scaling concepts (de Gennes 1979).

The two extreme concentration regions described above are separated by a region of transition in properties at a concentration $c = c^*$ where the chain coils begin to overlap (Fig. 7-7). The scaling properties of c^* are important. It is comparable to the local segment concentration within a single coil:

$$c^* \simeq x/(\overline{r^2})^{3/2} \sim x^{-0.80} \tag{7-20}$$

In terms of the volume fraction, $v_2^* \sim x^{-0.80}$ and is quite small ($\sim 10^{-3}$) for large x.

When c is small compared to c^* (the dilute-solution region), Flory (1953) showed that the equation of state takes the form

$$\frac{\pi}{kT} \simeq \frac{c}{x} + (\text{const.}) \left(\frac{c}{x}\right)^2 (\overline{r^2})^{3/2} + \cdots \tag{7-21}$$

which leads by scaling concepts to $A_2 \sim x^{-0.20}$, in good agreement with experiment. In the transition region, where v_2 is in the broad region between v_2^* and 1, application of scaling concepts to the osmotic pressure leads (des Cloizeaux 1975) to $\pi/kT \sim v_2^{2.25}$, in contrast to the Flory–Huggins prediction of v_2^2. This has been confirmed experimentally (Daoud 1975).

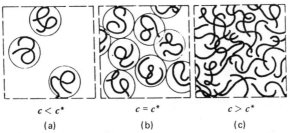

$$c < c^* \qquad\qquad c = c^* \qquad\qquad c > c^*$$

$$\text{(a)} \qquad\qquad\qquad \text{(b)} \qquad\qquad\qquad \text{(c)}$$

FIG. 7-7. Schematic of chain conformations in the (a) dilute, (b) transition, and (c) semi-dilute concentration regions. (From de Gennes 1979. Copyright 1979 by Cornell University. Used by permission of the publisher, Cornell University Press.)

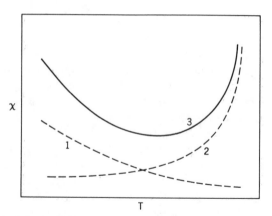

FIG. 7-8. Qualitative behavior of the parameter χ_1 according to the corresponding-state theories: curve 1, contribution from enthalpic polymer–solvent interactions (as predicted by the Flory–Huggins theory); curve 2, contribution due to free-volume dissimilarity between polymer and solvent; curve 3, total χ_1, sum of 1 and 2 (Patterson 1969).

GENERAL REFERENCES

Flory 1953; Huggins 1958; Carpenter 1970; Tager 1978; de Gennes 1979, Chapter 3; Kwei 1979, Chapter 4.2.

D. PHASE EQUILIBRIUM IN POLYMER SOLUTIONS

Polymer–Solvent Miscibility

This section is concerned with the equilibrium between two liquid phases, both of which contain amorphous polymer and one or more solvents. The treatment of cases involving a crystalline polymer phase is more complex and is given in part in Section E and in part in Chapter 10C.

When the temperature of a polymer solution is raised or lowered, the solvent eventually becomes thermodynamically poorer. Finally a temperature is reached beyond which polymer and solvent are no longer miscible in all proportions. At more extreme temperatures the mixture separates into two phases. Such phase separation may also be brought about by adding a nonsolvent liquid to the solution. In either case it takes place when the interaction parameter χ_1 exceeds a critical value near $\frac{1}{2}$ (see below).

It was shown in Section C that the Flory–Huggins theory attributes χ_1 to polymer–solvent interactions alone, and predicts it to increase monotonically as the temperature decreases (Fig. 7-8). Thus phase separation is predicted to take place only on lowering the temperature, the phase diagram looking like that in the lower portion of Fig. 7-9. The maximum temperature for phase separation is designated the *upper critical solution temperature*. Although it was not frequently observed

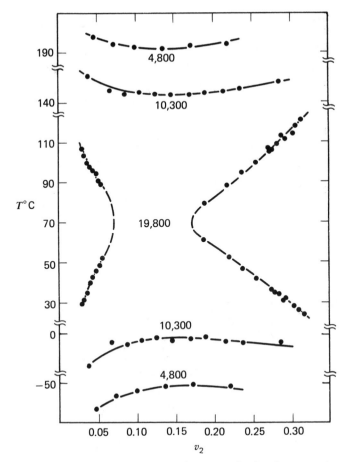

FIG. 7-9. Phase diagram of the system polystyrene in acetone, showing phase separation at upper and lower critical solution temperatures for polymer of the molecular weights indicated. Reprinted with permission from Siow 1972. Copyright 1972 by the American Chemical Society.)

until recent years (Freeman 1960), it is now recognized that phase separation invariably occurs also when the temperature is raised until a *lower critical solution temperature*† (Fig. 7-9, upper area) is reached. This phenomenon is explained by the corresponding-state theories of polymer solutions described qualitatively in Section C. The contribution to χ_1 from the free-volume dissimilarity between polymer and solvent is an increasing function of the temperature (Fig. 7-8). The total interaction parameter thus goes through a minimum, and two critical values of χ_1 are accessible. Each has the character of a Θ temperature (see below); at each the

---

†The perverse names of the upper and lower critical solution temperatures relate more directly to an alternative case (the common example is nicotine in water) in which the upper critical solution temperature is higher than the lower one and the phase diagram consists of a region of immiscibility completely surrounded by a one-phase region.

second virial coefficient approaches zero and the polymer chain approaches its unperturbed dimensions.

For the remainder of this section reference is made only to phase separation at an upper critical solution temperature. The predictions of the Flory–Huggins theory remain unchanged except for the sign of the temperature dependence in the case of a lower critical solution temperature.

Binary Polymer–Solvent Systems. The condition for equilibrium between two phases in a binary system is that the partial molar free energy of each component be equal in each phase. This condition corresponds to the requirement that the first and second derivatives of $\Delta \bar{G}_1$ (Eq. 7-10) with respect to v_2 be zero. Application of this condition leads to the critical concentration at which phase separation first appears:

$$v_{2c} = \frac{1}{1 + x^{1/2}} \simeq \frac{1}{x^{1/2}} \tag{7-22}$$

This is a rather small volume fraction; for a typical polymer ($x \simeq 10^4$), $v_{2c} \simeq 0.01$. The critical value of χ_1 is given by

$$\chi_{1c} = \frac{(1 + x^{1/2})^2}{2x} \simeq \frac{1}{2} + \frac{1}{x^{1/2}} \tag{7-23}$$

The critical value of χ_1 exceeds $\frac{1}{2}$ by a small increment, depending on molecular weight, and at infinite molecular weight equals $\frac{1}{2}$. The temperature at which phase separation begins is given by

$$\frac{1}{T_c} = \frac{1}{\Theta}\left[1 + \frac{1}{\psi_1}\left(\frac{1}{x^{1/2}} + \frac{1}{2x}\right) \right] \simeq \frac{1}{\Theta}\left(1 + \frac{C}{M^{1/2}}\right) \tag{7-24}$$

where C is a constant for the polymer–solvent system. Thus $1/T_c$ (K) varies linearly with the reciprocal square root of molecular weight. The Flory temperature Θ is the critical miscibility temperature in the limit of infinite molecular weight.

The qualitative features of the theory are in agreement with experiment. The dependence of precipitation temperature on polymer concentration is shown in Fig. 7-10. The critical values T_c and v_{2c} correspond to the maxima in the curves. In the neighborhood of the critical point the polymer concentration in the dilute phase is extremely small. The phenomenon of the coexistence of two liquid phases, one of which is a dilute solution and the other nearly pure solvent, is called *coacervation* (Section E).

The discrepancy between theory and experiment results (Koningsveld 1968) from the effect of finite molecular-weight distribution breadth in the polymer samples used.

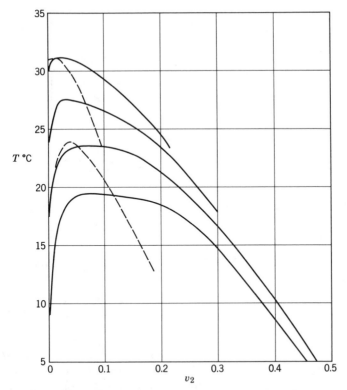

FIG. 7-10. Phase diagram showing precipitation temperature as a function of polymer concentration for fractions of polystyrene in diisobutyl ketone (Shultz 1952). Higher-molecular-weight fractions precipitate at higher temperatures (solid lines, experimental; dotted lines, theoretical).

The dependence of T_c upon molecular weight is shown in Fig. 7-11. The curves are accurately linear, and the Flory temperatures Θ obtained from the intercepts agree within experimental error ($<1°$) with those derived from osmotic measurements, taking Θ to be the temperature where A_2 is zero. Precipitation measurements on a series of sharp fractions offer perhaps the best method of determining Θ.

Ternary Systems. Although a thorough discussion of ternary systems is outside the scope of this book, a few cases have particular interest. The most commonly encountered system is that of polymer, solvent, and nonsolvent. The phase relations are conveniently displayed in the usual ternary diagram (Fig. 7-12). The position of the *binodal curve,* along which two phases are in equilibrium, depends upon molecular weight; the limiting critical point at infinite molecular weight is the analog of Θ in a two-component system.

Multicomponent Systems. The theory of phase separation in systems comprising a heterogeneous polymer in a single solvent is developed with the simplifying assumption that the interaction constants χ for all values of x are identical; only

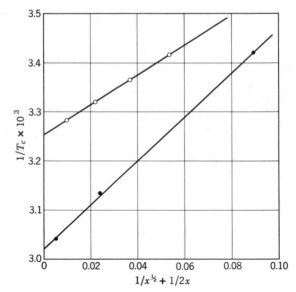

FIG. 7-11. Dependence of critical temperature for phase separation on molecular weight for (○) polystyrene in cyclohexane, and (●) polyisobutylene in diisobutyl ketone (Shultz 1952).

the size parameter x itself varies from one molecular species to another. In theory, all details of the system could be calculated from a knowledge of the size distribution of the polymer and the equilibrium conditions. In practice, this would be an enormous task; but the principal point of interest, namely, the efficiency with which molecular species are separated between the two phases, is simply derived. It is given by the equation

$$\ln\left(\frac{v_x{}'}{v_x}\right) = \sigma x \tag{7-25}$$

where $v_x{}'$ and v_x are, respectively, the concentrations of species x in the precipitated or more concentrated phase (primed) and the dilute phase (unprimed), and

$$\sigma = v_2\left(1 - \frac{1}{\bar{x}_n}\right) - v_2{}'\left(1 - \frac{1}{\bar{x}_n{}'}\right) + \chi_1[(1 - v_2)^2 - (1 - v_2{}')^2] \tag{7-26}$$

Here $v_2{}'$ and v_2 are the total polymer concentrations in the two phases, and $\bar{x}_n$ is the number average of x. The parameter σ thus depends in a complicated way on the relative amounts of all the polymer species in each phase. The following conclusions may be drawn from the theory:

a. A part of any given polymer species is always present in each phase. In fact, every species is actually more soluble in the precipitated phase, that is, $v_x{}' > v_x$ for all values of x.

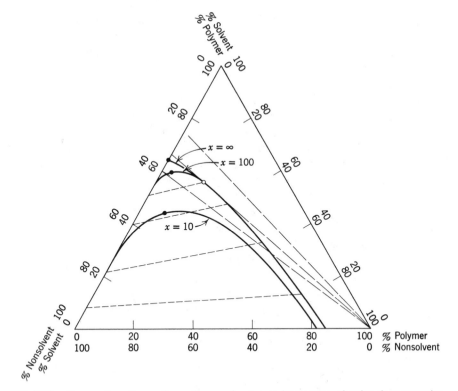

FIG. 7-12. Ternary phase diagram for a polymer–solvent–nonsolvent system, showing phase separation boundaries for the indicated values of x. Dotted lines are tie lines for $x = 100$, connecting points representing the compositions of pairs of phases in equilibrium. Dashed lines represent constant solvent–nonsolvent ratios (●, critical points; ○, precipitation threshold for $x = 100$).

b. Since σ depends on the concentrations of all species and on the details of their weight distribution, the result of a fractionation cannot be predicted in advance unless the details of the distribution are known. The Flory–Huggins theory predicts that only the number average $\bar{x}_n$ for each phase appears in σ, while an alternative analysis (Stockmayer 1949) expresses σ in terms of higher moments of the distribution for the whole polymer.

c. Fractionation is significantly more efficient in very dilute solutions: If $R = v'/v$, the ratio of the volumes of the precipitated and dilute phases, the fraction f_x of the constituent x in the dilute phase is

$$f_x = \frac{1}{1 + Re^{\sigma x}} \tag{7-27}$$

and that in the precipitated phase is

$$f_x' = \frac{Re^{\sigma x}}{1 + Re^{\sigma x}} \tag{7-28}$$

If $R \ll 1$, most of the smaller species remain in the dilute phase because of its greater volume. The larger partition factor v_x'/v_x for the higher-molecular-weight species causes them to appear selectively in the precipitated phase despite its smaller volume.

The resulting inefficient nature of the fractionation process is illustrated in Figs. 7-13 and 7-14. Figure 7-13 shows in the three lower curves the distribution of polymer species left in the dilute phase after precipitation at various values of R. The initial distribution is shown in the upper curve. The increase in efficiency as R becomes smaller is apparent. Figure 7-14 shows the overlapping distribution curves resulting even when R is unusually small. It is clear that separation into fractions that are close to monodisperse is a most difficult and time-consuming process. It should be noticed that the distribution curves of the fractions in Fig. 7-14 become narrower at lower molecular weights.

The foregoing discussion refers to fractionation by cooling from a single solvent. Similar considerations apply to fractionation at constant temperature in a solvent–nonsolvent system. A difference arises in that the solvent composition is different in the two phases. This difference should lead to more efficient fractionation in solvent–precipitant mixtures. Experimental evidence for such an effect has not been reported.

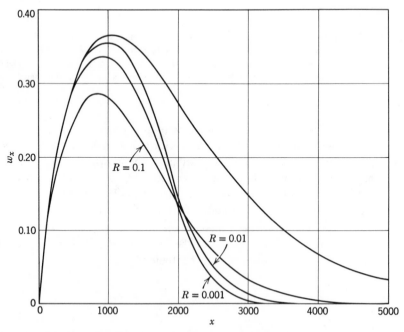

FIG. 7-13. Calculated separation of a polymer having the molecular-weight distribution shown in the upper curve with $R = 0.1$, 0.01, and 0.001, f_x chosen to be $\frac{1}{2}$ at $x = 200$ in each case (Flory 1953).

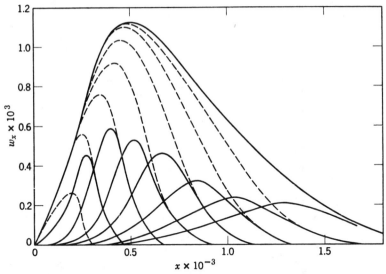

FIG. 7-14. Calculated distributions of a series of eight fractions separated from the distribution of the upper curve (Flory 1953); $R = 0.001$ in each case. Dashed curves represent the polymer remaining in the dilute phase after each successive precipitation.

Scaling Concepts. Figure 7-15 shows one important result of the application of scaling concepts to polymer physics. It is a typical phase diagram of a polymer solution, with the coordinates concentration c and reduced temperature $\tau = (T - \Theta)/\Theta$, showing the regions in which the application of scaling concepts becomes important. The dashed curve in the negative region of τ (region IV) is a coexistence or phase-separation curve of the type shown in Fig. 7-10. Region I' is the Θ region, of dilute solutions near $T = \Theta$ and $\tau = 0$. Region I is the dilute region, limited by a value of τ proportional to $(\overline{r^2})^{-1/2}$, and the line of c^* the critical concentration at which the chains begin to overlap; c^* is also proportional to $(\overline{r^2})^{-1/2}$. Region II is the semidilute region within which chain overlap becomes more and more important. The concentration c^{**} is, as seen, proportional to τ. Region III is the semidilute and concentrated Θ region, not yet studied in detail.

In several of these regions the Flory–Huggins theory describes the situation adequately. The critical point for phase separation, for example, is located correctly, but the shape of the coexistence curve in that region is not properly described by that theory. The properties of the polymer-rich phase are adequately described by the Flory–Huggins theory.

Polymer–Polymer Miscibility

Because of the commercially interesting properties of compatible polymer blends, interest in phase separation in polymer–polymer systems has increased in recent

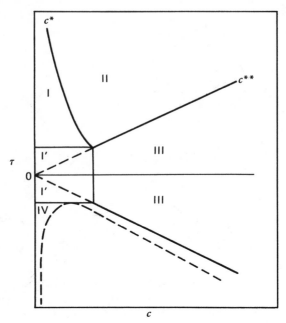

FIG. 7-15. Typical phase diagram of a polymer solution. See the text for explanation. (Reprinted with permission from Daoud 1976, Figure 3, page 974.)

years. In general, polymers mixed in pairs tend to be totally immiscible, but a significant number of exceptions exist and have been listed by Krause (1978).

Each of the several theories of polymer solution thermodynamics contributes to an understanding of polymer–polymer miscibility. Use of the solubility parameter predicts, for example, that for miscibility over the entire composition range the difference in solubility parameters of the two polymers cannot exceed 0.7 at $M = 10,000$, 0.2 at $M = 100,000$, or 0.08 at $M = 1,000,000$. The Flory–Huggins theory describes all the observed behavior, provided that interaction parameters are determined experimentally, but it does not provide a derivation of them, an understanding of the origin of the observed behavior, or predictive powers. The corresponding-state theory overcomes many of these limitations and leads to several conclusions at variance with the predictions of the Flory–Huggins theory.

The thermodynamics of phase separation in block copolymers has been developed by Krause (1970) and Meier (1969).

GENERAL REFERENCES

Flory 1953; Casassa 1977; Elias 1977, Chapter 6.6; Paul 1978; de Gennes 1979, Chapter 4; Kwei 1979, Chapter 4.3; Olabisi 1979.

E. FRACTIONATION OF POLYMERS BY SOLUBILITY

In this section are discussed only the most important and widely practiced techniques for fractionating polymers by solubility differences. Many variations and less useful techniques exist and are reviewed by Hall (1959), Guzmán (1961), and Cantow (1967).

Bulk Fractionation by Nonsolvent Addition

Fractional precipitation in bulk (Kotera 1967) is carried out by adding the nonsolvent to a dilute solution of the polymer until a slight turbidity develops at the temperature of fractionation. To ensure the establishment of equilibrium the mixture may be warmed until it is homogeneous and allowed to cool slowly back to the required temperature, which should thereafter be carefully maintained. The precipitated phase is allowed to settle to a coherent layer, and the supernatant phase is removed. A further increment of precipitant is added and the process repeated. The polymer is isolated from the precipitated phase, which may still be relatively dilute (perhaps 10% polymer).

Refractionation is often used to achieve better separation. A more efficient procedure is to restrict the initial experiments to very dilute solutions, or to refine each initial fraction by a single reprecipitation at higher dilution, returning the unprecipitated polymer from this separation to the main solution before the next fraction is removed.

The solvent and the precipitant should be chosen so that precipitation occurs over a relatively wide range of solvent composition yet is complete before too high a ratio of precipitant to solvent is reached. Other important considerations are the stability and volatility of the liquids and their ability to form a highly swollen, mobile gel phase. Relatively simple equipment, for example, a three-neck flask of several liters capacity (Fig. 7-16), is adequate for bulk precipitation fractionation.

Column Elution

Solvent–Gradient Elution. In elution methods (Elliott 1967), polymer is placed in contact with a series of liquids of gradually increasing solvent power. Species of lowest molecular weight, and thus highest solubility, dissolve in the first liquid, and successively higher molecular-weight fractions in subsequent liquids. To ensure rapid equilibration, the polymer must be present as a very thin film. For convenience, the polymer is often applied to a finely divided substrate such as sand or glass beads, which is subsequently used to pack a column. By the use of mixing vessels, a continuous gradient of solvent–nonsolvent composition can be produced and eluted through the column.

Thermal-Gradient Elution. A modification of the solvent–gradient method utilizes a small temperature gradient from one end to the other of the column, in addition to the gradient of solvent composition (Baker 1956, Porter 1967). The

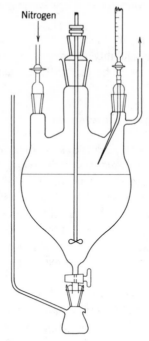

Nitrogen

FIG. 7-16. An apparatus suitable for bulk fractional precipitation (Hall 1959).

sample is initially confined to a small zone of the packing at the warm end of the column. Each species undergoes a series of solution and precipitation steps as it progresses down the column.

Analytical Precipitation Techniques

Analytical fractionation techniques are those in which fractions are not isolated, in contrast to the methods discussed above. They may be but are not necessarily carried out on a smaller scale than their preparative counterparts.

Summative Fractionation.   In this method (Billmeyer 1950, 1973), a series of polymer solutions is used. In each, part of the polymer is precipitated, the point of fractionation being varied throughout the series. Weight fractions w_x and molecular weights M_x of the precipitates are obtained; the product $w_x M_x$ may be related to the *integral* of the cumulative distribution curve. Since the two differentiations required to obtain the differential distribution would introduce intolerably large errors into the resulting differential curve, the data are properly analyzed to provide only information equivalent to the position of the maximum and breadth of the molecular-weight distribution curve. The method is advantageous for obtaining limited information about the molecular-weight distribution with a minimum of time, preliminary calibration effort, and equipment expense.

Turbidimetric Titration. In this technique (Giesekus 1967) a precipitant is added slowly to a polymer solution, and the turbidity due to precipitated polymer is measured by the decrease in intensity of a beam of transmitted light or by the increase in intensity of scattered light. Selection of an appropriate solvent–nonsolvent system for a given polymer requires much preliminary study, and it is usually not possible to obtain accurate information on molecular-weight distribution from the experimental data. Nonetheless, the method is attractive for control purposes once the proper operating conditions are established.

Effect of Polymer Structure on Solubility-Based Fractionation

Effect of Chemical Type. It must be reemphasized (Section A) that the solubility of polymers is determined primarily by their chemical composition, and in the usual case molecular weight is only a secondary variable. If chemical composition differences exist among the polymer species in a sample, fractionation is likely to occur primarily as a result of these differences and to yield information on them rather than on the distribution of molecular weights (Fuchs 1967). Turbidimetric titration has often been used to explore qualitatively the effects of chemical composition in copolymer systems (Giesekus 1967).

Effect of Chain Branching. On the basis of limited information (Schneider 1968), it appears that branching increases the solubility of high polymers. Thus at any given stage in a solubility-based fractionation process, the branched polymer being separated will consist of a mixture of species, some with low branching and low molecular weight, others more branched but with compensating higher molecular weight, all having the same solubility. Additional information is needed to characterize such a system in terms of either variable separately.

Effect of Crystallinity. Precipitation of polymer to a crystalline, rather than an amorphous, phase can lead to fractionation by molecular weight (with control of the other variables just mentioned) only if the (depressed) crystalline melting point is a strong function of molecular weight. This is the case only for polymers of rather low molecular weight (say, less than 20,000), but in these cases the technique works rather well for the production of narrow fractions in large amounts (Pennings 1968).

Treatment of Data

Information about the distribution of molecular weights in a polymer is often the goal of a fractionation experiment. Such information requires not only that the variables mentioned in the previous paragraphs be controlled, but also the measurement of the weight and molecular weight of each fraction. For convenience, intrinsic viscosity is often used for the molecular-weight measurement, but this must be limited to experiments with linear polymers.

The data are treated (Tung 1967) by plotting against molecular weight the cumulative or integral distribution curve of the combined weight of all fractions having molecular weight up to and including M. The differential distribution curve is obtained by differentiation. Examples are given in Chapter 8F. Experimental errors, indicated by the scatter of points in the integral curve, are magnified in this process. Only the gross features of the differential curve, such as the position of the peak and the breadth of the curve, are experimentally significant. Unless the fractionation is performed with extreme care, spurious features, such as double peaks in the differential distribution curve, may appear.

GENERAL REFERENCES

Hall 1959; Guzmán 1961; Schneider 1965; Cantow 1967; Johnson 1967; Tung 1977.

DISCUSSION QUESTIONS AND PROBLEMS

1. List and discuss the variables affecting the solubility of polymers.

2. Describe the two stages of the process of dissolving a polymer and suggest how to speed each of them up.

3. Calculate the solubility parameter δ of *bis*-phenol A polycarbonate (Chapter 15B) from molar attraction constants.

4. It is observed that a styrene–butadiene copolymer ($\delta = 16.5$) is insoluble in pentane ($\delta = 14.5$) and ethyl acetate ($\delta = 18.6$), but soluble in a 1:1 mixture of the two. Explain.

5. Describe the conformation of a polymer chain in solution. Define and relate various measures of the size of the chain, show the distribution of sizes by graphs, and discuss the effect of molecular structure on size.

6. For polystyrene of molecular weight 416,000 dissolved in toluene, the chain expansion factor α is 3.2. Calculate (a) the contour length, (b) the random-flight end-to-end distance, (c) the random-flight radius of gyration, (d) the unperturbed end-to-end distance, (e) the actual end-to-end distance, (f) the weight concentration of polymer at which spheres with radius 2.5 times the actual end-to-end distance would just touch (this radius is that at which the chance of finding chain segments outside the sphere becomes small; for simplicity, pack the spheres as if they were cubes), and (g) the volume fraction of polymer at this concentration (the effective density of the polystyrene chain in toluene is 1.10).

7. Define ideal solutions and show why polymer solutions can never be ideal, even at extreme dilutions.

8. Define excluded volume and discuss the difficulty of its evaluation.

9. Discuss the basis and contributions to polymer science of (a) the Flory–Huggins theory, (b) the Flory–Krigbaum theory, (c) the corresponding-state theories, and (d) scaling concepts.

10. Define the Flory temperature Θ and describe two ways of determining it from thermodynamic considerations.

11. Derive Eqs. 7-10 and 7-11 from Eq. 7-9.

12. Calculate the free energy of mixing 100 g of polystyrene, of molecular weight 100,000, with 1000 g of benzene. Take $\chi_1 = 0.23$ and assume that the polymer and the solvent have the same density. Comment on the sign and magnitude of the answer.

13. In the fractional precipitation of polymer in a mixture of good and poor solvents, how do the compositions of the two phases depend on polymer chain length?

14. Explain why poly(methyl methacrylate) and polystyrene are immiscible whereas their monomers are miscible in all proportions. Under what circumstances would this miscibility carry over to a copolymer of the two monomers?

15. The following data were synthesized to represent the fractionation of a polymer of relatively narrow molecular-weight distribution. Calculate $\bar{M}_w$ and $\bar{M}_n$ and plot the integral and differential distribution curves.

Weight Percent in Fraction	Molecular Weight
3	50,000
5	75,000
6	95,000
13	125,000
8	150,000
25	185,000
16	220,000
9	270,000
10	325,000
5	400,000

BIBLIOGRAPHY

Asmussen 1962. Frithjof Asmussen and Kurt Ueberreiter, "Velocity of Dissolution of Polymers. Part II," *J. Polym. Sci.* **57**, 199–208 (1962).

ASTM D3132. Standard Test Method for *Solubility Range of Resins and Polymers*, ANSI/ASTM Designation: D3132, American Society for Testing and Materials, Philadelphia, Pennsylvania, 1976.

Baker 1956. C. A. Baker and R. J. P. Williams, "A New Chromatographic Procedure and Its Application to High Polymers," *J. Chem. Soc.* **1956**, 2352–2362 (1956).

Bawn 1950. C. E. H. Bawn, R. F. J. Freeman, and A. R. Kamaliddin, "High Polymer Solutions. Part I. Vapour Pressure of Polystyrene Solutions," *Trans. Faraday Soc.* **46**, 677–684 (1950).

Billmeyer 1950. F. W. Billmeyer, Jr., and W. H. Stockmayer, "Method of Measuring Molecular Weight Distribution," *J. Polym. Sci.* **5**, 121–137 (1950).

Billmeyer 1969. Fred W. Billmeyer, Jr., "Molecular Structure and Polymer Properties," *J. Paint Technol.* **41**, 3–16, 209 (1969).

Billmeyer 1972. Fred W. Billmeyer, Jr., *Synthetic Polymers: Building the Giant Molecule,* Doubleday, Garden City, New York, 1972.

Billmeyer 1973. Fred W. Billmeyer, Jr., and Leonard R. Siebert, "Application of the Summative Fractionation Method to the Determination of $\bar{M}_w/\bar{M}_n$ for Narrow Distribution Polymers," Chapter 2 in Myer Ezrin, ed., *Polymer Molecular Weight Methods,* American Chemical Society, Washington, D.C., 1973.

Billmeyer 1977. Fred W. Billmeyer, Jr., "The Size and Weight of Polymer Molecules," Chapter 4 in Herman S. Kaufman and Joseph J. Falcetta, eds., *Introduction to Polymer Science and Technology: An SPE Textbook,* John Wiley & Sons, New York, 1977.

Birshtein 1966. T. M. Birshtein and O. B. Ptitsyn, *Conformations of Macromolecules* (translated by Serge N. Timasheff and Marina J. Timasheff), John Wiley & Sons, New York, 1966.

Bovey 1969. Frank A. Bovey, *Polymer Conformation and Configuration,* Academic Press, New York, 1969.

Bovey 1979. F. A. Bovey and T. K. Kwei, "Microstructure and Chain Conformation of Macromolecules," Chapter 3 in F. A. Bovey and F. H. Winslow, eds., *Macromolecules: An Introduction to Polymer Science,* Academic Press, New York, 1979.

Burrell 1955. Harry Burrell, "Solubility Parameters for Film Formers," *Off. Dig. Fed. Soc. Paint Technol.* **27**, 726–758 (1955).

Burrell 1975. H. Burrell, "Solubility Parameter Values," pp. IV-337–IV-359 in J. Brandrup and E. H. Immergut, eds., with the collaboration of W. McDowell, *Polymer Handbook,* 2nd ed., Wiley-Interscience, New York, 1975.

Cantow 1967. Manfred J. R. Cantow, ed., *Polymer Fractionation,* Academic Press, New York, 1967.

Carpenter 1970. D. K. Carpenter, "Solution Properties," pp. 627–678 in Herman F. Mark, Norman G. Gaylord, and Norbert F. Bikales, eds., *Encyclopedia of Polymer Science and Technology,* Vol. 12, Wiley-Interscience, New York, 1970.

Casassa 1977. Edward F. Casassa, "Phase Equilibrium in Polymer Solutions," Chapter 1 in L. H. Tung, ed., *Fractionation of Synthetic Polymers: Principles and Practices,* Marcel Dekker, New York, 1977.

Collins 1973. Edward A. Collins, Jan Bareš, and Fred W. Billmeyer, Jr., *Experiments in Polymer Science,* Wiley-Interscience, New York, 1973.

Daoud 1975. M. Daoud, J. P. Cotton, B. Farnoux, G. Jannink, G. Sarma, H. Benoit, R. Duplessix, C. Picot, and P. G. de Gennes, "Solutions of Flexible Polymers. Neutron Experiments and Interpretation," *Macromolecules* **8**, 804–818 (1975).

Daoud 1976. M. Daoud and G. Jannink, "Temperature–Concentration Diagram of Polymer Solutions," *J. Phys. Paris* **37**, 973–979 (1976).

de Gennes 1979. Pierre-Gilles de Gennes, *Scaling Concepts in Polymer Physics,* Cornell University Press, Ithaca, New York, 1979.

des Cloizeaux 1975. J. des Cloizeaux, "The Lagrangian Theory of Polymer Solutions at Intermediate Concentrations," *J. Phys. Paris* **36**, 281–291 (1975).

Eichinger 1968a. B. E. Eichinger and P. J. Flory, "Thermodynamics of Polymer Solutions. Part I. Natural Rubber and Benzene," *Trans. Faraday Soc.* **64**, 2035–2052 (1968).

Eichinger 1968b. B. E. Eichinger and P. J. Flory, "Thermodynamics of Polymer Solutions. Part 2. Polyisobutylene and Benzene," *Trans. Faraday Soc.* **64**, 2053–2060 (1968).

Eichinger 1968c. B. E. Eichinger and P. J. Flory, "Thermodynamics of Polymer Solutions. Part 3. Polyisobutylene and Cyclohexane," *Trans. Faraday Soc.* **64**, 2061–2065 (1968).

Eichinger 1968d. B. E. Eichinger and P. J. Flory, "Thermodynamics of Polymer Solutions. Part 4. Polyisobutylene and *n*-Pentane," *Trans. Faraday Soc.* **64**, 2066–2072 (1968).

Elias 1977. Hans-Georg Elias, *Macromolecules · 1 · Structure and Properties,* (translated by John W. Stafford), Plenum Press, New York, 1977.

Elliott 1967. John H. Elliott, "Fractional Solution," Chapter B.2 in Manfred J. R. Cantow, ed., *Polymer Fractionation,* Academic Press, New York, 1967.

Flory 1942. Paul J. Flory, "Thermodynamics of High Polymer Solutions," *J. Chem. Phys.* **10**, 51–61 (1942).

Flory 1950. P. J. Flory and W. R. Krigbaum, "Statistical Mechanics of Dilute Polymer Solutions," *J. Chem. Phys.* **18**, 1086–1094 (1950).

Flory 1953. Paul J. Flory, *Principles of Polymer Chemistry,* Cornell University Press, Ithaca, New York, 1953.

Flory 1965. P. J. Flory, "Statistical Thermodynamics of Liquid Mixtures," *J. Am. Chem. Soc.* **87**, 1833–1838 (1965).

Flory 1969. Paul J. Flory, *Statistical Mechanics of Chain Molecules,* Wiley-Interscience, New York, 1969.

Freeman 1960. P. I. Freeman and J. S. Rowlinson, "Lower Critical Points in Polymer Solutions," *Polymer* **1**, 20–26 (1960).

Fuchs 1967. O. Fuchs and W. Schmieder, "Chemical Inhomogeneity and Its Determination," Chapter D in Manfred J. R. Cantow, ed., *Polymer Fractionation,* Academic Press, New York, 1967.

Gardon 1965. J. L. Gardon, "Cohesive-Energy Density," pp. 833–862 in Herman F. Mark, Norman G. Gaylord, and Norbert M. Bikales, eds., *Encyclopedia of Polymer Science and Technology,* Vol. 3, Wiley-Interscience, New York, 1965.

Gee 1946. Geoffrey Gee and W. J. C. Orr, "The Interaction between Rubber and Liquids. VIII. A New Examination of the Thermodynamic Properties of the System Rubber + Benzene," *Trans. Faraday Soc.* **42**, 507–517 (1946).

Gee 1947. Geoffrey Gee, Part 5, "Equilibrium Properties of High Polymers and Gels," in "Discussion on 'Some Aspects of the Chemistry of Macromolecules,'" *J. Chem. Soc.* **1947**, 280–288 (1947).

Giesekus 1967. Hanswalter Giesekus, "Turbidimetric Titration," Chapter C.1 in Manfred J. R. Cantow, ed., *Polymer Fractionation,* Academic Press, New York, 1967.

Guzmán 1961. G. M. Guzmán, "Fractionation of High Polymers," pp. 113–183 in J. C. Robb and F. W. Peaker, eds., *Progress in High Polymers,* Vol. 1, Academic Press, New York, 1961.

Hall 1959. R. W. Hall, "The Fractionation of High Polymers," Chapter 2 in P. W. Allen, ed., *Techniques of Polymer Characterization,* Butterworths, London, 1959.

Hildebrand 1950. Joel H. Hildebrand and Robert L. Scott, *The Solubility of Nonelectrolytes,* 3rd ed., Reinhold, New York, 1950 (reprinted by Dover Publications, New York, 1964).

Hopfinger 1973. A. J. Hopfinger, *Conformational Properties of Macromolecules,* Academic Press, New York, 1973.

Hoy 1970. K. L. Hoy, "New Values of the Solubility Parameters from Vapor Pressure Data," *J. Paint Technol.* **42**, 76–118 (1970).

Huggins 1942a. Maurice L. Huggins, "Thermodynamic Properties of Solutions of Long-Chain Compounds," *Ann. N.Y. Acad. Sci.* **42**, 1–32 (1942).

Huggins 1942b. Maurice L. Huggins, "Some Properties of Solutions of Long-Chain Compounds," *J. Phys. Chem.* **46**, 151–158 (1942).

Huggins 1942c. Maurice L. Huggins, "Theory of Solutions of High Polymers," *J. Am. Chem. Soc.* **64**, 1712–1719 (1942).

Huggins 1958. Maurice L. Huggins, *Physical Chemistry of High Polymers,* John Wiley & Sons, New York, 1958.

Johnson 1967. Julian F. Johnson, Manfred J. R. Cantow, and Roger S. Porter, "Fractionation," pp. 231–260 in Herman F. Mark, Norman G. Gaylord, and Norbert M. Bikales, eds., *Encyclopedia of Polymer Science and Technology,* Vol. 7, Wiley-Interscience, New York, 1967.

Koningsveld 1968. R. Koningsveld, "Liquid–Liquid Separation of Polymer Solutions," pp. 172–213 in Donald McIntyre, ed., *Characterization of Macromolecular Structure,* Publication No. 1573, National Academy of Sciences, Washington, D.C., 1968.

Kotera 1967. Akira Kotera, "Fractional Precipitation," Chapter B.1 in Manfred J. R. Cantow, ed., *Polymer Fractionation,* Academic Press, New York, 1967.

Krause 1970. Sonja Krause, "Microphase Separation in Block Copolymers. Zeroth Approximation Including Surface Free Energies," *Macromolecules* **3,** 84–86 (1970).

Krause 1978. Sonja Krause, "Polymer–Polymer Compatibility," pp. 16–113 in D. R. Paul and Seymour Newman, eds., *Polymer Blends,* Vol. 1, Academic Press, New York, 1978.

Kwei 1979. T. K. Kwei, "Macromolecules in Solution," Chapter 4 in F. A. Bovey and F. H. Winslow, eds., *Macromolecules: An Introduction to Polymer Science,* Academic Press, New York, 1979.

Lewis 1961. Gilbert N. Lewis and Merle Randall, *Thermodynamics,* revised by Kenneth S. Pitzer and Leo Brewer, 2nd ed., McGraw-Hill, New York, 1961.

Lowry 1970. George G. Lowry, ed., *Markov Chains and Monte Carlo Calculations in Polymer Science,* Marcel Dekker, New York, 1970.

Meier 1969. D. J. Meier, "Theory of Block Copolymers. I. Domain Formation in A–B Block Copolymers," *J. Polym. Sci.* **C26,** 81–98 (1969).

Morawetz 1975. Herbert Morawetz, *Macromolecules in Solution,* 2nd ed., Wiley-Interscience, New York, 1975.

Newing 1950. M. J. Newing, "Thermodynamic Studies of Silicones in Benzene Solution," *Trans. Faraday Soc.* **46,** 613–620 (1950).

Olabisi 1979. Olagoke Olabisi, Lloyd M. Robeson, and Montgomery T. Shaw, *Polymer–Polymer Miscibility,* Academic Press, New York, 1979.

Patterson 1969. D. Patterson, "Free Volume and Polymer Solubility: A Qualitative View," *Macromolecules* **2,** 672–679 (1969).

Paul 1978. D. R. Paul and Seymour Newman, eds., *Polymer Blends,* Academic Press, New York, 1978.

Pennings 1968. A. J. Pennings, "Liquid–Crystal Phase Separation in Polymer Solutions," pp. 214–244 in Donald McIntyre, ed., *Characterization of Macromolecular Structure,* Publication No. 1573, National Academy of Sciences, Washington, D.C., 1968.

Porter 1967. Roger S. Porter and Julian F. Johnson, "Chromatographic Fractionation," Chapter B.3 in Manfred J. R. Cantow, ed., *Polymer Fractionation,* Academic Press, New York, 1967.

Prigogine 1957. I. Prigogine (with the collaboration of A. Bellemans and V. Mathot), *The Molecular Theory of Solutions,* Interscience, New York, 1957.

Rayleigh 1919. Lord Rayleigh, "On the Problem of Random Vibrations, and of Random Flights in One, Two, or Three Dimensions," *Philos. Mag.* **37,** 321–347 (1919).

Schneider 1965. Nathaniel S. Schneider, "Review of Solution Methods and Certain Other Methods of Polymer Fractionation," *J. Polym. Sci.* **C8,** 179–204 (1965).

Schneider 1968. N. S. Schneider, R. T. Traskos, and A. S. Hoffman, "Characterization of Branched Polyethylene Fractions from the Elution Column," *J. Appl. Polym. Sci.* **12,** 1567–1587 (1968).

Shultz 1952. A. R. Shultz and P. J. Flory, "Phase Equilibrium in Polymer–Solvent Systems," *J. Am. Chem. Soc.* **74,** 4760–4767 (1952).

Siow 1972. K. S. Siow, G. Delmas, and D. Patterson, "Cloud-Point Curves in Polymer Solutions with Adjacent Upper and Lower Critical Solution Temperatures," *Macromolecules* **5,** 29–34 (1972).

Snyder 1980. Lloyd Snyder, "Solutions to Solution Problems—2," *Chemtech* **10,** 188–193 (1980).

Stockmayer 1949. W. H. Stockmayer, "Solubility of Heterogeneous Polymers," *J. Chem. Phys.* **17**, 588 (1949).

Tager 1978. A. Tager, *Physical Chemistry of Polymers* (translated by David Sobolev and Nicholas Bobrov), Mir, Moscow, 1978 (Imported Publications, Chicago).

Taylor 1948. William J. Taylor, "Average Length and Radius of Normal Paraffin Hydrocarbon Molecules," *J. Chem. Phys.* **16**, 257–267 (1948).

Treloar 1958. L. R. G. Treloar, *The Physics of Rubber Elasticity,* 2nd ed., Clarendon Press, Oxford, 1958.

Tung 1967. L. H. Tung, "Treatment of Data," Chapter E in Manfred J. R. Cantow, ed., *Polymer Fractionation,* Academic Press, New York, 1967.

Tung 1977. L. H. Tung, ed., *Fractionation of Synthetic Polymers: Principles and Practices,* Marcel Dekker, New York, 1977.

Ueberreiter 1962. Kurt Ueberreiter and Frithjof Asmussen, "Velocity of Dissolution of Polymers. Part I," *J. Polym. Sci.* **57**, 187–198 (1962).

Volkenstein 1963. M. V. Volkenstein, *Configurational Statistics of Polymeric Chains,* translated by Serge N. Timasheff and M. J. Timasheff, Wiley-Interscience, New York, 1963.

Vollmert 1973. Bruno Vollmert, *Polymer Chemistry* (translated by Edmund H. Immergut), Springer, New York, 1973.

Zimm 1949. Bruno H. Zimm and Walter H. Stockmayer, "The Dimensions of Chain Molecules Containing Branches and Rings," *J. Chem. Phys.* **17**, 1301–1314 (1949).

CHAPTER EIGHT

MEASUREMENT OF MOLECULAR WEIGHT AND SIZE

The molecular weights of polymers can be determined by chemical or physical methods of functional-group analysis, by measurement of the colligative properties, light scattering, or ultracentrifugation, or by measurement of dilute-solution viscosity. All these methods except the last are, in principle, absolute: Molecular weights can be calculated without reference to calibration by another method. Dilute-solution viscosity, however, is not a direct measure of molecular weight. Its value lies in the simplicity of the technique and the fact that it can be related empirically to molecular weight for many systems.

With the exception of some types of end-group analysis, all molecular-weight methods require solubility of the polymer, and all involve extrapolation to infinite dilution or operation in a Θ solvent in which ideal-solution behavior is attained. The term *molar mass* is preferred by some authors (e.g., Billingham 1977) to molecular weight.

It is regrettable that over the years far more use is being made of empirical rather than absolute methods for determining molecular weight. Such methods require the use of calibration samples to give them meaning. Absolute methods for molecular-weight measurement, providing such samples, are practiced far less than a decade ago, and much special apparatus developed for these techniques is no longer commercially available. These methods are vital, and their continuing development and practice must be encouraged (Billmeyer 1976).

A. END-GROUP ANALYSIS

Molecular-weight determination through group analysis requires that the polymer contain a known number of determinable groups per molecule. The long-chain nature of polymers limits such groups to end groups. Thus the method is called *end-group analysis*. Since methods of end-group analysis count the number of molecules in a given weight of sample, they yield the number-average molecular weight (Section *B*) for polydisperse materials. The methods become insensitive at high molecular weight, as the fraction of end groups becomes too small to be measured with precision. Loss of precision often occurs at molecular weights above 25,000, the limitation being as much or more due to the inability to purify samples and reagents as to lack of sensitivity in the methods.

The discussion of end-group methods is conveniently divided to cover condensation and addition polymers because of the difference in the types of end groups usually found.

Condensation Polymers. End-group analysis in condensation polymers usually involves chemical methods of analysis for functional groups. Carboxyl groups in polyesters and in polyamides are usually titrated directly with base in an alcoholic or phenolic solvent, while amino groups in polyamides may be titrated with acid under similar conditions. Hydroxyl groups are usually determined by reacting them with a titratable reagent, but infrared spectroscopy has been used. The chemical methods are often limited by insolubility of the polymer in solvents suitable for the titrations.

Addition Polymers. No general procedures can be given for end-group analysis in addition polymers because of the variety of type and origin of the end groups. When the polymerization kinetics is well known, analysis may be made for initiator fragments containing identifiable functional groups, elements, or radioactive atoms, or for end groups arising from transfer reactions with solvent or for unsaturated end groups in linear polyethylene and poly-α-olefins, as in the infrared spectroscopic analysis for vinyl groups.

GENERAL REFERENCES

Hellman 1962; Collins 1973, Exp. 10; Garmon 1975; Billingham 1977, Chapter 9.

B. COLLIGATIVE PROPERTY MEASUREMENT

The relations between the colligative properties and molecular weight for infinitely dilute solutions rest upon the fact that the activity of the solute in a solution becomes equal to its mole fraction as the solute concentration becomes sufficiently small. The activity of the solvent must equal its mole fraction under these conditions, and it follows that the depression of the activity of the solvent by a solute is equal to the mole fraction of the solute.

The colligative property methods are based on vapor-pressure lowering, boiling-point elevation (*ebulliometry*), freezing-point depression (*cryoscopy*), and the osmotic pressure (*osmometry*). The working equations for these methods are derived from Eq. 7-10. When Boltzmann's constant is replaced by the gas constant throughout to convert to a molar basis, the equations for the methods of interest for polymer solutions are

$$\lim_{c \to 0} \frac{\Delta T_b}{c} = \frac{RT^2}{\rho \, \Delta H_v} \frac{1}{\bar{M}_n}$$

$$\lim_{c \to 0} \frac{\Delta T_f}{c} = \frac{RT^2}{\rho \, \Delta H_f} \frac{1}{\bar{M}_n} \tag{8-1}$$

$$\lim_{c \to 0} \frac{\pi}{c} = \frac{RT}{\bar{M}_n}$$

where ΔT_b, ΔT_f, and π are the boiling-point elevation, freezing-point depression, and osmotic pressure, respectively; ρ is the density of the solvent; ΔH_v and ΔH_f are the enthalpies of vaporization and fusion, respectively, of the solvent per gram; and c is the solute concentration in grams per cubic centimeter. The *number-average molecular weight* $\bar{M}_n$, defined below, has been inserted to make the equations applicable to polydisperse solutes.

The relative applicability of the colligative methods to polymer solutions is demonstrated in Table 8-1. It is clear that direct measurement of vapor-pressure lowering in polymer solutions is unrewarding. It is possible, however, to utilize vapor-pressure lowering indirectly through the technique of *vapor-phase osmometry*, in which one measures a temperature difference relatable to vapor-pressure lowering through the Clapeyron equation. This temperature difference is of the same order of magnitude as those observed in cryoscopy and ebulliometry.

While such differences, of the order of $1 \times 10^{-3}°C$, can currently be measured with considerable precision, the larger effect observed in the osmotic experiment makes this technique more useful for polymer solutions, and it has been more widely used than other colligative techniques for polymer systems. To distinguish it from vapor-phase osmometry, we refer hereafter to membrane osmometry as the specific name for the last-named technique in Table 8-1.

Number-Average Molecular Weight

Since typical polymers consist of mixtures of many molecular species, molecular-weight methods always yield average values. The measurement of the colligative properties in effect counts the number of moles of solute per unit weight of sample. This number is the sum over all molecular species of the number of moles N_i of each species present:

$$\sum_{i=1}^{\infty} N_i$$

TABLE 8-1.   Colligative Properties of a Solution of Polystyrene of $M = 20,000$ at $c = 0.01$ g/cm³ in Benzene

Property	Value
Vapor-pressure lowering	4×10^{-3} mm Hg
Boiling-point elevation	1.3×10^{-3}°C
Freezing-point depression	2.5×10^{-3}°C
Osmotic pressure	15 cm solvent

The total weight w of the sample is similarly the sum of the weights of each molecular species,

$$w = \sum_{i=1}^{\infty} w_i = \sum_{i=1}^{\infty} N_i M_i \qquad (8\text{-}2)$$

The average molecular weight given by these methods is known as the *number-average molecular weight $\bar{M}_n$*. By the definition of molecular weight as weight of sample per mole,

$$\bar{M}_n = \frac{w}{\sum\limits_{i=1}^{\infty} N_i} = \frac{\sum\limits_{i=1}^{\infty} M_i N_i}{\sum\limits_{i=1}^{\infty} N_i} \qquad (8\text{-}3)$$

It is easy to show that the colligative methods measure $\bar{M}_n$. Taking membrane osmometry as an example, we may write Eq. 8-1 separately for each species i in the mixture present. Omitting the limit sign for convenience and transposing, we obtain

$$\pi_i = RT \frac{c_i}{M_i}$$

Summing over i, we may replace $\Sigma \pi_i$ by π, since each solute molecule contributes independently to the osmotic pressure, and note that $c = \Sigma c_i$. The mixture is then described by an unspecified average molecular weight $\bar{M}$:

$$\pi = RT \sum \frac{c_i}{M_i} = RT \frac{c}{\bar{M}} \qquad (8\text{-}4)$$

Solving, we have $\bar{M} = c/\Sigma(c_i/M_i)$. For unit volume the c's may be replaced by w's. Noting that $w_i = M_i N_i$ shows that $\bar{M} = \bar{M}_n$.

End-group methods as well as colligative methods give the number-average molecular weight. The number average is very sensitive to changes in the weight fractions of low-molecular-weight species, and relatively insensitive to similar changes for high-molecular-weight species.

Concentration Dependence of the Colligative Properties

For convenience, the concentration dependence of the colligative properties is developed in terms of the osmotic pressure; the results are directly applicable to cryoscopy and ebulliometry by appropriate substitution in Eq. 8-1. By Eq. 7-11 or 7-12 a general expression for the osmotic pressure at a finite concentration is

$$\frac{\pi}{RTc} = A_1 + A_2c + A_3c^2 + \cdots$$

$$= \frac{1}{\bar{M}_n} (1 + \Gamma c + g\Gamma^2c^2 + \cdots) \tag{8-5}$$

where $\Gamma = A_2/A_1$ and g is a slowly varying function of the polymer–solvent interaction with values near zero for poor solvents and values near 0.25 for good solvents (Krigbaum 1952, Stockmayer 1952).

In most cases the term in c^2 may be neglected; when dependence on c^2 is significant, it may be convenient to take $g = 0.25$ and write

$$\frac{\pi}{RTc} = \frac{1}{\bar{M}_n} \left(1 + \frac{\Gamma}{2} c\right)^2 \tag{8-6}$$

In terms of the polymer–solvent interaction constant χ_1 of the Flory–Huggins and Flory–Krigbaum theories, the second virial coefficient is given by

$$A_2 = \frac{\rho_1}{M_1\rho_2^2} \left(\frac{1}{2} - \chi_1\right) = \frac{\rho_1}{M_1\rho_2^2} \psi \left(1 - \frac{\Theta}{T}\right) \tag{8-7}$$

where subscript 1 indicates the solvent, and 2 the polymer. As noted in Chapter 7C, this expression does not account for the molecular-weight dependence of A_2 correctly.

As Eqs. 8-5 and 8-7 suggest, it is usual to plot π/c versus c. In general, a straight line results whose intercept at $c = 0$ is $A_1 = 1/\bar{M}_n$ and whose slope is A_2 and allows evaluation of the polymer–solvent interaction constant χ_1. If the solvent is good enough or the concentration high enough so that the c^2 term is significant, the points may deviate from a straight line. In such cases it is useful to plot $(\pi/RTc)^{1/2}$ versus c, as suggested by Eq. 8-6. Typical data are discussed in the section Membrane Osmometry.

Vapor-Phase Osmometry

An indirect measurement of vapor-pressure lowering by the technique known as vapor-phase osmometry is the method of choice for measuring $\bar{M}_n$ of samples too low in molecular weight to be measured in a membrane osmometer. In this method the property measured, in a so-called "vapor-pressure osmometer" (Fig. 8-1), is the small temperature difference resulting from different rates of solvent evaporation from and condensation onto droplets of pure solvent and polymer solution maintained in an atmosphere of solvent vapor. The basis of the method is that this temperature difference is proportional to the vapor-pressure lowering of the polymer solution at equilibrium and thus to the number-average molecular weight: Because of heat losses, the full temperature difference expected from theory is not attained. Measurements must be made at several concentrations and extrapolated to $c = 0$. Like ebulliometry and cryoscopy, the method is calibrated with low-molecular-weight standards and is useful for values of $\bar{M}_n$ at least up to 40,000 in favorable cases. It is rapid and has the additional advantage that only a few milligrams of sample are required. The lower limit of $\bar{M}_n$ that can be measured is that at which the solute becomes appreciably volatile. Since the method does not measure equilibrium vapor-pressure lowering but depends on the development of quasi-steady-

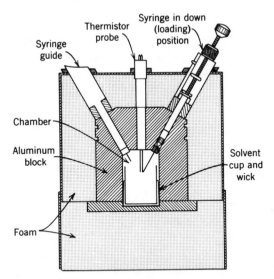

FIG. 8-1. Measurement chamber of a vapor-phase osmometer (Pasternak 1962). Droplets of solvent and polymer solution are placed, with the aid of hypodermic syringes, on the "beads" of two thermistors used as temperature-sensing elements and maintained in equilibration with an atmosphere of solvent vapor. The lower activity of the solvent in the solution droplet leads to an excess of solvent condensation over evaporation there compared to the droplet of pure solvent. The excess heat of vaporization thus liberated leads to a rise in temperature, usually somewhat less than that predicted from thermodynamic considerations because of heat losses.

state phenomena, care must be taken to standardize such variables as time of measurement and drop size between calibration and sample measurement.

Ebulliometry

In the boiling-point elevation experiment the boiling point of the polymer solution is compared directly with that of the (condensing) pure solvent in a vessel known as an ebulliometer. Sensing devices include differential thermometers and multi-junction thermocouples or thermistors arranged in a Wheatstone bridge circuit. It is customary to calibrate the ebulliometer with a substance of known molecular weight, for example, octacosane ($M = 396$) or tristearin ($M = 892$).

Although with care accurate molecular weights can be obtained up to $\bar{M}_n = 30,000$ or more, the ebulliometric experiment occasionally suffers from a limitation in the tendency of polymer solutions to foam on boiling. Not only may this lead to unstable operation, but also the polymer may concentrate in the foam because of its greater surface, thus rendering uncertain the actual concentration of the solution. No equipment suitable for ebulliometry in the high-polymer range is commercially available, and this method remains largely in the category of a referee technique.

Cryoscopy

The freezing-point depression or cryoscopic method is similar to the ebulliometric technique in several respects. The preferred temperature-sensing element is the thermistor used in a bridge circuit. The freezing points of solvent and solution are often compared sequentially. Calibration with a substance of known molecular weight is customary. Although the limitations of the method appear to be somewhat

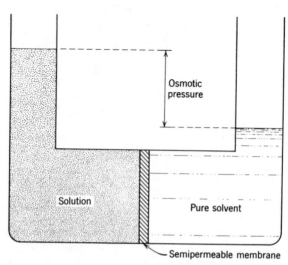

FIG. 8-2. The principle of operation of the membrane osmometer.

less severe than those of ebulliometry, care must be taken that supercooling be controlled. The use of a nucleating agent to provide controlled crystallization of the solvent is helpful in this respect. Reliable results are obtained for molecular weights at least as high as 30,000. Again, lack of commercial equipment relegates this method largely to referee status, despite the appeals of simplicity and ease of operation.

Membrane Osmometry

The principle of membrane osmometry is illustrated in Fig. 8-2. The two compartments of an *osmometer* are separated by a *semipermeable membrane*, through which only solvent molecules can penetrate, and which is closed except for capillary tubes. With the polymer solute confined to one side of the osmometer, the activity of the solvent is different in the two compartments. The thermodynamic drive toward equilibrium results in a difference in liquid level in the two capillaries. The resulting

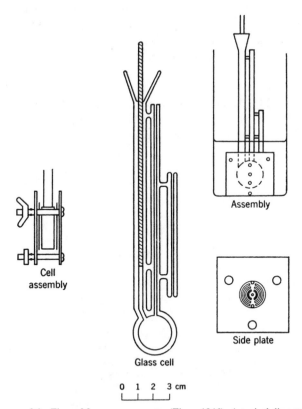

Cell
assembly

Assembly

Side plate

Glass cell

0 1 2 3 cm

FIG. 8-3. Diagram of the Zimm–Myerson osmometer (Zimm 1946). A typical diameter for the measuring and reference (solvent cell) capillary is 0.5–1 mm. The closure of the filling tube is a 2-mm metal rod. A mercury seal is used at the top to ensure tightness.

hydrostatic pressure increases the solvent activity on the solution side until, when the applied pressure equals the osmotic pressure, equilibrium is achieved.

Osmometers. Two types of osmometer have been used extensively in the past. One, the block osmometer, is a relatively large and bulky metal instrument. The membrane area is large and the solution volume small—advantages for rapid equilibration. More popular are relatively small and simple osmometers based on the Zimm–Myerson design (Zimm 1946), in which two membranes are held against a glass solution cell by means of perforated metal plates, as shown in Fig. 8-3. The assembled instrument is suspended in a large tube partly filled with solvent. In the Stabin (1954) modification the membranes are supported on both sides by channeled metal plates to ensure their rigidity. Advantages of these osmometers are small size and low cost, making multiple installations feasible, allowing the possibility of complete immersion in constant-temperature baths, and providing ease of filling and adjusting osmotic height.

Modern membrane osmometers are rapid-equilibrating automatic devices in which the solvent compartment is completely closed and fitted with a sensitive pressure-sensing device rather than a capillary (Fig. 8-4). The instruments can be brought to balance before an appreciable amount of solvent has to pass through the membrane. As a result of this rapid action, the osmometers come to equilibrium in 1–5 min instead of 10–20 h, as typically required in conventional instruments.

Membranes. The usual materials used as osmotic membranes include collodion (nitrocellulose of 11–13.5% nitrogen); regenerated cellulose, made by denitration of collodion; gel cellophane that has never been allowed to dry after manufacture; bacterial cellulose, made by the action of certain strains of bacteria; rubber; poly(vinyl

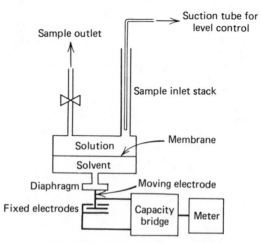

FIG. 8-4. Schematic of a modern rapid automatic osmometer. Osmotic pressure is read directly (after suitable calibration) as the change in capacitance of a condenser caused by the deflection of the flexible diaphragm (Reprinted with permission from Billingham 1977).

TABLE 8-2. Apparent Molecular Weights of Branched Polyethylenes by Membrane Osmometry[a]

	Apparent $\bar{M}_n$ of Sample		
Method	75	76	77
Average of cryoscopy, ebulliometry, and vapor-phase osmometry (Table 8-3)	11,000	15,300	18,500
Rapid membrane osmometer, first observable values (5–7 min after sample introduction)	19,900	22,100	27,900
Rapid membrane osmometer, 36 min after sample introduction	29,400	25,000	32,800
Conventional membrane osmometer, 1000 min after sample introduction	39,500	30,300	36,400

[a]Holleran (1968).

alcohol); polyurethanes; poly(vinyl butyral); and polychlorotrifluoroethylene. Of these, gel cellophane is probably most widely used at present.

Permeability of Solute Through the Membrane. The success of the osmotic experiment depends on the availability of a membrane through which solvent but not solute molecules can pass freely. Existing membranes only approximate ideal semipermeability, and the chief limitation of the osmotic method is the diffusion of low-molecular-weight species through the membrane. As currently practiced, the method is reliable and accurate only for experiments in which diffusion is absent, such as measurement of unfractionated polymers with $\bar{M}_n > 50{,}000$, or polymers from which low-molecular-weight species have been removed by fractionation or extraction, with $\bar{M}_n > 20{,}000$. The upper limit to molecular weights that can be measured by the osmotic method is about 10^6, set by the precision with which small osmotic heights can be read. When diffusion is present, the apparent osmotic pressure is always less than the true osmotic pressure and falls with time. Extrapolation back to zero time still gives too low an osmotic pressure and hence too high a molecular weight. The magnitude of the error resulting from solute diffusion and its dependence on the type of measurement are illustrated in Table 8-2.

Typical Data. Figures 8-5 and 8-6 show typical osmotic data. In both cases concentrations were low enough so that dependence of π/RTc on c^2 was negligible. The parallelism of the lines in Fig. 8-5, representing samples of different molecular weight in the same solvent, illustrates the usual result that the polymer–solvent interactions are slowly varying functions of molecular weight. The constant intercept of the lines in Fig. 8-6 obtained for one sample in several different solvents illustrates the independence of $(\pi/RTc)_{c=0} = 1/\bar{M}_n$ of solvent type and the range of A_2 or χ_1 that can be obtained with a given polymer and a range of solvents.

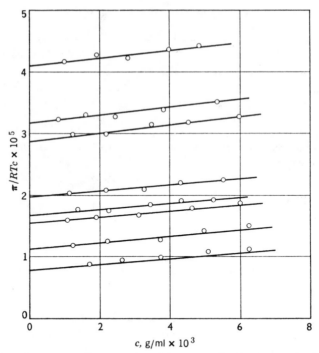

FIG. 8-5. Plots of $\pi/RTc = 1/\bar{M}_n$ versus c for cellulose acetate fractions in acetone solution (Badgley 1949).

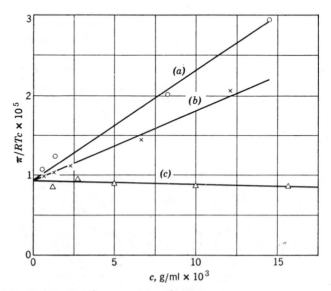

FIG. 8-6. Plot of $\pi/RTc = 1/\bar{M}_n$ versus c for nitrocellulose in (a) acetone, (b) methanol, and (c) nitrobenzene (Gee 1944, data of Dobry 1935).

196

TABLE 8-3. Number-Average Molecular Weight of Polyethylene by Various Methods[a]

	Sample						
	Branched Polyethylene			Linear Polyethylene			
Method	75	76	77	99	99H	99L	101
End groups by infrared	—	—	—	8,100	37,400	4,800	11,700
Ebulliometry	11,150	—	18,400	—	—	—	—
Cryoscopy	10,700	13,300	19,100	8,100	—	3,320	11,800
Membrane osmometry	(30,300)	(26,600)	(31,400)	—	40,600	—	—
Vapor-phase osmometry	11,000	16,300	18,800	8,600	38,400	3,500	11,900

[a]Billmeyer (1964b).

The advantage of selecting a poor solvent for osmotic measurements in order to minimize the error in extrapolation is evident. At the Flory temperature Θ the slope of π/RTc versus c is zero to concentrations of several percent.

Comparison of Data

Although it is conservative to state that most methods for measuring polymer molecular weights have an accuracy of only 5–10%, significantly closer agreement among methods can be achieved in favorable cases. An extensive intercomparison of methods for determining $\bar{M}_n$ gave the data in Table 8-3. Here, despite additional difficulties imposed by high-temperature operation, several methods gave excellent agreement for samples of polyethylene. The osmotic molecular weights in parentheses show the effects of diffusion of low-molecular-weight species through the membranes. They were obtained with conventional osmometers; data on these samples obtained with rapid automatic osmometers are given in Table 8-2.

GENERAL REFERENCES

Bonnar 1958; McIntyre 1968; Collins 1973, Chapter 7A; Glover 1975; Morawetz 1975, Chapter 4A; Billingham 1977, Chapters 1, 3, and 4; Billmeyer 1977; and specifically the following:

Vapor-Phase Osmometry. Collins 1973, Exp. 12; Billingham 1977, Chapter 4; Burge 1977; ASTM D3592.
Ebulliometry. Glover 1966, Ezrin 1968.
Membrane Osmometry. Krigbaum 1967; Coll 1968; Collins 1973, Exp. 11; Ulrich 1975; Billingham 1977, Chapter 3; Burge 1977; ASTM D3750.

C. LIGHT SCATTERING

Light scattering occurs whenever a beam of light encounters matter. The nuclei and electrons undergo induced vibrations in phase with the incident light wave and act as sources of light that is propagated in all directions, aside from a polarization effect, with the same wavelength as the exciting beam. Light scattering accounts for many natural phenomena, including the colors of the sky and the rainbow, and of most white materials.

In 1871 Lord Rayleigh applied classical electromagnetic theory to the problem of the scattering of light by the molecules of a gas. He showed that the quantity of light scattered, for particles small compared to the wavelength of the light, is inversely proportional to the number of scattering particles per unit volume and to the fourth power of the wavelength; the latter dependence accounts for the blue color of the sky.

In Rayleigh's treatment it was assumed that each particle scattered as a point source independent of all others. This is equivalent to assuming that the relative positions of the particles are random. In liquids this is not the case, and the scattered light intensity is reduced about 50-fold because of destructive interference of the light scattered from different particles. In the calculation of scattered intensity in liquids, due to Einstein and Smoluchowski in the early 1900's, the scattering is considered to be due to local thermal fluctuations in density that make the liquid optically inhomogeneous.

Light Scattering From Solutions of Particles Small Compared to the Wavelength.   In solutions and in mixtures of liquids, additional light scattering arises from irregular changes in density and refractive index due to fluctuations in composition. Debye (1944, 1947) calculated the effect of these fluctuations, relating them to the change in concentration c associated with the osmotic pressure π per mole of solute:

$$\Delta\tau = \frac{32\pi^3}{3\lambda^4}\frac{RTc}{N_0}\left(n\frac{dn}{dc}\right)^2 \bigg/ \left(\frac{d\pi}{dc}\right) \tag{8-8}$$

where τ is the turbidity, $\Delta\tau$ represents the excess turbidity of the solution over that of the pure solvent, λ is the wavelength, and n is the refractive index. In the absence of absorption, τ is related to the primary-beam intensities before and after passing through the scattering medium. If the incident intensity I_0 is reduced to I in a length l of sample, then

$$\frac{I}{I_0} = e^{-\tau l} \tag{8-9}$$

The turbidity, which is the total scattering integrated over all angles, is often replaced by the *Rayleigh ratio* R_θ, which relates the scattered intensity at angle θ to the incident beam intensity. For particles small compared to λ, $\tau = (16\pi/3) R_{(\theta = 90^\circ)}$. Inserting the relation between osmotic pressure and molecular weight yields the *Debye equation:*

$$K \frac{c}{\Delta R_{90}} = H \frac{c}{\Delta \tau} = \frac{1}{M} + 2A_2 c + \cdots$$

where (8-10)

$$K = \frac{2\pi^2 n^2}{N_0 \lambda^4} \left(\frac{dn}{dc}\right)^2 \quad \text{and} \quad H = \frac{32\pi^3 n^2}{3 N_0 \lambda^4} \left(\frac{dn}{dc}\right)^2$$

Equation 8-10 forms the basis of the determination of polymer molecular weights by light scattering. Beyond the measurement of τ or R_θ, only the refractive index n and the *specific refractive increment* dn/dc require experimental determination. The latter quantity is a constant for a given polymer, solvent, and temperature and is measured with an interferometer or a differential refractometer (Brice 1951).

Equation 8-10 is correct only for vertically polarized incident light and for optically isotropic particles. The use of unpolarized light requires that τ be multiplied by $1 + \cos^2 \theta$. This term arises from the fact that light can be propagated only at right angles to the direction of oscillation of the electric moment. Whether polarized or unpolarized light is used, the scattered intensity is symmetrical about the angle of observation, 90°. If the scattering particles are anisotropic, a correction for depolarization is required (Cabannes 1929).

Weight-Average Molecular Weight

In the derivation of Eq. 8-10 it is shown that the amplitude of the scattered light is proportional to the polarizability and hence to the mass of the scattering particle. Thus the intensity of scattering is proportional to the square of the particle mass. If the solute is polydisperse, the heavier molecules contribute more to the scattering than the light ones. The total scattering at zero concentration is

$$\Delta \tau = \sum_{i=1}^{\infty} \Delta \tau_i = H \sum_{i=1}^{\infty} c_i M_i = Hc\bar{M}_w \tag{8-11}$$

defining the *weight-average molecular weight* according to any of the relationships

$$\bar{M}_w = \frac{\sum_{i=1}^{\infty} c_i M_i}{c} = \sum_{i=1}^{\infty} w_i M_i = \frac{\sum_{i=1}^{\infty} N_i M_i^2}{\sum_{i=1}^{\infty} N_i M_i} \tag{8-12}$$

where w_i is the weight fraction and

$$c = \sum_{i=1}^{\infty} c_i$$

as in the corresponding definition of number-average molecular weight.

The significance of the weight- and number-average molecular weights is noteworthy. The quantity $\bar{M}_w$ is always greater than $\bar{M}_n$, except for a monodisperse system. The ratio $\bar{M}_w/\bar{M}_n$ is a measure of the polydispersity of the system. $\bar{M}_w$ is particularly sensitive to the presence of high-molecular-weight species, whereas $\bar{M}_n$ is influenced more by species at the lower end of the molecular-weight distribution.

Light Scattering in Solutions of Larger Particles

When the size of a scattering particle exceeds about $\lambda/20$, different parts of the particle are exposed to incident light of different amplitude and phase. The scattered light is made up of waves coming from different parts of the particle and interfering with one another. Consequently, the scattered light intensity varies with the angle θ. Debye and others described this variation with a *particle scattering factor* $P(\theta)$ that depends upon the model selected to describe the scattering system. For spheres

$$P(\theta) = \left[\left(\frac{3}{u^3} \right) (\sin u - u \cos u) \right]^2 \tag{8-13}$$

where

$$u = 2\pi \left(\frac{d}{\lambda_s} \right) \sin \left(\frac{\theta}{2} \right)$$

d is the diameter of the spheres, and $\lambda_s = \lambda/n$ is the wavelength in the solution.

For a monodisperse system of randomly coiling polymers the radial density function is the Gaussian distribution, and the intensity of scattering is given by

$$P(\theta) = \left(\frac{2}{v^2} \right) [e^{-v} - (1 - v)] \tag{8-14}$$

where

$$v = (16\pi^2) \left(\frac{\bar{s^2}}{\lambda_s^2} \right) \sin^2 \left(\frac{\theta}{2} \right)$$

and $\bar{s^2}$ is the mean-square radius of gyration of the molecule.

Similar functions can be derived for other models, such as rods and discs. The functions $P(\theta)$ are plotted in Fig. 8-7. It can be seen from the curves that the

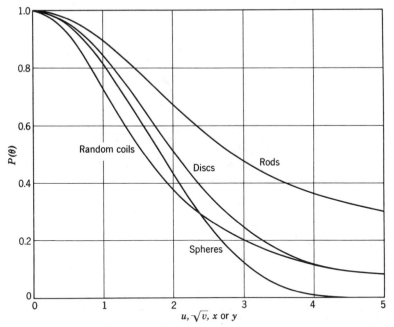

FIG. 8-7. The variation of the particle scattering function $P(\theta)$ with the size parameter u (spheres), v (monodisperse random coils), x (rods), or y (discs).

scattered intensity is greater in the forward than in the backward direction, falling off smoothly with angle except for rather large spheres. It is sometimes convenient to characterize the size of the scattering particles by the *dissymmetry* of light scattering, $z = i_{45°}/i_{135°}$, measured at two angles symmetrical about 90°. The dissymmetry increases rapidly with increasing particle size.

If the scattering particles are not small compared to the wavelength, as is usually the case with polymer solutions, Eq. (8-10) (in which M is replaced by $\bar{M}_w$ for application to polydisperse systems) is modified to insert $P(\theta)$:

$$K\frac{c}{\Delta R_{90}} = H\frac{c}{\Delta\tau} = \frac{1}{\bar{M}_w P(\theta)} + 2A_2c + \cdots \tag{8-15}$$

It has been shown (Zimm 1948a,b) that a z-average dimension is obtained from $P(\theta)$ for polydisperse systems.

In practice, the correction for dissymmetry can be made in two ways. In Debye's original method, the dissymmetry was measured and compared with the functions of Fig. 8-7 to evaluate particle size and thence $P(\theta)$. In more common current use is Zimm's method, in which the left-hand side of Eq. 8-15 is plotted (*Zimm plot*) against $\sin^2(\theta/2) + kc$, where k is an arbitrary constant. A rectilinear grid results (Fig. 8-8), allowing extrapolation to both $c = 0$ and $\theta = 0$, where $P(\theta) = 1$.

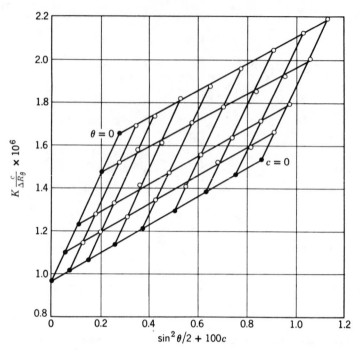

FIG. 8-8. Zimm plot showing the light scattering from a sample of polystyrene in butanone (Billmeyer 1955). Molecular parameters are $\bar{M}_w = 1,030,000$; $A_2 = 1.29 \times 10^{-4}$ ml-mole/g^2; $(\overline{s_z^2})^{1/2} = 460$ Å; $z = 1.44$.

Determination of Molecular Weight by Light Scattering

Photometers. Modern light-scattering instruments use either a mercury arc or a laser as a source and detect the scattered light photoelectrically. Many are interfaced to computers for control, data handling, and computation of results. The essential features of a "classical" light-scattering photometer are described with respect to the widely used Brice-Phoenix instrument (Brice 1950) (Fig. 8-9). Light from a mercury-arc source (S) passes through a lens (L), a polarizer (P₁), and a mono-chromatizing filter (F). It then strikes either a calibrated reference standard or a glass cell (C) holding the polymer solution. After passing through the cell, the primary light beam is absorbed in a light-trap tube (T). Scattered light from cell or standard is viewed by a multiplier phototube (R) after passing through a slit system D₁–D₂ and polarizer (P₂). The phototube is powered by high voltage from a regulated electronic power supply. Its output signal is transmitted either to a strip-chart recorder or to an analogue-to-digital converter for computer use.

A typical scattering cell consists of a glass cylinder having flat entrance and exit windows. The scattering cell is centered on the axis of rotation of the receiver phototube.

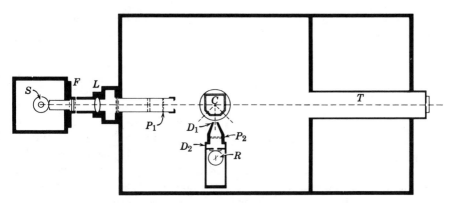

FIG. 8-9. Sketch of the essential components of a light-scattering photometer.

Sample Preparation. The proper preparation of a sample for light-scattering measurement is of major importance. Polymer molecules are usually small compared to particles of dust or other extraneous material. The scattering of these extraneous substances, while usually confined to low angles, may easily outweigh the scattering from the polymer solution. Consequently, solvents and polymer solutions must be clarified by filtration or by ultracentrifugation. Proper clarification, as evidenced by obtaining a rectilinear Zimm plot, is essential for the unambiguous interpretation of light-scattering data.

Accuracy in light-scattering measurement is favored by the proper choice of solvent. The difference in refractive index between polymer and solvent should be as large as possible. In addition, the solvent should itself have relatively low scattering, and the polymer-solvent system should not have too high a second virial coefficient, since the extrapolation to zero polymer concentration becomes less certain for high A_2. Mixed solvents must be avoided unless both components have the same refractive index.

Calibration. Although, in principle, observed scattered intensity can be related to turbidity through knowledge of the geometry of the photometer, calibration with substances of known turbidity is common practice. These substances include reflecting standards, colloidal suspensions, and simple liquids. Tungstosilicic acid, $H_4SiW_{12}O_{40}$, $M = 2879$, has been recommended as a primary calibration standard in preference to pure liquids (too sensitive to impurities) or uniform-particle-size latexes (too sensitive to residual polydispersity) (Kratohvil 1966, Billmeyer 1971, Levine 1976).

Treatment of Data. The weight-average molecular weight is the inverse of the intercept at $c = 0$ and $\theta = 0$ in the Zimm plot (Fig. 8-8). The second virial coefficient is calculated from the slope of the lines at constant angle by Eq. 8-15, and to a first approximation is independent of angle. The radius of gyration is derived from the slope of the zero-concentration line as a function of angle. The appropriate relationship, derived from Eq. 8-14, is

$$\overline{s_z^2} = \frac{3\lambda_s^2}{16\pi^2} \times \frac{\text{initial slope}}{\text{intercept}} \tag{8-16}$$

Range of Applicability. Light scattering from solutions has been used to measure molecular weights as low as that of sucrose and as high as those of proteins. In practice, polymer molecular weights on the order of 10,000–10,000,000 can readily be measured, with the possibility of extending the range in either direction in favorable cases. Applications include determination of absolute values of $\bar{M}_w$, especially for proteins and other biological materials, and calibration of intrinsic viscosity–molecular weight relations (Section D). It is not, however, possible to obtain information on particle shape or polydispersity from light-scattering measurements except under most unusual conditions. The method is applicable to copolymers only with severe restriction, owing to the difference in refractive index that generally exists between the two types of chain repeat units. Branched polymers can be measured without restriction.

Other Scattering Methods

Dynamic Light Scattering. This term, among others (Brillouin scattering, laser light scattering, laser spectroscopy, light-beating spectroscopy, photon correlation spectroscopy), is applied to a variety of related methods in which the spectrum of scattered light (secondary peaks, linewidths, etc.) is determined by both static and dynamic means. The most valuable additional information obtained is the rotational diffusion coefficient of the polymer (Cummins 1973, Chu 1974, Kinsinger 1975, Berne 1976, Shimoda 1976).

Neutron Scattering. In experiments entirely analogous to light scattering, the angular dependence of scattering at small angles can be observed when a monochromatic collimated beam of coherent "cold" neutrons is directed at a solid polymer sample (Allen 1976). Advantage is taken of the much greater scattering power of deuterons compared to protons to provide the equivalent of the refractive-index difference between polymer and solvent in light scattering: Fully deuterated polymer samples are prepared and studied in mixtures with normal, protonated samples. The major results are in the area of polymer conformations in the solid state (Benoit 1974, Richards 1978).

Light Scattering From Very Large Particles. The measurement by light scattering of the size of particles considerably larger than polymer molecules is discussed by van de Hulst (1957), Billmeyer (1964a), Kerker (1969), and Livesey (1969).

GENERAL REFERENCES

Billmeyer 1964a; Eskin 1964; McIntyre 1964; Kerker 1969; Huglin 1972; Collins 1973, Chapter 7B, Exps. 13 and 14; Casassa 1975; Morawetz 1975, Chapter 5C; Billingham 1977, Chapter 5; Elias 1977, Chapters 9.5 and 9.6; Hyde 1978; ASTM D4001.

D. ULTRACENTRIFUGATION

Ultracentrifugation techniques are the most intricate of the existing methods for determining the molecular weights of high polymers. They are far more successful in application to compact protein molecules than to random-coil polymers, where extended conformations increase deviations from ideality and the chances of mechanical entanglement of the polymer chains. Despite recent theoretical and experimental developments that increase the utility of the techniques for random-coil polymer systems, the methods are used overwhelmingly for biological materials.

Experimental Techniques

The ultracentrifuge consists of an aluminum rotor several inches in diameter that is rotated at high speed in an evacuated chamber. The solution being centrifuged is held in a small cell within the rotor near its periphery. The rotor may be driven electrically or by an oil or air turbine. The cell is equipped with windows, and the concentration of polymer along its length is determined by optical methods based on measurements of refractive index or absorption. Solvents for ultracentrifugation experiments must be chosen for difference from the polymer in both density (to ensure sedimentation) and refractive index (to allow measurement). In addition, mixed solvents must usually be avoided, and low solvent viscosity is desirable.

Sedimentation Equilibrium

In the sedimentation equilibrium experiment, the ultracentrifuge is operated at a low speed of rotation for times up to 1 or 2 weeks under constant conditions. A thermodynamic equilibrium is reached in which the polymer is distributed in the cell solely according to its molecular weight and molecular-weight distribution, the force of sedimentation on each species being just balanced by its tendency to diffuse back against the concentration gradient resulting from its movement in the centrifugal field. The force on a particle of mass m at a distance r from the axis of rotation is $\omega^2 r(1 - \bar{v}\rho)m$, where ω is the angular velocity of rotation, $\bar{v}$ is the partial specific volume of the polymer, and ρ is the density of the solution. Writing the partial molar free energy in terms of this force and applying the conditions for equilibrium gives the result, for ideal solutions,

$$\bar{M}_w = \frac{2RT \ln (c_2/c_1)}{(1 - \bar{v}\rho)\omega^2(r_2^2 - r_1^2)} \tag{8-17}$$

where c_1 and c_2 are the concentrations at two points r_1 and r_2 in the cell; r_1 and r_2 must be taken at the meniscus and the cell bottom to include all the molecular species in the average. If the data are treated somewhat differently, it can be shown that the *z-average molecular weight* $\bar{M}_z$ is obtained:

$$\bar{M}_z = \frac{\sum_{i=1}^{\infty} N_i M_i^3}{\sum_{i=1}^{\infty} N_i M_i^2} = \frac{\sum_{i=1}^{\infty} w_i M_i^2}{\sum_{i=1}^{\infty} w_i M_i} \tag{8-18}$$

The problem of nonideality is best met by working at the Θ temperature. In nonideal solutions the apparent molecular weight is a linear function of concentration at temperatures near Θ, the slope depending primarily on the second virial coefficient.

The major disadvantage to the sedimentation equilibrium experiment lies in the fact that several days may be required to reach equilibrium, even if the time is shortened by the use of short cells or a synthetic boundary cell to reduce the distance the various species must travel as equilibrium is established.

The distribution of molecular weights can be obtained directly from sedimentation-equilibrium measurements. The solution of an integral equation is involved, however, and the analysis remains a difficult one.

Equilibrium Sedimentation in a Density Gradient. In this approach a mixed solvent is chosen so that the relative sedimentation of the two components gives rise to a density gradient. The solute forms a band centering at the point where its effective density equals that of the solvent mixture. The band is Gaussian in shape with respect to solute concentration, the half-width being inversely proportional to the solute molecular weight. A feature that is of major importance is the sensitivity of the method to small differences in effective density among the solute species.

Approach to Equilibrium

Archibald (1947) showed that measurement of c and dc/dr at the cell boundaries allows molecular weight to be determined at any stage in the equilibrium process. If measurements are made early enough, before the molecular species have time to redistribute in the cell, the weight-average molecular weight and second virial coefficient can be evaluated. In practice, measurements can be made about 10–60 min after the start of the experiment.

Sedimentation Transport

In the *sedimentation transport* or *sedimentation velocity* experiment, the ultracentrifuge is operated at high speed so that the solute is transported to the bottom of the cell. The rate of sedimentation is given by the *sedimentation constant*

$$s = \frac{1}{\omega^2 r} \frac{dr}{dt} = \frac{m(1 - \bar{v}\rho)}{f} \tag{8-19}$$

and related to particle mass through a *frictional coefficient* f. For rigid particles f is well defined (for solid spheres of diameter d, f is given by Stokes' law, $f = 3\pi\eta d$), but for random-coil polymers the evaluation of f is more difficult. It can be related to the diffusion coefficient D: At infinite dilution $D = kT/f$, where

$$\frac{D}{s} = \frac{kT}{M(1 - \bar{v}\rho)} \tag{8-20}$$

However, even for a single solute species the equations have not been extended to take into account simultaneously the effects of diffusion in the ultracentrifuge cell, deviations from ideality, and compressibility of solvent and solute in the high centrifugal fields required. For polydisperse solutes additional complications arise in assigning average molecular weights, since D and s must be averaged separately. In general, simple averages such as $\bar{M}_w$ and $\bar{M}_z$ are not obtained (Elias 1975).

The most fruitful result of the sedimentation transport experiment now appears to be examination of the distribution of sedimentation constants. By working in a Θ solvent, the concentration dependence of s and D is reduced enough to avoid extrapolation to $c = 0$. The observed distribution of s over the cell must still be extrapolated to infinite time to eliminate the effects of diffusion relative to sedimentation, and correction for compression must be made. The resulting distribution of s can be converted to a distribution of molecular weight, either through detailed knowledge of the dependence of s and D on M or from the relation

$$s = KM^{1/2} \tag{8-21}$$

which is very nearly true at $T = \Theta$. The number of approximations and the complex treatment of the data required in this procedure suggest that its use be limited to qualitative observations. In this respect, however, the experiment can be extremely valuable, for it provides directly qualitative information about the nature of the distribution of species present.

Preparative Separations. The usefulness of the ultracentrifuge in the preparation of samples rather than in the production of analytic data should not be overlooked. Preparative ultracentrifuges have utility in fractionating polymer samples and in freeing them from easily sedimented impurities.

GENERAL REFERENCES

Svedberg 1940; Fujita 1962, 1975; Bowen 1970; Williams 1972; McCall 1973; Morawetz 1975, Chapters 4*B* and 6A; Scholte 1975; Billingham 1977, Chapter 6; Elias 1977, Chapter 9.7.

E. SOLUTION VISCOSITY AND MOLECULAR SIZE

The usefulness of solution viscosity as a measure of polymer molecular weight has been recognized ever since the early work of Staudinger (1930). Solution viscosity is basically a measure of the size or extension in space of polymer molecules. It is empirically related to molecular weight for linear polymers; the simplicity of the measurement and the usefulness of the viscosity–molecular weight correlation are so great that viscosity measurement constitutes an extremely valuable tool for the molecular characterization of polymers.

Experimental Methods

Measurements of solution viscosity are usually made by comparing the *effux time* t required for a specified volume of polymer solution to flow through a capillary tube with the corresponding effux time t_0 of the solvent. From t, t_0, and the solute concentration are derived several quantities whose defining equations and names are given in Table 8-4. Two sets of nomenclature are in use for these quantities; one (Cragg 1946) has had long and widespread application, the other (International Union 1952) was proposed for greater clarity and precision. In this book the common nomenclature is retained. In this system the concentration c is expressed in grams per deciliter (g/dl or g/100 ml).

The intrinsic viscosity $[\eta]$ is independent of concentration by virtue of extrapolation to $c = 0$, but is a function of the solvent used. The inherent viscosity at

TABLE 8-4. Nomenclature of Solution Viscosity

Common Name	Recommended Name	Symbol and Defining Equation
Relative viscosity	Viscosity ratio	$\eta_r = \eta/\eta_0 \simeq t/t_0$
Specific viscosity	—	$\eta_{sp} = \eta_r - 1 = (\eta - \eta_0)/\eta_0 \simeq (t - t_0)/t_0$
Reduced viscosity	Viscosity number	$\eta_{red} = \eta_{sp}/c$
Inherent viscosity	Logarithmic viscosity number	$\eta_{inh} = (\ln \eta_r)/c$
Intrinsic viscosity	Limiting viscosity number	$[\eta] = (\eta_{sp}/c)_{c=0} = [(\ln \eta_r)/c]_{c=0}$

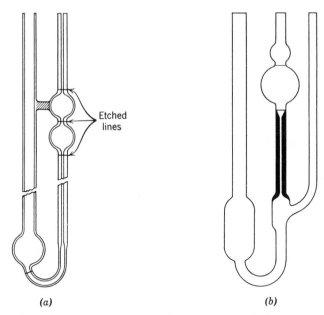

FIG. 8-10. Capillary viscometers commonly used for measurement of polymer solution viscosities: (a) Ostwald–Fenske and (b) Ubbelohde.

a specified concentration, usually 0.5 g/dl, is sometimes used as an approximation to [η].

Dilute-solution viscosity is usually measured in capillary viscometers of the Ostwald–Fenske or Ubbelohde type (Fig. 8-10a,b, respectively). The latter has the advantage that the measurement is independent of the amount of solution in the viscometer; measurements at a series of concentrations can easily be made by successive dilutions.

For highest precision, the following precautions are usually observed. The viscosity measurement is made in a constant-temperature bath regulated to at least ±0.02°C. The effux time is kept long (preferably more than 100 sec) to minimize the need for applying corrections to the observed data. For accuracy in extrapolating to $c = 0$, the solution concentration is restricted to the range that gives relative viscosities between 1.1 and 1.5.

Treatment of Data.   Viscosity data as a function of concentration are extrapolated to infinite dilution by means of the Huggins (1942) equation

$$\frac{\eta_{sp}}{c} = [\eta] + k'[\eta]^2 c \tag{8-22}$$

where k' is a constant for a series of polymers of different molecular weights in a given solvent. The alternative definition of the intrinsic viscosity leads to the equation (Kraemer 1938)

$$\frac{\ln \eta_r}{c} = [\eta] + k''[\eta]^2 c \qquad (8\text{-}23)$$

where $k' - k'' = \frac{1}{2}$. Typical data are shown in Fig. 8-11. The slopes of the lines vary as $[\eta]^2$, as required by Eqs. 8-22 and 8-23 for polymer species of different molecular weight in the same solvent.

At intrinsic viscosities above about 2, and even lower in some cases, there may be an appreciable dependence of viscosity on rate of shear in the viscometer. This dependence is not eliminated by extrapolation to infinite dilution; measurement as a function of shear rate and extrapolation to zero shear rate are also required. Special apparatus and extreme care are required (Zimm 1962).

Empirical Correlations Between Intrinsic Viscosity and Molecular Weight for Linear Polymers

Staudinger's prediction in 1930 that the reduced viscosity of a polymer is proportional to its molecular weight has needed only slight modification: The intrinsic viscosity has been substituted for the reduced viscosity, and it has been recognized that the proportionality is to a power of the molecular weight somewhat less than 1. The relation is expressed in the equation

$$[\eta] = K'M^a \qquad (8\text{-}24)$$

where K' and a are constants determined from a double logarithmic plot of intrinsic viscosity and molecular weight (Fig. 8-12). Such plots are usually found to be

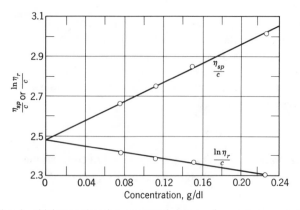

FIG. 8-11.   Reduced and inherent viscosity–concentration curves for a polystyrene in benzene (Ewart 1946).

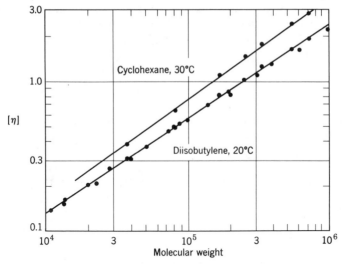

FIG. 8-12.   Intrinsic viscosity–molecular weight relationships for polyisobutylene in diisobutylene and cyclohexane (Flory 1943; Krigbaum 1953).

straight lines within experimental error over large ranges of the variables. For randomly coiled polymers the exponent a varies from 0.5 in a Θ solvent to a maximum of about 1.0. For many systems a lies between 0.6 and 0.8. Typical values of K' range between 0.5 and 5×10^{-4}. Both K' and a are functions of the solvent as well as of the polymer type. This empirical relation between viscosity and molecular weight is valid only for linear polymers.

Intrinsic-viscosity measurement leads to the *viscosity-average molecular weight*, defined as

$$\bar{M}_v = \left(\sum_{i=1}^{\infty} w_i M_i^a \right)^{1/a} = \left(\frac{\sum_{i=1}^{\infty} N_i M_i^{1+a}}{\sum_{i=1}^{\infty} N_i M_i} \right)^{1/a} \tag{8-25}$$

Thus $\bar{M}_v$ depends on a as well as on the distribution of molecular species. For many polymers $\bar{M}_v$ is 10–20% below $\bar{M}_w$. For $a = 1$, $\bar{M}_v = \bar{M}_w$. If fractionated polymers, for which $\bar{M}_n \simeq \bar{M}_v \simeq \bar{M}_w$, are available, any absolute molecular-weight measurement may be combined with a viscosity measurement to evaluate the constants in Eq. 8-24. If the molecular-weight distribution of the samples is known in sufficient detail to calculate $\bar{M}_v$ from a molecular-weight average that has been measured, the constants in the equation

$$[\eta] = K' \bar{M}_v^a \tag{8-26}$$

may be evaluated. Since $\bar{M}_v$ is nearer to $\bar{M}_w$ than to $\bar{M}_n$, the weight-average molecular weight is preferred for this calibration. It can be shown that polymers with fairly broad but well-known distributions are preferred to fractions for this purpose. A less precise relation,

$$[\eta] = K' \bar{M}_w{}^a \tag{8-27}$$

is sometimes used to relate weight-average molecular weight directly to intrinsic viscosity for a limited series of samples. Extensive tabulations of K' and a are available (Kurata 1975).

Equation 8-27 is known to become inaccurate for molecular weights below about 50,000, because deviations from the linear relationship set in. For better results in this region, and with some theoretical justification (Stockmayer 1963), use of one of several expressions of the form

$$[\eta] = KM^{1/2} + K''M \tag{8-28}$$

is recommended. Here the first term is determined by short-range interactions, as in Eq. 8-31, and the second term by long-range interactions. The form and value of K'' vary among the several theories available (Cowie 1966).

Intrinsic Viscosity and Molecular Size

Theories (Flory 1949, 1953) of the frictional properties of polymer molecules in solution show that the intrinsic viscosity is proportional to the effective hydrodynamic volume of the molecule in solution divided by its molecular weight. The effective volume is proportional to the cube of a linear dimension of the randomly coiling chain. If $(\bar{r^2})^{1/2}$ is the dimension chosen,

$$[\eta] = \Phi \frac{(\bar{r^2})^{3/2}}{M} \tag{8-29}$$

where Φ is a universal constant. Replacing $(\bar{r^2})^{1/2}$ by $\alpha(\bar{r_0^2})^{1/2}$ and recalling that $\bar{r_0^2}/M$ is a function of chain structure independent of its surroundings or molecular weight, it follows that

$$[\eta] = \Phi \left(\frac{r_0^2}{M} \right)^{3/2} M^{1/2} \alpha^3 = KM^{1/2} \alpha^3 \tag{8-30}$$

where $K = \Phi(\bar{r_0^2}/M)^{3/2}$ is a constant for a given polymer, independent of solvent and molecular weight. The dimensions $(\bar{r^2})^{1/2}$ and $(\bar{r_0^2})^{1/2}$ and the expansion factor α used here are usually identified with those discussed in Chapter 7. Extensions of the Flory–Krigbaum treatment (Chapter 7C) imply that Φ is not a universal constant, and that the power of α in Eq. 8-30 is somewhat less than 3 (Kurata 1960, 1963).

It follows from the properties of α that, at $T = \Theta$, $\alpha = 1$ and

$$[\eta]_\Theta = KM^{1/2} \tag{8-31}$$

There is ample experimental evidence for the validity of this equation. Values of K are near 1×10^{-3} for a number of polymer systems.

An important feature of Flory's viscosity theory is that it furnishes confirmation (if not evaluation) of the Θ temperature as that at which $a = \frac{1}{2}$, and it allows the determination of the unperturbed dimensions of the polymer chain. Even if a Θ solvent is not available, several extrapolation techniques (summarized by Cowie 1966) are available for estimating the unperturbed dimensions from viscosity data in good solvents. The best and simplest of these techniques appears to be that of Stockmayer (1963) based on Eq. 8-28.

Determination of Chain Branching

The size of a dissolved polymer molecule, which can be estimated by means of its intrinsic viscosity, depends upon a number of factors. These include the molecular weight, the local chain structure as reflected in the factor $\overline{r_0^2}/M$, the perturbations reflected in α, and finally the gross chain structure, in which the degree of long-chain branching is an important variable. [The presence of short branches, such as ethyl or butyl groups in low-density polyethylene (Chapter 13A), has little effect on solution viscosity.] By evaluating or controlling each of the other factors, viscosity measurements may be used to estimate the degree of long-chain branching of a polymer.

The relation between molecular size and the number and type of branch points has been calculated (Zimm 1949), leading to a relation g between the radii of gyration for branched and linear chains of the same numbers of segments:

$$g = \frac{\overline{s^2} \text{ (branched)}}{\overline{s^2} \text{ (linear)}} \tag{8-32}$$

Several theories have been proposed relating g to $[\eta]$ (branched)/$[\eta]$ (linear). Only recently has it been possible to decide which of these yields the best fit to experimental data; the situation is well summarized by Graessley (1968). At present it still seems appropriate to consider the ratio $[\eta]$ (branched)/$[\eta]$ (linear) evaluated at constant molecular weight as a qualitative indication of chain branching rather than to attempt to assign numerical values to the degree of chain branching in a given polymer sample.

GENERAL REFERENCES

Kurata 1963; Lyons 1967; Moore 1967; ASTM D2857; Collins 1973, Chapter 7C, Exp. 15; Carpenter 1975; Billingham 1977, Chapter 7; Elias 1977, Chapter 9.9.

F. GEL PERMEATION CHROMATOGRAPHY

Gel permeation chromatography, more correctly termed *size exclusion chromatography*, is a separation method for high polymers, similar to but advanced in practice over *gel filtration* as carried out by biochemists, that has become a prominent and widely used method for estimating molecular-weight distributions since its discovery (Moore 1964) just over two decades ago in 1961. The separation takes place in a chromatographic column filled with beads of a rigid porous ''gel''; highly cross-linked porous polystyrene and porous glass are preferred column-packing materials. The pores in these gels are of the same size as the dimensions of polymer molecules.

A sample of a dilute polymer solution is introduced into a solvent stream flowing through the column. As the dissolved polymer molecules flow past the porous beads

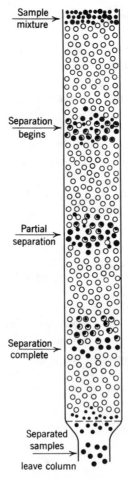

Sample
mixture

Separation
begins

Partial
separation

Separation
complete

Separated
samples

leave column

FIG. 8-13.   Principle of the separation of molecules according to size by gel permeation chromatography (after Cazes 1966 and Billmeyer 1969).

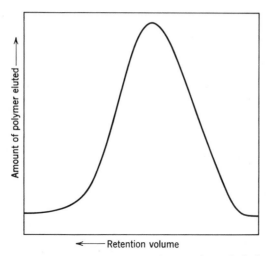

FIG. 8-14. A typical gel permeation chromatogram: polystyrene in tetrahydrofuran, with $\bar{M}_w/\bar{M}_n$ = 2.9.

(Fig. 8-13), they can diffuse into the internal pore structure of the gel to an extent depending on their size and the pore-size distribution of the gel. Larger molecules can enter only a small fraction of the internal portion of the gel, or are completely excluded; smaller polymer molecules penetrate a larger fraction of the interior of the gel.

The larger the molecule, therefore, the less time it spends inside the gel, and the sooner it flows through the column. The different molecular species are eluted from the column in order of their molecular size (Benoit 1966) as distinguished from their molecular weight, the largest emerging first.

A complete theory predicting retention times or volumes as a function of molecular size has not been formulated for gel permeation chromatography. A specific column or set of columns (with gels of different pore sizes) is calibrated empirically to give such a relationship, by means of which a plot of amount of solute versus retention volume (the chromatogram, Fig. 8-14) can be converted into a molecular-size-distribution curve. For convenience, commercially available narrow-distribution polystyrenes are often used. If the calibration is made in terms of a molecular-size parameter, for example, $[\eta]M$ (whose relation to size is given by Eq. 8-29), it can be applied to a wide variety of both linear and branched polymers (Fig. 8-15).

As in all chromatographic processes, the band of solute emerging from the column is broadened by a number of processes, including contributions from the apparatus, flow of the solution through the packed bed of gel particles, and the permeation process itself (Kelley 1970). Corrections for this zone broadening can be made empirically; it usually becomes unimportant when the sample has $\bar{M}_w/\bar{M}_n > 2$.

Gel permeation chromatography is extremely valuable for both analytic and preparative work with a wide variety of systems ranging from low to very high

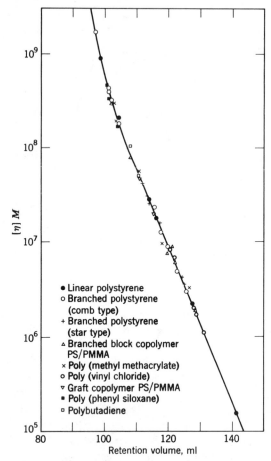

FIG. 8-15. Calibration curve for gel permeation chromatography based on hydrodynamic volume as expressed by the product $[\eta]M$ (Grubisic 1967). Among the polymer types shown are linear polystyrene, two types of branched polystyrene, poly(methyl methacrylate), poly(vinyl chloride), polybutadiene, poly(phenyl siloxane), and two types of copolymer.

molecular weights. The method can be applied to a wide variety of solvents and polymers, depending on the type of gel used. With polystyrene gels, relatively nonpolar polymers can be measured in solvents such as tetrahydrofuran, toluene, or (at high temperatures) *o*-dichlorobenzene; with porous glass gels, more polar systems, including aqueous solvents, can be used. A few milligrams of sample suffices for analytic work, and the determination is complete in as short a time as a few minutes using modern high-pressure, high-speed equipment.

The results of careful gel permeation chromatography experiments for molecular-weight distribution agree so well with results from other techniques that there is serious doubt as to which is correct when residual discrepancies occur. Figure 8-16 shows the extent of agreement between this method and a solvent-gradient elution

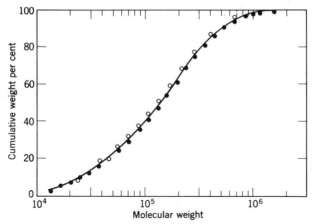

FIG. 8-16. Typical cumulative molecular-weight distribution curve for a sample of polypropylene (Crouzet 1969): gradient-elution data (○) and data from gel permeation chromatography (●). Molecular weight is plotted logarithmically because of the very broad distribution in this sample.

fractionation, while Fig. 8-17 demonstrates the degree of fit between the experiment and a distribution curve calculated from polymerization kinetics.

GENERAL REFERENCES

Altgelt 1971; Collins 1973, Chapter 7D, Exp. 15; Ouano 1975; ASTM D3536; Billingham 1977, Chapter 8; Tung 1977; Dawkins 1978; Yau 1979; ASTM D3593.

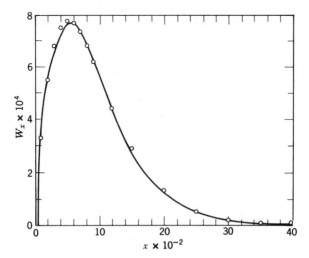

FIG. 8-17. Fit of gel permeation chromatography data for polystyrene to a molecular-weight distribution curve calculated from polymerization kinetics (May 1968).

G. POLYELECTROLYTES

Polymers with ionizable groups along the chain, termed polyelectrolytes, normally exhibit properties in solution that are quite different from those with nonionizable structures. There are many examples of polyelectrolytes, including polyacids such as poly(acrylic acid) and hydrolyzed copolymers of maleic anhydride, polybases such as poly(vinyl amine) and poly(4-vinyl pyridine), polyphosphates, nucleic acids, and proteins.

When they are soluble in nonionizing solvents—for example, poly(acrylic acid) in dioxane—polyelectrolytes behave in completely normal fashion, but in aqueous solution they are ionized, with three major results. First, the mutual repulsion of their charges causes expansions of the chain far beyond those resulting from changes from good to poor solvents with ordinary polymers. The size of the polyelectrolyte random coil is, moreover, a function of the concentrations of polymer and added salt, if any, since both influence the degree of ionization.

Second, the ionization of the electrolyte groups leads to a variety of unusual effects in the presence of small amounts of added salt. The intensity of light scattering decreases because of the ordering of the molecules in solution, while the osmotic pressure and ultracentrifugation behavior are determined predominantly by the total charge on the molecule (Donnan effect).

Finally, the ionic charges attached to the chains create regions of high local charge density, affecting the activity coefficients and properties of small ions in these localities. Although the various effects cannot be separated completely, the results of chain expansion are of primary interest for the measurement of molecular weight and size.

Those properties depending on the size of the chain, such as viscosity and angular dependence of light scattering, are strongly affected by chain expansion. The viscosity may even increase markedly as polymer concentration *decreases*, with consequent increase in the degree of ionization of the polymer. When very high chain extensions are reached (up to half of the fully extended chain length), the effect reverses, but it does not disappear at infinite dilution. On the other hand, the addition of low-molecular-weight electrolyte (salt) to the aqueous solution increases the ionic strength of the solution outside the polymer coil relative to that inside, and also reduces the thickness of the layer of "bound" counterions around the chain. Both effects cause the chain to contract, and when the concentration of added salt reaches, say, 0.1 M, behavior is again normal. With some special precautions, molecular weights may be measured by light scattering and equilibrium ultracentrifugation, and intrinsic viscosity–molecular weight relations may be established for polyelectrolytes in the presence of added salt.

Because of the preponderance of small ions, the colligative properties of polyelectrolytes in ionizing solvents measure counterion activities rather than molecular weight. In the presence of added salt, however, correct molecular weights of polyelectrolytes can be measured by membrane osmometry, since the small ions can equilibrate across the membrane. The second virial coefficient differs from that previously defined, since it is determined by both ionic and nonionic polymer–solvent interactions.

The transport and electrophoretic properties of polyelectrolytes are beyond the scope of this book.

GENERAL REFERENCES

Armstrong 1969; Morawetz 1975, Chapter 7; Anderson 1982.

DISCUSSION QUESTIONS AND PROBLEMS

1. Discuss the value of knowledge of the molecular weight and distribution of a polymer to the plastics engineer.

2. Which methods would you use to obtain this information on a routine basis, as in process control? Why?

3. Which methods would you use to obtain this information for a new polymer type not previously studied? Why?

4. Of what value is the ability to make measurements by such absolute methods as membrane osmometry or light scattering? Review your answers to the previous questions, assuming that these absolute methods were not available, and discuss.

5. You wish to determine $\bar{M}_n$ for a polymer, but find it too high to be measured by vapor-phase osmometry, yet, when using membrane osmometry, there is significant diffusion through the membrane. Outline two different ways of obtaining the desired information, and discuss their relative merits.

6. If light scattering and osmotic pressure are measured in the same solvent, compare the slopes of the plots of $1/M$ versus c.

7. Over what molecular-weight range can light scattering be used to measure $\bar{M}_w$, and what limitations set this range?

8. What types of ultracentrifugation experiment are suitable for random-coil polymers?

9. What molecular-weight method would be suitable for a polymer with an intrinsic viscosity of about 10 dl/g?

10. What kind of information would light scattering give you about the polymer of Question 9?

11. What molecular-weight technique is best for calibrating $[\eta]-M$ relations?

12. What kind of information does one obtain from measuring $[\eta]$?

13. How can you make the best guess at M from measurements of $[\eta]$ on an entirely new kind of polymer?

14. Under what conditions can you use $[\eta]$ to measure $\overline{r_0^2}/M$?

15. How can $\overline{r_0^2}/M$ be used to measure chain branching?

16. Calculate the end-to-end distance of a polymer with molecular weight 1,000,000 and intrinsic viscosity of 2.10 dl/g; take $\Phi = 2.1 \times 10^{21}$. What can you say about the light-scattering behavior of such a polymer?

17. Light scattering and osmometry were performed on a solution of a polymer in a Θ solvent at 27°C. Concentration was 0.01 g/liter. Take $R = 1 \times 10^5$, $K = 0.4 \times 10^{-5}$ for these units, $\pi = 3.5$ cm of solvent, and $R_\theta = 0.01$ cm^{-1}. Calculate $\bar{M}_w$ and $\bar{M}_n$ and comment on their ratio.

18. A polymer with $M = 100,000$ obeys the Mark–Houwink equation with $K = 1 \times 10^{-4}$ and $a = 0.80$. Huggins' constant is 0.33. Calculate the relative viscosity at $c = 0.30$ g/dl.

19. A sample of polyethylene was studied by vapor-phase osmometry in o-dichlorobenzene at 130°C. Data for the polymer and for tristearin, $M = 891$, are given below. Calculate $\bar{M}_n$ and A_2 for the polyethylene sample.

Polyethylene		Tristearin	
c (g/liter)	Δr	c (g/liter)	Δr
12.40	0.61	4.41	3.41
20.36	1.44	10.27	7.83
30.54	2.89	15.29	11.84
40.17	4.90	19.88	14.97

20. The following freezing-point data were obtained for a sample of polyethylene in hexamethylbenzene at 165°C:

Polymer Added (g)	Chart Divisions From Solvent Freezing Point
0.428	60
0.739	103
1.055	148
1.348	189

For convenience, a calibration was made with octacosane, $M = 395$, that gave a freezing-point depression of 1223 divisions per gram in the same solvent. The density of hexamethylbenzene at the freezing point may be taken as 0.85, to a first approximation. Calculate $\bar{M}_n$ and A_2.

21. The following data were obtained by membrane osmometry of solutions of a polyethylene sample in xylene at 90°C. At this temperature the density of xylene is 0.8014 and that of polyethylene is 0.8173. Calculate the molecular weight of the sample and the second virial coefficient.

c (g/liter):	2.00	4.00	6.00	8.00
π (cm xylene):	2.585	5.965	9.660	13.935

22. Angular-light-scattering measurements on solutions of a nonionic polymer in water at $\lambda = 4359$ Å gave the following results for ΔR_θ as a function of the angle of observation θ and concentration c:

	$\Delta R_\theta \times 10^4$ for c (g/liter)			
θ	0.315	0.690	1.115	1.620
25	1.87	3.86	5.79	7.75
30	1.77	3.64	5.45	7.32
35	1.64	3.39	5.10	6.85
45	1.39	2.88	4.36	5.90
60	1.05	2.18	3.32	4.54
75	0.810	1.69	2.59	3.52
90	0.685	1.44	2.22	3.05
105	0.665	1.41	2.19	3.04
120	0.720	1.52	2.37	3.31
135	0.813	1.72	2.70	3.76

If $n = 1.340$ and $dn/dc = 0.192$ ml/g, calculate $\bar{M}_w$, A_2, and $\overline{s_z^2}$.

23. Determination of the sedimentation coefficient of a polymer in aqueous solution at a concentration c led to the following data, where r is the distance of the boundary from the axis of rotation and t is the time of centrifugation in minutes:

t (min)	r (cm)
5	6.560
10	6.575
15	6.591
20	6.609
25	6.622
30	6.640
35	6.654
40	6.670

The speed of centrifugation was 59,780 rpm, the temperature 20°C. Calculate the sedimentation coefficient.

24. The following solution-viscosity data were obtained for samples of polystyrene in 2-butanone at 25°C, in which successive increments of a stock solution of polymer were added to solvent in a Ubbelohde viscometer. Finally, the total weight of polymer in the viscometer was measured.

	Sample				
	A	B	C	D	E
Efflux time (s) for					
20 ml of solvent	209.3	223.6	209.0	213.7	208.2
20 ml solvent + 5 ml stock	225.0	241.7	235.0	223.2	234.3
20 ml solvent + 10 ml stock	236.3	255.9	254.3	229.8	253.3
20 ml solvent + 15 ml stock	245.1	265.2	268.7	234.4	267.8
20 ml solvent + 20 ml stock	251.7	273.5	280.4	238.1	279.2
Total grams of polymer in viscometer at end of run	0.7644	0.6326	0.5802	0.2693	0.4033
$\bar{M}_w$ from light scattering	17,000	30,000	70,000	83,000	136,000

Is there evidence for chain branching in any of the samples? Explain and discuss.

25. The following data were obtained by gel permeation chromatography in tetrahydrofuran at 25°C of a sample of poly(methyl methacrylate):

Retention Volume (ml)	Recorder Chart Divisions Above Base Line
130	0.5
135	6.0
140	25.7
145	44.5
150	42.0
155	25.6
160	8.9
165	2.2

Samples of narrow-distribution polystyrene chromatogrammed under the same conditions gave a linear calibration curve with $M = 98,000$ eluting at 130 ml and $M = 1800$ eluting at 165 ml.

a. Calculate $\bar{M}_n$ and $\bar{M}_w$, assuming the above calibration applies.

b. Comment on possible systematic errors in the calibration and outline in detail a better method. Specify what additional information you would need to apply it.

c. Derive the equation defining the type of molecular-weight average obtained in the "universal" calibration method in gel permeation chromatography.

BIBLIOGRAPHY

Allen 1976. G. Allen, "A Review of Neutron Scattering with Special Reference to the Measurement of the Unperturbed Dimensions in Macromolecules," Chapter 1 in K. J. Ivin, ed., *Structural Studies of Macromolecules by Spectroscopic Methods,* John Wiley & Sons, New York, 1976.

Altgelt 1971. Klaus H. Altgelt and Leon Segal, eds., *Gel Permeation Chromatography,* Marcel Dekker, New York, 1971.

Anderson 1982. Charles F. Anderson and Herbert Morawetz, "Polyelectrolytes," pp. 495–530 in Martin Grayson, ed., *Kirk–Othmer Encyclopedia of Chemical Technology,* 3rd ed., Vol. 18, Wiley-Interscience, New York, 1982.

Archibald 1947. W. J. Archibald, "A Demonstration of Some New Methods of Determining Molecular Weights from the Data of the Ultracentrifuge," *J. Phys. Colloid Chem.* **51,** 1204–1214 (1947).

Armstrong 1969. R. W. Armstrong and U. P. Strauss, "Polyelectrolytes," pp. 781–861 in Herman F. Mark, Norman G. Gaylord, and Norbert M. Bikales, eds., *Encyclopedia of Polymer Science and Technology,* Vol. 10, Wiley–Interscience, New York, 1969.

ASTM D2857. Standard Test Method for *Dilute Solution Viscosity of Polymers,* ASTM-ANSI Designation: D2857, American Society for Testing and Materials, Philadelphia, Pennsylvania, 1970.

ASTM D3536. Standard Test Method for *Molecular Weight Averages and Molecular Weight Distribution of Polystyrene by Liquid Exclusion Chromatography (Gel Permeation Chromatography—GPC),* ASTM-ANSI Designation: D3536, American Society for Testing and Materials, Philadelphia, Pennsylvania, 1976.

ASTM D3592. Standard Recommended Practice for *Determining Molecular Weight by Vapor Pressure Osmometry,* ASTM-ANSI Designation: D3592, American Society for Testing and Materials, Philadelphia, Pennsylvania, 1977.

ASTM D3593. Standard Test Method for *Molecular Weight Averages and Molecular Weight Distribution of Certain Polymers by Liquid Size-Exclusion Chromatography (Gel Permeation Chromatography—GPC) Using Universal Calibration,* ASTM Designation: D3593, American Society for Testing and Materials, Philadelphia, Pennsylvania, 1980.

ASTM D3750. Standard Practice for *Determination of Number-Average Molecular Weight of Polymers by Membrane Osmometry,* ASTM-ANSI Designation: D3750, American Society for Testing and Materials, Philadelphia, Pennsylvania, 1979.

ASTM D4001. Standard Practice for *Determination of Weight-Average Molecular Weight of Polymers by Light Scattering,* ASTM-ANSI Designation: D4001, American Society for Testing and Materials, Philadelphia, Pennsylvania, 1981.

Badgley 1949. W. J. Badgley and H. Mark, "Osmometry and Viscometry of Polymer Solutions," pp. 75–112 in R. E. Burk and Oliver Grummit, eds., *High Molecular Weight Compounds (Frontiers in Chemistry),* Vol. 6, Interscience, New York, 1949.

Benoit 1966. Henri Benoit, Zlatka Grubisic, Paul Rempp, Danielle Decker, and Jean-Georges Zilliox, "Study by Liquid-Phase Chromatography of Linear and Branched Polystryrenes of Known Structure" (in French), *J. Chem. Phys.* **63**, 1507–1514 (1966).

Benoit 1974. H. Benoit, J. P. Cotton, D. Decker, B. Farnoux, J. S. Higgins, G. Jannink, R. Ober, C. Picot, and J. des Cloizeaux, "Conformation of Polymer Chains in the Bulk," *Macromolecules* **7**, 863–871 (1974).

Berne 1976. Bruce J. Berne and Robert Pecora, *Dynamic Light Scattering,* John Wiley & Sons, New York, 1976.

Billingham 1977. N. C. Billingham, *Molar Mass Measurements in Polymer Science,* Kogan Page Ltd. and Halsted Press, John Wiley & Sons, New York, 1977.

Billmeyer 1955. F. W. Billmeyer, Jr., and C. B. de Than, "Dissymmetry of Molecular Light Scattering in Polymethyl Methacrylate," *J. Am. Chem. Soc.* **77**, 4763–4767 (1955).

Billmeyer 1964a. Fred W. Billmeyer, Jr., "Principles of Light Scattering," Chapter 56 in I. M. Kolthoff and Philip J. Elving, eds., with the assistance of Ernest B. Sandell, *Treatise on Analytical Chemistry,* Part I, Vol. 5, Wiley-Interscience, New York, 1964.

Billmeyer 1964b. F. W. Billmeyer, Jr., and V. Kokle, "The Molecular Structure of Polyethylene. XV. Comparison of Number-Average Molecular Weights by Various Methods," *J. Am. Chem. Soc.* **86**, 3544–3546 (1964).

Billmeyer 1969. Fred W. Billmeyer, Jr., "Molecular Structure and Polymer Properties," *J. Paint Technol.* **41**, 3–16, 209 (1969).

Billmeyer 1971. Fred W. Billmeyer, Jr., Harold I. Levine, and Peter J. Livesey, "The Refraction Correction in Light Scattering," *J. Colloid Interface Sci.* **35**, 204–214 (1971).

Billmeyer 1976. Fred W. Billmeyer, Jr., "Trends in Polymer Characterization," *J. Polym. Sci. Symp.* **55**, 1–10 (1976).

Billmeyer 1977. Fred W. Billmeyer, Jr., "The Size and Weight of Polymer Molecules," Chapter 4 in Herman S. Kaufman and Joseph J. Falcetta, eds., *Introduction to Polymer Science and Technology—An SPE Textbook,* John Wiley & Sons, New York, 1977.

Bonnar 1958. R. U. Bonnar, M. Dimbat, and F. H. Stross, *Number-Average Molecular Weights,* Interscience, New York, 1958.

Bowen 1970. T. J. Bowen, *An Introduction to Ultracentrifugation,* Wiley-Interscience, New York, 1970.

Brice 1950. B. A. Brice, M. Halwer, and R. Speiser, "Photoelectric Light-Scattering Photometer for Determining High Molecular Weights," *J. Opt. Soc. Am.* **40**, 768–778 (1950).

Brice 1951. B. A. Brice and M. Halwer, "A Differential Refractometer," *J. Opt. Soc. Am.* **41**, 1033–1037 (1951).

Burge 1977. David E. Burge, "Molecular Weight Determination by Osmometry," *Am. Lab.* **9** (b), 41–51 (1977).

Cabannes 1929. J. Cabannes and Y. Rocard, *La diffusion moléculaire de la lumière,* Presses Universitaires de France, Paris, 1929.

Carpenter 1975. Dewey K. Carpenter and Lowell Westerman, "Viscometric Methods of Studying Molecular Weight and Molecular Weight Distribution," Chapter 7 in Philip E. Slade, Jr., ed., *Polymer Molecular Weights,* Part I, Marcel Dekker, New York, 1975.

Casassa 1975. Edward F. Casassa and Guy C. Berry, "Light Scattering from Solutions of Macromolecules," Chapter 5 in Philip E. Slade, Jr., ed., *Polymer Molecular Weights,* Part I, Marcel Dekker, New York, 1975.

Cazes 1966. Jack Cazes, "Topics in Chemical Instrumentation. XXIX. Gel Permeation Chromatography," *J. Chem. Educ.* **43**, A567–A582, A625–A642 (1966).

Chu 1974. Benjamin Chu, *Laser Light Scattering,* Academic Press, New York, 1974.

Coll 1968. Hans Coll and F. H. Stross, "Determination of Molecular Weights by Equilibrium Osmotic-Pressure Measurements," pp. 10–27 in Donald McIntyre, ed., *Characterization of Macromolecular Structure,* Publication No. 1573, National Academy of Sciences, Washington, D.C., 1968.

Collins 1973. Edward A. Collins, Jan Bareš, and Fred W. Billmeyer, Jr., *Experiments in Polymer Science,* Wiley-Interscience, New York, 1973.

Cowie 1966. J. M. G. Cowie, "Estimation of Unperturbed Polymer Dimensions from Viscosity Measurements in Non-Ideal Solvents," *Polymer* **7**, 487–495 (1966).

Cragg 1946. L. H. Cragg, "The Terminology of Intrinsic Viscosity and Related Functions," *J. Colloid Sci.* **1**, 261–269 (1946).

Crouzet 1969. P. Crouzet, F. Fine, and P. Magnin, "Comparison of the Gel Permeation Chromatographic and Gradient Elution Fractionation Methods for the Determination of the Molecular Weight Distribution of Polypropylene" (in French), *J. Appl. Polym. Sci.* **13**, 205–213 (1969).

Cummins 1973. H. Z. Cummins and E. R. Pike, eds., *Photon Correlation and Light Beating Spectroscopy,* Plenum Press, New York, 1973.

Dawkins 1978. J. V. Dawkins and G. Yeadon, "High Performance Gel Permeation Chromatography," Chapter 3 in J. V. Dawkins, ed., *Developments in Polymer Characterisation—1,* Applied Science, London, 1978.

Debye 1944. P. Debye, "Light Scattering in Solutions," *J. Appl. Phys.* **15**, 338–342 (1946).

Debye 1947. P. Debye, "Molecular-Weight Determination by Light Scattering," *J. Phys. Colloid Chem.* **51**, 18–32 (1947).

Dobry 1935. A. Dobry, "Apparatus for the Measurement of Very Small Osmotic Pressures in Colloidal Solutions" (in French), *J. Chim. Phys.* **32**, 46–49 (1953).

Elias 1975. Hans-Georg Elias, "Polymolecularity and Polydispersity in Molecular Weight Determinations," *Pure Appl. Chem.* **43**, 115–147 (1975).

Elias 1977. Hans-Georg Elias, *Macromolecules·1·Structure and Properties* (translated by John W. Stafford), Plenum Press, New York, 1977.

Eskin 1964. V. E. Eskin, "Light Scattering as a Method of Studying Polymers," *Sov. Phys. Usp.* **7**, 270–304 (1964).

Ewart 1946. R. H. Ewart, "Significance of Viscosity Measurements on Dilute Solutions of High Polymers," pp. 197–251 in H. Mark and G. S. Whitby, eds., *Scientific Progress in the Field of Rubber and Synthetic Elastomers (Advances in Colloid Science)* Vol. 2, Interscience, New York, 1946.

Ezrin 1968. Myer Ezrin, "Determination of Molecular Weight by Ebulliometry," pp. 3–9 in Donald McIntyre, ed., *Characterization of Macromolecular Structure*, Publication No. 1573, National Academy of Sciences, Washington, D.C., 1968.

Flory 1943. Paul J. Flory, "Molecular Weights and Intrinsic Viscosities of Polyisobutylenes," *J. Am. Chem. Soc.* **65**, 372–382 (1943).

Flory 1949. Paul J. Flory, "The Configuration of Real Polymer Chains," *J. Chem. Phys.* **17**, 303–310 (1949).

Flory 1953. Paul J. Flory, *Principles of Polymer Chemistry*, Cornell University Press, Ithaca, New York, 1953.

Fujita 1962. Hiroshi Fujita, *Mathematical Theory of Sedimentation Analysis*, Academic Press, New York, 1962.

Fujita 1975. Hiroshi Fujita, *Foundations of Ultracentrifugal Analysis*, Wiley-Interscience, New York, 1975.

Garmon 1975. Ronald G. Garmon, "End Group Determinations," Chapter 3 in Philip E. Slade, Jr., ed., *Polymer Molecular Weights*, Part I, Marcel Dekker, New York, 1975.

Gee 1944. Geoffrey Gee, "The Molecular Weights of Rubber and Related Materials. V. The Interpretation of Molecular Weight Measurements on High Polymers," *Trans. Faraday Soc.* **40**, 261–266 (1944).

Glover 1966. Clyde A. Glover, "Determination of Molecular Weights by Ebulliometry," pp. 1–67 in Charles N. Reilley and Fred W. McLafferty, eds., *Advances in Analytical Chemistry and Instrumentation*, Vol. 5, Wiley-Interscience, New York, 1966.

Glover 1975. Clyde A. Glover, "Absolute Colligative Property Methods," Chapter 4 in Philip E. Slade, Jr., ed., *Polymer Molecular Weights*, Part I, Marcel Dekker, New York, 1975.

Graessley 1968. William W. Graessley, "Detection and Measurement of Branching in Polymers," pp. 371–388 in Donald McIntyre, ed., *Characterization of Macromolecular Structure*, Publication No. 1573, National Academy of Sciences, Washington, D.C., 1968.

Grubisic 1967. Z. Grubisic, P. Rempp, and H. Benoit, "A Universal Calibration for Gel Permeation Chromatography," *J. Polym. Sci.* **B5**, 753–759 (1967).

Hellman 1962. Max Hellman and Leo A. Wall, "End-Group Analysis," Chapter 5 in Gordon M. Kline, ed., *Analytical Chemistry of Polymers*, Part III, Wiley-Interscience, New York, 1962.

Holleran 1968. Peter M. Holleran and Fred W. Billmeyer, Jr., "Rapid Osmometry with Diffusible Polymers," *J. Polym. Sci.* **B6**, 137–140 (1968).

Huggins 1942. Maurice L. Huggins, "The Viscosity of Dilute Solutions of Long-Chain Molecules. IV. Dependence on Concentration," *J. Am. Chem. Soc.* **64**, 2716–2718 (1942).

Huglin 1972. M. B. Huglin, ed., *Light Scattering from Polymer Solutions*, Academic Press, New York, 1972.

Hyde 1978. A. J. Hyde, "Light Scattering in Synthetic Polymer Systems," Chapter 4 in J. V. Dawkins, ed., *Developments in Polymer Characterisation—1*, Applied Science, London, 1978.

International Union 1952. International Union of Pure and Applied Chemistry, "Report on Nomenclature in the Field of Macromolecules," *J. Polym. Sci.* **8**, 255–277 (1952).

Kelley 1970. Richard N. Kelley and Fred W. Billmeyer, Jr., "A Review of Peak Broadening in Gel Chromatography," *Sep. Sci.* **5**, 291–316 (1970).

Kerker 1969. Milton Kerker, *The Scattering of Light and Other Electromagnetic Radiation*, Academic Press, New York, 1969.

Kinsinger 1975. Jack B. Kinsinger, "Macromolecules: Their Masses, Sizes, and Related Distributions," pp. 291–306 in John J. Burke and Volker Weiss, eds., *Characterization of Materials in Research: Ceramics and Polymers*, Syracuse University Press, Syracuse, New York, 1975.

Kraemer 1938. Elmer O. Kraemer, "Molecular Weights of Cellulose and Cellulose Derivatives," *Ind. Eng. Chem.* **30**, 1200–1203 (1938).

Kratohvil 1966. J. P. Kratohvil, L. E. Oppenheimer, and M. Kerker, "Correlation of Turbidity and Activity Data. III. The System Tungstosilicic Acid–Sodium Chloride–Water," *J. Phys. Chem.* **70**, 2834–2839 (1966).

Krigbaum 1952. W. R. Krigbaum and P. J. Flory, "Treatment of Osmotic Pressure Data," *J. Polym. Sci.* **9**, 503–588 (1952).

Krigbaum 1953. W. R. Krigbaum and P. J. Flory, "Statistical Mechanics of Dilute Polymer Solutions. IV. Variation of the Osmotic Second Coefficient with Molecular Weight," *J. Am. Chem. Soc.* **75**, 1775–1784 (1953).

Krigbaum 1967. W. R. Krigbaum and R. J. Roe, "Measurement of Osmotic Pressure," Chapter 79 in I. M. Kolthoff and Philip J. Elving, eds., with the assistance of Ernest B. Sandell, *Treatise on Analytical Chemistry,* Part I, Vol. 7, Wiley-Interscience, New York, 1967.

Kurata 1960. Michio Kurata, Walter H. Stockmayer, and Antonio Roig, "Excluded Volume Effect of Linear Polymer Molecules," *J. Chem. Phys.* **33**, 151–155 (1960).

Kurata 1963. M. Kurata and W. H. Stockmayer, "Intrinsic Viscosities and Unperturbed Dimensions of Long Chain Molecules," *Adv. Polym. Sci.* **3**, 196–312 (1963).

Kurata 1975. M. Kurata, Y. Tsunashima, M. Iwama, and K. Kamada, "Viscosity–Molecular Weight Relationships and Unperturbed Dimensions of Linear Chain Molecules," pp. IV-1–IV-60 in J. Brandrup and E. H. Immergut, eds., with the collaboration of W. McDowell, *Polymer Handbook,* 2nd ed., Wiley-Interscience, New York, 1975.

Levine 1976. Harold I. Levine, Robert J. Fiel, and Fred W. Billmeyer, Jr., "Very Low-Angle Light Scattering. A Characterization Method for High-Molecular-Weight DNA," *Biopolymers* **15**, 1267–1281 (1976).

Livesey 1969. P. J. Livesey and F. W. Billmeyer, Jr., "Particle-Size Determination by Low-Angle Light Scattering: New Instrumentation and a Rapid Method of Interpreting Data," *J. Colloid Interface Sci.* **30**, 447–472 (1969).

Lyons 1967. John W. Lyons, "Measurement of Viscosity," Chapter 83 in I. M. Kolthoff and Philip J. Elving, eds., with the assistance of Ernest B. Sandell, *Treatise on Analytical Chemistry,* Part I, Vol. 7, Wiley-Interscience, New York, 1967.

McCall 1973. J. S. McCall and B. J. Potter, *Ultracentrifugation,* Macmillan, New York, 1973.

McIntyre 1964. D. McIntyre and F. Gornick, eds., *Light Scattering from Dilute Polymer Solutions,* Gordon and Breach, New York, 1964.

McIntyre 1968. Donald McIntyre, ed., *Characterization of Macromolecular Structure,* Publication No. 1573, National Academy of Sciences, Washington, D.C., 1968.

May 1968. James A. May, Jr., and William B. Smith, "Polymer Studies by Gel Permeation Chromatography. II. The Kinetic Parameters for Styrene Polymerization," *J. Phys. Chem.* **72**, 216–221 (1968).

Morawetz 1975. Herbert Morawetz, *Macromolecules in Solution,* 2nd ed., Wiley-Interscience, New York, 1975.

Moore 1964. J. C. Moore, "Gel Permeation Chromatography. I. A New Method for Molecular Weight Distribution of High Polymers," *J. Polym. Sci.* **A2**, 835–843 (1964).

Moore 1967. W. R. Moore, "Viscosities of Dilute Polymer Solutions," Chapter 1 in A. D. Jenkins, ed., *Progress in Polymer Science,* Vol. 1, Pergamon Press, New York, 1967.

Ouano 1975. Augustus C. Ouano, Edward M. Barrall, II, and Julian F. Johnson, "Gel Permeation Chromatography," Chapter 2 in Philip E. Slade, Jr., ed., *Polymer Molecular Weights,* Part II, Marcel Dekker, New York, 1975.

Pasternak 1962. R. A. Pasternak, P. Brady, and H. C. Ehrmantraut, "Apparatus for the Rapid Determination of Molecular Weight," *DECHEMA Monogr.* **44**, 205–207 (1962).

Richards 1978. R. W. Richards, "Molecular Dimensions of Amorphous Polymers by Neutron Scattering," Chapter 5 in J. V. Dawkins, ed., *Developments in Polymer Characterisation—1,* Applied Science, London, 1978.

Scholte 1975. Th. G. Scholte, "Sedimentation Techniques," Chapter 8 in Philip E. Slade, Jr., ed., *Polymer Molecular Weights*, Part II, Marcel Dekker, New York, 1975.

Shimoda 1976. K. Shimoda, ed., *High-Resolution Laser Spectroscopy*, Springer-Verlag, New York, 1967.

Stabin 1954. J. V. Stabin and E. H. Immergut, "A High-Speed Glass Osmometer," *J. Polym. Sci.* **14**, 209–212 (1954).

Staudinger 1930. H. Staudinger and W. Heuer, "Highly Polymerized Compounds. XXXIII. A Relation Between the Viscosity and the Molecular Weight of Polystyrenes" (in German), *Ber. Dtsch. Chem. Ges. B* **63**, 222–234 (1930).

Stockmayer 1952. Walter H. Stockmayer and Edward F. Casassa, "The Third Virial Coefficient in Polymer Solutions," *J. Chem. Phys.* **20**, 1560–1566 (1952).

Stockmayer 1963. W. H. Stockmayer and Marshall Fixman, "On the Estimation of Unperturbed Dimensions from Intrinsic Viscosities," *J. Polym. Sci.* **C1**, 137–141 (1963).

Svedberg 1940. The Svedberg and Kai O. Pedersen, *The Ultracentrifuge*, Clarendon Press, Oxford, 1940; Johnson Reprint Corp., New York, 1959.

Tung 1977. L. H. Tung and John C. Moore, "Gel Permeation Chromatography," Chapter 6 in L. H. Tung, ed., *Fractionation of Synthetic Polymers: Principles and Practices*, Marcel Dekker, New York, 1977.

Ulrich 1975. Robert D. Ulrich, "Membrane Osmometry," Chapter 2 in Philip E. Slade, Jr., ed., *Polymer Molecular Weights*, Part I, Marcel Dekker, New York, 1975.

van de Hulst 1957. H. C. van de Hulst, *Light Scattering by Small Particles*, John Wiley & Sons, New York, 1957.

Williams 1972. J. W. Williams, *Ultracentrifugation of Macromolecules: Modern Topics*, Academic Press, New York, 1972.

Yau 1979. W. W. Yau, J. J. Kirkland, and D. D. Bly, *Modern Size-Exclusion Liquid Chromatography: Practice of Gel Permeation and Gel Filtration Chromatography*, John Wiley & Sons, New York, 1979.

Zimm 1946. B. H. Zimm and I. Myerson, "A Convenient Small Osmometer," *J. Am. Chem. Soc.* **68**, 911–912 (1946).

Zimm 1948a. Bruno H. Zimm, "The Scattering of Light and the Radial Distribution Function of High Polymer Solutions," *J. Chem. Phys.* **16**, 1093–1099 (1948).

Zimm 1948b. Bruno H. Zimm, "Apparatus and Methods for Measurement and Interpretation of the Angular Variation of Light Scattering: Preliminary Results on Polystyrene Solutions," *J. Chem. Phys.* **16**, 1099–1116 (1948).

Zimm 1949. Bruno H. Zimm and Walter H. Stockmayer, "The Dimensions of Chain Molecules Containing Branches and Rings," *J. Chem. Phys.* **17**, 1301–1314 (1949).

Zimm 1962. Bruno H. Zimm and Donald M. Crothers, "Simplified Rotation Cylinder Viscometer for DNA," *Proc. Nat. Acad. Sci. U.S.A.* **48**, 905–911 (1962).

CHAPTER NINE

ANALYSIS AND TESTING
OF POLYMERS

The purpose of this chapter is to assemble information on the applicability to polymers of various standard methods of physical and chemical analysis. A background is thus established for the uninterrupted discussion in succeeding chapters of the results of such analyses and tests. In no case is the treatment in this chapter exhaustive; it is intended to provide, with minimum detail, a broad picture of the applicability of each method to polymer systems and of the type of information obtained.

A. CHEMICAL ANALYSIS OF POLYMERS

The chemical analysis of polymers is not basically different from the analysis of low-molecular-weight organic compounds, provided that appropriate modification is made to ensure solubility or the availability of sites for reaction (e.g., insoluble specimens should be ground to expose a large surface area). The usual methods for functional group and elemental analysis are generally applicable, as are many other standard analytic techniques. Several books on the identification and analysis of plastics have appeared (Haslam 1965, Saunders 1966, Lever 1968, Wake 1969, Braun 1982, Krause 1983); the analytical chemistry of the more common plastics is treated by Kline (1959). Chemical reactions of polymers provide additional means of chemical analysis, as do their reactions of degradation (Chapter 6D). Collins (1973) discusses various analyses as means of following the course of polymerization. Brief descriptions follow of two powerful techniques for analyzing low-molecular-weight products of reactions (usually degradation) of polymers.

Mass Spectrometry (Wall 1962, Sedgwick 1978). In the most common applications of mass spectrometry to polymer systems, the polymer is allowed to react (as in thermal degradation, Chapter 6D) to form low-molecular-weight fragments that are condensed at liquid-air temperature. They are then volatilized, ionized, and separated according to mass and charge by the action of electric and magnetic fields in a typical mass spectrometer analysis. From the abundance of the various ionic species found, the structures of the low-molecular-weight species can be inferred.

Gas Chromatography (Cassel 1962, Stevens 1969). Gas chromatography is a method of separation in which gaseous or vaporized components are distributed between a moving gas phase and a fixed liquid phase or solid adsorbent. By a continuous succession of adsorption or elution steps, taking place at a specific rate for each component, separation is achieved. The components are detected by one of several methods as they emerge successively from the chromatographic column. From the detector signal, proportional to the instantaneous concentration of the dilute component in the gas stream, is obtained information about the number, nature, and amounts of the components present.

GENERAL REFERENCES

Kline 1959; Brauer 1965; Feinland 1965; Braun 1982; Krause 1983.

B. SPECTROSCOPIC METHODS

Infrared Spectroscopy

Emission or absorption spectra arise when molecules undergo transitions between quantum states corresponding to two different internal energies. The energy difference ΔE between the states is related to the frequency of the radiation emitted or absorbed by the quantum relation $\Delta E = h\nu$. Infrared frequencies in the wavelength range 1–50 μm† are associated with molecular vibration and vibration–rotation spectra.

A molecule containing N atoms has $3N$ normal vibration modes, including rotational and translational motions of the entire molecule. For highly symmetrical molecules with very few atoms, the entire infrared and Raman‡ spectrum can be

---

† Infrared absorption wavelengths are often expressed in wave numbers (wavelength^{-1}) in centimeter^{-1}: 1 μm = 10,000 cm^{-1}, 50 μm = 200 cm^{-1}, and so on, where μm = micrometer, 10^{-6} meter, formerly (and still widely) called the micron.

‡ The Raman effect results when the frequency of visible light is altered in the scattering process by the absorption or emission of energy produced by changes in molecular vibration and vibration–rotation quantum states.

correlated to and explained by the vibrational modes, but even for most low-molecular-weight substances, N is too large for such analysis. Useful information can still be obtained, however, because some vibrational modes involve localized motions of small groups of atoms and give rise to absorption bands at frequencies characteristic of these groups and the type of motions they undergo.

In polymers the infrared absorption spectrum is often surprisingly simple, if one considers the large number of atoms involved. This simplicity results first from the fact that many of the normal vibrations have almost the same frequency and therefore appear in the spectrum as one absorption band and, second, from the strict selection rules that prevent many of the vibrations from causing absorptions.

Experimental Methods. General features of infrared spectroscopy are by now so well known that it suffices to comment briefly on special problems of sample preparation for polymers. One of the greatest experimental difficulties in work with substances that absorb heavily in the infrared, including many polymers, is obtaining sufficiently thin samples. Common methods of sample preparation include the following: compression molding, by far the most widely used preparation technique; dissolving the polymer in a solvent, such as carbon disulfide or tetrachloroethylene, whose spectrum is relatively free of intense absorption bands; preparing a thin film by microtoming or milling; casting a thin film from solution; and pressing a finely ground mixture of the sample with KBr to form a disc or wafer.

Detection of Chemical Groups in Typical Spectra. Figure 9-1 gives a few of the many chemical linkages or groups that can be detected in polymer spectra, together with the approximate wavelengths at which they occur. Among the infrared absorption bands of interest in common polymers are those in polyethylene corresponding to C—H stretching (3.4 μm), C—H bending of CH_2 groups (6.8 μm) and CH_3 groups (shoulder at 7.25 μm on an amorphous band at 7.30 μm), and CH_2 rocking of sequences of methylene groups in paraffin structures (13.9 μm). (The structure of polyethylene is discussed in more detail in Chapter 13.) Other absorption bands of interest are those due to C=C at 6.1 μm in natural rubber; carbonyl at 5.8 μm and ether at 8.9 μm in poly(methyl methacrylate); aromatic structures at 6.2, 6.7, 13.3, and 14.4 μm in polystyrene; C—Cl at 14.5 μm in poly(vinyl chloride); peptide groups at 3.0, 6.1, and 6.5 μm in nylon; and CF_2 at 8.2–8.3 μm in polytetrafluoroethylene.

Related Methods. It is desirable to supplement as much as possible the observations of infrared absorption usually obtained in the 2–15 μm wavelength range. For this purpose observations in the *far-infrared* region (say up to 200 μm) become important. Despite serious experimental difficulties, valuable information is available in this region: For some polymers, such as polytetrafluoroethylene, most of the absorption bands occur above 15 μm.

The application of *Fourier-transform* infrared spectroscopy to polymers has been reviewed (D'Esposito 1978). Since the requirements for activity of a vibration for causing absorption in the infrared and for causing *Raman scattering* are often

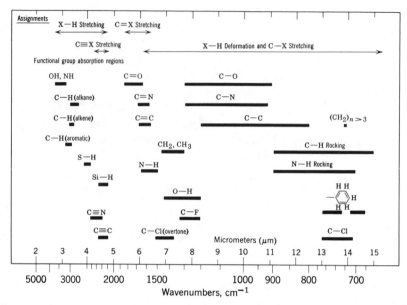

FIG. 9-1.   Infrared absorption bands of interest in polymers arranged by approximate wavelength and frequency.

different, information from Raman experiments in general supplements that obtainable from infrared absorption. Utilization of lasers as light sources has greatly decreased experimental difficulties in this technique, and the method has now become one of the standard tools for polymer analysis (Hendra 1969, Koenig 1971).

Dichroism.   For any molecular vibration leading to infrared absorption, there is a periodic change in electric dipole moment. If the direction of this change is parallel to a component of the electric vector of the infrared radiation, absorption occurs; otherwise it does not. In oriented bulk polymers, the dipole-moment change can be confined to specified directions. The use of polarized infrared radiation then leads to absorption that is a function of the orientation of the plane of polarization. This phenomenon is called *dichroism* and is usually measured as the *dichroic ratio,* the ratio of the optical densities of an absorption band measured with radiation polarized parallel and perpendicular, respectively, to a specified direction in the sample. Dichroic ratios depend upon both the degree of orientation and the angle between the direction of the transition moment and the selected direction in the sample (for example, the axial direction in a fiber). They usually range between 0.1 and 1.0 (see also Chapter 10*E*).

Crystallinity.   The infrared absorption spectra of the same polymer in the crystalline and amorphous states can differ for at least two reasons. First, specific intermolecular interactions may exist in the crystalline polymer that lead to sharpening or splitting of certain bands; and second, some specific conformations may exist in one but not the other phase, leading to bands characteristic exclusively of either crystalline or amorphous material. An example of the latter effect is poly(ethylene terephthalate), in which the —OCH_2CH_2O— portion of each repeat unit is restricted to the all-trans conformation in the crystal, but can exist in part in the gauche form in the melt. Several bands characteristic of each conformation have been identified (Ward 1957). In favorable cases, such as 66-nylon (Starkweather 1956), percent crystallinity can be determined in absolute terms from infrared absorption data.

Geometric Isomerism.   The determination of the various types of geometric isomers associated with unsaturation in polymer chains is of great importance, for example, in the study of the structure of modern synthetic rubbers (Chapter 13). Table 9-1 lists some of the important infrared absorption bands resulting from olefinic groups. In synthetic "natural" rubber, *cis*-1,4-polyisoprene, relatively small amounts of 1,2- and 3,4-addition can easily be detected, though it is more difficult to distinguish between the cis and trans configurations. Nuclear magnetic resonance spectroscopy (see below) is also useful for this analysis.

Nuclear Magnetic Resonance Spectroscopy

Since about 1960 nuclear magnetic resonance (NMR) spectroscopy has become a major tool for the study of chain configuration, sequence distribution, and microstructure in polymers. Its use has evolved from early broad-line studies of the onset of molecular motion in solid polymers, through the widely practiced solution studies of proton NMR, to the application of the more difficult but more powerful carbon-13 NMR methods to both liquids and solids. Despite the widespread use of NMR, a brief summary of its origins and experimental methods is warranted.

Experimental Methods.   The NMR technique utilizes the property of spin (angular momentum and its associated magnetic moment) possessed by nuclei whose atomic number and mass number are not both even. Such nuclei include the isotopes of hydrogen and ^{13}C, ^{15}N, ^{17}O, and ^{19}F. Application of a strong magnetic field to material containing such nuclei splits the energy level into two, representing states

TABLE 9-1. Absorption Wavelengths of Olefinic Groups[a]

Group Containing C=C	Wavelength (μm)
Vinyl, R_1CH=CH_2	10.1 and 11.0
trans-R_1CH=CHR_2	10.4
Vinylidene, R_1R_2C=CH_2	11.3
R_1R_2C=CHR_3	12.0
cis-R_1CH=CHR_2	14.2 (variable)

[a]Cross (1950).

with spin parallel and antiparallel to the field. Transitions between the states lead to absorption or emission of an energy

$$E = h\nu_0 = 2\mu H_0 \tag{9-1}$$

where the frequency ν_0 is in the microwave region for fields of strength H_0 of the order of 10,000 gauss and up, and μ is the magnetic moment of the nucleus. The energy change is observed as a resonance peak or line in experiments where either H or ν_0 is varied, with the other held constant.

With an assembly of nuclei, the field on any one is modified by the presence of the others:

$$h\nu_0 = 2\mu(H_0 + H_L) \tag{9-2}$$

where H_L is the local field with a strength of 5–10 gauss. A distribution of local fields usually exists so that the resonance line becomes broadened. This broadening was once studied as a function of temperature to indicate the temperature of the onset of molecular motion, but it was soon found that with ample molecular motion present, as with liquids and solutions, quite narrow lines can be observed. The positions of these lines on the scale of frequency or magnetic field depend on the local fields, which in turn result from the nature and location of the atomic groups in the vicinity of the protons. Ths displacements in the resonances, called *chemical shifts,* are measured in parts per million in frequency (or the equivalent field strength) on a scale labeled δ. The zero of the δ scale is a reference point provided by the single resonance of the equivalent protons in a substance showing minimum chemical shift, such as tetramethyl silane. Tables of chemical shifts resulting from groups in the neighborhood of the nucleus being studied are now common and are not repeated here.

In proton NMR, additional complexity and additional information result from coupling of the resonances of protons on adjacent carbon atoms, resulting in the splitting of their resonances into $n + 1$ peaks, where n is the number of equivalent neighboring protons. To aid in interpretation, two experimental modifications are useful. The first of these is the use of high magnetic field strengths, in the range 60,000–220,000 gauss compared to the few tens of thousand gauss used in early broad-line work. Superconducting magnets are used to obtain the highest field strengths. As Fig. 9-2 shows, the increased resolution at the higher field strength aids in the unambiguous interpretation of the spectra.

The second development, double resonance or spin decoupling, effects great simplifications in the spectra. A second radiofrequency field is used that has the effect of removing the coupling and collapsing multiplet spectra to much simpler ones (Fig. 9-3). Mention should also be made of the use of ^2H NMR to provide spectra simplified by elimination of proton–proton coupling due to the lower frequency of occurrence of the deuteron.

A great advantage of ^{13}C NMR over proton NMR is the much greater range of chemical shifts exhibited by ^{13}C (and most other nuclei), some 200 ppm in contrast

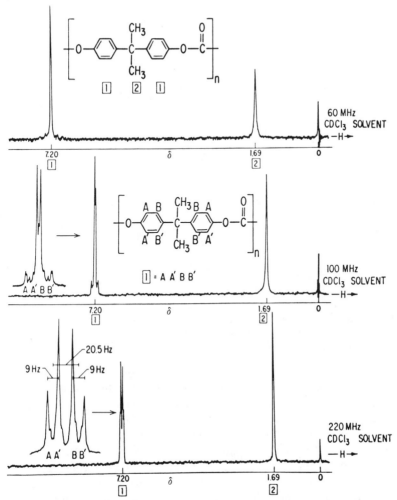

FIG. 9-2. High-resolution NMR spectra of bisphenol-A polycarbonate in deuterochloroform solution showing the effect of frequency on resolution (Sudol 1969).

to 10 ppm for ^{1}H. In addition, there is no carbon–carbon spin coupling due to the low chemical abundance of the ^{13}C nucleus, 1.1% relative to ^{12}C. Finally, the resonances of carbon nuclei are sensitive to configurational as well as chemical environments. Offsetting these advantages are the lower abundance of the ^{13}C nuclei and their lower nuclear magnetic moment, about one-quarter that of the proton; both effects reduce the intensity of the NMR signal.

This lower sensitivity is overcome by the use of pulsed Fourier-transform NMR, in which a high-power microsecond pulse of radiofrequency energy sets all the carbon nuclei into resonance at once, eliminating the need to sweep the frequency or magnetic field. The data are recorded as the subsequent decay of the resonance with time, which is the Fourier transform of the desired spectrum. Records from

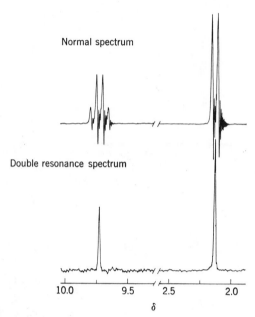

Normal spectrum

Double resonance spectrum

| 10.0 | 9.5 | 2.5 | 2.0 |

δ

FIG. 9-3. NMR spectra of propylene oxide in carbon tetrachloride solution: (top) conventional spectrum and (bottom) double-resonance spectrum showing simplification resulting from collapse of the splitting due to coupling (Bovey 1965).

repeat pulses are summed by computer, which transforms the data to the spectrum desired when the signal-to-noise ratio has reached a satisfactory level.

Interest in studying molecular motion in solid polymers has returned with the development of the theory of line broadening. It can be shown that the local field seen by a pair of like nuclei in a solid sample is given by an isotropic component (the origin of the chemical shift observed in solution) plus a broadening component proportional to $3 \cos^2\theta - 1$, where θ is the angle between the line connecting the nuclei and the direction of the magnetic field. If the structure of the polymer solid is random, this component broadens the resonance to a width greater than the total ^{13}C spectrum; but if the specimen can be oriented and spun around an axis at $\theta = 54.7°$, for which $3 \cos^2\theta - 1 = 0$—the so-called "magic angle"—the line width is reduced enough so that all the carbon resonances can be resolved.

Stereochemical Configuration.   A major application of NMR to polymer systems has been the elucidation of the stereochemical configurations of polymer chains; this topic is discussed further in Chapter 10A. Poly(methyl methacrylate) was first studied over 20 years ago (Bovey 1960). It is now possible to analyze for the statistical frequency of occurrence of all possible combinations of up to four successive pairs of units (dyads) that can occur with either the same (*meso*) or opposite (*racemic*) configurations.

Geometric Isomerism. Both proton and ^{13}C NMR have proven very useful in supplementing infrared spectroscopy as a means of determining isomerism around the carbon–carbon bond in polymer chains (Duch 1970).

Copolymer Sequences. The principles of analyzing short sequences of monomers in a copolymer are not different from those of determining stereochemical configuration. NMR results have contributed extensively in this field.

Electron Paramagnetic Resonance Spectroscopy

Although electron paramagnetic resonance (EPR) and NMR spectroscopy are similar in basic principles and in experimental techniques, they detect different phenomena and therefore yield different information. The major use of EPR spectroscopy lies in the detection of free radicals (Chapter 3A). These species are uniquely characterized by their magnetic moment, arising from the presence of an unpaired electron. Measurement of a magnetic property of a material containing free radicals, such as its magnetic susceptibility, gives the concentration of free radicals, but the method lacks sensitivity and cannot reveal the structure of the radicals. Electron paramagnetic resonance spectroscopy is essentially free from these limitations.

Experimental Methods. As in NMR spectroscopy, the action of a strong magnetic field on a material containing free radicals removes the degeneracy of their ground-state energy levels. For low radical concentrations the new energy levels are given by two terms. The first is

$$E = h\nu_0 = g\beta\mu_0 H_0 \tag{9-3}$$

where g is a tensor relating the field direction and the symmetry directions in the radical, β is the magnetic moment of the electron spin, and μ_0 is the magnetic permeability of a vacuum. The second term represents coupling of the electron spin with the nuclear spins in the molecule. This coupling results in the splitting of the resonance line into a symmetrical group of lines whose positions and amplitudes depend on, and therefore give information about, the structure of the radical.

Applications to Polymers. The investigation of free radicals produced by high-energy irradiation (Chapter 6D) of polytetrafluoroethylene (Rexroad 1958, Lebedev 1960) serves as an example of an early application of EPR to polymer radicals.

The EPR spectrum of irradiated polytetrafluoroethylene is interpreted as arising from radicals of the type

$$—(CF_2)_x\overset{\cdot}{C}F(CF_2)_y—$$

remaining trapped in the polymer after a C—F bond has been broken and the fluorine atom has diffused away. The spectrum shows fine structure due to coupling between the unpaired electron and neighboring ^{19}F nuclei. The fine structure is lost below 270 K as the motion of the ^{19}F nuclei is slowed down. The spectrum does not change at higher temperatures, indicating that the radical is stable to at least 550 K. Secondary products, resulting from the reaction of the primary radical with such substances as O_2 and NO, were also studied.

It may be noted that, as discussed in Chapters 1C and 6D, the strength of the C—F bond and the tendency of polytetrafluoroethylene to degrade rather than crosslink on irradiation would suggest the breaking of a C—C rather than a C—F bond as a likely source of radicals in this polymer.

GENERAL REFERENCES

Infrared spectroscopy: Zbinden 1964; Hummel 1966; Kössler 1967; Elliott 1969; Boerio 1972; Collins 1973, Exp. 32; Siesler 1980.

NMR spectroscopy: Bovey 1968, 1969, 1972, 1976; Schaefer 1974, 1976, 1979; Cunliffe 1978.

EPR spectroscopy: Bresler 1966; Bullock 1976; Rånby 1977; Boyer 1980.

C. X-RAY DIFFRACTION ANALYSIS

The *x-ray diffraction* method is a powerful tool for investigating orderly arrangements of atoms or molecules through the interaction of electromagnetic radiation to give interference effects with structures comparable in size to the wavelength of the radiation. If the structures are arranged in an orderly array or lattice, the interferences are sharpened so that the radiation is scattered or diffracted only under specific experimental conditions. Knowledge of these conditions gives information regarding the geometry of the scattering structures. The wavelengths of x-rays are comparable to interatomic distances in crystals; the information obtained from scattering at wide angles describes the spatial arrangements of the atoms. Low-angle x-ray scattering is useful in detecting larger periodicities, which may arise from lamellar crystallites (Chapter 10C) or from voids.

The results of the application of the x-ray method in determining the crystal structures of polymers are described in Chapter 10B. Collected crystallographic data for polymers have been tabulated (Miller 1975).

Experimental Methods

X-rays are usually produced by bombarding a metal target with a beam of high-voltage electrons. This is done inside a vacuum tube, the x-rays passing out through a beryllium or polyester film window in the tube in a well-defined beam. Choice of the target metal and the applied voltage determines the wavelength or wavelengths of x-rays produced. Experiments in which nearly monochromatic x-rays are used are of the greatest interest.

The diffracted x-rays may be detected by their action on photographic films or plates, or by means of a radiation counter and electronic equipment feeding data to a computer. Each method has its advantages. Qualitative examination of the diffraction pattern and accurate measurement of angles and distances are best made from a photographic record, whereas for precise measurement of the intensity of the diffracted beam the counting technique is preferred.

X-rays of a given wavelength are diffracted only for certain specific orientations of the sample. If the sample is a single crystal, it must be placed in all possible orientations during the experiment, usually by rotating or oscillating the sample about one of its axes to achieve the desired orientations with respect to the x-ray beam. Alternatively, a sample made up of a powder of very small crystals may be used. If the minute powdered particles are randomly oriented, all orientations will be included within the sample. The powder method is more convenient, but gives less information than the single-crystal method, since the latter allows orientations about one crystal axis at a time to be investigated.

Application to Polymers

Since polymer single crystals as now prepared (Chapter 10C) are too small for x-ray diffraction experiments, the crystal structure of a polymer is usually determined from x-ray patterns of a fiber drawn from the polymer. Because of the alignment of the crystalline regions with the long axes of the molecules parallel to the fiber axis (Chapter 10E), the pattern is essentially identical to a rotation pattern from a single crystal. In such a pattern diffraction maxima occur in rows perpendicular to the fiber axis, called *layer lines* (seen, for example, in Fig. 10-22b).

Chain Conformations. As the periodicity of molecular structure of a polymer is characterized by the existence of a *repeat unit,* so the periodicity of its crystal is characterized by a *repeat distance.* The repeat distance is directly determined by measuring the distance between the layer lines: The greater the repeat distance, the closer together are the layer lines. Determination of the repeat unit is considerably more difficult. Classically, it proceeds through derivation, from the positions of the diffracted x-ray beams, of the dimensions of the *unit cell* (usually the simplest geometric volume unit that, by repetition, builds up the three-dimensional crystalline array). The positions of the atoms in the unit cell are then derived from the relative intensities of the diffracted beams. The diffraction patterns of polymers, however, do not provide sufficient information to allow such analyses to be carried to completion. Additional structural information is utilized from other sources, such as normal bond lengths and angles and atomic arrangements along the chain suggested by the chemical structure of the sample. For example, the repeat distance of 2.55 Å in the crystals of polyethylene is readily identified with a single repeat unit in the planar zigzag conformation.

Many polymer conformations can be described in terms of atoms regularly spaced along helices (Chapter 10B). Methods for analyzing the diffraction patterns from helical structures provide a highly versatile technique for the determination of the repeat unit in such polymers.

Chain Packing.   The packing of the chains is described most completely in terms of the unit cell and its contents (shown for polyethylene in Fig. 10-3). The volume of the unit cell, and hence the volume occupied in the crystal by a single repeat unit, can be obtained from the repeat distance and the positions of the diffraction spots on the layer lines. This volume measures the density of the crystals, which is useful in determining the degree of crystallinity, as outlined in Chapter 10D.

The atomic arrangement within the unit cell is much more difficult to determine than the cell dimensions. Trial structures can often be deduced from these dimensions and a knowledge of the chain conformation, and can be tested in terms of calculated and observed diffraction intensities, using well-established crystallographic techniques.

Disorder in the Crystal Structure.   In the discussion so far it has been assumed that polymer crystals consist of perfect geometrical arrays of atoms. However, defects and distortions are present in crystals of all materials, and, together with grosser types of disorder, they play an important role in the crystal structures of polymers. X-ray diffraction data give a great deal of information about the qualitative aspects of disorder in the crystal structure, since disordering results in the broadening of the diffraction maxima. Since different classes of diffraction maxima are sensitive in different ways to particular types of disorder, a systematic study of the broadening yields detailed information on the types of disturbances in the crystals (Wunderlich 1973).

Quantitative measurement of the disorder is difficult because broadening of the diffraction maxima is caused both by small crystallite size and by distortions within larger crystals. As discussed in Chapter 10C, the former explanation was traditionally adopted in interpreting the diffraction pattern from polymers. However, the distortion of perfect crystals into *paracrystals* (Hosemann 1962) is now considered important in many polymers.

Orientation.   Unless the crystalline regions in polymers are oriented, as by drawing a fiber, the diffraction maxima merge into rings, made up of the maxima from a large number of crystallites in many different orientations (Chapter 10E, especially Fig. 10-22). If orientation is intermediate between these extremes, the rings split into arcs, and a quantitative evaluation of the crystallite orientation is made by measuring the angular spread and intensity of these arcs.

GENERAL REFERENCES

Alexander 1969; Kakudo 1972; Collins 1973, Chapter 8B, Exp. 18; Wunderlich 1973; Herglotz 1975; Brown 1978.

D. MICROSCOPY

While the techniques of light and electron microscopy are well described in the literature, their application to polymer analysis involves sufficient extension beyond the ordinary techniques to warrant brief comments.

Light Microscopy. Reflected-light microscopy is valuable for examining the texture of solid opaque polymers. For materials that can be prepared as thin films (often, for example, by casting on the microscope slide), examination by transmitted light is usual, but little detail can be seen without some type of enhancement. Two common techniques are used. One is polarized-light microscopy, in which advantage is taken of the ability of crystalline materials to rotate the plane of polarized light. Thus the structure of spherulites (Chapter 10C) is studied with the sample between crossed polarizers, and the crystalline melting point is taken as the temperature of disappearance of the last traces of crystallinity when using a hot-stage polarizing microscope. A second useful technique is phase-contrast microscopy, which allows observations of structural features involving differences in refractive index rather than absorption of light as in the conventional case. Interference microscopy, allowing measurement of thicknesses as low as a few angstrom units, has proved valuable in the study of polymer single crystals (Chapter 10C). Resolution, of course, is limited to object sizes about half the wavelength; practically, about 2000 Å.

Electron Microscopy and Electron Diffraction. Resolution of smaller objects can be achieved in electron microscopy, though the theoretical resolving power mentioned above is scarcely approached, the practical limit of resolution being a few angstrom units. Electron microscopy has been a powerful tool in the study of the morphology of crystalline polymers (Chapter 10C). The usual techniques of replication, heavy-metal shadowing, and solvent etching are widely used. The direct observation of thin specimens, such as polymer single crystals, is also possible and allows the observation of the electron-diffraction pattern of the same specimen area, invaluable for determining crystallographic directions and relating them to morphology. A severe problem, however, is damage of the specimen by the electron beam. Typically, polymer single crystals are severely damaged in times of a few seconds to a minute. This problem can be alleviated by maintaining the sample well below room temperature with a cold stage, by the use of accelerating voltages several times higher than the usual 50,000–100,000 V, or by the use of an image intensifier providing a cathode-ray tube image with far less beam current than that ordinarily required.

Scanning Electron Microscopy. In scanning electron microscopy a fine beam of electrons is scanned across the surface of an opaque specimen to which a light conducting film has been applied by evaporation. Secondary electrons, backscattered electrons, or (in the *electron microprobe*) x-ray photons emitted when the beam hits the specimen are collected to provide a signal used to modulate the intensity of the electron beam in a television tube, scanning in synchronism with the microscope beam. Because the latter maintains its small size over large distances relative to the specimen, the resulting images have great depth of field and a remarkable three-dimensional appearance (conventional stereoscopic pairs of photographs can be made). Resolution is currently limited to the order of 100 Å.

GENERAL REFERENCES

Birbeck 1961; Hartshorne 1964; Fischer 1966; Geil 1966; Sjöstrand 1967; Thornton 1968; Collins 1973, Chapter 8C,D, Exps. 19 and 20; Hemsley 1978.

E. THERMAL ANALYSIS

The field of the thermal analysis of polymers has expanded greatly since the introduction of simple, inexpensive instruments (compared to classical instrumentation) for several types of thermal measurements just over 20 years ago. In addition to the traditional calorimetric and differential thermal analysis, the field now includes equipment for thermogravimetric analysis, thermomechanical analysis, electrical thermal analysis, and effluent gas analysis. Not only can one study the enthalpy changes associated with heating, annealing, crystallizing, or otherwise thermally treating polymers, but one can now study a wide variety of responses of the system to temperature, including polymerization, degradation, or other chemical changes.

Differential Scanning Calorimetry

Experimental Methods. In contrast to earlier use of a large, expensive adiabatic calorimeter for measurements of specific heat and enthalpies of transition, these measurements are now usually carried out on quite small samples in a *differential scanning calorimeter (DSC)*. The term is applied to two different modes of analysis, of which the one more closely related to traditional calorimetry is described here. In DSC an average-temperature circuit measures and controls the temperature of sample and reference holders to conform to a predetermined time–temperature program. This temperature is plotted on one axis of an x–y recorder. At the same time, a temperature-difference circuit compares the temperatures of the sample and reference holders and proportions power to the heater in each holder so that the temperatures remain equal. When the sample undergoes a thermal transition, the power to the two heaters is adjusted to maintain their temperatures, and a signal proportional to the power difference is plotted on the second axis of the recorder. The area under the resulting curve is a direct measure of the heat of transition.

Although the DSC is less accurate than a good adiabatic calorimeter (1–2% versus 0.1%), its accuracy is adequate for most uses and its advantages of speed and low cost make it the outstanding instrument of choice for most modern calorimetry.

Application to Polymers. As a typical result, specific heat–temperature curves obtained (by adiabatic calorimetry) (Smith 1956) on heating quenched (amorphous) specimens of poly(ethylene terephthalate) are shown in Fig. 9-4. Each curve rises linearly with temperature at low temperatures and then rises more steeply at the glass transition, 60–80°C. With the onset of mobility of the molecular chains above this transition, crystallization takes place, as indicated by the sharp drop in the specific heat curve. At still higher temperatures, 220–270°C, the crystals melt with a corresponding rise in the specific heat curve.

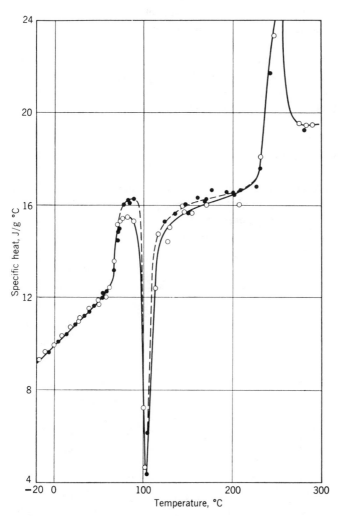

FIG. 9-4. Curves of specific heat as a function of increasing temperature for quenched (amorphous) poly(ethylene terephthalate) (Smith 1956).

Differential Thermal Analysis

Experimental Methods. In differential thermal analysis (DTA) the sample and an inert reference substance, undergoing no thermal transition in the temperature range of interest, are heated at the same rate. The temperature difference between sample and reference is measured and plotted as a function of sample temperature. The temperature difference is finite only when heat is being evolved or absorbed because of exothermic or endothermic activity in the sample, or when the heat capacity of the sample is changing abruptly. Since the temperature difference is directly proportional to the heat capacity, the curves resemble specific heat curves, but are inverted because, by convention, heat evolution is registered as an upward peak and heat absorption as a downward peak.

Application to Polymers. A typical DTA result (Ke 1960) is the differential thermal analysis curve for poly(ethylene terephthalate) shown in Fig. 9-5. Features of the curve may readily be identified by comparison with Fig. 9-4. The lower crystalline melting range in the specimen of Fig. 9-5 is attributed to impurities in the polymer.

Other Thermal Methods

Thermogravimetric Analysis. In thermogravimetric analysis (TGA) a sensitive balance is used to follow the weight change of the sample as a function of temperature. Typical applications include the assessment of thermal stability and decomposition temperature (Fig. 9-6), extent of cure in condensation polymers, com-

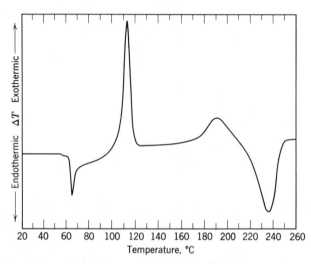

FIG. 9-5. Differential thermal analysis curve for amorphous poly(ethylene terephthalate) (Ke 1960); compare with Fig. 9-4.

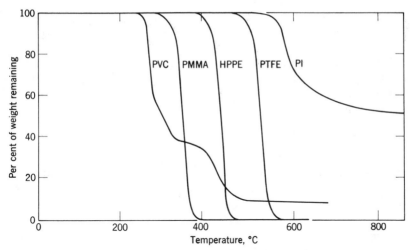

FIG. 9-6. Relative thermal stability of polymers as determined by weight loss on heating at 5°C/min in nitrogen in thermogravimetric analysis (Chiu 1966). Poly(vinyl chloride) (PVC) first loses HCl; later the mixture of unsaturated carbon–carbon backbone and unchanged poly(vinyl chloride) partly chars and partly degrades to small fragments. Poly(methyl methacrylate) (PMMA), branched polyethylene (HPPE), and polytetrafluoroethylene (PTFE) degrade completely to volatile fragments (Chapter 6D), while a polyimide (PI) partially decomposes, forming a char above 800°C.

position and some information on sequence distribution in copolymers, and composition of filled polymers, among many others.

Thermomechanical Analysis. Thermomechanical analysis (TMA) measures the mechanical response of a polymer system as the temperature is changed. Typical measurements include dilatometry, penetration or heat deflection, torsion modulus, and stress–strain behavior.

GENERAL REFERENCES

Ke 1966; Slade 1966–1970; Smothers 1966; Miller 1969; Schwenker 1969; Mackenzie 1970; Wunderlich 1970, 1971; Reich 1971; Collins 1973, Chapter 9C,E, Exps. 21, 22, and 24; Richardson 1978; Turi 1981.

F. PHYSICAL TESTING

The purpose of this section is to provide brief descriptions of the more important test methods for measuring the physical properties of polymers. The discussion is not intended to be comprehensive or detailed. Test methods related to rheology and viscoelasticity are discussed in Chapter 11.

The literature related to the physical testing of polymers is extensive. Several compilations are useful: the series edited by Schmitz (1965, 1966, 1968) and Brown (1969) and those volumes of methods of test and recommended practices dealing with plastics issued by the American Society for Testing and Materials (ASTM). There are also many pertinent articles in the *Encyclopedia of Polymer Science and Technology* (Mark 1964–1970). Beyond this listing, specific references have largely been omitted from this section.

Mechanical Properties

Stress–Strain Properties in Tension.   One of the most informative mechanical experiments for any material is the determination of its *stress–strain curve in tension*. This is usually done by measuring continuously the force developed as the sample is elongated at constant rate of extension.

The generalized stress–strain curve for plastics shown in Fig. 9-7 serves to define several useful quantities, including *modulus* or *stiffness* (the slope of the curve), *yield stress,* and *strength* and *elongation at break*. This type of curve is typical of a plastic such as polyethylene. Figure 9-8 shows stress–strain curves typical of some other classes of polymeric materials. The properties of these polymer types are related to the characteristics of their stress–strain curves in Table 9-2.

Tensile properties are usually measured at rates of strain of 1–100%/min. At higher rates of strain—up to 10^6%/min—tensile strength and modulus usually increase severalfold, while elongation decreases. The interpretation of these results is complicated by large temperature rises in the test specimen.

In addition to tensile measurements, tests may also be performed in *shear, flexure, compression,* or *torsion*. For materials in film form, flexural tests are often

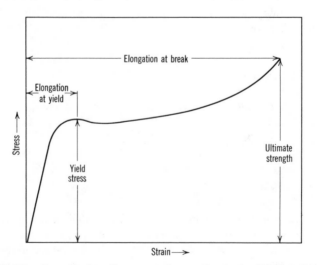

FIG. 9-7.   Generalized tensile stress–strain curve for plastics (Winding 1961).

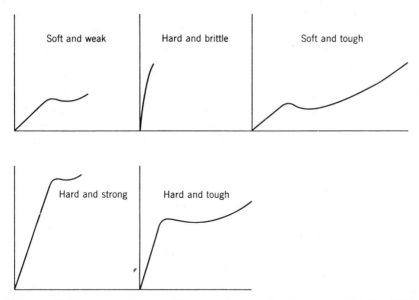

FIG. 9-8. Tensile stress–strain curves for several types of polymeric materials (Winding 1961).

used. These may include (for stiffer materials) measurement of *flexural modulus* or (for less stiff materials) *flexural* or *folding endurance* tests (see next pargraph).

Fatigue Tests. When subjected to cyclic mechanical stresses, most materials fail at a stress considerably lower than that required to cause rupture in a single stress cycle. This phenomenon is called *fatigue*. Various modes of fatigue testing in common use include alternating tensile and compressive stress and cyclic flexural stress. Results are reported as plots of stress versus number of cycles to fail. Many materials show a fatigue endurance limit, or a maximum stress below which fatigue failure never takes place.

TABLE 9-2. Characteristic Features of Stress–Strain Curves as Related to Polymer Properties[a]

Description of Polymer	Characteristics of Stress–Strain Curve			
	Modulus	Yield Stress	Ultimate Strength	Elongation at Break
Soft, weak	Low	Low	Low	Moderate
Soft, tough	Low	Low	Yield stress	High
Hard, brittle	High	None	Moderate	Low
Hard, strong	High	High	High	Moderate
Hard, tough	High	High	High	High

[a]Winding (1961).

Fatigue failure may arise from the absorption of energy in a material that is not perfectly elastic. This energy is manifested as heat, leading to a temperature rise, a lower modulus, and rapid failure. Energy absorption is accompanied and measured by a difference in phase between stress and strain in the cyclic test.

Impact Tests.   In most cases rupture in polymer samples may be divided into two classes: brittle and ductile. Brittle rupture occurs if the material behaves elastically up to the point of failure, that is, does not yield or draw, whereas in ductile rupture the specimen is permanently distorted near the point of failure. Brittle failure is characterized by lack of distortion of the broken parts. Two aspects of brittle rupture are of interest: the temperature below which brittle failure occurs under a given set of experimental conditions, and measures of the toughness of materials at ambient temperature. The test methods used to obtain both types of information are highly empirical. Since the specimen is destroyed, multiple testing is usually employed to allow statistical evaluation of the results.

The *brittle point*, or temperature at the onset of brittleness, is usually determined by subjecting a specimen to impact in a standardized but empirical way. The temperature of the test is varied until that temperature is found where, statistically, half the specimens fail by brittle rupture. The brittle point is roughly related to the glass transition temperature (Chapter 11*D*).

Impact strength of plastics is commonly measured by tests in which a pendulum with a massive striking edge is allowed to hit the specimen. From the travel of the pendulum after breaking the specimen can be calculated the energy required to cause the break. The specimen is often notched in an effort to improve the reproducibility of the mode of failure.

Other forms of impact test include a large variety in which falling objects (balls, darts, etc.) strike the specimen. Usually the height from which the missile must be dropped to cause failure in half the specimens is taken as a measure of the toughness of the material.

Tear Resistance.   When plastics are used as films, particularly in packaging applications, their resistance to tearing is an important property. In one test of *tear strength* a specimen is torn apart at a cut made by a sharp blade. Energy is provided by a falling pendulum, and the work done is measured by the residual energy of the pendulum. Tear strength and tensile strength are closely related.

Hardness.   As usually conceived, *hardness* is a composite property combining concepts of resistance to *penetration, scratching, marring,* and so on. Most hardness tests for plastics are based on resistance to penetration by an indentor pressed into the plastic under a constant load.

Abrasion Resistance.   Abrasion resistance in plastics usually takes the form of a scratch test, in which the material is subjected to many scratches, usually from contact with an abrasive wheel or a stream of falling abrasive material. The degree of abrasion can be determined by loss of weight for severe damage, but is more usually measured by evidence of surface marring, such as loss of gloss or development of haze in transparent specimens.

Friction, hardness, and abrasion resistance are related closely to the viscoelastic properties of polymers (Chapter 11C).

Thermal Properties

Softening Temperature.   In addition to the rheological tests discussed in Chapter 11A, a thermal property of great interest is the *softening temperature* of a plastic. Various ways of measuring this property include observation of the temperatures at which (a) an indentor under fixed load penetrates a specified distance into the material (*Vicat test*; see Chapter 12D); (b) a bar, held in flexure under constant load, deforms a specified amount (*deflection temperature* or *heat distortion test*); (c) a polymer sample becomes molten and leaves a trail when moved across a hot metal surface with moderate pressure (*polymer melt or stick temperature test*); and (d) a polymer specimen fails in tension under its own weight (*zero-strength temperature test*).

Flammability.   The flammability of plastics is usually tested as the burning rate of a specified sample. The self-extinguishing tendency of the material on the removal of an external flame is also important.

Optical Properties

Transmittance and Reflectance.   A major determinant of the appearance of a *transparent* material (one that does not scatter light) is its *transmittance,* the ratio of the intensities of light passing through and light incident on the specimen. Similarly, the appearance of an *opaque* material (one that may reflect light but does not transmit it) is characterized by its *reflectance*, the ratio of the intensities of the reflected and the incident light. A *translucent* substance is one that transmits part and reflects part of the light incident on it.

Transmittance and reflectance may be measured as a function of the wavelength of light in a spectrophotometer. When adjusted to correspond to visual perception, by weighting the effects at various wavelengths according to the power incident from a specified light source (often daylight) and the response of the human eye to light flux, these quantities are called *luminous transmittance* and *luminous reflectance*.

Color.   Color is the subjective sensation in the brain resulting from the perception of those aspects of the appearance of objects that result from the spectral composition of the light reaching the eye. Other aspects of appearance (gloss, haze, transparency; see below) are properly not part of the phenomenon of color. Since color is subjective, it cannot be described completely in physical terms, though to a first approximation color depends largely on the spectral power distribution of a light source, the spectral reflectance of the illuminated object, and the spectral response curves of the eye. In terms of visually perceived quantities, the description of color requires specification of three variables, a common set of which are *hue*, the attribute that determines whether the color is red, green, blue, and so on; *lightness,* the attribute that permits a color to be classified equivalent to some member of a scale of grays ranging from white to black; and *saturation,* the attribute of any color possessing a hue that determines the degree of difference of the color from the gray of the same lightness. The instrumental "measurement" of color consists in determining sets of numbers, correlating approximately with visually perceived quantities for sufficiently restrictive conditions, that allow one to judge whether two colors are alike or different. Despite the many limitations involved, the ability to make such determinations quantitatively and objectively is of great commercial importance. The principles of color technology, including visual perception, the measurement of color and color difference, color matching, and the coloring of plastics and other materials, have been discussed by Billmeyer (1981).

Gloss. Gloss is the geometrically selective reflectance of a surface responsible for its shiny or lustrous appearance. Surface reflectance is commonly at a maximum in or near the specular direction, that is, the direction at which a mirror would reflect light. Photoelectric instruments are available for measuring gloss at a variety of angles of incidence and reflection.

Haze. For transparent materials, haze is that percentage of transmitted light that in passing through the specimen deviates from the incident beam by forward scattering. In commercial hazemeters only light deviating more than 2.5° from the transmitted beam direction is considered haze. The effect of haze is to impart a cloudy or milky appearance to the sample, but its transparency (defined in the next paragraph) need not be reduced.

Transparency. For a transparent plastic material transparency is defined as the state permitting perception of objects through or beyond the specimen. A sample of low transparency may not exhibit haze, but objects seen through it will appear blurred or distorted. Transparency can be measured as that fraction of the normally incident light that is transmitted without deviation from the primary beam direction of more than 0.1° (Webber 1957).

Electrical Properties

Dielectric Constant and Loss Factor. The dielectric constant of an (insulating) material is the ratio of the capacities of a parallel plate condenser measured with and without the dielectric material placed between the plates. The difference is, of course, due to the polarization of the dielectric. If the field applied to the condenser is time dependent (as in alternating current), the polarization is time dependent also. However, because of the resistance to motion of the atoms in the dielectric, there is a delay between changes in the field and changes in the polarization. This delay is often expressed as a phase difference or *loss angle* δ. The *power factor* is then defined as sin δ, and the *dissipation factor* as tan δ. The product of the dielectric constant and the power factor is called the *loss factor*. It is proportional to the energy absorbed per cycle by the dielectric from the field. Dielectric constant and loss factor are usually measured over a frequency range from 60 cps to thousands of megacycles per second.

Resistivity. The resistance of most polymers to the flow of direct current is very high, and conductivity probably results from the presence of ionic impurities whose mobility is limited by the high viscosity of the medium. Both surface and volume resistivity are important properties for applications of polymers as insulating materials.

Dielectric Strength. Insulators will not sustain an indefinitely high voltage: As the applied voltage is increased, a point is reached where a catastrophic decrease in resistance takes place, accompanied by a physical breakdown of the dielectric. As in mechanical fatigue experiments, application of lower voltages causes eventual breakdown: Curves of time to fail versus voltage can be plotted. Such curves show two distinct regions of failure. At short times failure is presumed to occur as a result of the inability of the electrons conducting the current to dissipate rapidly enough the energy they receive from the field. Breakdown at longer times appears to be due to corona attack, which is sensitive to the atmosphere surrounding the dielectric and to the presence of mechanical strains.

Arc Resistance. The surfaces of some polymers may become carbonized and conduct current readily when exposed to an electrical discharge. This property is important for such applications as insulation for gasoline-engine ignition systems.

Electronic Properties. Recognition of the possibility of producing unusual electronic properties in polymers has led to considerable research activity and several new products in recent years. The phenomena involved include electrical conductivity, charge storage, energy transfer and contact electrification or triboelectricity. Both real and polarization charges can be stored for long periods of time by polyethylene, poly(ethylene terephthalate), and a variety of fluorine-containing polymers, the most important of which is poly(vinylidene fluoride). In the latter polymer, thermal or sonic energy can be converted to electrical energy and vice versa by pyroelectric and piezoelectric phenomena, leading to electret microphones and loudspeakers with remarkable fidelity among other new products. The electrical conductivity of such polymers as polyacetylene, $-(CH{=}CH)_x-$, and poly-p-phenylene can be no less than 20 orders of magnitude greater than that of the usual polymer insulators.

Chemical Properties

The following properties may not involve chemical treatment or attack exclusively, but are conveniently grouped together as different from those considered in previous sections.

Resistance to Solvents. The effect of solvents (broadly defined as liquids in general) on polymers may take several forms: *solubility* (Chapter 2A); *swelling,* including the absorption of water; *environmental stress cracking,* in which the specimen fails by breaking when exposed to mechanical stress in the presence of an organic liquid or an aqueous solution of a soap or other wetting agent; and *crazing,* in which a specimen fails by the development of a multitude of very small cracks in the presence of an organic liquid or its vapor, with or without the presence of mechanical stress.

Vapor Permeability. The *permeability* of a polymer to a gas or vapor is the product of the *solubility* of the gas or vapor in the polymer and its *diffusion coefficient.* Permeability is directly measured as the rate of transfer of vapor through unit thickness of the polymer in film form, per unit area and pressure difference across the film.

Weathering. For convenience and reproducibility, the behavior of materials on long exposure to weather is often simulated by exposure to *artificial weathering* sources, such as filtered carbon-arc lamps in controlled atmospheres. Such tests are usually not appreciably accelerated over natural weathering, except that they may be applied continuously.

GENERAL REFERENCES

Baer 1964; Mark 1964–1970; Ritchie 1965; Schmitz 1965, 1966, 1968; Lever 1968; Brown 1969; Collins 1973, Chapter 10B–D, Exps. 27–29; Allcock 1981, Chapter 21; Duke 1982; Horowitz 1982; Mort 1982; Seanor 1982; ASTM.

DISCUSSION QUESTIONS AND PROBLEMS

1. Which of the techniques described in this chapter could contribute to the solution of each of the following problems? Discuss the advantages and disadvantages of each. (a) Locate a crystalline melting temperature; (b) determine degree of orientation (see also Chapter 10*E*); (c) ascertain the arrangement of molecular chains in a polymer crystal; (d) locate a glass-transition temperature; (e) characterize the double bonds in a diene polymer; (f) measure an enthalpy of fusion; (g) investigate the mechanism of oxidation of a polymer; (h) study molecular motion in polymer chains; (i) estimate degree of crystallinity; (j) measure the amide content of an ester–amide copolymer; (k) determine the mechanism of polymerization of

$$CH_2CH{=}CHO$$

2. Explain how infrared spectroscopy can be used to determine copolymer reactivity ratios. Give an illustrative example and discuss limitations to the method.

3. Show how NMR can be used to (a) distinguish between head-to-head and head-to-tail polymerization in polymers and (b) distinguish between a random copolymer and a mixture of homopolymers.

4. How may x-ray diffraction data be used to estimate the sizes of polymer crystallites, and what alternative interpretations of the data exist? (See also Chapter 10C.)

5. What microscopic methods would be suitable for studying (a) the morphology of crystalline polymers (Chapter 10C), (b) surface oxidation in a hydrocarbon polymer, (c) the structure of the beads used in a GPC column (Chapter 8F), (d) suspected gel particles in branched polyethylene (Chapter 13A)?

6. With the aid of sketches of the instruments, describe the difference between DTA and DSC, and discuss the determinations for which each method is best suited.

7. Draw typical DSC and DTA thermograms for a crystalline polymer, showing the glass transition, crystallization, crystalline melting, and thermal degradation.

8. Which of the stress–strain curves in Fig. 9-8 would you expect to find for a polymer useful for (a) a gear in a machine, (b) the housing for an automatic pencil, (c) a garden hose, (d) the packing material to cushion delicate instruments, (e) the outer covering of a basketball, (f) a decorative paperweight, (g) a fiber useful for rope.

BIBLIOGRAPHY

Alexander 1969. Leroy E. Alexander, *X-Ray Diffraction Methods in Polymer Science,* John Wiley & Sons, New York, 1969.

Allcock 1981. Harry R. Allcock and Frederick W. Lampe, *Contemporary Polymer Chemistry,* Prentice-Hall, Englewood Cliffs, New Jersey, 1981.

ASTM. Annual Book of ASTM Standards, Section 8—Plastics (Vols. 08.01—08.03), American Society for Testing and Materials, Philadelphia, Pennsylvania, 1983, and annually thereafter.

Baer 1964. Eric Baer, ed., *Engineering Design for Plastics,* Reinhold, New York, 1964.

Billmeyer 1981. Fred W. Billmeyer, Jr., and Max Saltzman, *Principles of Color Technology,* 2nd ed., John Wiley & Sons, New York, 1981.

Birbeck 1961. M. S. C. Birbeck, "Techniques for the Electron Microscopy of Proteins," Chapter 1 in P. Alexander and R. J. Block, eds., *Analytical Methods of Protein Chemistry,* Vol. 3, Pergamon Press, New York, 1961.

Boerio 1972. F. J. Boerio and J. L. Koenig, "Vibrational Spectroscopy of Polymers," *J. Macromol. Sci. Rev. Macromol. Chem.* **C7,** 209–249 (1972).

Bovey 1960. F. A. Bovey and G. V. D. Tiers, "Polymer NSR Spectroscopy. II. The High Resolution Spectra of Methyl Methacrylate Polymers Prepared with Free Radical and Anionic Initiators," *J. Polym. Sci.* **44,** 173–182 (1960).

Bovey 1965. Frank A. Bovey, "Nuclear Magnetic Resonance," *Chem. Eng. News* **43** (35), 98–121 (1965).

Bovey 1968. F. A. Bovey, "Nuclear Magnetic Resonance," pp. 356–396 in Herman F. Mark, Norman

G. Gaylord, and Norbert M. Bikales, eds., *Encyclopedia of Polymer Science and Technology,* Vol. 9, Wiley-Interscience, New York, 1968.

Bovey 1969. Frank A. Bovey, *Nuclear Magnetic Resonance Spectroscopy,* Academic Press, New York, 1969.

Bovey 1972. F. A. Bovey, *High Resolution NMR of Macromolecules,* Academic Press, New York, 1972.

Bovey 1976. F. A. Bovey, "High-Resolution Carbon-13 Studies of Polymer Structure," Chapter 10 in K. J. Ivin, ed., *Structural Studies of Macromolecules by Spectroscopic Methods,* John Wiley & Sons, New York, 1976.

Boyer 1980. R. F. Boyer and S. E. Keinath, eds., *Molecular Motion in Polymers by ESR,* Harwood Academic, New York, 1980.

Brauer 1965. G. M. Brauer and G. M. Kline, "Chemical Analysis," pp. 632–665 in Herman F. Mark, Norman G. Gaylord, and Norbert M. Bikales, eds., *Encyclopedia of Polymer Science and Technology,* Vol. 3, Wiley-Interscience, New York, 1965.

Braun 1982. Dietrich Braun, *Simple Methods for Identification of Plastics,* revised ed., Macmillan, New York, 1982.

Bresler 1966. S. E. Bresler and E. N. Kazbekow, "Electron-Spin Resonance," pp. 669–692 in Herman F. Mark, Norman G. Gaylord, and Norbert M. Bikales, eds., *Encyclopedia of Polymer Science and Technology,* Vol. 5, Wiley-Interscience, New York, 1966.

Brown 1969. W. E. Brown, ed., *Testing of Polymers,* Vol. 4, Wiley-Interscience, New York, 1969.

Brown 1978. D. S. Brown and R. E. Wetton, "Recent Advances in the Study of Polymers by Small Angle X-Ray Scattering," Chapter 6 in J. V. Dawkins, ed., *Developments in Polymer Characterisation—1,* Applied Science, London, 1978.

Bullock 1976. A. T. Bullock and G. G. Cameron, "E.s.r. Studies of Spin-Labeled Synthetic Polymers," Chapter 15 in K. J. Ivin, ed., *Structural Studies of Macromolecules by Spectroscopic Methods,* John Wiley & Sons, New York, 1976.

Cassel 1962. James M. Cassel, "Chromatography," Chapter 10 in Gordon M. Kline, ed., *Analytical Chemistry of Polymers,* Part II, Wiley-Interscience, New York, 1962.

Chiu 1966. Jen Chiu, "Applications of Thermogravimetry to the Study of High Polymers," *Appl. Polym. Symp.* **2,** 25–42 (1966).

Collins 1973. Edward A. Collins, Jan Bareš, and Fred W. Billmeyer, Jr., *Experiments in Polymer Science,* Wiley-Interscience, New York, 1973.

Cross 1950. L. H. Cross, R. B. Richards, and H. A. Willis, "The Infrared Spectrum of Ethylene Polymers," *Disc. Faraday Soc.* **9,** 235–256 (1950).

Cunliffe 1978. A. V. Cunliffe, "^{13}C NMR Spectroscopy of Polymers," Chapter 1 in J. V. Dawkins, ed., *Developments in Polymer Characterisation—1,* Applied Science, London, 1978.

D'Esposito 1978. L. D'Esposito and J. L. Koenig, "Applications of Fourier Transform IR to Synthetic Polymers and Biological Macromolecules," Chapter 2 in John R. Ferraro and Louis J. Basile, eds., *Fourier Transform IR Spectroscopy; Applications to Chemical Systems,* Vol. 1, Academic Press, New York, 1978.

Duch 1970. Michael W. Duch and David M. Grant, "Carbon-13 Chemical Shift Studies of the 1,4-Polybutadienes and the 1,4-Polyisoprenes," *Macromolecules* **3,** 165–174 (1970).

Duke 1982. Charles B. Duke and Harry W. Gibson, "Polymers, Conductive," pp. 755–793 in Martin Grayson, ed., *Kirk–Othmer Encyclopedia of Chemical Technology,* 3rd ed., Vol. 18, Wiley-Interscience, New York, 1982.

Elliott 1969. Arthur Elliott, *Infrared Spectra and Structure of Organic Long-Chain Polymers,* St. Martin's Press, New York, 1969.

Feinland 1965. R. Feinland, "Chromatography," pp. 731–762 in Herman F. Mark, Norman G. Gaylord, and Norbert M. Bikales, eds., *Encyclopedia of Polymer Science and Technology,* Vol. 3, Wiley-Interscience, New York, 1965.

Fischer 1966. E. W. Fischer and H. Goddar, "Electron-Diffraction Analysis," pp. 641–661 in Herman F. Mark, Norman G. Gaylord, and Norbert M. Bikales, eds., *Encyclopedia of Polymer Science and Technology,* Vol. 5, Wiley-Interscience, New York, 1966.

Geil 1966. Phillip H. Geil, "Electron Microscopy," pp. 662–669 in Herman F. Mark, Norman G. Gaylord, and Norbert M. Bikales, eds., *Encyclopedia of Polymer Science and Technology,* Vol. 5, Wiley-Interscience, New York, 1966.

Hartshorne 1964. N. H. Hartshorne and A. Stuart, *Practical Optical Crystallography,* 2nd ed., American Elsevier, New York, 1964.

Haslam 1965. John Haslam and Harry A. Willis, *Identification and Analysis of Plastics,* D. Van Nostrand, Princeton, New Jersey, 1965.

Hemsley 1978. D. Hemsley, "Microscopy of Polymer Surfaces," Chapter 8 in J. V. Dawkins, ed., *Developments in Polymer Characterisation—1,* Applied Science, London, 1978.

Hendra 1969. P. J. Hendra, "Laser-Raman Spectra of Polymers," *Adv. Polym. Sci.* **6,** 151–169 (1969).

Herglotz 1975. H. K. Herglotz, "Characterization of Polymers by Unconventional X-Ray Techniques," Chapter 5 in John J. Burke and Volker Weiss, eds., *Characterization of Materials in Research: Ceramics and Polymers,* Syracuse University Press, Syracuse, New York, 1975.

Horowitz 1982. Emanuel Horowitz, "Plastics Testing," pp. 207–228 in Martin Grayson, ed., *Kirk–Othmer Encyclopedia of Chemical Technology,* 3rd ed., Vol. 18, Wiley-Interscience, New York, 1982.

Hosemann 1962. R. Hosemann and S. N. Bagchi, *Direct Analysis of Diffraction by Matter,* North-Holland, Amsterdam, 1962.

Hummel 1966. Dieter O. Hummel, *Infrared Spectra of Polymers: In the Medium and Long Wavelength Region,* John Wiley & Sons, New York, 1966.

Kakudo 1972. M. Kakudo and N. Kasai, *X-Ray Diffraction by Polymers,* Elsevier, New York, 1972.

Ke 1960. Bacon Ke, "Application of Differential Thermal Analysis to High Polymers," pp. 361–392 in John Mitchell, Jr., I. M. Kolthoff, E. S. Proskauer, and A. Weissberger, eds., *Organic Analysis,* Vol. 4, Interscience, New York, 1960.

Ke 1966. Bacon Ke, "Differential Thermal Analysis," pp. 37–65 in Herman F. Mark, Norman G. Gaylord, and Norbert M. Bikales, eds., *Encyclopedia of Polymer Science and Technology,* Vol. 5, Wiley-Interscience, New York, 1966.

Kline 1959. Gordon M. Kline, ed., *Analytical Chemistry of Polymers,* Part I, Interscience, New York, 1959.

Koenig 1971. J. L. Koenig, "Raman Scattering of Synthetic Polymers—A Review," *Appl. Spec. Rev.* **4,** 233–306 (1971).

Kössler 1967. Ivo Kössler, "Infrared-Absorption Spectroscopy," pp. 620–642 in Herman F. Mark, Norman G. Gaylord, and Norbert M. Bikales, eds., *Encyclopedia of Polymer Science and Technology,* Vol. 7, Wiley-Interscience, New York, 1967.

Krause 1983. Anneliese Krause, Anfried Lange, and Myer Ezrin, *Plastics Analysis Guide,* Macmillan, New York, 1983.

Lebedev 1960. Ya. S. Lebedev, Iu. D. Tsvetkov, and V. V. Voevodskii, "Paramagnetic Resonance Spectra of Fluoroalkyl and Nitrose-Fluoroalkyl Radicals in Irradiated Teflon," *Opt. Spectrosc.* **8,** 426–428 (1960).

Lever 1968. A. E. Lever and J. A. Rhys, *Properties and Testing of Plastics Materials,* 3rd ed., Chemical Rubber Co., New York, 1968.

Mackenzie 1970. R. C. Mackenzie, *Differential Thermal Analysis, Fundamental Aspects,* Vol. 1, Academic Press, New York, 1970.

Mark 1964–1970. Herman F. Mark, Norman G. Gaylord, and Norbert M. Bikales, eds., *Encyclopedia of Polymer Science and Technology,* Wiley-Interscience, New York, 14 volumes + suppl., 1964–1970.

Miller 1969. Gerald W. Miller, "The Thermal Characterization of Polymers," *Appl. Polym. Symp.* **10**, 35–72 (1969).

Miller 1975. Robert L. Miller, "Crystallographic Data for Various Polymers," pp. III-1–III-137 in J. Brandrup and E. H. Immergut, eds., with the collaboration of W. McDowell, *Polymer Handbook*, 2nd ed., Wiley-Interscience, New York, 1975.

Mort 1982. J. Mort and G. Pfister, eds., *Electronic Properties of Polymers*, Wiley-Interscience, New York, 1982.

Rånby 1977. B. Rånby and J. F. Rabek, *ESR Spectroscopy in Polymer Research*, Springer, New York, 1977.

Reich 1971. Leo Reich and David W. Levi, "Thermogravimetric Analysis," pp. 1–41 in Herman F. Mark, Norman G. Gaylord, and Norbert M. Bikales, eds., *Encyclopedia of Polymer Science and Technology*, Vol. 14, Wiley-Interscience, New York, 1971.

Rexroad 1958. Harvey N. Rexroad and Walter Gordy, "Electron Spin Resonance Studies of Irradiated Teflon: Effects of Various Gases," *J. Chem. Phys.* **30**, 399–403 (1958).

Richardson 1978. M. J. Richardson, "Quantitative Differential Scanning Calorimetry," Chapter 7 in J. V. Dawkins, ed., *Developments in Polymer Characterisation—1*, Applied Science, London, 1978.

Ritchie 1965. P. D. Ritchie, ed., *Physics of Plastics*, D. Van Nostrand, Princeton, New Jersey, 1965.

Saunders 1966. K. J. Saunders, *The Identification of Plastics and Rubbers*, Chapman and Hall, London, 1966.

Schaefer 1974. Jacob Schaefer, "The Carbon-13 NMR Analysis of Synthetic High Polymers," Chapter 4 in George C. Levy, ed., *Topics in Carbon-13 NMR Spectroscopy*, Vol. 1, Wiley-Interscience, New York, 1974.

Schaefer 1976. J. Schaefer, "The Analysis of ^{13}C n.m.r. Relaxation Experiments on Polymers," Chapter 11 in K. J. Ivin, ed., *Structural Studies of Macromolecules by Spectroscopic Methods*, John Wiley & Sons, New York, 1976.

Schaefer 1979. Jacob Schaefer and E. O. Stejskal, "High-Resolution ^{13}C NMR of Solid Polymers," Chapter 4 in George C. Levy, ed., *Topics in Carbon-13 NMR Spectroscopy*, Vol. 3, Wiley-Interscience, New York, 1979.

Schmitz 1965. John V. Schmitz, ed., *Testing of Polymers*, Vol. 1, Wiley-Interscience, New York, 1965.

Schmitz 1966. John V. Schmitz, ed., *Testing of Polymers*, Vol. 2, Wiley-Interscience, New York, 1966.

Schmitz 1968. John V. Schmitz, ed., *Testing of Polymers*, Vol. 3, Wiley-Interscience, New York, 1968.

Schwenker 1969. Robert F. Schwenker, Jr. and Paul D. Garn, eds., *Thermal Analysis, Instrumentation, Organic Materials and Polymers*, Vol. 1, Academic Press, New York, 1969.

Seanor 1982. Donald A. Seanor, ed., *Electrical Properties of Polymers*, Academic Press, New York, 1982.

Sedgwick 1978. R. D. Sedgwick, "Mass Spectrometry," Chapter 2 in J. V. Dawkins, ed., *Developments in Polymer Characterisation—1*, Applied Science, London, 1978.

Siesler 1980. H. W. Siesler and K. Holland-Moritz, *IR and Raman Spectroscopy of Polymers*, Marcel Dekker, New York, 1980.

Sjöstrand 1967. Fritiof S. Sjöstrand, *Electron Microscopy of Cells and Tissues, Instrumentation and Techniques*, Vol. 1, Academic Press, New York, 1967.

Slade 1966–1970. Philip E. Slade, Jr. and Lloyd T. Jenkins, eds., *Techniques and Methods of Polymer Evaluation*, Vol. 1, *Thermal Analysis*, 1966; Vol 2, *Thermal Characterization Techniques*, 1970; Marcel Dekker, New York.

Smith 1956. Carl W. Smith and Malcolm Dole, "Specific Heat of Synthetic High Polymers. VII. Polyethylene Terephthalate," *J. Polym. Sci.* **20**, 37–56 (1956).

Smothers 1966. W. J. Smothers and Y. Chiang, *Handbook of Differential Thermal Analysis*, Chemical Publishing Co., New York, 1966.

Starkweather 1956. Howard W. Starkweather and Robert E. Moynihan, "Density, Infrared Absorption, and Crystallinity in 66 and 610 Nylons," *J. Polym. Sci.* **22,** 363–368 (1956).

Stevens 1969. Malcolm P. Stevens, *Characterization and Analysis of Polymers by Gas Chromatography,* Marcel Dekker, New York, 1969.

Sudol 1969. Robert S. Sudol, "Determination of Polymer Structure by High-Resolution Nuclear Magnetic Resonance Spectroscopy," *Anal. Chim. Acta* **46,** 231–237 (1969).

Thornton 1968. P. R. Thornton, *Scanning Electron Microscopy; Applications to Materials and Device Science,* Chapman and Hall, London, 1968.

Turi 1981. Edith A. Turi, ed., *Thermal Characteristics of Polymeric Materials,* Academic Press, New York, 1981.

Wake 1969. William C. Wake, *The Analysis of Rubber and Rubber-Like Polymers,* 2nd ed., John Wiley & Sons, New York, 1969.

Wall 1962. Leo A. Wall, "Mass Spectrometry," Chapter 6 in Gordon M. Kline, ed., *Analytical Chemistry of Polymers,* Part II, Wiley-Interscience, New York, 1962.

Ward 1957. I. M. Ward, "Configurational Changes in Polyethylene Terephthalate," *Chem. Ind.* **1957,** 1102 (1957).

Webber 1957. Alfred C. Webber, "Method for the Measurement of Transparency of Sheet Materials" *J. Opt. Soc. Am.* **47,** 785–789 (1957).

Winding 1961. Charles C. Winding and Gordon D. Hiatt, *Polymeric Materials,* McGraw-Hill, New York, 1961.

Wunderlich 1970. B. Wunderlich and H. Baur, "Heat Capacities of Linear High Polymers," *Adv. Polym. Sci.* **7,** 151–368 (1970).

Wunderlich 1971. Bernhard Wunderlich, "Differential Thermal Analysis," Chapter 8 in Arnold Weissberger and Bryant W. Rossiter, eds., *Physical Methods of Chemistry,* Vol. 1 of A. Weissberger, ed., *Techniques of Chemistry,* Part 5, Wiley-Interscience, New York, 1971.

Wunderlich 1973. Bernhard Wunderlich, *Macromolecular Physics, Crystals, Structure, Morphology, and Defects,* Vol. 1, Academic Press, New York, 1973.

Zbinden 1964. Rudolf Zbinden, *Infrared Spectroscopy of High Polymers,* Academic Press, New York, 1964.

PART FOUR

STRUCTURE AND PROPERTIES

CHAPTER TEN

MORPHOLOGY AND ORDER IN CRYSTALLINE POLYMERS

A. CONFIGURATIONS OF POLYMER CHAINS

In this book the word *configuration* is used to describe those arrangements of atoms that cannot be altered except by breaking and reforming primary chemical bonds. In contrast, as discussed in Chapter 7*B*, arrangements that can be altered by rotating groups of atoms around single bonds are called *conformations*. Examples involving conformations of polymer chains include trans versus gauche arrangements of consecutive carbon–carbon single bonds and the helical arrangements found in some polymer crystal structures (Section *B*). Examples involving configurations include head-to-head, tail-to-tail, and head-to-tail arrangements in vinyl polymers (Chapter 3*A*) and the several *stereoregular* arrangements described in this section, including 1,2- and 1,4-addition, cis and trans isomers, and arrangements around asymmetric carbon atoms.

Configurations Involving an Asymmetric Carbon Atom

Staudinger (1932) recognized early that polymers of monosubstituted olefins should contain a series of asymmetric carbon atoms along the chain, and much speculation followed over the possibility of synthesizing such polymers in stereoregular forms. As discussed in Chapter 4*D*, such syntheses have been achieved by several routes. The regular structure of the resulting polymers, notably poly(α-olefins), was recognized by Natta (1955, 1959*a*), who devised the nomenclature now accepted (Huggins 1962) to describe stereoregular polymers of this type. As indicated in Fig.

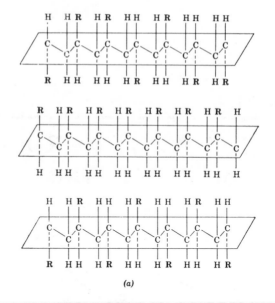

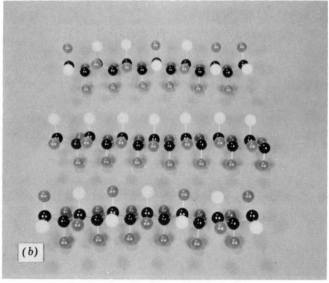

FIG. 10-1. Diagrams (a) and models (b) showing the irregular atactic (top) and the stereoregular isotactic (center) and syndiotactic (bottom) configurations in a vinyl polymer (Natta 1959b). The main carbon–carbon chain is depicted in the fully extended (all trans) planar zigzag conformation.

10-1, where the polymer chain is depicted in the fully extended (all trans) planar zigzag conformation, the configuration resulting when all the substituent groups R on the vinyl polymer lie above (or below) the plane of the main chain is called *isotactic*. If substituent groups lie alternately above and below the plane, the configuration is called *syndiotactic,* whereas a random sequence of positions is said to lead to the *atactic* configuration. More complicated arrangements and their nomenclature are beyond the scope of this book.

The synthesis of isotactic and syndiotactic polymers has now become almost commonplace, and many examples of each type are known. In addition to poly(α-olefins), polystyrene and poly(methyl methacrylate) are examples of polymers that can be made in the isotactic configuration, while poly(methyl methacrylate) and polybutadiene in which 1,2-addition occurs exclusively, giving the structure

$$
\begin{array}{c}
CH{=}CH_2 \\
| \\
-CH_2-CH-CH_2-CH- \\
| \\
CH{=}CH_2
\end{array}
$$

are examples of polymers that can be prepared in the syndiotactic configuration. It has been shown that some stereoregular polymers had been made earlier but were not recognized at the time, a notable example being poly(vinyl isobutyl ether) (Schildknecht 1948), which is isotactic.

As shown in Section *B*, whether a polymer is isotactic or syndiotactic usually determines its crystal structure, and the assignment of an all-isotactic or all-syndiotactic structure to a polymer can be made from its crystal structure. The use of NMR spectroscopy (Chapter 9*B*) is a more powerful technique, however, since it allows determination of the stereoregular configuration of successive monomers in sequences up to about 5 units long. For example, examination of the resonance absorption of the methylene protons in poly(methyl methacrylate) allows one to distinguish between and determine the relative numbers of sequences of two monomers (*dyads*) that have syndiotactic (*racemic, r*) and isotactic (*meso, m*) symmetry. The α-methyl proton resonance allows estimation of the numbers of 3-monomer sequences (*triads*) with configurations *mm, mr,* and *rr*. Use of ^{13}C-NMR allows the configurations of tetrads and pentads to be determined. The data can be interpreted to indicate whether the probability P_m of adding the next monomer with the same (meso) configuration as the last one added is independent of the configuration of the growing chain, or not, and if the former, to determine this probability. Typical of the results are that for predominantly syndiotactic poly(methyl methacrylate), $P_m = 0.24$; for free-radical poly(vinyl chloride), $P_m = 0.43$, corresponding to a slightly syndiotactic but nearly atactic structure (for which $P_m = 0.50$); and for predominantly isotactic poly(methyl methacrylate) and polypropylene, the probability of stereoregularity of propagation is more complex (Bovey 1979). For most polymers tacticity changes very little with the temperature of polymerization.

Head-to-Head, Tail-to-Tail Configurations

Nuclear magnetic resonance spectroscopy (Chapter 9B) has provided the first detailed information on the frequency of occurrence of a monomer adding in the
reverse fashion, to provide an occasional head-to-head, tail-to-tail instead of the
normal head-to-tail sequence. An example is the use of ^{19}F-NMR to study triad
sequences in poly(vinylidine fluoride), in which it has been found that a monomer
adds in reverse fashion 5–6% of the time, but the "mistake" is immediately rectified
with the next addition.

Configurations Involving a Carbon–Carbon Double Bond

As discussed in Chapter 6D, polymers of 1,3-dienes, containing one residual double
bond per repeat unit after polymerization, can contain sequences with several different configurations. For a monosubstituted butadiene such as isoprene, the following structures are possible:

$$
\begin{array}{cc}
\underset{R}{\overset{-CH_2}{\diagdown}}C=C\underset{H}{\overset{CH_2-}{\diagup}} & \underset{R}{\overset{-CH_2}{\diagdown}}C=C\underset{CH_2-}{\overset{H}{\diagup}} \\
\textit{cis-}1,4 & \textit{trans-}1,4
\end{array}
$$

$$
\begin{array}{cc}
\overset{R}{\underset{CH=CH_2}{-CH_2-C-}} & \overset{-CH_2-CH-}{\underset{R}{C=CH_2}} \\
1,2 & 3,4
\end{array}
$$

Stereoregular polymers with several of these configurations are known (Chapter
13D,E). Their structures are confirmed by x-ray diffraction and infrared spectroscopy.

Other Stereoregular Configurations

There is no doubt that stereoregularity plays an important role in the structures of
proteins, nucleic acids, and other substances of biological importance. Deoxyri-
bonucleic acid (DNA), for example, has a highly stereoregular double helix structure, now famous (Watson 1953) (Fig. 10-2). Other stereoregular structures in
nonbiological polymers are also well known (Chapter 15).

Optically Active Polymers

Many polymers, both biological and synthetic, are capable of rotating the plane of
polarization of light and are said to be optically active. While in simple low-
molecular-weight compounds optical activity is associated with the presence of

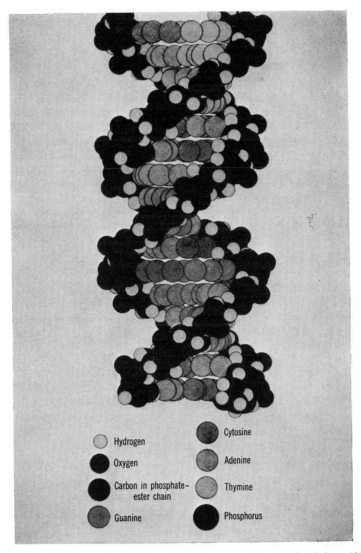

FIG. 10-2. The famous double helix of DNA (Kornberg 1962, after Feughelman 1955).

asymmetric carbon atoms, this is not universally true in polymers. Staudinger (1932) pointed out that every second chain carbon atom in vinyl polymers is formally asymmetric. Yet such polymers are not usually optically active, even when isotactic or syndiotactic, because of intramolecular compensation.

Optically active polymers can be prepared, however, in several ways (Schulz 1968): (1) polymerization of optically active monomers in such a way that their asymmetric centers are preserved; (2) introduction of activity into optically inactive polymers by reactions that place asymmetric centers on side chains; (3) stereospecific polymerization of optically inactive monomers with optically active catalysts in such a manner that the repeat units have one or more asymmetric carbon atoms of a given configuration; or (4) polymerization of racemic monomer mixtures to yield polymer from only one of the monomers (as by using a selective initiator) or from both polymerizing simultaneously and independently to yield a mixture of dextrorotatory and levorotatory polymer which can be separated.

An example of an optically active polymer is that made from L-propylene oxide (Price 1956):

$$-CH_2-\underset{\underset{CH_3}{|}}{\overset{\overset{H}{|}}{C^*}}-O-CH_2-\underset{\underset{CH_3}{|}}{\overset{\overset{H}{|}}{C^*}}-O-$$

GENERAL REFERENCES

Natta 1967; Farina 1968; Schulz 1968; Morawetz 1975, Chapter 3; Tadokoro 1979.

B. CRYSTAL STRUCTURES OF POLYMERS

Structural Requirements for Crystallinity

Many polymers, including most fibers, are partially crystalline. The most direct evidence of this fact is provided by x-ray diffraction studies. The x-ray patterns of crystalline polymers show both sharp features associated with regions of three-dimensional order, and more diffuse features characteristic of molecularly disordered substances like liquids. The occurrence of both types of feature is evidence that ordered and disordered regions coexist in most crystalline polymers. Additional evidence to this effect comes from other polymer properties, such as densities, which are intermediate between those calculated for completely crystalline and amorphous species.

The close relation between regularity of molecular structure and crystallizability has long been recognized. Typical crystalline polymers are those whose molecules are chemically and geometrically regular in structure. Occasional irregularities, such as chain branching in polyethylene or copolymerization of a minor amount of isoprene with polyisobutylene as in butyl rubber, limit the extent of crystallization but do not prevent its occurrence. Typical noncrystalline polymers, on the other hand, include those in which irregularity of structure occurs: copolymers with significant amounts of two or more quite different monomer constituents, or atactic polymers.

Before the recognition of the importance of stereoregularity to crystallinity in polymers, it was thought that the bulkiness of the side groups in some atactic polymers, such as polystyrene and poly(methyl methacrylate), alone prevented their crystallization. The argument was based in part on the fact that poly(vinyl alcohol) can be made to crystallize, although it is derived by hydrolysis from poly(vinyl acetate), which has never been crystallized. The crystallinity of poly(vinyl alcohol) was once taken as evidence that it was isotactic. Therefore poly(vinyl acetate) should have been similarly isotactic and would be expected to crystallize—perhaps not in a completely extended conformation—despite the bulkiness of the acetate groups. Poly(vinyl alcohol) is in fact atactic. Other irregular polymers, such as copolymers of ethylene and vinyl alcohol, ethylene and tetrafluoroethylene, and ethylene and carbon monoxide, are also crystalline. All these polymers have very similar structures. It is now clear that the CH_2, CHOH, CF_2, and C=O groups are near enough to the same size to fit into similar crystal lattices despite the stereochemical irregularity of the polymers. On the other hand, the CHCl group is apparently too large, since chlorinated polyethylene is noncrystalline. The absence of crystallinity in poly(vinyl acetate) and similar polymers is due to the combination of their atactic structure and the size of their substituent groups.

When the disordering features of atactic configuration and bulky side groups are combined with strong interchain forces, the polymer may have a structure that is intermediate in kind between crystalline and amorphous. An example is polyacrylonitrile, in which steric and intramolecular dipole repulsions lead to a stiff, irregularly twisted, backbone chain conformation. The stiff molecular chains pack like rigid rods in a lattice array; lateral order between chains is present, but longitudinal order along chains is absent.

Structures Based on Extended Chains

There is little doubt that the fully extended planar zigzag is the conformation of minimum energy for an isolated section of a hydrocarbon chain, the energy of the trans conformation being some 3.3 kJ/mole less than that of the gauche. It is to be expected, therefore, that fully extended chain conformations will be favored in polymer crystal structures unless substituents on the chains cause steric hindrance relieved by the assumption of other forms. Fully extended chains are found in the crystal structures of polyethylene, poly(vinyl alcohol), syndiotactic polymers including poly(vinyl chloride) and poly(1,2-butadiene), most polyamides, and cellulose.

Polyethylene. Except for end groups, the arrangement of chains in the usual orthorhombic structure of polyethylene is essentially the same as that in crystals of linear paraffin hydrocarbons containing 20–40 carbon atoms. The unit cell of polyethylene is rectangular with dimensions $a = 7.41$ Å, $b = 4.94$ Å, and $c = 2.55$ Å. The latter is the chain repeat distance, which is identical with the fully extended zigzag repeat distance. The arrangement of the chains with respect to one another is shown in Fig. 10-3. The positions of the hydrogen atoms are not detectable by the x-ray method (since the scattering power of an atom is proportional to its atomic weight) but have been calculated assuming tetrahedral bonds and a C—H distance of 1.10 Å.

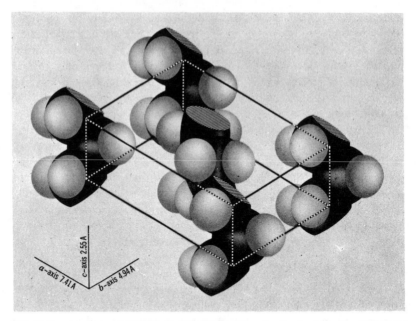

FIG. 10-3. Arrangement of chains in the orthorhombic unit cell of polyethylene (Geil 1963).

Polyethylene, like many other polymers, exhibits *polymorphism,* the ability to exist in more than one crystal structure, depending on conditions. When crystalline polyethylene is stressed, as in cold drawing (Section *E*), the rectangular face of the unit cell perpendicular to the chain direction is distorted into a parallelogram with one included angle, designated β, becoming 107.9°. The lengths of the axes are altered to $a = 8.09$ Å and $b = 4.79$ Å. The chain formerly in the center of the unit cell now lies in the center of the a axis. This unit cell is monoclinic. At high pressures another, hexagonal, structure is formed in which the order associated with the all-trans conformation is lost and the molecules pack like rods (Bassett 1976). Similar structures exist in normal paraffins at temperatures close to their melting points.

Poly(vinyl alcohol). The crystal structure of poly(vinyl alcohol) is similar to that of polyethylene, as might be expected, since the CHOH group is small enough to fit into the polyethylene structure in place of a CH_2 group with little difficulty. The unit cell is monoclinic with $a = 7.81$ Å, $b = 2.52$ Å (chain axis), $c = 5.51$ Å; $\beta = 91°42'$. Pairs of chains are linked together by hydrogen bonds, and similarly linked into sheets, insofar as the stereochemical irregularity allows.

Poly(vinyl chloride). Crystallinity in free-radical poly(vinyl chloride) is not well developed; the polymer is primarily syndiotactic but with considerable irregularity. Its crystal structure appears to be typical of syndiotactic polymers, with a repeat distance corresponding to four chain carbon atoms. Poly(1,2-butadiene) has a similar structure, but the carbon–carbon chains are slightly distorted from the trans conformation to relieve repulsion between side-chain and neighboring-chain atoms.

Polyamides. The structures of poly(hexamethylene adipamide) (66-nylon), poly(hexamethylene sebacamide) (610-nylon), and polycaprolactam (6-nylon) are made up of fully extended chains linked together by hydrogen bonds to form sheets that may be packed together in two different ways, giving two crystal modifications. Oxygen atoms of one molecule are always found opposite NH groups of a neighboring molecule, with the N—H---O distance of 2.8 Å smaller than normal because of hydrogen bond formation. Other nylons (99-, 106-, 1010-, and 11-nylon) contain chains slightly distorted from the planar zigzag form.

Cellulose. The crystal structure of cellulose is of historic interest, since it represents the first instance in which the concepts of a crystal unit cell and a chain structure were reconciled by the concept of the chain extending through successive unit cells. The basic feature of the crystal structure of cellulose is the repeating segment of two cellobiose units, joined in 1,4-linkages, with the —CH₂OH side chains in successive units occurring on opposite sides of the chain. Many forms and derivatives of cellulose have crystal structures in which these chains are packed in slightly different ways.

Distortions From Fully Extended Chains

Polyesters. In most aliphatic polyesters and in poly(ethylene terephthalate) the polymer chains are shortened by rotation about the C—O bonds to allow close packing. As a result, the main chains are no longer planar. The group

$$
\begin{array}{ccc}
\mathrm{O} & & \mathrm{O} \\
\| & & \| \\
\mathrm{C} & \langle\bigcirc\rangle & \mathrm{C}
\end{array}
$$

in poly(ethylene terephthalate) is planar, however, as required by resonance.

Polyisoprenes and Polychloroprene. Polymers of isoprene and chloroprene in which 1,4-addition predominates can occur in all-cis or all-trans configurations; natural rubber and gutta-percha are naturally occurring *cis*-1,4-polyisoprene and *trans*-1,4-polyisoprene, respectively. All these polymers have similar crystal structures except that the repeat distance of the trans polymers corresponds to one monomer unit and that of the cis polymers to two.

The repeat distance of *trans*-1,4-polyisoprene (gutta-percha) (4.72–4.77 Å) is smaller than that corresponding to a fully extended chain (5.04 Å).

The structure of natural rubber (*cis*-polyisoprene) is similar to that of gutta-percha, except as required by the longer repeat distance due to the cis configuration.

The chain is slightly out of plane at the single bonds and in plane around the double bond, the observed repeat distance of 8.1 Å being less than the extended distance of 10.08 Å.

The structure of polychloroprene is similar to that of gutta-percha except that the molecules are oriented differently with respect to one another in the crystal because of differences in polarity between the chlorine and methyl groups.

Polypeptides. Proteins, including the natural fibers wool and silk, are made up of α-amino acids joined into long polypeptide chains with the planar structure

$$
\begin{array}{ccc}
\text{H} & & \text{CHR—} \\
& \text{N—C} & \\
\text{—CHR} & & \text{O}
\end{array}
$$

where R represents the amino acid chain. These polymer chains can exist in two quite different types of crystal structure.

One of these forms, the β-keratin structure, is made up of nearly extended polypeptide chains arranged into sheets through hydrogen bonding. The sheets are not quite planar and have been described as pleated or rippled. A number of possible structures have been proposed, differing in the detailed arrangement of the chains into sheets. One of these, the antiparallel-chain, pleated sheet structure, is now considered the most likely structure for silk fibroin.

The helical α-keratin structure of polypeptides is described below.

Helical Structures

Polymers with bulky substituents closely spaced along the chain often take on a helical conformation in the crystalline phase, since this allows the substituents to pack closely without appreciable distortion of chain bonds. Most isotactic polymers, as well as polymers of some 1,1-disubstituted ethylenes such as isobutylene, fall in this class. Other helical structures of interest are those of polytetrafluoroethylene and the α-keratin structure.

Isotactic Polymers. Many isotactic polymers crystallize with a helical conformation in which alternate chain bonds take trans and gauche positions. For the gauche positions the rotation is always in that direction that relieves steric hindrance by placing R and H groups in juxtaposition, generating either a left-hand or a right-hand helix.

If the side group is not too bulky, the helix has exactly three units per turn and the arrangement is similar to that in Fig. 10-4a. This arrangement is found in isotactic polypropylene, one form of poly(1-butene), polystyrene, and others. More bulky side groups require more space, resulting in the formation of looser helices as shown in Fig. 10-4b–d. Isotactic poly(methyl methacrylate) forms a helix with five units in two turns, while polyisobutylene forms a helix with eight units in five turns.

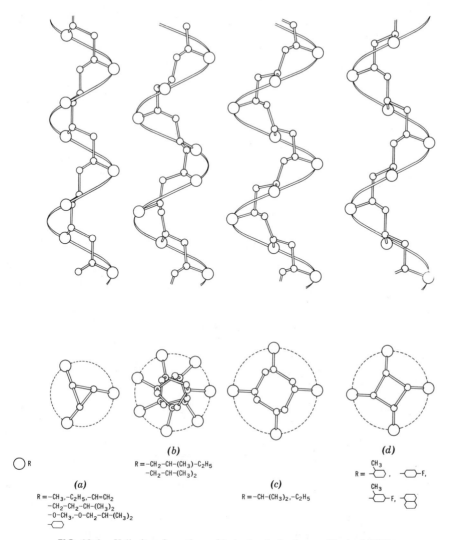

FIG. 10-4. Helical conformations of isotactic vinyl polymers (Gaylord 1959).

Polypeptides. Helical structures are of great importance in biological materials. In addition to helical conformations proposed for DNA and other polymers of biological interest, the α-keratin structure assumed by many polypeptides is a helix, in which about 3.6–3.7 polypeptide residues occur per turn (Fig. 10-5). Adjacent turns are held together by intramolecular hydrogen bonding. Several variations of this structure have been proposed.

Polytetrafluoroethylene. This polymer exists in two helical conformations that may be described as twisted ribbons in which the fully extended planar form is distorted to have a 180° twist in 13 CF_2 units in the more stable (low-temperature)

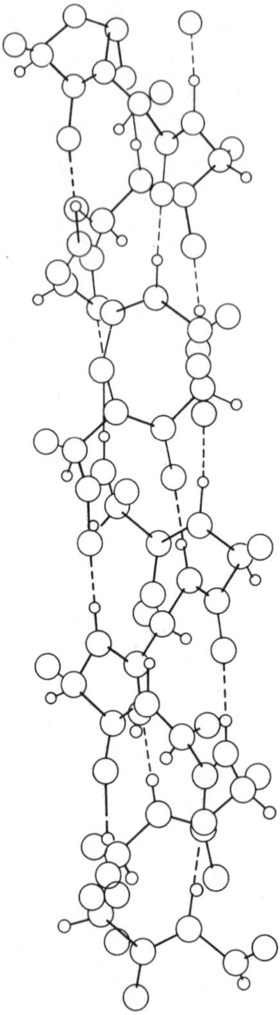

FIG. 10-5. Helical structure of α-keratin proteins (Pauling 1951).

form. Above 19°C this form is replaced by a slightly untwisted conformation with 15 CF_2 units per half-twist. Above 19°C the x-ray diffraction pattern shows diffuse streaks that are interpreted as resulting from small angular displacements of molecular segments about their long axes. Above 30°C more diffuseness occurs in the x-ray pattern, increasing as the temperature is raised to and above the melting point, 327°C. This additional diffuseness is attributed to random angular displacement of the molecules about their long axes.

In terms of molecular motion, it is considered that the molecules undergo slight torsional motion—that is, twisting and untwisting of the helices—that causes the lattice to expand laterally, and at 19°C is strong enough to cause the untwisting from 13 to 15 CF_2 groups in each 180° twist. Between 19 and 30°C the molecules are "locked" into the threefold helix corresponding to the 15 CF_2 repeat distance; their motion is restricted to small-angle oscillations about their "rest positions." Above 30°C the molecules lose their preferred angular orientation, becoming randomly misaligned about their axes. Nearly cylindrical in shape, they pack like parallel rods. Above 30°C torsional motion increases, with the twisting and untwisting becoming more pronounced, until at T_m all crystalline order is lost.

In this respect, polytetrafluoroethylene is an example of a newly recognized type of polymer mesophase, termed a *condis* (for conformationally disordered) crystal. In such a mesophase cooperative motion between various conformational isomers is permitted (Wunderlich 1983).

Helix-Coil Transitions in Polypeptides   (Poland 1970). The α-keratin helix structure that has been described as characteristic of polypeptides and many polymers of biological origin can undergo several phase transitions, including melting to random coils at T_m or in the presence of a solvent. A transition of particular interest results from the fact that the helix can exist as a stable conformation in solution, so that the helix-coil transition can occur without involving a solid phase.

The stability of the helix in solution results from intramolecular hydrogen bonding. Since disruption of the helix requires breaking about three such bonds per turn, some 20–40 kJ of enthalpy must be expended to realize the greater entropy of the random-coil state. Once the transition is initiated, by a change in temperature or composition, the preferred conformation is perpetuated over many bonds. In the simplest case the helical conformation occurs either not at all or in long sequences with infrequent interruption if not throughout the chain. The change is fairly abrupt and resembles a phase transition.

GENERAL REFERENCES

Geil 1963; Miller 1966; Corradini 1968; Bryan 1969; Wunderlich 1973; Miller 1975; Tadokoro 1979; Fava 1980.

C. MORPHOLOGY OF CRYSTALLINE POLYMERS

Polymer Single Crystals

Although the formation of single crystals of polymers was observed during polymerization many years ago, it was long believed that such crystals could not be produced from polymer solutions because of molecular entanglement. In 1953 (Schlesinger 1953) and in several laboratories in 1957 (Fischer 1957, Keller 1957, Till 1957), the growth of such crystals was reported. The phenomenon has been reported for so many polymers, including gutta-percha, polyethylene, polypropylene and other poly(α-olefins), polyoxymethylene, polyamides, and celluose and its derivatives, that it appears to be quite general and universal (Geil 1963, Blackadder 1967).

Lamellae.   All the structures described as polymer single crystals have the same general appearance, being composed of thin, flat platelets (lamellae) about 100 Å thick and often many micrometers in lateral dimensions. They are usually thickened by the spiral growth of additional lamellae from screw dislocations. A typical lamellar crystal is shown in Fig. 10-6. The size, shape, and regularity of the crystals depend on their growth conditions, such factors as solvent, temperature, and growth rate being important. The thickness of the lamellae depends on crystallization temperature and any subsequent annealing treatment (Hendra 1976).

Electron-diffraction measurements, performed in almost all the researches cited above, indicate that the polymer chains are oriented normal or very nearly normal to the plane of the lamellae. Since the molecules in the polymers are at least 1000 Å long and the lamellae are only about 100 Å thick, the only plausible explanation is that the chains are folded (Keller 1957). This arrangement has been shown to be sterically possible. In polyethylene, for example, the molecules can fold in such a way that only about five chain carbon atoms are involved in the fold itself.

Many single crystals of essentially linear polyethylene show secondary structural features, including corrugations (Fig. 10-7) and pleats (Fig. 10-8). Both these

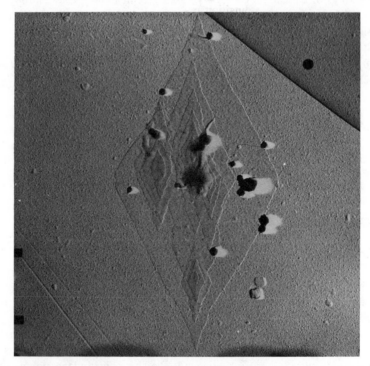

FIG. 10-6.   Electron micrograph of a single crystal of 6-nylon grown by precipitation from dilute glycerine solution (Geil 1960). The lamellae are about 60 Å thick.

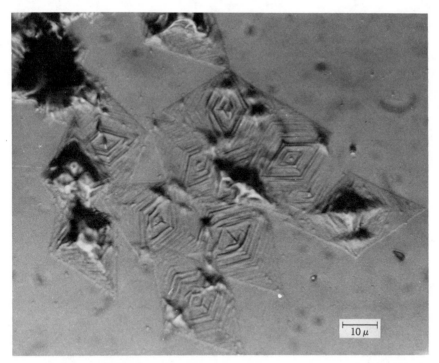

FIG. 10-7. Optical micrograph showing corrugations in single crystals of linear polyethylene grown from a solution in perchlorethylene (Reneker 1960).

features result from the fact that many crystals of polyethylene, and perhaps other polymers too, grow in the form of hollow pyramids. When solvent is removed during preparation of the crystals for microscopy, surface-tension forces cause the pyramids to collapse.

The hollow pyramidal structure is related to the packing of the folded chains. If successive planes of folded molecules are displaced from their neighbors by an integral (or, with rotation of 180°, by a half-integral) number of repeat distances, a pyramidal structure with a slope of 32° (or 17°) results.

More Complex Structures. The growth of crystal structures from polymer solutions can result in more complex structures sometimes reminiscent of features found in polymers crystallized from the melt. Typical of these are sheaflike arrays corresponding to the nuclei and initial growth habit of spherulites. Each ribbon in the array is composed of a number of lathlike lamellae with the usual 50–100 Å thickness. Twinned crystals, dendritic growths, clusters of hollow pyramids, spiral growths, steps, Bragg extinction lines, dislocation networks, moiré patterns, and epitaxial growths have been described, among others.

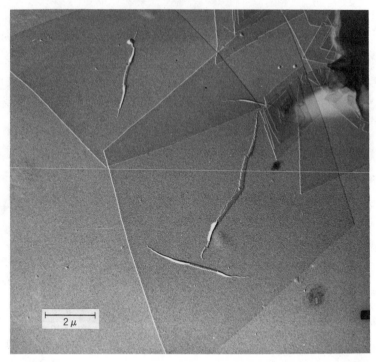

FIG. 10-8.   Electron micrograph showing pleats in a crystal like that of Fig. 10-7 (Reneker 1960).

Disorder and Nature of the Fold Surfaces.   The major unsolved problem (Keller 1969) in polymer single-crystal studies has for some time centered on the nature of the fold surfaces. Although there is strong structural and kinetic evidence favoring regular folding with immediate adjacent reentry of the chain into the crystal, there is even more convincing evidence that considerable molecular disorder exists in polymer single crystals. The extent of this disorder appears too great to be compatible with intercrystalline defects and to require some sort of amorphous region or layer at the fold surface. Such a region might involve loose or irregular folds or a more random "switchboard" model as suggested by Flory (1962), and as indicated schematically in Fig. 10-9. Although there appears to be agreement that the fold surfaces in polymer single crystals are rather regular and tight, the controversy remains unresolved into the 1980s.

Structure of Polymers Crystallized From the Melt

The Fringed Micelle Concept.   Although x-ray diffraction studies show recognizable crystalline features in some high polymers, the Bragg reflections appear broad and diffuse compared to those from well-developed simple crystals. Diffraction theory indicates that this broadening can arise from either small crystallite size or the presence of lattice defects (Bunn 1961), but the diffraction patterns from

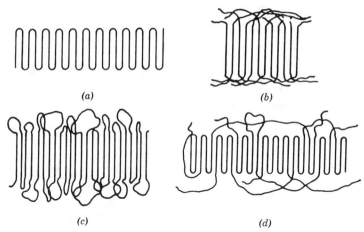

FIG. 10-9. Schematic two-dimensional representations of models of the fold surface in polymer lamellae (Ingram 1968): (a) sharp folds, (b) "switchboard" model, (c) loose loops with adjacent reentry, (d) a combination of several features.

polymers are usually too weak to permit discrimination between these possibilities. Historically, the hypothesis of small crystallite size was selected as more probable.

With this assumption, rough estimates from the width of diffraction rings indicated that crystallite size rarely exceeds a few hundred angstroms. A substantial background of diffuse scattering suggested the coexistence of an appreciable amorphous fraction. It was suggested that the polymer chains are precisely aligned over distances corresponding to the dimensions of the crystallites, but include more disordered segments that do not crystallize, and hence are included in amorphous regions. Since the chains are very long, they were visualized as contributing segments to several crystalline and amorphous regions, leading to a *composite single-phase* structure known as the *fringed micelle* or *fringed crystallite* model (Fig. 10-10), the fringes representing transition material between the crystalline and amorphous phases.

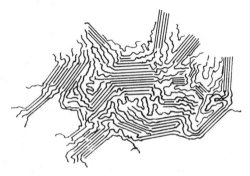

FIG. 10-10. Fringed micelle model of the crystalline–amorphous structure of polymers (Bryant 1947).

The fringed micelle concept enjoyed popularity for many years, largely because it led to simple and appealing consequences such as the strong bonding of crystalline and amorphous regions into a composite structure of good mechanical properties, and the simple interpretation of degree of crystallinity in terms of the percentages of well-defined crystalline and amorphous regions. However, it tends to draw attention away from the details of fine structure and gives little insight into the structures of larger entities such as spherulites.

Almost simultaneously with the discovery of polymer single crystals with a lamellar structure involving chain folding, it was found that spherulites are complex structures of lamellae and that individual lamellae in polymers crystallized from the melt are almost macroscopic in size. These features, incompatible with the interpretation of x-ray diffraction-pattern broadening in terms of small crystallites, suggested that reexamination of the interpretation of the x-ray data was needed.

The Defect Structure of Crystalline Polymers. It is rewarding to recognize the existence of defects in crystalline polymers just as in any other crystalline material. The presence of several kinds of such defects is now well recognized. Among them are the following:

a. Point defects, such as vacant lattice sites and interstitial atoms. Chain ends, which themselves must be considered defects because they differ chemically from the rest of the chain, are usually accompanied by vacancies. Interstitial atoms or groups may be foreign material or may be associated with the chain, for example, as certain kinds of side chains.

b. Dislocations, primarily screw dislocations and edge dislocations. Screw dislocations resulting in growth spirals have been observed many times in both polymer single crystals and bulk polymers. Edge dislocations have been discovered more recently through the study of moiré patterns.

c. Two-dimensional imperfections, those of particular interest being the fold surfaces (Fig. 10-9).

d. Chain-disorder defects, including folds, changes of alignment, and the like.

e. Amorphous defects, defined as disorders large enough to disrupt the lattice surrounding their own immediate area by forcing other atoms out of normal lattice positions.

The net effect of all these defects, in particular the fold surfaces and surrounding regions, is to provide localized amorphous regions contributing to diffuse x-ray scattering, and deformed or distorted lattices, termed paracrystalline (Hosemann 1962), which contribute in part to x-ray line broadening. In summary, the fringed micelle concept has largely been supplanted, especially for highly crystalline polymers, by a defect-crystal concept accounting equally well for the observable facts, and utilizing, in addition, experimentally verified defect structures. Much still remains to be learned, however, about the structure of polymers with low or intermediate crystallinity; here the fringed micelle concept may give way to a chain-folding model with partial adjacent and partial random reentry (Peterlin 1980).

Extended-Chain Crystals. In addition to producing folded-chain structures, polymers can also crystallize from the melt in the form of extended-chain crystals (Fig. 10-11) (Geil 1964, Wunderlich 1968). This morphology is more easily obtained with low-molecular-weight polymers, but this depends on crystallization conditions, including applied pressure and supercooling. The extended-chain morphology, in which the size of the crystal in the chain direction is essentially equal to the extended chain length, is now considered the equilibrium morphology for polyethylene and probably many other crystalline polymers (Bassett 1976*b*).

Structure of Spherulites. The most prominent organization in polymers on a scale larger than lamellae is the *spherulite,* (ideally) a spherical aggregate ranging from submicroscopic in size to millimeters in diameter in extreme cases. Spherulites are recognized (Fig. 10-12) by their characteristic appearance in the polarizing microscope, where they are seen as (ideally) circular birefringent areas possessing a dark Maltese cross pattern. The birefringence effects are associated with molecular

FIG. 10-11. Extended-chain crystals of linear polyethylene (Prime 1969). Fracture surface of a sample crystallized at 5000 atm.

FIG. 10-12.   Ringed spherulites of poly(trimethylene glutarate) observed in the optical microscope between crossed polarizers (Keller 1959).

orientation resulting from the characteristic lamellar morphology now recognized in spherulites.

Relation of Spherulites to Crystallites.   Although there has not been general agreement on whether spherulites are formed by rearrangement of previously crystallized material or as products of primary crystallization, the latter interpretation is considered more probable. It has been shown by examination of polymers that are partially spherulitic that in the usual case the spherulites are crystalline, whereas the intermediate nonspherulitic material is amorphous. Spherulites thus appear to represent the crystalline portions of the sample, growing at the expense of the noncrystalline melt. The point of initiation of spherulite growth, its nucleus, may be a foreign particle (heterogeneous nucleation) or may arise spontaneously in the melt (homogeneous nucleation).

Morphology of Spherulites.   Electron-microscopic evidence indicates that spherulites have a lamellar structure for almost all polymers. Crystallization spreads by the growth of individual lamellae. Growth may begin from screw dislocations at the nucleus or at spiral growths originating at crystal defects farther out in the spherulite. When two spherulites meet during crystallization, lamellae from both extend across the boundary into any uncrystallized material available. This results in interleaving, which holds the material together. Evidence that the lamellar structure persists throughout the body of spherulites, and that lamellae thus are basic structural elements of most solid polymers, is furnished by electron microscopic examination of fracture surfaces.

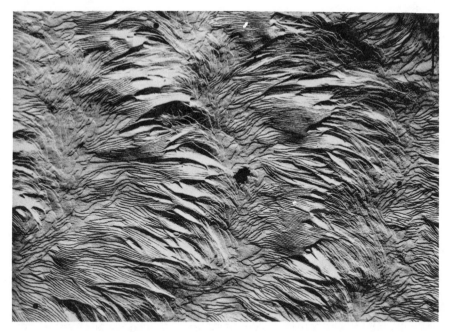

FIG. 10-13.   Electron micrograph of a portion of a ringed spherulite in linear polyethylene (photograph by E. W. Fischer, from Geil 1963).

The morphology of spherulites in some polymers is complicated by the development of a ring structure (Fig. 10-12). In the electron microscope it is seen (Fig. 10-13) that the ring structure is associated with a periodic twisting of the lamellae; the cause of the twisting is unknown.

GENERAL REFERENCES

Geil 1963; Ingram 1968; Wunderlich 1973; Schultz 1974, Chapters 2 and 3; Mandelkern 1975; Bassett 1976a, Chapters 2–4; Khoury 1976; Fava 1980; Peterlin 1980; Hobbs 1982.

D. CRYSTALLIZATION AND MELTING

Crystallization Kinetics

Experimental observations of the development of crystallinity in polymers are of two types. In one case a property varying with total amount of crystallinity, such as specific volume, is observed as a function of time at constant temperature. Alternatively the rate of formation and growth of spherulites is observed directly with the microscope. The two techniques lead to essentially the same results.

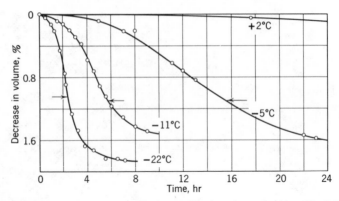

FIG. 10-14. Volume–time relations during the crystallization of natural rubber (Wood 1946). Arrows indicate points where crystallization is half complete.

The development of crystallinity in polymers is not instantaneous. Curves of specific volume as a function of time at temperatures below the crystalline melting point (Fig. 10-14) show that crystallization sometimes cannot be considered complete for long periods. Since the time for complete crystallization is somewhat indefinite, it is customary to define the rate of crystallization at a given temperature as the inverse of the time needed to attain one-half of the total volume change.

The rate so defined is a characteristic function of temperature (Fig. 10-15). As the temperature is lowered, the rate increases, goes through a maximum, and then decreases as the mobility of the molecules decreases and crystallization becomes diffusion controlled. At temperatures at which the rate is very low, the polymer may be supercooled and maintained in the amorphous state.

If data for specific volume as a function of time during crystallization, such as those for natural rubber in Fig. 10-14, are plotted against log time, all the curves have the same shape and can be superposed by a shift along the time axis. This behavior is also typical of the crystallization of low-molecular-weight substances, and curves of this type can be fitted with an equation due to Avrami (1939):

$$\ln \left(\frac{V_\infty - V_t}{V_\infty - V_0} \right) = - \frac{1}{w_c} kt^n \qquad (10\text{-}1)$$

where V_∞, V_t, and V_0 are specific volumes at the times indicated in the subscripts, w_c is the weight fraction of material crystallized, k is the constant describing the rate of crystallization, and n is an exponent varying with the type of nucleation and growth process.

Although this analysis provides a satisfying analogy between crystallization in polymers and that in low-molecular-weight substances, quantitative aspects of the agreement with theory are not completely satisfactory. The shape of the curves, as in Fig. 10-14, is not very sensitive to values of the exponent n in the expected range of 2–4, and it is not always possible to determine n or indeed to ascertain

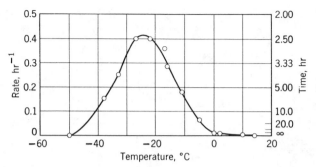

FIG. 10-15. Rate of crystallization of rubber as a function of temperature (Wood 1946).

whether n is constant throughout the crystallization process. It is generally found that crystallization continues in polymer systems for much longer times than Eq. 10-1 predicts. This results from the fact that crystallization in polymers involves the steps of (primary) nucleation and relatively rapid spherulite growth, followed by a slow, kinetically difficult improvement in crystal perfection. "Amorphous" chains remaining between the spherulite lamellae after the first stage also crystallize in the slow second step.

Extension of these kinetics to the case of polymers was carried out by Lauritzen (1960) and Hoffman (1961), who calculated the rate at which a nucleus of critical size to sustain growth into a crystallite is formed by the addition of strands of polymer chain. They evaluated the driving force for crystallization as the difference in free energy between the mass of the nucleus and a similar mass in the melt. This expression involved terms for the free-energy change due to supercooling and for surface free energy. Their final equation, though complicated, predicts the correct order of magnitude of the thickness of the resulting lamellar crystal, and the temperature dependence of crystallization rate of the form shown in Fig. 10-15. An alternative thermodynamic approach (Peterlin 1960) has received less general acceptance.

The concept of the lamellar structure of polymers implies large-scale molecular motion in both the melt and the solid crystalline polymer. Either prefolding and some local orientation exists in the crystallizing polymer, or the molecules must undergo a considerable degree of motion during crystallization. Extensive melting and crystallization with an extremely rapid, severalfold increase in fold period occur during annealing of polymer single crystals and melt-crystallized lamellae; the extent to which the crystal morphology is disrupted, even at temperatures below T_m, is illustrated in Fig. 10-16. The development of spherulites in polymers quenched from the melt and then held at temperatures between T_m and T_g is additional evidence of high chain mobility below T_m. The existence of mobile lattice defects provides a plausible mechanism for this mobility in crystalline material. It has been found, as might be expected, that the ease with which these molecular motions take place decreases as molecular weight increases.

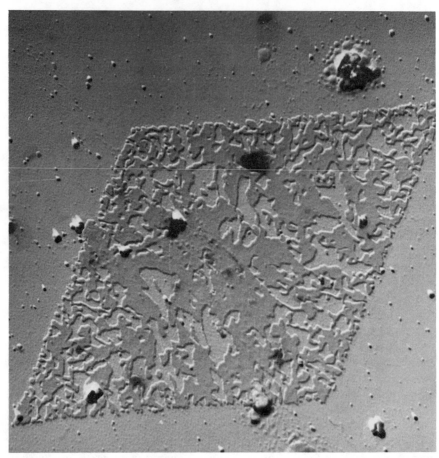

FIG. 10-16.   A single crystal of linear polyethylene crystallized from perchlorethylene solution and then annealed for 30 min at 125°C, at least 10°C below T_m (Statton 1960). The fold period increased from about 100 Å to almost 200 Å during annealing. Compare Fig. 10-8.

Determination of T_m

The disappearance of a polymer crystalline phase at the melting point is accompanied by changes in physical properties: The material becomes a (viscous) liquid, with discontinuous changes in density, refractive index, heat capacity, transparency, and other properties. Measurement of any of these properties may be used to detect the *crystalline melting point* T_m. Since melting occurs over a temperature range, the melting point determined by a particular method may depend to some extent on the sensitivity of the method. For example, crystalline melting points may be determined by noting the temperature of disappearance of the last traces of crystallinity as evidenced by birefringence observed between crossed polarizers on a hot-stage microscope. This melting point is usually several degrees higher than those obtained from methods depending on gross changes in properties such as specific volume and specific heat. Because of retention of order above T_m and the existence of a melting range in polymers, properties sensitive to the amount of crystalline material present undergo change over a relatively wide range in temperature, as evidenced by experimental data for specific volume or specific heat (see, for example, Fig. 9-4).

Today almost all measurements of T_m are made by differential thermal analysis (Chapter 9E). Other methods that could be used to detect T_m are x-ray diffraction and infrared and NMR spectroscopy, each of these being suited to the detection of crystalline material as described in Chapter 9. As the experiments are usually performed, they are not adapted to detecting T_m but, rather, are carried out at constant temperature.

Thermodynamics of Crystalline Melting

For a polymer homogeneous in molecular weight, a statistical thermodynamic analysis based on a lattice model yields the relation

$$\frac{1}{T} - \frac{1}{T_m} = \frac{R}{\Delta H_m} \left(\frac{1}{xw_a} + \frac{1}{x - \zeta + 1} \right) \tag{10-2}$$

where w_a is the weight fraction of material that is amorphous at temperature T (expressed in Kelvins as required in all thermodynamic equations). The equilibrium melting point of the pure polymer at infinite chain length is $T_m = \Delta H_m / \Delta S_m$. The degree of polymerization is x, and ζ is a parameter characterizing the crystallite size. Equation 10-2 indicates that even for a polymer of uniform chain length, melting occurs over a finite temperature range. The physical reason for this behavior is the higher surface free energy of small crystallites. Equation 10-2 also shows that the temperature interval for fusion decreases with increasing molecular weight, and further analysis indicates that it broadens with increasing breadth of molecular-weight distribution. On the other hand, as w_a approaches unity, its derivative $\partial w_a / \partial T$ does not vanish; in consequence, the last traces of crystallinity disappear at a well-defined temperature, as confirmed by experiment.

The extended-chain crystals described in Section C are now considered to be the equilibrium state for high polymers, and thus the only state whose melting is properly described by Eq. 10-2. The more common folded-chain crystals are not in their most stable state, and as T_m is approached, they can undergo several possible changes in structure, exemplified by the reorganization and recrystallization seen during annealing. The melting point in systems of this kind occurs at the point where the free-energy change ΔG is zero, even though the system is not at equilibrium. The process is termed *zero-entropy-production melting*. The melting point is a measure of the degree of imperfection of the metastable system, and is greatly dependent on the thermal history of the sample and in particular the heating rate (Jaffe 1967).

For polymer–diluent systems, use of the Flory–Huggins expression for the free energy of mixing (Chapter 7C) leads to the expression

$$\frac{1}{T_m} - \frac{1}{T_m^0} = \frac{R}{\Delta H_m} \frac{V_2}{V_1} (v_1 - \chi_1 v_1^2) \qquad (10\text{-}3)$$

where T_m and T_m^0 are the melting points of the polymer–diluent mixture and the pure polymer, respectively, V is molar volume, v is volume fraction, and χ_1 is the polymer–diluent interaction parameter defined in Chapter 7C. Thus the melting-point depression depends on the volume fraction of diluent added and its interaction with the polymer. For thermodynamically good solvents, large depressions (40–50°C) are sometimes observed in dilute solutions. Since the melting temperature is only slightly dependent on molecular weight, crystallization does not provide an efficient method of molecular-weight fractionation.

Heats and Entropies of Fusion

Measurement of the heat and entropy of fusion yields important experimental data for assessing the effect of polymer structure on T_m, as discussed in Chapter 12A.

Heats of fusion are readily derived from several experimental sources, at present most often differential scanning calorimetry (Chapter 9E). The observed heat of fusion ΔH^* calculable from specific-heat measurements is not ΔH_m, the heat of fusion per mole of repeating unit, but is related to it by the degree of crystallinity w_c: $\Delta H^* = w_c \Delta H_m$. It is ΔH_m that appears directly in Eq. 10-3, however, and in the related equation for the melting points of copolymers,

$$\frac{1}{T_m} - \frac{1}{T_m^0} = -\frac{R}{\Delta H_m} \ln n \qquad (10\text{-}4)$$

where n is the mole fraction of crystallizing units. Thus ΔH_m can be determined from melting-point data for these two types of systems even if w_c is not known.

In some cases ΔH_m can be inferred from the heats of fusion of low-molecular-weight homologues. It is found, for example, that the increment per CH_2 group in the heat of fusion of n-paraffins becomes constant for chain lengths greater than six carbon atoms. With certain assumptions, this increment may be taken as the heat of fusion of crystalline polyethylene. The method can be extended to the calculation of the lattice or sublimation energy at 0 K, a measure of the intermolecular forces in the crystal.

A comparison is made in Table 10-1 of the values of ΔH_m for polyethylene determined from specific-heat measurements, the melting-point depressions of polymer–diluent mixtures, and extrapolation of data on paraffins. Heats of fusion for several other polymers are included.

The total entropy change on fusion can be calculated from T_m and the heat of fusion, since $\Delta G_m = \Delta H_m - T \Delta S_m = 0$ at the melting point. The total change

TABLE 10-1.   Heat and Entropy of Fusion of Crystalline Polymers

Polymer	ΔH_m (J/g)	ΔS_m (J/g-deg)	$(\Delta S_m)_v$ (J/g-deg)
Polyethylene		0.68[d]	0.53[d]
From specific heat	277[a]		
From diluents	280[b]		
From extrapolation	276[c]		
Polyoxymethylene	249[d]	0.55[d]	0.33[d]
Polytetrafluoroethylene	57[d]	0.10[d]	0.06[d]
Natural rubber	64[e]	0.21[f]	0.10[f]
Gutta percha	190[f]	0.54[f]	0.31[f]

[a] Wunderlich 1957.   [c] Billmeyer 1957.   [e] Roberts 1955.
[b] Quinn 1958.   [d] Starkweather 1960.   [f] Mandelkern 1956.

in entropy is made up of the change in entropy on fusion at constant volume $(\Delta S_m)_v$ plus the change in entropy due to expansion to the volume of the liquid phase. The latter can be evaluated by a simple thermodynamic relation to yield the equation

$$(\Delta S_m)_v = \Delta S_m - (V_L - V_s)\left(\frac{\partial P}{\partial T}\right)_{v,n} \tag{10-5}$$

where V_S and V_L are the volumes of the solid and liquid phases, respectively. Values of $(\Delta S_m)_v$ so calculated, as well as of ΔS_m, are included in Table 10-1.

Degree of Crystallinity

The determination of the degree of crystallinity of a polymer is affected by the conceptual difficulties discussed in Section C. The interpretation of measurements of crystallinity depends on the model used in designing the experiments: Each of the methods to be described involves a simplifying assumption, usually concerned with the degree to which defects and disordered regions in paracrystalline material are weighted in the determination of the amorphous fraction. Fortunately, independent techniques lead to unexpectedly good experimental agreement, implying that differences in their resolution, specifically the difference in the lateral extent of lattice order detected, are relatively minor (Richardson 1969).

The three major methods of determining crystallinity, discussed in the following paragraphs, are based on specific volume, x-ray diffraction, and infrared spectroscopy. Polyethylene is used throughout as an example, although each method is applicable, at least in principle, to other polymers as well.

Another well-established method, based on measurement of heat content as a function of temperature through the fusion range, is now easily carried out using differential scanning calorimetric measurements. Other methods are based on increased swelling or increased chemical reactivity of the amorphous regions. Nuclear magnetic resonance spectroscopy (Chapter 9B) has been correlated with crystallinity, but this technique places more emphasis on other properties than on differences between crystalline and amorphous character.

Specific Volume.   The crystallinity of a material is given in terms of the specific volumes of the specimen (V), the pure crystals (V_c), and the completely amorphous material (V_a), as

$$w_c = \frac{V_a - V}{V_a - V_c} \tag{10-6}$$

This relation assumes additivity of the specific volumes, but this requirement is fulfilled for polymers of homogeneous structure, as far as is known. It is implicit that the sample be free of voids. Since specific volume can be determined to 1 or 2 parts in 10^4, the method can attain high precision. Its accuracy depends on the uncertainty in V_a and V_c, which are known to about 1 part in 10^3 for polyethylene. The crystalline specific volume is determined from x-ray unit cell dimensions, while V_a is usually obtained by extrapolation of the specific volumes of polymer melts.

X-Ray Diffraction.   The x-ray method allows calculation of the relative amounts of crystalline and amorphous material in a sample if it is possible to resolve the contributions of the two types of structure to the x-ray diffraction pattern. A favorable situation is depicted in Fig. 10-17, where the scattering envelope of a polyethylene sample (outer line) has been resolved into contributions of two crystalline peaks (with indices 110 and 200) and a broad amorphous peak. The estimation

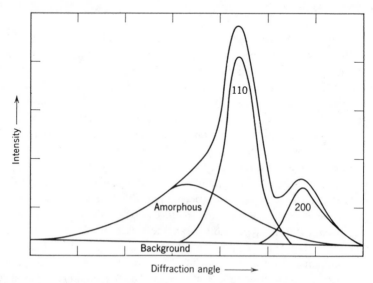

FIG. 10-17.   Resolution of the x-ray scattering curve of a polyethylene into contributions from two crystalline peaks (indices 110 and 200), the amorphous peak, and background.

of amount of crystallinity is usually based on comparison of the areas under the peaks, but other measures, such as peak height, may be used. With proper attention to experimental detail, this method provides one of the fundamental measures of crystallinity in polymers.

Infrared Absorption.   Suitable infrared absorption bands can be used as primary measures of crystallinity in polymers. The application requires that the band bear a simple and unambiguous relationship to the crystalline or amorphous character of the polymer, and that absorption data can be obtained or inferred for the pure crystalline and amorphous polymers. These conditions are satisfied, for example, for an infrared band at 7.67 μm in polyethylene identified with a methylene wagging motion in the amorphous regions, and a band at 10.6 μm in 610-nylon whose specific absorbance varies linearly with specific volume.

GENERAL REFERENCES

Mandelkern 1964; Sharples 1966; Price 1968; Sanchez 1974; Schultz 1974, Chapter 9; Bassett 1976*a*, Chapters 5–8; Hoffman 1976; Wunderlich 1976, 1980; Hobbs 1982.

E. STRAIN-INDUCED MORPHOLOGY

When a polymer is crystallized from solution in the absence of external forces, the single crystals described in Section *C* form. If the solution is subjected to an external stress such as stirring, an entirely different morphology can result. When a polymer mass is crystallized in the absence of external forces, there is no preferred direction in the specimen along which the polymer chains lie. If such an unoriented crystalline polymer is subjected to an external stress, it undergoes a rearrangement of the crystalline material. Changes in the x-ray diffraction pattern suggest that the polymer chains align in the direction of the applied stress. At the same time the physical properties of the sample change markedly.

Fibrillar Crystallization

If a dilute (~1%) polymer solution is stirred during crystallization, crystalline aggregates consisting of a central rod or ribbon with lamellar overgrowths can be formed (Fig. 10-18). The morphology has, for obvious reasons, been called the "shish kebab" morphology. The backbone, or "shish," consists of relatively but not fully extended chains, with a significant number of defects, probably chain folds. The "kebabs" form by epitaxial growth using the backbone as a nucleus. This is perhaps the most spectacular of a number of morphologies produced by crystallization under unusual conditions.

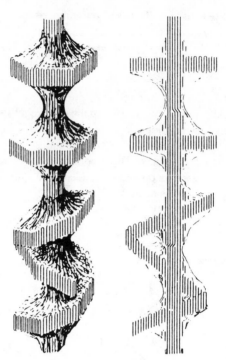

FIG. 10-18. Schematic model of "shish kebab" morphology formed by crystallization of a stirred polymer solution. (Reprinted with permission from Pennings 1977. Copyright 1977 by John Wiley & Sons.)

Cold Drawing

When the orientation process is carried out below T_m but above the glass-transition temperature T_g, as when an unoriented fiber is stretched rapidly, the sample does not become gradually thinner, but suddenly becomes thinner at one point, in a process known as "necking down" (Fig. 10-19). As the stretching is continued, the thin or drawn section increases in length at the expense of the undrawn portion of the sample. The diameters of the drawn and undrawn portions remain about the same throughout the process. The *draw ratio,* or ratio of the length of the drawn fiber to that of the undrawn, is about 4 or 5 to 1 for a number of polymers, including branched polyethylene, polyesters, and polyamides, but is much higher (10 to 1 or more) in linear polyethylene.

In general, the degree of crystallinity in the specimen does not change greatly during drawing if crystallinity was previously well developed. If the undrawn polymer was amorphous or only partially crystallized, crystallinity is likely to increase during cold drawing.

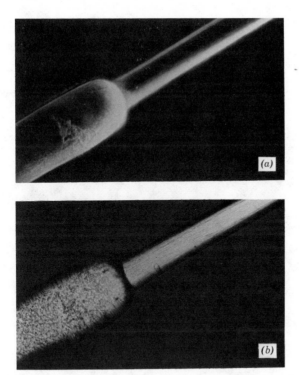

FIG. 10-19. (a) Cold-drawn fiber of polyethylene, showing sections of drawn and undrawn material (Bunn 1953). (b) Cold-drawn fiber of polyethylene observed between crossed polarizers, showing evidence of random orientation in the undrawn section and orientation parallel to the direction of stress in the drawn section.

Morphology Changes During Orientation

Observations in the electron microscope indicate that when polymer single crystals are stressed, fibrils 50–100 Å in diameter are drawn across the break (Fig. 10-20). These fibrils must contain many molecules; smaller fibrils have not been resolved in the microscope. The fibrils appear to come from individual lamellae, and the sharp discontinuity between the drawn and the undrawn material suggests that the molecules are unfolding. In some cases the fibrils appear to have a periodicity of 100–400 Å along their length; a similar period is observed by low-angle x-ray diffraction.

Examination of melt or solution-cast specimens shows that the lamellae tend to rotate toward the draw direction and on further elongation break up into microlamellae and finally into submicroscopic units. Since the elongation at break in cold-drawn specimens is less than could be obtained by complete unfolding, it has been suggested that the smallest units are microcrystallites in which the chains remain folded. Periodic spacing of these crystallites within the fibrils would account for their observed periodicity.

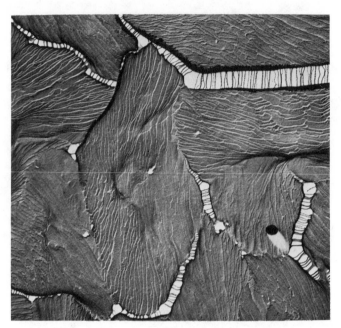

FIG. 10-20.   Electron micrograph of a film of linear polyethylene crystallized by evaporation of the
solvent (Keller 1957). Fibrils drawn across the breaks are 50–100 Å in diameter and
appear to have a 100–300 Å periodicity along their length.

Spherulites tend to remain intact during the first stages of drawing, often elongating to markedly ellipsoidal shapes. Rupture of the sample usually occurs at
spherulite boundaries.

The structure of fibers, whether formed by drawing or crystallized from an
oriented melt, is still unsettled. X-ray evidence indicates that the molecules are
aligned and shows that a long periodicity is present. Microfibrils are undoubtedly
important structural elements. Among the models proposed are fringed fibrils, a
development of the fringed micelle model (Section C) in which the crystalline
regions are aligned in the direction of the fiber axis; a "string" model (Statton
1959), in which the microfibrils have alternating crystalline and amorphous regions,
but there are fewer ties molecules between them than in the fringed fibril concept;
a "folded fibril" model in which the microfibrils consist of small stacked sections
of lamellae; and a paracrystalline model in which the imperfect crystalline regions
are thought to be much larger than required by the other models.

Perhaps the final stages in the formation of an oriented fiber by drawing can be
imagined (Peterlin 1965, 1967) as shown in Fig. 10-21. A multilayer lamellar
crystal is destroyed with tilting and slipping of the chains. At points of concentration
of defects, blocks of folded chains break off and are incorporated along with
unfolded chains into the new fiber structure. In consequence of these processes,
the number of chain folds decreases and the number of tie molecules increases.

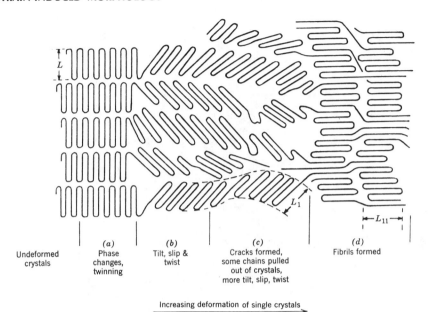

	(a)	(b)	(c)	(d)
Undeformed crystals	Phase changes, twinning	Tilt, slip & twist	Cracks formed, some chains pulled out of crystals, more tilt, slip, twist	Fibrils formed

Increasing deformation of single crystals

FIG. 10-21. Suggested model for fiber formation by chain tilting and slipping followed by breaking off of blocks of lamellae (Peterlin 1967).

The latter and the partial interpenetration of lamellae increase the mechanical stability, tensile strength, and stiffness of the fiber.

Degree of Orientation

As in the measurement of degree of crystallinity (Section D), the determination of degree of orientation is complicated both by the complexity of the types of orientation that can occur and by the conceptual difficulties involved in the current picture of the structure of an oriented specimen. The interpretation of the measurements depends on the model and simplifying assumptions in each method, but these differences are relatively minor, and the agreement among results by different methods is in general satisfactory.

X-Ray Diffraction. As indicated in Chapter 4C, x-ray diffraction patterns of unoriented polymers resemble small-molecule powder photographs, characterized by rings rather than diffraction spots. As the specimen is oriented, these rings break into arcs, and at high degrees of orientation approach the relatively sharp patterns characteristic of paracrystals (Fig. 10-22). From the distribution of scattered intensity in the arcs or spots, the degree of orientation can be measured in terms of distribution functions. These functions describe the amount of material occurring in crystalline regions having angular coordinates describing their orientation that fall within certain limits. For practical purposes, only average values of the functions are important.

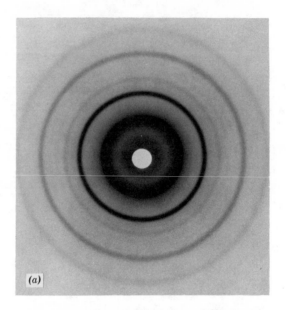

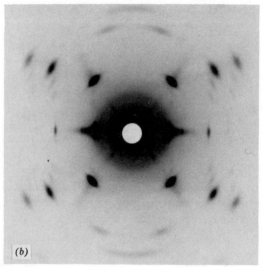

FIG. 10-22.   X-ray diffraction patterns for unoriented (a) and oriented (b) polyoxymethylene (courtesy of E. S. Clark).

Birefringence.   The birefringence, or change in index of refraction with direction, of a crystalline polymer is made up of contributions from the crystalline and amorphous regions plus a contribution, *form birefringence,* resulting from the shape of the crystals or the presence of voids. For completely unoriented material, the contributions of both the crystalline and the amorphous regions are zero. The increase in birefringence on orientation is due primarily to the crystalline regions and is proportional to the degree of crystallinity, the intrinsic birefringence of the crystals (which can be calculated), and an orientation factor that can be identified with that determined by the x-ray technique.

Infrared Dichroism. Infrared absorption is dependent upon change in dipole moment. Such a change is a vector quantity, that is, one that is confined to definite directions for certain groups. In oriented samples the amount of absorption of plane-polarized infrared radiation may vary markedly with change in direction of the plane of polarization. Thus the stretching vibrations of the C=O and N—H groups in polyamides involve changes in dipole moment that are almost exactly perpendicular to the chain axis; the corresponding absorption bands are weak for polarized radiation vibrating along the chain axis and strong for that vibrating perpendicular to the axis. When separate absorptions can be found for both the crystalline and the amorphous regions, the dichroism of these bands gives information about orientation in both regions of the specimen.

Orientation in Noncrystalline Polymers

The presence of crystalline material is not essential for the development of molecular orientation. Even polymers such as polystyrene, which do not crystallize, undergo considerable molecular orientation when subjected to stress at temperatures above T_g. This orientation (which is often seen in injection-molded pieces) is stable below T_g and is accompanied by the development of birefringence and of enhanced strength properties in the direction of the applied stress. This type of orientation can be measured by birefringence, by a property change associated with the orientation, or by a change on subsequent heating above T_g. The force with which the specimen tends to retract on heating above T_g is a suitable and readily measured property characterizing orientation in amorphous polymers.

GENERAL REFERENCES

Peterlin 1967; Szabolcs 1967; Ingram 1968; Wilchinsky 1968; Desper 1975.

DISCUSSION QUESTIONS AND PROBLEMS

1. Draw structural formulas indicating the stereoregular chain configuration in (a) atactic polystyrene, (b) isotactic polypropylene, and (c) syndiotactic poly(vinyl chloride).

2. Name two examples of crystallizable atactic polymers.

3. Polyisobutylene has unit cell dimensions of approximately $18 \times 12 \times 7$ Å and contains two chains of eight repeat units each per unit cell. What would

be the density of 100% crystalline polyisobutylene? Why is your answer different from the observed density of 0.915?

4. Polytetrafluoroethylene has a first-order phase transition at 19°C. The crystal structures above and below the transition are hexagonal, with the following parameters:

Temperature	Interchain Distance (Å)	Repeat Length (Å)	CF_2 per Repeat Length
Low	5.62	16.9	13
High	5.66	19.5	15

Calculate the density of the polymer above and below the transition and discuss the phase change in terms of molecular motion and possible effects on the mechanical properties of polytetrafluoroethylene.

5. Contrast the fringed micelle and lamellar models of the structure of crystalline polymers with respect to (a) description and sketch of the basic units, (b) observation basis, (c) interpretation of x-ray evidence, (d) spherulite structure, (e) correlation between density and crystallinity, and (f) changes on cold drawing.

6. Discuss, with sketches, the structure of solution-grown polyethylene single crystals.

7. Discuss the evidence for chain folding in polymers and the conditions under which it occurs.

8. Describe the growth and structure, including molecular arrangement, of spherulites in polymers.

9. Sketch graphs of the following: (a) specific volume versus time during crystallization at several temperatures below T_m, indicating the definition of rate of crystallization; (b) rate of crystallization versus temperature, showing the locations of T_m and T_g; and (c) equilibrium specific volume versus temperature over a range including both T_m and T_g.

10. On the basis of the data in Table 10-1, compare the following on the basis of intermolecular forces, energy barriers to rotation about bonds, and other structural features: (a) $(\Delta S_m)_v$ for polyethylene and polyoxymethylene, (b) $(\Delta S_m)_v$ for polyethylene and polytetrafluoroethylene, and (c) ΔH_m for natural rubber and gutta-percha (see also Chapter 13D).

11. Using the data of Table 10-1, calculate T_m for natural rubber and gutta-percha. Predict the properties of these polymers at room temperature. Do your predictions agree with the actual properties of the polymers? Can you reconcile any differences?

12. From the data plotted in Figs. 10-14 and 10-15, predict the behavior of natural

rubber subjected to the following thermal treatments: cool rapidly to $-40°C$, hold for 24 hr; raise rapidly to $-5°C$, hold for 24 hr; raise rapidly to $+20°C$, hold for 24 hr.

13. Draw a sketch of a fiber during the process of cold drawing. Indicate the orientation of the chains in the drawn and undrawn regions in terms of the probable structural elements present.

14. Discuss (a) extended-chain crystals, (b) "shish kebab" morphology, and (c) structural changes in single crystals on annealing.

BIBLIOGRAPHY

Avrami 1939. Melvin Avrami, "Kinetics of Phase Change. I. General Theory," *J. Chem. Phys.* **7**, 1103–1112 (1939).

Bassett 1976a. D. C. Bassett, *Principles of Polymer Morphology*, Cambridge University Press, Cambridge, 1976.

Bassett 1976b. D. C. Bassett, "Chain-Extended Polyethylene in Context: A Review," *Polymer* **17**, 460–470 (1976).

Billmeyer 1957. Fred W. Billmeyer, Jr., "Lattice Energy of Crystalline Polyethylene," *J. Appl. Phys.* **28**, 1114–1118 (1957).

Blackadder 1967. D. A. Blackadder, "Ten Years of Polymer Single Crystals," *J. Macromol. Sci. Rev. Macromol. Chem.* **C1** (2), 297–326 (1967).

Bovey 1979. F. A. Bovey and T. K. Kwei, "Microstructure and Chain Conformation of Macromolecules," Chapter 3 in F. A. Bovey and F. H. Winslow, eds., *Macromolecules: An Introduction to Polymer Science*, Academic Press, New York, 1979.

Bryan 1969. W. P. Bryan, G. E. Hein, and M. F. Perutz, "Proteins," pp. 620–677 in Herman F. Mark, Norman G. Gaylord, and Norbert M. Bikales, eds., *Encyclopedia of Polymer Science and Technology*, Vol. 11, Wiley-Interscience, New York, 1969.

Bryant 1947. W. M. D. Bryant, "Polythene Fine Structure," *J. Polym. Sci.* **2**, 547–564 (1947).

Bunn 1953. C. W. Bunn, "Polymer Texture," Chapter 10 in Rowland Hill, ed., *Fibres from Synthetic Polymers*, Elsevier, New York, 1953.

Bunn 1961. Charles W. Bunn, *Chemical Crystallography*, 2nd ed., Oxford University Press, London, 1961.

Corradini 1968. P. Corradini, "Chain Conformation and Crystallinity," Chapter 1 in A. D. Ketley, ed., *The Stereochemistry of Macromolecules*, Vol. 1, Marcel Dekker, New York, 1968.

Desper 1975. C. Richard Desper, "Characterization of Molecular and Crystalline Orientation of Anisotropic Solid Polymers," Chapter 16 in John J. Burke and Volker Weiss, eds., *Characterization of Materials in Research: Ceramics and Polymers*, Syracuse University Press, Syracuse, New York, 1975.

Farina 1968. Mario Farina and Giancarlo Bressan, "Optically Active Stereoregular Polymers," Chapter 4 in A. D. Ketley, ed., *The Stereochemistry of Macromolecules*, Vol. 3, Marcel Dekker, New York, 1968.

Fava 1980. R. A. Fava, ed., *Polymers, Part B: Crystal Structure and Morphology*, Vol. 16 of L. Marton and C. Marton, eds., *Methods of Experimental Physics*, Academic Press, New York, 1980.

Feughelman 1955. M. Feughelman, R. Langridge, W. E. Seeds, A. R. Stokes, H. R. Wilson, C. W. Hooper, M. H. F. Wilkins, R. K. Barclay, and L. D. Hamilton, "Molecular Structure of Deoxyribose Nucleic Acid and Nucleoprotein," *Nature* **175**, 834–838 (1955).

Fischer 1957. E. W. Fischer, "Step and Spiral Crystal Growth of High Polymers" (in German), *Z. Naturforsch*. **12a**, 753–754 (1957).

Flory 1962. P. J. Flory, "On the Morphology of the Crystalline State in Polymers," *J. Am. Chem. Soc*. **84**, 2857–2867 (1962).

Gaylord 1959. Norman G. Gaylord and Herman F. Mark, *Linear and Stereoregular Addition Polymers*, Interscience, New York, 1959.

Geil 1960. P. H. Geil, "Nylon Single Crystals," *J. Polym. Sci*. **44**, 449–458 (1960).

Geil 1963. Phillip H. Geil, *Polymer Single Crystals*, Wiley-Interscience, New York, 1963.

Geil 1964. Phillip H. Geil, Franklin R. Anderson, Bernhard Wunderlich, and Tamio Arakawa, "Morphology of Polyethylene Crystallized from the Melt Under Pressure," *J. Polym. Sci*. **A2**, 3707–3720 (1964).

Hendra 1976. P. J. Hendra, "The Measurement of Lamellar Thickness by Raman Methods," Chapter 6 in K. J. Ivin, ed., *Structural Studies of Macromolecules by Spectroscopic Methods*, John Wiley & Sons, New York, 1976.

Hobbs 1982. S. Y. Hobbs, "Polymer Morphology," Chapter 7 in Mahendra D. Baijal, ed., *Plastics Polymer Science and Technology*, John Wiley & Sons, New York, 1982.

Hoffman 1961. John D. Hoffman and John I. Lauritzen, Jr., "Crystallization of Bulk Polymers with Chain Folding: Theory of Growth of Lamellar Spherulites," *J. Res. Nat. Bur. Stand*. **65A**, 297–336 (1961).

Hoffman 1976. John D. Hoffman, G. Thomas Davis, and John I. Lauritzen, Jr., "The Rate of Crystallization of Linear Polymers with Chain Folding," Chapter 7 in N. B. Hannay, ed., *Treatise on Solid State Chemistry, Crystalline and Noncrystalline Solids*, Vol. 3, Plenum Press, New York, 1976.

Hosemann 1962. R. Hosemann and S. N. Bagchi, *Direct Analysis of Diffraction by Matter*, North-Holland, Amsterdam, 1962.

Huggins 1962. M. L. Huggins, G. Natta, V. Desreux, and H. Mark, "Report on Nomenclature Dealing with Steric Regularity in High Polymers," *J. Polym. Sci*. **56**, 152–161 (1962).

Ingram 1968. P. Ingram and A. Peterlin, "Morphology," pp. 204–274 in Herman F. Mark, Norman G. Gaylord, and Norbert M. Bikales, eds., *Encyclopedia of Polymer Science and Technology*, Vol. 9, Wiley-Interscience, New York, 1968.

Jaffe 1967. M. Jaffe and B. Wunderlich, "Melting of Polyoxymethylene," *Kolloid Z. Z. Polym*. **216–217**, 203–216 (1967).

Keller 1957. A. Keller, "A Note on Single Crystals in Polymers: Evidence for a Folded Chain Configuration," *Philos. Mag*. [8] **2**, 1171–1175 (1957).

Keller 1959. A. Keller, "Investigations on Banded Spherulites," *J. Polym. Sci*. **39**, 151–173 (1959).

Keller 1969. A. Keller, "Solution Grown Polymer Crystals—A Survey of Some Problematic Issues," *Kolloid Z. Z. Polym*. **231**, 386–421 (1969).

Khoury 1976. F. Khoury and E. Passaglia, "The Morphology of Crystalline Synthetic Polymers," Chapter 6 in N. B. Hannay, ed., *Treatise on Solid State Chemistry, Crystalline and Noncrystalline Solids*, Vol. 3, Plenum Press, New York, 1976.

Kornberg 1962. Arthur Kornberg, *Enzymatic Synthesis of DNA*, John Wiley & Sons, New York, 1962.

Lauritzen 1960. John I. Lauritzen, Jr., and John D. Hoffman, "Theory of Formation of Polymer Crystals with Folded Chains in Dilute Solution," *J. Res. Nat. Bur. Stand*. **64A**, 73–102 (1960).

Mandelkern 1956. L. Mandelkern, F. A. Quinn, Jr., and D. E. Roberts, "Thermodynamics of Crystallization in High Polymers: Gutta-Percha," *J. Am. Chem. Soc*. **78**, 926–932 (1956).

Mandelkern 1964. Leo Mandelkern, *Crystallization of Polymers*, McGraw-Hill, New York, 1964.

Mandelkern 1975. L. Mandelkern, "Morphology of Semicrystalline Polymers," Chapter 13 in John J. Burke and Volker Weiss, eds., *Characterization of Materials in Research: Ceramics and Polymers*, Syracuse University Press, Syracuse, New York, 1975.

Miller 1966. Robert L. Miller, "Crystallinity," pp. 449–528 in Herman F. Mark, Norman G. Gaylord, and Norbert M. Bikales, eds., *Encyclopedia of Polymer Science and Technology,* Vol 4, Wiley-Interscience, New York, 1966.

Miller 1975. Robert L. Miller, "Crystallographic Data for Various Polymers," pp. III-1–III-137 in J. Brandrup and E. H. Immergut, eds., with the collaboration of W. McDowell, *Polymer Handbook,* 2nd ed., Wiley-Interscience, New York, 1975.

Morawetz 1975. Herbert Morawetz, *Macromolecules in Solution,* 2nd ed., Wiley-Interscience, New York, 1975, Chapter 3.

Natta 1955. G. Natta, Piero Pino, Paolo Corradini, Ferdinando Danusso, Enrico Mantica, Giorgio Mazzanti, and Giovanni Moranglio, "Crystalline High Polymers of α-Olefins," *J. Am. Chem. Soc.* **77,** 1708–1710 (1955).

Natta 1959a. G. Natta and F. Danusso, "Nomenclature Relating to Polymers Having Sterically Ordered Structure," *J. Polym. Sci.* **34,** 3–11 (1959).

Natta 1959b. Giulio Natta and Paolo Corradini, "Conformation of Linear Chains and Their Mode of Packing in the Crystal State," *J. Polym. Sci.* **39,** 29–46 (1959).

Natta 1967. Giulio Natta and Ferdinando Danusso, eds., *Stereoregular Polymers: Stereoregular Polymers and Stereospecific Polymerization,* Pergamon Press, New York, 1967.

Pauling 1951. Linus Pauling, Robert B. Corey, and H. R. Branson, "The Structure of Proteins: Two Hydrogen-Bonded Helical Configurations of the Polypeptide Chain," *Proc. Nat. Acad. Sci. U.S.A.* **37,** 205–211 (1951).

Pennings 1977. A. J. Pennings, "Bundle-like Nucleation and Longitudinal Growth of Fibrillar Polymer Crystals from Flowing Solutions," *J. Polym. Sci. Polym. Symp.* **59,** 55–86 (1977).

Peterlin 1960. Anton Peterlin, "Chain Folding and Free Energy Density in Polymer Crystals," *J. Appl. Phys.* **31,** 1934–1938 (1960).

Peterlin 1965. A. Peterlin, "Crystalline Character in Polymers," *J. Polym. Sci.* **C9,** 61–89 (1965).

Peterlin 1967. A. Peterlin, "The Role of Chain Folding in Fibers," pp. 283–340 in H. F. Mark, S. M. Atlas, and E. Cernia, eds., *Man-Made Fibers, Science and Technology,* Wiley-Interscience, New York, 1967.

Peterlin 1980. A. Peterlin, "Chain Folding in Lamellar Crystals," *Macromolecules* **13,** 777–782 (1980).

Poland 1970. Douglas Poland and Harold A. Scheraga, *Theory of Helix-Coil Transitions in Biopolymers,* Academic Press, New York, 1970.

Price 1956. Charles C. Price and Maseh Osgan, "The Polymerization of *l*-Propylene Oxide," *J. Am. Chem. Soc.* **78,** 4787–4792 (1956).

Price 1968. Fraser P. Price, "Kinetics of Crystallization," pp. 63–83 in Herman F. Mark, Norman G. Gaylord, and Norbert M. Bikales, eds., *Encyclopedia of Polymer Science and Technology,* Vol. 8, Wiley-Interscience, New York, 1968.

Prime 1969. R. Bruce Prime, Bernhard Wunderlich, and Louis Melillo, "Extended-Chain Crystals. V. Thermal Analysis and Electron Microscopy of the Melting Process in Polyethylene," *J. Polym. Sci. A-2* **7,** 2091–2097 (1969).

Quinn 1958. F. A. Quinn, Jr., and L. Mandelkern, "Thermodynamics of Crystallization in High Polymers: Poly-(ethylene)," *J. Am. Chem. Soc.* **80,** 3178–3182 (1958).

Reneker 1960. D. H. Reneker and P. H. Geil, "Morphology of Polymer Single Crystals," *J. Appl. Phys.* **31,** 1916–1925 (1960).

Richardson 1969. M. J. Richardson, "Crystallinity Determination in Polymers and a Quantitative Comparison for Polyethylene," *Br. Polym. J.* **1,** 132–137 (1969).

Roberts 1955. Donald E. Roberts and Leo Mandelkern, "Thermodynamics of Crystallization in High Polymers: Natural Rubber," *J. Am. Chem. Soc.* **77,** 781–786 (1955).

Sanchez 1974. Isaac Sanchez, "Modern Theories of Polymer Crystallization," *J. Macromol. Sci. Rev. Macromol. Chem.* **C10,** 113–148 (1974).

Schildknecht 1948. C. E. Schildknecht, S. T. Gross, H. R. Davidson, J. M. Lambert, and A. O. Zoss, "Polyvinyl Isobutyl Ethers—Properties and Structures," *Ind. Eng. Chem.* **40,** 2104–2115 (1948).

Schlesinger 1953. Walter Schlesinger and H. M. Leeper, "Gutta. I. Single Crystals of Alpha-Gutta," *J. Polym. Sci.* **11,** 203–213 (1953).

Schultz 1974. Jerold Schultz, *Polymer Materials Science,* Prentice-Hall, New York, 1974.

Schulz 1968. Rolf C. Schulz, "Optically Active Polymers," pp. 507–524 in Herman F. Mark, Norman G. Gaylord, and Norbert M. Bikales, eds., *Encyclopedia of Polymer Science and Technology,* Vol. 9, Wiley-Interscience, New York, 1968.

Sharples 1966. Allan Sharples, *Introduction to Polymer Crystallization,* St. Martin's Press, New York, 1966.

Starkweather 1960. Howard W. Starkweather, Jr., and Richard H. Boyd, "The Entropy of Melting of Some Linear Polymers," *J. Phys. Chem.* **64,** 410–414 (1960).

Statton 1959. W. O. Statton, "Polymer Texture: The Arrangement of Crystallites," *J. Polym. Sci.* **41,** 143–155 (1959).

Statton 1960. W. O. Statton and P. H. Geil, "Recrystallization of Polyethylene during Annealing," *J. Appl. Polym. Sci.* **3,** 357–361 (1960).

Staudinger 1932. Hermann Staudinger, *Die Hochmolecularen Organischen Verbindungen, (High Molecular Weight Organic Compounds)* (in German), Springer-Verlag, Berlin, 1932.

Szabolcs 1967. O. Szabolcs and I. Szabolcs, "Current Ideas on the Morphology of Synthetic Fibers," pp. 341–374 in H. F. Mark, S. M. Atlas, and E. Cernia, eds., *Man-Made Fibers, Science and Technology,* Vol. 1, Wiley-Interscience, New York, 1967.

Tadokoro 1979. Hiroyuki Tadokoro, *Structure of Crystalline Polymers,* John Wiley & Sons, New York, 1979.

Till 1957. P. H. Till, Jr., "The Growth of Single Crystals of Linear Polyethylene," *J. Polym. Sci.* **24,** 301–306 (1957).

Watson 1953. J. D. Watson and F. H. C. Crick, "A Structure of Deoxyribose Nucleic Acid," *Nature* **171,** 737–738 (1953).

Wilchinsky 1968. Zigmond W. Wilchinsky, "Orientation," pp. 624–648 in Herman F. Mark, Norman G. Gaylord, and Norbert M. Bikales, eds., *Encyclopedia of Polymer Science and Technology,* Vol. 9, Wiley-Interscience, New York, 1968.

Wood 1946. Lawrence A. Wood, "Crystallization Phenomena in Natural and Synthetic Rubbers," pp. 57–95 in H. Mark and G. S. Whitby, eds., *Scientific Progress in the Field of Rubber and Synthetic Elastomers (Advances in Colloid Science),* Vol. 2, Interscience, New York, 1946.

Wunderlich 1957. Bernhard Wunderlich and Malcolm Dole, "Specific Heat of Synthetic High Polymers. VII. Low Pressure Polyethylene," *J. Polym. Sci.* **24,** 201–213 (1957).

Wunderlich 1968. Bernhard Wunderlich and Louis Melillo, "Morphology and Growth of Extended Chain Crystals of Polyethylene," *Makromol. Chem.* **118,** 250–264 (1968).

Wunderlich 1973. Bernhard Wunderlich, *Macromolecular Physics, Volume 1: Crystal Structure, Morphology, Defects,* Academic Press, New York, 1973.

Wunderlich 1976. Bernhard Wunderlich, *Macromolecular Physics, Volume 2: Crystal Nucleation, Growth, Annealing,* Academic Press, New York, 1976.

Wunderlich 1980. Bernhard Wunderlich, *Macromolecular Physics, Volume 3: Crystal Melting,* Academic Press, New York, 1980.

Wunderlich 1983. Bernhard Wunderlich and Janusz Grebowicz, "Thermotropic Mesophases and Mesophase Transitions of Linear, Flexible Macromolecules," *Adv. Polym. Sci.* **60–61,** 1–59, 1983.

CHAPTER ELEVEN

RHEOLOGY AND THE MECHANICAL PROPERTIES OF POLYMERS

Rheology is, by definition, the science of deformation and flow of matter. The rheological behavior of polymers involves several widely different phenomena, which can be related to some extent to different molecular mechanisms. These phenomena and their associated major mechanisms are as follows:

a. *Viscous flow,* the irreversible bulk deformation of polymeric material, associated with irreversible slippage of molecular chains past one another.

b. *Rubberlike elasticity,* where the local freedom of motion associated with small-scale movement of chain segments is retained, but large-scale movement (flow) is prevented by the restraint of a diffuse network structure.

c. *Viscoelasticity,* where the deformation of the polymer specimen is reversible but time dependent and associated (as in rubber elasticity) with the distortion of polymer chains from their equilibrium conformations through activated segment motion involving rotation about chemical bonds.

d. *Hookean elasticity,* where the motion of chain segments is drastically restricted and probably involves only bond stretching and bond angle deformation: The material behaves like a glass.

These four phenomena are discussed in Sections A–D, respectively. Together they form the basis for a description of the mechanical properties of amorphous polymers. The mechanical properties of semicrystalline polymers, however, depend intimately on the restraining nature of their crystalline regions and can be inferred only in part from the rheological behavior of amorphous polymers. The mechanical properties of crystalline polymers are therefore discussed separately in Section *E*.

A. VISCOUS FLOW

Phenomena of Viscous Flow

If a force per unit area *s* causes a layer of liquid at a distance *x* from a fixed boundary wall to move with a velocity *v*, the viscosity η is defined as the ratio between the shear stress *s* and the velocity gradient $\partial v/\partial x$ or rate of shear $\dot\gamma$:

$$s = \eta \frac{\partial v}{\partial x} = \eta\dot\gamma \tag{11-1}$$

If η is independent of the rate of shear, the liquid is said to be *Newtonian* or to exhibit ideal flow behavior (Fig. 11-1*a*). Two types of deviation from Newtonian flow are commonly observed in polymer solutions and melts (Bauer 1967). One is *shear thinning* or *pseudoplastic* behavior, a reversible decrease in viscosity with increasing shear rate (Fig. 11-1*b*). Shear thinning results from the tendency of the applied force to disturb the long chains from their favored equilibrium conformation (Chapter 7*B*), causing elongation in the direction of shear. An opposite effect, *shear*

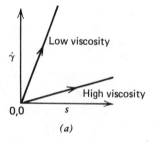
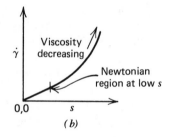

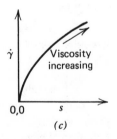
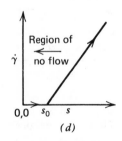

FIG. 11-1. Dependence of shear rate $\dot\gamma$ on shear stress *s* for (a) Newtonian, (b) pseudoplastic, and (c) dilatant behavior, and (d) the presence of a yield stress s_0 followed by Newtonian behavior.

thickening or *dilatant* behavior (Fig. 11-1*c*), in which viscosity increases with increasing shear rate, is not observed in polymers.

A second deviation from Newtonian flow is the exhibition of a *yield value,* a critical stress below which no flow occurs. Above the yield value, flow may be either Newtonian (as indicated in Fig. 11-1*d*) or non-Newtonian. For most polymer melts, only an apparent yield value is observed.

The above effects are shear dependent but time independent. Some fluids also exhibit reversible time-dependent changes in viscosity when sheared at constant stress. Viscosity decreases with time in a *thixotropic fluid,* and increases with time in a *rheopectic* fluid, under constant shear stress.

For low-molecular-weight liquids, the temperature dependence of viscosity is found to follow the simple exponential relationship

$$\eta = Ae^{E/RT} \tag{11-2}$$

where E is an *activation energy for viscous flow* and A is a constant.

These features of the flow of liquids can be explained in terms of several molecular theories. That of Eyring (Glasstone 1941) is based upon a lattice structure for the liquid, containing some unoccupied sites or holes. These sites move at random throughout the liquid as they are filled and created anew by molecules jumping from one site to another. Under an applied stress the probability of such jumps is higher in the direction that relieves the stress. If each jump is made by overcoming an energy barrier of height E, the theory leads to Eq. 11-2. The energy of activation E is expected to be related to the latent heat of vaporization of the liquid, since the removal of a molecule from the surroundings of its neighbors forms a part of both processes. Such a relation is indeed found and is taken as evidence that the particle that moves from site to site is probably a single molecule.

As molecular weight is increased in a homologous series of liquids up to the polymer range, the activation energy of flow E does not increase proportionally with the heat of vaporization but levels off at a value independent of molecular weight. This is taken to mean that in long chains the unit of flow is considerably smaller than the complete molecule. It is rather a segment of the molecule whose size is of the order of 5–50 carbon atoms. Viscous flow takes place by successive jumps of segments (with, of course, some degree of coordination) until the whole chain has shifted.

Dynamics of Polymer Melts. It is now accepted that polymer chains are strongly intertwined and entangled in the melt; the dynamic behavior of such a system has been reviewed (Graessley 1974), but is only poorly understood. Thermodynamically, the chains are essentially ideal, as was first realized by Flory (1949). Their freedom of motion results from the presence of a *correlation hole* around each flow unit, within which the concentration of similar units from other chains is reduced. The presence of these correlation holes, and of the ideal but entangled nature of chains in the melt, has been confirmed by neutron-scattering experiments (Cotton 1974).

Flow Measurement

Methods commonly used for measuring the viscosity of polymer solutions and melts are listed in Table 11-1. The most important of these methods involve rotational and capillary devices (Van Wazer 1963, Whorlow 1979, Dealy 1981, 1983).

Rotational Viscometry. Rotational viscometers are available with several different geometries, including concentric cylinders, two cones of different angles, a cone and a plate, or combinations of these. Measurements with rotational devices become difficult to interpret at very high shear stresses, owing to the generation of heat in the specimen because of dissipation of energy, and to the tendency of the specimen to migrate out of the region of high shear. This phenomenon, the *Weissenberg effect,* arises because the stress in any material can always be analyzed into the components of a 3×3 stress tensor, in which the off-diagonal elements, called *normal stresses* because they act perpendicular to the surface of the specimen, are not negligible in viscoelastic fluids.

A simple rotational instrument used in the rubber industry is the *Mooney viscosimeter.* This empirical instrument measures the torque required to revolve a rotor at constant speed in a sample of the polymer at constant temperature. It is used to study changes in the flow characteristics of rubber during milling or mastication (Chapter 19). The *Brabender Plastograph* is a similar device.

Capillary Viscometry. Capillary rheometers, usually made of metal and operated either by dead weight or by gas pressure, or at constant displacement rate, have advantages of good precision, ruggedness, and ease of operation. They may be built to cover the range of shear stresses found in commercial fabrication operations. However, they have the disadvantage that the shear stress in the capillary varies from zero at the center to a maximum at the wall.

An elementary capillary rheometer (extrusion plastometer) is used to determine the flow rate of polyethylene in terms of *melt index,* defined as the mass rate of

TABLE 11-1. Summary of Methods for Measuring
Viscosity

Method	Approximate Useful Viscosity Range (Poise)
Capillary pipette	$10^{-2}-10^{3}$
Falling sphere	$1-10^{5}$
Capillary extrusion	$1-10^{8}$
Parallel plate	$10^{4}-10^{9}$
Falling coaxial cylinder	$10^{5}-10^{11}$
Stress relaxation	$10^{3}-10^{10}$
Rotating cylinder	$1-10^{12}$
Tensile creep	$10^{5}->10^{12}$

flow of polymer through a specified capillary under controlled conditions of temperature and pressure.

Experimental Results

Molecular Weight and Shear Dependence.   As discussed further in Chapter 12, the most important structural variable determining the flow properties of polymers is molecular weight or, alternatively, chain length, Z (the number of atoms in the chain). Although early data (Flory 1940) suggested that log η was proportional to $Z^{1/2}$, it is now well established (Fox 1956) for essentially all polymers studied that, for values of Z above a critical value Z_c,

$$\log \eta = 3.4 \log \bar{Z}_w + k \tag{11-3}$$

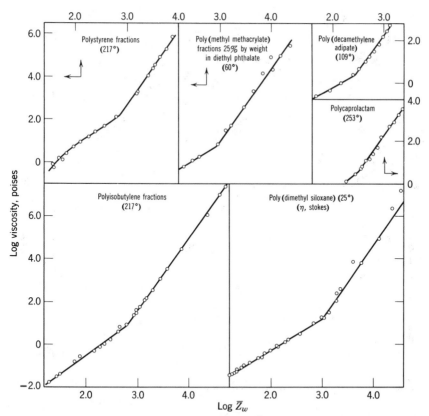

FIG. 11-2. Dependence of melt viscosity η on chain length $\bar{Z}_w$ for low shear rate, showing the regions below $Z_c \simeq 600$, where η is approximately proportional to $\bar{Z}_w^{1.75}$, and above Z_c, where $\eta \sim \bar{Z}_w^{3.4}$ (Fox 1956).

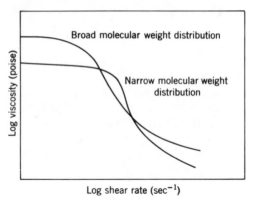

FIG. 11-3.   Generalized melt viscosity–shear rate curves for polymers with broad and narrow distributions of molecular weight.

where k is temperature dependent. This equation is valid only for shear stress
sufficiently low (10^2–10^3 dynes/cm^2) that the viscosity is Newtonian. The weight-
average chain length $\bar{Z}_w$ is usually assumed to be the appropriate average for the
above conditions. de Gennes (1979) discusses the unusual exponent 3.4 in terms
of scaling concepts, considering it a major unsolved problem. Theories to explain
it, including his own *reptation* model (de Gennes 1971), based on the wriggling
motion of a chain inside a "tube" formed by its neighbors, fail to predict the
exponent closer than 3.0 with no obvious reason for the discrepancy.

For chain lengths below Z_c, which is about 600 for many polymers, the viscosity
is found to depend upon a power of $\bar{Z}_w$ (and hence $\bar{M}_w$) in the range 1.75–2.0. In
this range, shear rate has little effect on viscosity. Typical experimental data for
the two regions described above are shown in Fig. 11-2.

While the Newtonian melt viscosity is determined by $\bar{M}_w$ as described in Eq.
11-3, the dependence of viscosity on shear rate also depends upon the molecular-
weight distribution. As depicted schematically in Fig. 11-3, the drop in melt viscosity below its Newtonian value begins at a lower shear rate and continues over
a broader range of shear rates for polymers with broader distributions of molecular
weight. At sufficiently high shear rates, the melt viscosity appears to depend primarily on $\bar{M}_n$ rather than $\bar{M}_w$. Qualitative information, at least, about the molecular-
weight distribution can be obtained from melt viscosity–shear rate studies.

Flow Instabilities.   At shear stresses in the neighborhood of 2×10^6 dynes/cm^2
for many polymers, instabilities in the flow appear, with the result that the upper
Newtonian region is rarely realized in bulk polymers. These instabilities are manifested as a striking and abrupt change in the shape of the polymer stream emerging
from the capillary of the rheometer. At and above a critical stress, the shape of
the emerging stream changes from that of a regular cylinder to a rough or distorted one.

Temperature Dependence of Viscosity.   On close examination of polymer systems, the activation energy for viscous flow of Eq. 11-2 is found to be constant only for small ranges of temperature. For pure liquids, it was found many years ago that most of the change in viscosity with temperature is associated with the concurrent change in volume. This observation led to theories of viscosity based on the concept of free volume, whose application to polymers is discussed in Section D. A major result of these theories is the WLF equation (Eqs. 11-13 and 11-22), in which the temperature dependence of melt viscosity is expressed in terms of the glass transition temperature T_g (or another reference temperature) and universal constants. Since the terms describing the variation of melt viscosity with temperature and with molecular weight are independent, the WLF equation can be combined with Eq. 11-3 to yield the relation, for low shear rates,

$$\log \eta = 3.4 \log \bar{Z}_w - \frac{17.44(T - T_g)}{51.6 + T - T_g} + k' \qquad (11\text{-}4)$$

where k' is a constant depending only on polymer type. This equation holds over the temperature range from T_g to about $T_g + 100$ K.

GENERAL REFERENCES

Eirich 1956–1969; Van Wazer 1963; Carley 1975; Walters 1975; Nielsen 1977; White 1978; Whorlow 1979; Dealy 1981; Hull 1981; Pierce 1982; Rosen 1982, Chapters 15 and 17.

B. KINETIC THEORY OF RUBBER ELASTICITY

Rubberlike elasticity is in many respects a unique phenomenon, involving properties markedly different from those of low-molecular-weight solids, liquids, or gases. The properties of typical elastomers are defined by the following requirements:

a.   They must stretch rapidly and considerably under tension, reaching high elongations (500–1000%) with low damping, that is, little loss of energy as heat.

b.   They must exhibit high tensile strength and high modulus (stiffness) when fully stretched.

c.   They must retract rapidly, exhibiting the phenomenon of *snap* or *rebound*.

d.   They must recover their original dimensions fully on the release of stress, exhibiting the phenomena of *resilience* and *low permanent set*.

Although the thermodynamics associated with rubber elasticity was developed in the middle of the nineteenth century, the molecular requirements for the exhibition of rubbery behavior were not recognized until 1932. Theories of the mechanism relating these molecular-structure requirements to the phenomena of rubber elasticity were developed soon after.

As discussed further in Chapter 12E, the molecular requirements of elastomers may be summarized as follows:

a. The material must be a high polymer.

b. It must be above its glass transition temperature T_g to obtain high local segment mobility.

c. It must be amorphous in its stable (unstressed) state for the same reason.

d. It must contain a network of crosslinks to restrain gross mobility of its chains.

Thermodynamics of Rubber Elasticity

The Ideal Elastomer.   In discussions of the mechanical properties of a polymer, parameters related to its distortion must be included, as well as the variables defining the state of the system, such as pressure and temperature. In particular, when a sample of rubber is stretched, work is done on it and its free energy is changed. By restricting the development to stretching in one direction, for simplicity, the elastic work W_{el} equals $f\, dl$, where f is the retractive force and dl the change in length. If the change in free energy is G,

$$f = \left(\frac{\partial G}{\partial l}\right)_{T,p} = \left(\frac{\partial H}{\partial l}\right)_{T,p} - T\left(\frac{\partial S}{\partial l}\right)_{T,p} \qquad (11\text{-}5)$$

In analogy with an ideal gas, where

$$\left(\frac{\partial E}{\partial V}\right)_T = 0, \qquad p = T\left(\frac{\partial S}{\partial V}\right)_T \qquad (11\text{-}6)$$

an *ideal elastomer* is defined by the condition that

$$\left(\frac{\partial H}{\partial l}\right)_{T,p} = 0, \qquad f = -T\left(\frac{\partial S}{\partial l}\right)_{T,p} \qquad (11\text{-}7)$$

The negative sign results since work is done on the specimen to increase its length.

Entropy Elasticity.   Equation 11-7 shows that the retractive force in an ideal elastomer is due to its decrease in entropy on extension. The molecular origin of this *entropy elasticity* is the distortion of the polymer chains from their most probable conformations in the unstretched sample. As described in Chapter 7B, the distribution of these conformations is Gaussian, the probability of finding a chain end in a unit volume of the space coordinates x, y, z at a distance r from the other end being

$$W(x, y, z) = \left(\frac{b}{\sqrt{\pi}}\right)^3 e^{-b^2 r^2} \tag{11-8}$$

where $b^2 = \frac{3}{2} x l^2$, x being the number of links, with length l. Since the entropy of the system is proportional to the logarithm of the number of configurations it can have,

$$S = \text{const.} - k b^2 r^2 \tag{11-9}$$

where k is Boltzmann's constant. It follows that for a single polymer chain the retractive force f' for an extension of magnitude dr is

$$f' = -T \frac{dS}{dr} = 2kTb^2 r \tag{11-10}$$

It is customarily assumed that the retractive force f (Eq. 11-7) for a bulk polymer sample can be identified as the sum of the forces f' for all the chains in the specimen. This assumption, which is inaccurate in detail but justified in most cases, implies that individual chains contribute additively and without interaction to the elasticity of the macroscopic sample, and that the distribution of end-to-end distances undergoes transformation identically with the sample dimensions (affine transformation).

Stress–Strain Behavior of Elastomers

The stress–strain curves of typical elastomers (Fig. 11-4) show marked deviations from the straight line required by Hooke's law. The relatively low slope of the curve (defined near the origin as the modulus) decreases to about one-third its original value over the first hundred percent elongation, and later increases, often to quite high values at high elongations. To explain this behavior, it is convenient first to examine the stress–strain properties of a simple model, and then to consider the relation of the model to actual elastomers.

A Simple Model of an Elastomeric Network (Guth 1946). It is assumed that the actual tangled mass of polymer chains may be represented by an idealized network of flexible chains, irregular in detail but homogeneous and isotropic, extending throughout the sample. The network consists of m chains of average length $(\overline{r^2})^{1/2}$ per unit volume directed along each of three perpendicular axes. To obtain the proper space-filling properties of the model it is assumed that the space between the chains is filled with an incompressible fluid exerting a hydrostatic pressure p outward against the elastic tension of the chains. By considering the equilibration of inward and outward forces when a cube of this material is stretched

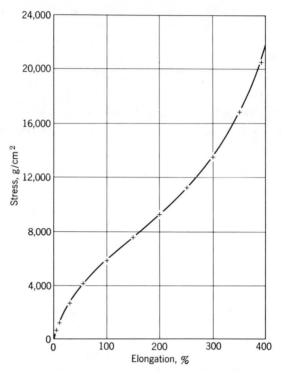

FIG. 11-4.   Stress–strain curve for a typical elastomer (Guth 1946).

into a parallelepiped, using Eq. 11-10, it can be shown that the stress s is related to the strain γ by

$$s = 2mkTb^2\left(\gamma - \frac{1}{\gamma^2}\right) \tag{11-11}$$

The modulus† is given by

$$G = \frac{ds}{d\gamma} = 2mkTb^2\left(1 + \frac{2}{\gamma^3}\right) \tag{11-12}$$

As the strain increases from zero ($\gamma = 1$) to large values of γ, the slope decreases to one-third its initial value as required by the experimental facts. Equation 11-12 predicts the stress–strain curve of actual elastomers very well up to elongations of 300% or more (Fig. 11-5).

----

†Note that the symbol G is conventionally used to represent both the Gibbs free energy (Eq. 11-5) and, throughout the remainder of this chapter, the modulus of elasticity.

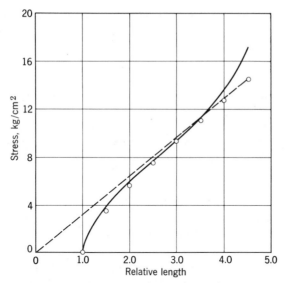

FIG. 11-5. Experimental and theoretical stress–strain curves (Guth 1946): experimental (——), computed from Eq. 11-12 (○), and asymptote of Eq. 11-12 for high elongations (- - -).

Behavior at High Extensions.   Figure 11-5 suggests that at elongations greater than about 300%, the stress–strain curves of actual elastomers have a higher slope than that predicted by Eq. 11-12. This is far below the ultimate elongation of 1000% for a good elastomer. One reason for the failure of the theory is that the Gaussian distribution of chain lengths that holds at low elongations fails in the region of interest for elastomers, as the actual lengths approach those of fully extended chains. A better approximation than the Gaussian is available. With its use the theory predicts the actual stress–strain curves rather well for elastomers that do not crystallize on stretching. In the cases where crystallinity does develop, the slope still increases faster than predicted. As is true in general for the problem of the mechanical behavior of crystalline polymers (Section *E*), a quantitative theory of the effect of crystallinity on the modulus of elastomers has not been developed.

GENERAL REFERENCES

Treloar 1975; Smith 1977; Brydson 1978, Chapter 3; Gent 1978; Shen 1978.

C. VISCOELASTICITY

Sections *A* and *B* have dealt with the equilibrium response of linear and network polymer structures to external stress. In this section are considered the time-dependent mechanical properties of amorphous polymers. As in the studies of rate phenomena, theories of dynamic mechanical phenomena in polymers are less thoroughly developed than those referring to the equilibrium states. They are also more dependent on the details of models. The present problem is, first, to find suitable models or mathematical functions with which to describe the several types of molecular motion postulated earlier to be associated with the processes of elasticity, viscoelasticity, and viscous flow in polymers; and, second, to utilize these models in relating the behavior of the polymers to their molecular structure.

The description of the viscoelastic response of amorphous polymers to small stresses is greatly simplified by the application of the following two general principles that are widely applicable to these systems.

The Boltzmann Superposition Principle. This principle states that strain is a linear function of stress, so that the total effect of applying several stresses is the sum of the effects of applying each one separately. Application of the superposition principle makes it possible to predict the mechanical response of an amorphous polymer to a wide range of loading conditions from a limited amount of experimental data. The principle applies to both static and time-dependent stresses.

Time–Temperature Equivalence. An increase in temperature accelerates molecular and segmental motion, bringing the system more rapidly to equilibrium or apparent equilibrium and accelerating all types of viscoelastic processes. A convenient way of formulating this effect of temperature is in terms of the ratio a_T of the time constant (*relaxation time*) of a particular response τ at temperature T to its value τ_0 at a convenient reference temperature T^0. For many cases, including most nonpolar amorphous polymers, a_T does not vary with τ, so that changes in temperature shift the distribution of relaxation times, representing all possible molecular responses of the system, to smaller or greater values of τ but do not otherwise alter it. Time and temperature affect viscoelasticity only through the product of a_T and actual time, and a_T is called a *shift factor*. Its application to experimental data is described in the following section.

Despite the marked dependence on molecular structure of the relation between a_T and absolute temperature, nearly general empirical relations have been derived by expressing the temperature for each material in terms of its glass-transition temperature T_g or some nearly equivalent reference temperature. Among the most successful of these relations is the Williams–Landel–Ferry (WLF) equation (Williams 1955):

$$\log a_T = \frac{-17.44(T - T_g)}{51.6 + T - T_g} \tag{11-13}$$

This equation holds over the temperature range from T_g to about $T_g + 100$ K. The constants are related to the free volume as described in Section D.

Experimental Methods

Stress Relaxation. If elongation is stopped during the determination of the stress–strain curve of a polymer (Chapter 9F), the force or stress decreases with time as the specimen approaches equilibrium or quasi-equilibrium under the imposed strain. The direct observation and measurement of this phenomenon constitutes the *stress relaxation* experiment. Usually the sample is deformed rapidly to a specified strain, and stress at this strain is observed for periods ranging from several minutes to several days or longer.

It has been shown both experimentally and theoretically that the stress-relaxation behavior of rubbery polymers can be factored into independent functions of strain and time. At small strains, the stress–strain function is almost linear and can be represented by a time-dependent modulus of elasticity $G(t)$.

Typical stress-relaxation data for a well-studied sample of polyisobutylene are shown in Fig. 11-6, where $G(t)$ is plotted against time for experiments performed at several temperatures. By means of time–temperature equivalence these data may be shifted to produce a "master curve," as shown in Fig. 11-7. The curve indicates that any specified modulus can be observed after a period that depends, through

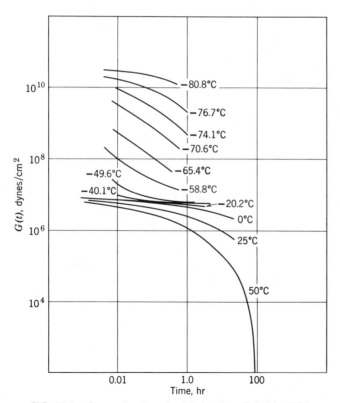

FIG. 11-6. Stress relaxation of polyisobutylene (Tobolsky 1956).

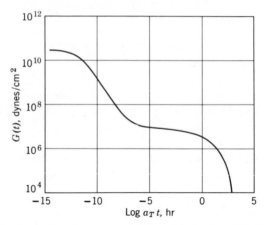

FIG. 11-7. Stress relaxation master curve for polyisobutylene (Tobolsky 1956).

the factor a_T, on temperature. As temperature is increased, a given modulus is observed at shorter times.

Master curves such as that of Fig. 11-7 show the different types of viscoelastic behavior usually observed with amorphous materials. At small times the high modulus and low slope are characteristic of glassy behavior. The following region of rapidly decreasing modulus represents the glass transition, described further in Section D. This is followed by a flat region of rubbery behavior produced either through entanglements among relatively long molecules or through permanent crosslinks. The ultimate slope at long times represents the region of viscous flow in uncrosslinked polymers.

Creep. Creep is studied by subjecting a sample rapidly to a constant stress and observing the resulting time-dependent strain for relatively long periods of time, frequently for a week or more or even for a year or more. Creep and stress relaxation are complementary aspects of plastics behavior and in many cases may provide equivalent information for studies of both fundamental viscoelastic properties and performance in practical applications. Creep experiments are usually easier, more economical, and more feasible for long periods of time.

Dynamic Methods. The delayed reaction of a polymer to stress and strain also affects its dynamic properties. If a simple harmonic stress of angular frequency ω is applied to the sample, the strain lags behind the stress by a phase angle whose tangent measures the *internal friction* $\Delta E/E$, where ΔE is the energy dissipated in taking the sample through a stress cycle and E is the energy stored in the sample when the strain is a maximum. The internal friction is a maximum when the dynamic modulus of the material (the ratio of the stress to that part of the strain which is in phase with the stress) is in a region of relaxation. In terms of frequency the relaxation behavior of the dynamic modulus and the internal friction is illustrated in Fig. 11-8.

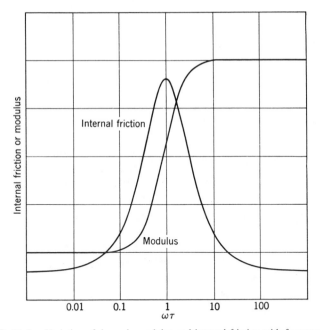

FIG. 11-8. Variation of dynamic modulus and internal friction with frequency.

At low frequencies (about 1 sec^{-1}) it is convenient to measure the internal friction of polymers by observing the free decay of torsional oscillations of a thin fiber or film loaded with a large moment of inertia (*torsion pendulum*). The internal friction is the logarithm of the ratio of the amplitudes of successive free oscillations, while the modulus is calculated from the frequency of the oscillations.

A natural extension of this technique is to force oscillation with an external driving force. In such techniques the specimen is driven in flexural, torsional, or longitudinal oscillations by an oscillating force of constant amplitude but variable frequency, and the displacement amplitude of the resulting oscillation is observed. These techniques are particularly useful in the kilocycle frequency range. At very high frequencies (above 1 megacycle), the dimensions of the sample become awkwardly small for mechanical techniques, and it is convenient to measure the attenuation of sound waves in the sample. These methods are discussed fully by Ferry (1980).

Sperling (1982) describes a simple classroom experiment illustrating the above principles.

Models of Viscoelastic Behavior

Before attempting to devise a model to duplicate the viscoelastic behavior of an actual polymer, it is well to examine the response γ of ideal systems to a stress s. An ideal elastic element is represented by a spring which obeys Hooke's law, with

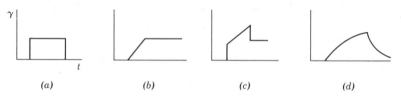

FIG. 11-9.   Strain–time relationships at constant stress for simple models (Alfrey 1948): (a) ideal elastic spring, (b) Newtonian fluid (dashpot), (c) Maxwell element, (d) Voigt element.

a modulus of elasticity G. The elastic deformation is instantaneous and independent of time: $\gamma = (1/G)s$ (Fig. 11-9a). A completely viscous response is that of a Newtonian fluid, whose deformation is linear with time while the stress is applied and is completely irrecoverable: $d\gamma/dt = (1/\eta)s$, where η is the viscosity of the fluid (Fig. 11-9b). A simple mechanical analogy of a Newtonian fluid is a dashpot.

The two elements of spring and dashpot can be combined in two ways. If they are placed in series, the resulting *Maxwell element* (Fig. 11-10a) exhibits flow plus elasticity on the application of stress (Fig. 11-9c); when the stress is applied, the spring elongates while the dashpot slowly yields. On the removal of the stress the spring recovers but the dashpot does not. The strain is given by the equation

$$\frac{d\gamma}{dt} = \frac{1}{\eta}s + \frac{1}{G}\frac{ds}{dt} \tag{11-14}$$

The relation of creep to stress relaxation may be seen by considering the experiment in which a strain is obtained and then held by fixing the ends of the system: $d\gamma/dt = 0$ in Eq. 11-14. The equation can then be solved:

$$s = s_0 e^{-(G/\eta)t} = s_0 e^{-t/\tau} \tag{11-15}$$

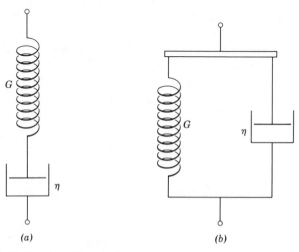

FIG. 11-10.   (a) A Maxwell element and (b) a Voigt element.

where the stress s relaxes from its initial value s_0 exponentially as a function of time. The time η/G after which the stress reaches $1/e$ of its initial value is the relaxation time τ.

The response of a tangled mass of polymer chains to a stress is better represented by a parallel combination of spring and dashpot in a *Kelvin* or *Voigt element* (Fig. 11-10b). This element shows a *retarded elastic* or *viscoelastic* response (Fig. 11-9d). The dashpot acts as a damping resistance to the establishment of the equilibrium of the spring. The equation for the strain is

$$\eta \frac{d\gamma}{dt} + G\gamma = s \tag{11-16}$$

If a stress is applied and after a time removed (Fig. 11-9d), the deformation curve is given by

$$\gamma = \frac{s}{G} (1 - e^{-(G/\eta)t}) = \frac{s}{G} (1 - e^{-t/\tau}) \tag{11-17}$$

where τ is a *retardation time*. When the stress is removed the sample returns to its original shape along the exponential curve:

$$\gamma = \gamma_0 e^{-t/\tau} \tag{11-18}$$

General Mechanical Models for an Amorphous Polymer. The three components that make up the simplest behavior of an actual polymer sample in creep can be represented by a mechanical model which combines a Maxwell and a Voigt element in series. If a stress is suddenly applied to this model, the strain changes with time as shown in Fig. 11-11; the corresponding behavior of the model is shown in Fig. 11-12.

The model departs from the initial conditions (a) at time t_1 by an elastic deformation s/G_1 (b). A viscoelastic response approaching s/G_2 as an equilibrium value

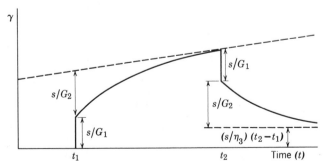

FIG. 11-11.   Strain–time relationship for a generalized mechanical model for creep combining elasticity, viscoelasticity, and flow (Alfrey 1948).

and a viscous flow at the rate s/η_3 follow (c). On the removal of the stress at time t_2, the elastic element relaxes immediately (d) and the viscoelastic one slowly (e), but the viscous flow is never recovered.

Alternatively, the creep and stress-relaxation experiments can be described with a generalized model consisting of a Maxwell and a Voigt element arranged in parallel. The two generalized models are entirely equivalent.

Although these models exhibit the chief characteristics of the viscoelastic behavior of polymers, they are nevertheless very much oversimplified. The flow of the polymer is probably not Newtonian, and its elastic response may not be Hookean. Moreover, the behavior of a real polymer cannot be characterized by a single relaxation time, but requires a spectrum of relaxation times to account for all phases of its behavior.

Treatment of Experimental Data

Distribution of Relaxation Times. Equation 11-17 indicates that most of the relaxation associated with a single element takes place within one cycle of log time. In contrast, experiments indicate (as in Fig. 11-7) that relaxation phenomena in polymers extend over much wider ranges of time. Thus the actual behavior of polymers can be described only in terms of a distribution of model elements and an associated distribution of relaxation or retardation times:

$$\gamma(t) = s \int_{-\infty}^{\infty} \bar{J}\,(\log \tau)(1 - e^{-t/\tau})d \log \tau \qquad (11\text{-}19)$$

where $J = 1/G$ is the *elastic compliance* and $\bar{J}(\log \tau)$ the distribution of retardation times.

In principle, knowledge of $\gamma(t)$ over the entire range of time allows $\bar{J}(\log \tau)$ to be evaluated explicitly. This is difficult in practice, however, and it is convenient for mathematical simplicity to adopt simple empirical forms of the distribution

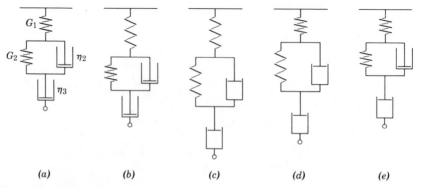

(a)   (b)   (c)   (d)   (e)

FIG. 11-12. Interpretation of strain–time curve of Fig. 11-11 in terms of the generalized mechanical model for creep.

function $\bar{J}(\log \tau)$ and evaluate their parameters from the more limited experimental data at hand. Alternatively, the experimental data may be fitted graphically by summing the theoretical curves corresponding to a small number (three or four) of discrete relaxation times. The usefulness of the distribution of relaxation times arises from its identification with certain molecular parameters of the specimen, as described below.

Molecular Structure and Viscoelasticity

The phenomena of relaxation processes can be thought of in terms of the effect of thermal motion on the orientation of polymer molecules. When a mechanical stress is applied to a polymer, introducing deformations of the chains, the entropy of the system decreases as less probable conformations are taken up. The free energy correspondingly increases. If the sample is kept in the deformed state, stress relaxation takes place as a result of the thermal motions of the chains, the molecular deformations are obliterated, and the excess free energy is dissipated as heat. The details of the stress-relaxation process depend upon the multiplicity of ways in which the polymer molecules can regain their most probable conformations through thermal motion. These complex motions of a polymer molecule can be expressed as a series of characteristic modes requiring various degrees of long-range cooperation among the segments of the chain. Thus the first mode corresponds to translation of the entire molecule, requiring maximum cooperation, the second corresponds to motion of the ends of the chain in opposite directions, requiring somewhat less cooperation, and so on. With each of these modes is associated a characteristic relaxation time; there are so many modes that over most of the time scale the discrete spectrum of relaxation times can be approximated by a continuous distribution.

Theories based on the above considerations, derived for dilute solutions of polymer molecules, have been combined (Ferry 1980) with the concepts of entanglement of long molecules successful in predicting the molecular-weight dependence of viscous flow (Section A). Although exact numerical agreement is not achieved, for example, in calculating the dynamic modulus of polyisobutylene over a wide range of frequency and molecular weight, the qualitative situation seems extremely satisfactory.

In the rubbery region, the maximum relaxation time is strongly dependent on molecular weight. In this region the motions of the molecules are long range in nature, involving motions of units of the order of the length of the molecule itself. This is the region where entanglements are important.

In the region around the glass transition, however, only vibrations of the parts of the molecule between entanglements are important. So long as the molecules are long enough for entanglements to exist, this portion of the curve is expected, and found, to be essentially independent of the molecular weight and to depend primarily upon the local structure of the polymer.

GENERAL REFERENCES

Alfrey 1948, 1967; Eirich 1956–1969; Tobolsky 1960; Van Wazer 1963; McCrum
    1967; Ward 1971; Nielsen 1974; Sauer 1975; Sternstein 1977; Kramer 1978;
    Tager 1978, Chapter 7; Ferry 1980; Young 1981, Chapter 5; Rosen 1982,
    Chapter 18.

D. THE GLASSY STATE AND THE GLASS TRANSITION

As anticipated in the previous sections, and in particular in Fig. 11-7, all amorphous
polymers assume at sufficiently low temperatures the characteristics of glasses,
including hardness, stiffness, and brittleness. Many aspects of the glassy state and
the glass transition were mentioned in Section C; what remains is to discuss measurement of the transition temperature and its molecular interpretation in more detail.

One property associated with the glassy state is a low volume coefficient of
expansion. This low coefficient occurs as the result of a change in slope of the
curve of volume versus temperature at the point called the *glass-transition temperature T_g*. This behavior is shown for natural rubber in Fig. 11-13. In the high-
temperature region, the slope of the curve (expansion coefficient) is characteristic
of a rubber; below T_g at about $-70°C$, it is characteristic of a glass.

Figure 11-13 illustrates another general phenomenon: The amorphous regions in
partially crystalline polymers also assume a glassy state, T_g being independent of
degree of crystallinity to a first approximation. The magnitude of the phenomena
associated with T_g decreases with decreasing amorphous content, however. As a
result, T_g is sometimes difficult to detect in highly crystalline polymers. In terms
of the lamellar model (Chapter 10C) the glass transition is considered to involve
defect regions within or at the boundaries of the lamellae.

In contrast to crystalline melting at a temperature T_m (about $+10°C$ in Fig. 11-
13), there is not an abrupt change in volume at T_g, but only a change in the slope

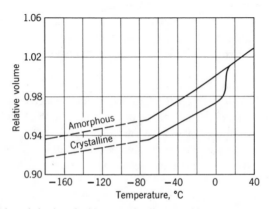

FIG. 11-13. Volume behavior of rubber near the glass-transition temperature (Bekkedahl 1934).

of the volume-temperature curve. In analogy to thermodynamic first- and second-order transitions, T_g is sometimes referred to as a second-order or apparent second-order transition. This nomenclature is considered poor, however, since it implies more thermodynamic significance than the nature of the transition warrants.

Measurement of T_g

The glass-transition temperature can be detected in a variety of experiments, which can be roughly classified into those dealing with bulk properties of the polymer, and those measuring the nature and extent of molecular motion. The classification is to some extent arbitrary, since, as indicated below, T_g is in fact the temperature of onset of extensive molecular motion.

Bulk Properties.   Perhaps the most common way of estimating T_g is by means of the volume expansion coefficient, as indicated above. Other bulk properties whose temperature coefficients undergo marked changes at T_g, and which can therefore be used for its determination, include heat content (Chapter 9E), refractive index, stiffness, and hardness.

Molecular Motion.   Experiments that are sensitive to the onset of molecular motion in polymer chains may be used to detect the glass transition. Such methods include the measurement of internal friction (Section C), dielectric loss in polar polymers, and NMR spectroscopy (Chapter 9B).

Phenomena Related to T_g.   The onset of brittleness, as measured in impact tests, and the softening of amorphous polymers, as measured in thermal tests of various sorts, take place at temperatures near T_g (see Chapter 9F).

Time Effects Near T_g.   If a polymer sample is cooled rapidly to a temperature just below T_g, its volume continues to decrease for many hours. In consequence, the value observed as T_g in a volume–temperature experiment depends on the time scale of the measurements. It is convenient to define T_g in terms of an arbitrary convenient (but not highly critical) time interval, such as 10 min to 1 hr.

For similar reasons, other tests for T_g give results somewhat dependent on the time scale of the experiments, with tests requiring shorter times yielding higher values of T_g. The brittle temperature as determined in an impact test is, for example, normally somewhat higher than T_g as otherwise measured.

Molecular Interpretation of T_g

In the glassy state, large-scale molecular motion does not take place, rather, atoms and small groups of atoms move against the local restraints of secondary bond forces, much as atoms vibrate around their equilibrium positions in a crystal lattice, except that the glassy state does not have the regularity of the crystalline state. The glass transition corresponds to the onset of liquidlike motion of much longer segments of molecules, characteristic of the rubbery state. This motion requires more free volume than the short-range excursions of atoms in the glassy state. The rise in the relative free volume with increasing temperature above T_g leads to the higher observed volume expansion coefficient in this region. Since the fully extended chain is the conformation of minimum energy (Chapter 10B), it tends to be assumed more frequently as the temperature is lowered. As the molecules thus straighten out, the free volume decreases. In consequence, flow becomes more difficult. The glass transition (observed at infinite time)—or, alternatively, the onset of crystallization where possible—is taken as the point where the number of possible conformations of the amorphous phase decreases sharply toward one.

The fraction f of "free" volume may be defined as

$$f = f_g + (T - T_g)\,\Delta\alpha, \qquad T \geq T_g \qquad (11\text{-}20)$$
$$f = f_g \qquad\qquad\qquad\qquad T < T_g$$

Thus f is constant at the value f_g for all temperatures below T_g. Here the volume expansion coefficient α is that resulting from the increase in amplitude of molecular vibrations with temperature. Above T_g new free volume is created as the result of an increase $\Delta\alpha$ in the expansion coefficient.

Williams, Landel, and Ferry (Williams 1955) proposed that log viscosity varies linearly with $1/f$ above T_g, so that

$$\ln\left(\frac{\eta}{\eta_g}\right) = \frac{1}{f} - \frac{1}{f_g} \qquad (11\text{-}21)$$

Substitution into Eq. 11-20 leads to

$$\log\left(\frac{\eta}{\eta_g}\right) = -\frac{a(T - T_g)}{b + T - T_g} \qquad (11\text{-}22)$$

which is the WLF equation presented in Section C (Eq. 11-13), the numerical constants for a and b given there being determined by fitting literature data on the viscosity–temperature behavior of many glass-forming substances. The shift factor a_T is seen to be just the ratio of the viscosity at T relative to that at T_g. The latter is about 10^{13} poise for many substances.

Equation 11-22 also implies that both the viscosity of the polymer and the activation energy for viscous flow $\Delta E = 2.3R\,d(\log \eta)/d(1/T)$ should become infinite at $T = T_g - b = T_g - 51.6$. Thus by extrapolating downward from behavior well above T_g, one would predict that all molecular motion should become completely frozen at $T < T_g - 51.6$. What happens, of course, is that new mechanisms of deformation take over more or less sharply as this critical range is approached, in fact at T_g.

The above discussion embodies elements of several theories of the glass transition, based on free-volume concepts, kinetics, and statistical thermodynamics.

These theories are summarized by Ward (1971) and Ferry (1980); important contributions were made by Doolittle (1951), Bueche (1953), Gibbs (1958), DiMarzio (1958), Kovacs (1958), Cohen (1959), and Adam (1965).

Molecular Motion Below T_g

The foregoing discussion is concerned with the transition involving the motion of long segments of the polymer chain. At lower temperatures, other transitions may occur, produced by the motion of short sections of the main chain or of side chains. Although some characteristics of the glassy state, such as brittleness, may occasionally occur only below one of these lower transitions, it is proposed that the transition of highest temperature be called T_g. Alternatively, the transitions may be denoted α, β, γ, and so on, in order of descending temperature.

Transitions due to the motion of short segments of the main polymer chain occur most prominently in crystalline polymers such as polyethylene, polypropylene, and polytetrafluoroethylene. Such polymers also typically exhibit an α-transition.

Side-chain transitions occur in methacrylate polymers as a result of the relaxation of the carbomethoxy side chain at about 20°C (torsion pendulum) and the relaxation of the aliphatic ester group below -150°C.

GENERAL REFERENCES

Ward 1971; Nielsen 1974; Armeniades 1975; Tager 1978, Chapters 8 and 9; Ferry 1980; Young 1981, Chapter 5.

E. THE MECHANICAL PROPERTIES OF CRYSTALLINE POLYMERS

The models developed in the preceding sections represent well the rheological and mechanical properties of amorphous polymers. The viscoelastic properties of crystalline polymers are much more complex, however, and are not amenable to adequate theoretical explanation for three reasons.

First, an amorphous polymer is isotropic. This means that models suitable for describing shear stress, for example, are adequate to describe tensile stress or other types. Since crystalline polymers are not isotropic, this universality does not hold and the range of application of any model is severely limited.

Second, the homogeneous nature of amorphous polymers ensures that an applied stress is distributed uniformly throughout the system, at least down to very small dimensions. In crystalline polymers the relatively large crystallites are bound together in such a way that large stress concentrations inevitably develop.

Finally, a crystalline polymer is a mixture of regions of different degrees of order ranging all the way from completely ordered crystallites to completely amorphous regions. As the stress on the sample changes, the amounts of these regions change continuously as the crystallites melt or grow. This change of composition with respect to ordering is the most difficult obstacle to overcome in formulating a theory of the mechanical behavior of crystalline polymers. Even in the simplest cases the necessity of having the mechanical model change continuously with the applied stress has led to serious difficulties.

In consequence, neither the Boltzmann superposition principle nor time–temperature equivalence apply to crystalline polymers. Without these simplifying principles, attempts to explain the viscoelastic response of crystalline polymers in terms of models become complex and are only qualitative.

Typical Behavior

An orderly, if qualitative, discussion of the mechanical properties of crystalline polymers requires their classification into several categories, as indicated in Table 11-2. It is in the range of intermediate degree of crystallinity that the properties, unique to polymers, of most importance for mechanical and engineering applications are found. Further, these properties are found for the most part in the temperature range between T_g and T_m, and at temperatures not far below T_g. Well below the glass transition, molecular motion is essentially absent and the material behaves as a hard, glassy solid with the presence or absence of crystallinity making little difference: for example, the properties of atactic and isotactic polystyrene are quite similar at room temperature. Above T_m, of course, crystallinity plays no part in the properties of the amorphous viscoelastic melt.

Polymers with low crystallinity include plasticized poly(vinyl chloride) and elastic polyamides. These materials behave like lightly crosslinked amorphous polymers, their crystalline regions acting like crosslinks which are stable with respect to time but unstable with respect to temperature. The viscoelastic properties of these polymers are much like those of amorphous polymers except that the transition region between glassy and rubbery or liquid behavior is very much broadened on the temperature scale.

At very low extensions ($< 1\%$), at temperatures well below T_m, and at not too long times, polymers with intermediate degrees of crystallinity (such as low-density polyethylene) behave much like those with very low crystallinity. Time–temperature equivalence is not applicable to such polymers. The transition regions of these polymers in modulus and the corresponding distributions of relaxation times are exceptionally broad, as indicated by the master curves of Fig. 11-14. (It should be

TABLE 11-2. Classification of Crystalline Polymers

Predominant Properties in Temperature Range	Degree of Crystallinity		
	Low (5–10%)	Intermediate (20–60%)	High (70–90%)
Above T_g	Rubbery	Leathery, tough	Stiff, hard (brittle)
Below T_g	Glassy, brittle	Hornlike, tough	Stiff, hard, brittle

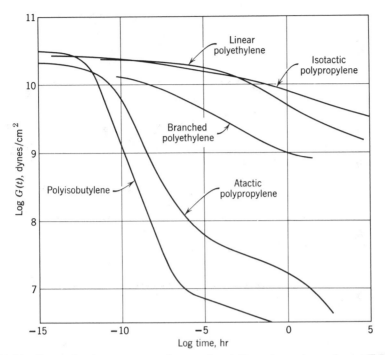

FIG. 11-14. Stress-relaxation master curves for several crystalline and amorphous polymers (Tobolsky 1960).

pointed out that these curves were themselves derived by application of time–temperature equivalence and must be considered only as idealizations.)

At higher extensions, these polymers exhibit the phenomena of a yield stress and cold drawing, with the accompanying changes in crystalline morphology described in Chapter 10E. No adequate theory relating rheological and mechanical behavior in this region to structure has yet been formulated. Polymers of intermediate degree of crystallinity are characteristically leathery or horny in texture, and exhibit good impact resistance which in many cases is retained even below T_g. The exact structural features responsible for this toughness have not been well defined.

The major effects of a further increase in crystallinity to very high values include (a) a further increase in modulus, as the high modulus characteristic of the crystalline regions is approached, and (b) the onset of a tendency toward brittleness, in the sense of failure at low strains. Such polymers, of which high-density polyethylene is typical, can be cold drawn only with difficulty. Tensile failure usually occurs at or slightly beyond the yield stress, accompanied by distortion or deformation that appears to occur at slip boundaries or dislocations, reminiscent of the viscoelastic behavior of metals.

Crystallization on Stressing. The application of a mechanical stress to a non-crystalline but crystallizable polymer can cause crystallinity to develop, either by raising T_m or by increasing the rate of crystallization. An example of the former effect is the crystallization of natural rubber on stretching described in Section *B*. At room temperature, unstretched natural rubber is above its crystalline melting point. As a tensile stress is applied, T_m is raised and the rubber crystallizes to an oriented structure. When the stress is released, T_m is reduced and the rubber melts as it retracts.

An example of the effect of stress on crystallization rate is seen in polymers such as poly(ethylene terephthalate) which can be quenched to a metastable amorphous state at temperatures well below T_m. Without an applied stress, the rate of nucleation and crystal growth is very low. When a tensile stress is applied, T_m is raised and the rate of crystallization is greatly increased so that crystallization takes place during stretching. These crystals do not melt when the stress is removed, of course.

GENERAL REFERENCES

Alfrey 1948, 1967; Tobolsky 1960; McCrum 1967; Ward 1971; Nielsen 1974; Schultz 1974, Chapter 11; McCullough 1977; Tager 1978, Chapter 9; Ferry 1980; Young 1981, Chapter 5.

DISCUSSION QUESTIONS AND PROBLEMS

1. Define and distinguish among the following terms: (a) shear thickening, (b) shear thinning, (c) thixotropic, (d) rheopectic, (e) Newtonian behavior.

2. State the equations relating the melt viscosity of a polymer to (a) molecular weight and (b) temperature.

3. List four physical properties characteristic of typical elastomers.

4. List four molecular-structure characteristics necessary for the development of typical elastomer physical properties.

5. Describe two quantitative aspects of the elastomeric state that are explained by the kinetic theory of rubber elasticity.

6. Why can the kinetic theory of rubber elasticity not predict the entire stress–strain curve for a typical elastomer?

7. Derive the stress–strain equation for the simple stretching of an elastomer, Eq. 11-11.

8. Derive an equation similar to Eq. 11-11 for the case of an elastomer with each dimension increased by a factor σ due to swelling with an organic solvent.

9. List three types of response of a typical polymer to stress and suggest a molecular mechanism for each type.

10. Sketch a strain–time curve for each of the mechanisms described in Question 9.

11. Describe briefly (a) creep and (b) stress relaxation.

12. Why cannot the theories of the viscoelasticity of amorphous polymers be extended to crystalline polymers?

BIBLIOGRAPHY

Adam 1965. Gerold Adam and Julian H. Gibbs, "On the Temperature Dependence of Cooperative Relaxation Properties in Glass-Forming Liquids," *J. Chem. Phys.* **43**, 139–146 (1965).

Alfrey 1948. Turner Alfrey, Jr., *Mechanical Behavior of High Polymers,* Interscience, New York, 1948.

Alfrey 1967. Turner Alfrey, Jr., and Edward F. Gurnee, *Organic Polymers,* Prentice-Hall, Englewood Cliffs, New Jersey, 1967.

Armeniades 1975. C. D. Armeniades and Eric Baer, "Transitions and Relaxations in Polymers," Chapter 6 in Herman S. Kaufman and Joseph J. Falcetta, eds., *Introduction to Polymer Science and Technology: An SPE Textbook,* Wiley-Interscience, New York, 1975.

Bauer 1967. Walter H. Bauer and Edward A. Collins, "Thixotropy and Dilatancy," Chapter 8 in Frederick R. Eirich, ed., *Rheology—Theory and Applications,* Vol. 4, Academic Press, New York, 1967.

Bekkedahl 1934. Norman Bekkedahl, "Forms of Rubber as Indicated by Temperature–Volume Relationship," *J. Res. Nat. Bur. Stand.* **13**, 411–431 (1934).

Brydson 1978. J. A. Brydson, *Rubber Chemistry,* Applied Science, London, 1978.

Bueche 1953. F. Bueche, "Segmental Mobility of Polymers Near Their Glass Temperatures," *J. Chem. Phys.* **21**, 1850–1855 (1953).

Carley 1975. James F. Carley, "Rheology," Chapter 8 in Herman S. Kaufman and Joseph J. Falcetta, eds., *Introduction to Polymer Science and Technology: An SPE Textbook,* Wiley-Interscience, New York, 1975.

Cohen 1959. Morrel H. Cohen and David Turnbull, "Molecular Transport in Liquids and Glasses," *J. Chem. Phys.* **31**, 1164–1169 (1959).

Cotton 1974. J. P. Cotton, D. Decker, H. Benoit, B. Farnoux, J. Higgins, G. Janniuk, R. Ober, C. Picot, and J. des Cloizeaux, "Conformation of Polymer Chains in the Bulk," *Macromolecules* **7**, 863–872 (1974).

Dealy 1981. John M. Dealy, *Rheometers for Molten Plastics: A Practical Guide to Testing & Property Measurement,* Van Nostrand Reinhold, New York, 1981.

Dealy 1983. J. M. Dealy, "Melt Rheometer Update," *Plastics Eng.* **34**, 57–61 (1983).

de Gennes 1971. P. G. de Gennes, "Reptation of a Polymer Chain in the Presence of Fixed Obstacles," *J. Chem. Phys.* **55**, 572–579 (1971).

de Gennes 1979. Pierre-Gilles de Gennes, *Scaling Concepts in Polymer Physics,* Cornell University Press, Ithaca, New York, 1979.

DiMarzio 1958. E. A. DiMarzio and J. H. Gibbs, "Chain Stiffness and the Lattice Theory of Polymer Phases," *J. Chem. Phys.* **28**, 807–813 (1958).

Doolittle 1951. Arthur K. Doolittle, "Studies in Newtonian Flow. II. The Dependence of the Viscosity of Liquids on Free-Space," *J. Appl. Phys.* **22**, 1471–1475 (1951).

Eirich 1956-1969. Frederick R. Eirich, ed., *Rheology—Theory and Applications,* Academic Press, New York, Vol. 1, 1956; Vol. 2, 1958; Vol. 3, 1960; Vol. 4, 1967; Vol. 5, 1969.

Ferry 1980. John D. Ferry, *Viscoelastic Properties of Polymers,* 3rd ed., John Wiley & Sons, New York, 1980.

Flory 1940. Paul J. Flory, "Viscosities of Linear Polyesters. An Exact Relationship between Viscosity and Chain Length," *J. Am. Chem. Soc.* **62**, 1057–1070 (1940).

Flory 1949. Paul J. Flory, "The Configuration of Real Polymer Chains," *J. Chem. Phys.* **17**, 303–310 (1949).

Fox 1956. T. G Fox, Serge Gratch, and S. Loshaek, "Viscosity Relationships for Polymers in Bulk and in Concentrated Solution," Chapter 12 in Frederick R. Eirich, ed., *Rheology—Theory and Applications,* Vol. 1, Academic Press, New York, 1956.

Gent 1978. A. N. Gent, "Rubber Elasticity: Basic Concepts and Behavior," Chapter 1 in Frederick R. Eirich, ed., *Science and Technology of Rubber,* Academic Press, New York, 1978.

Gibbs 1958. Julian H. Gibbs and Edmund A. DiMarzio, "Nature of the Glass Transition and the Glassy State," *J. Chem. Phys.* **28**, 373–383 (1958).

Glasstone 1941. Samuel Glasstone, Keith J. Laidler, and Henry Eyring, *The Theory of Rate Processes,* McGraw-Hill, New York, 1941.

Graessley 1974. William W. Graessley, "The Entanglement Concept in Polymer Rheology," *Adv. Polym. Sci.* **16**, 1–179 (1974).

Guth 1946. E. Guth, H. M. James, and H. Mark, "The Kinetic Theory of Rubber Elasticity," pp. 253–299 in H. Mark and G. S. Whitby, eds., *Scientific Progress in the Field of Rubber and Synthetic Elastomers (Advances in Colloid Science),* Vol. 2, Interscience, New York, 1946.

Hull 1981. Harry H. Hull, *An Approach to Rheology Through Multi-Variable Thermodynamics: Or, Inside the Thermodynamic Black Box,* Harry A. Hull, Sun City, Florida, 1981.

Kovacs 1958. A. J. Kovacs, "The Isothermal Volume Contraction of Amorphous Polymers" (in French), *J. Polym. Sci.* **30**, 131–147 (1958).

Kramer 1978. Ole Kramer and John D. Ferry, "Dynamic Mechanical Properties," Chapter 5 in Frederick R. Eirich, ed., *Science and Technology of Rubber,* Academic Press, New York, 1978.

McCrum 1967. N. G. McCrum, B. Read, and G. Williams, *Anelastic and Dielectric Effects in Polymeric Solids,* John Wiley & Sons, New York, 1967.

McCullough 1977. R. L. McCullough, "Anisotropic Elastic Behavior of Crystalline Polymers," in J. M. Schultz, ed., *Treatise on Materials Science and Technology,* Vol. 10, Part B (Herbert Herman, series ed.), Academic Press, New York, 1977.

Nielsen 1974. Lawrence E. Nielsen, *Mechanical Properties of Polymers and Composites,* Marcel Dekker, New York, 1974.

Nielsen 1977. Lawrence E. Nielsen, *Polymer Rheology,* Marcel Dekker, New York, 1977.

Pierce 1982. Percy E. Pierce and Clifford K. Schoff, "Rheological Measurements," pp. 259–319 in Martin Grayson, ed., *Kirk–Othmer Encyclopedia of Chemical Technology,* 3rd ed., Vol. 20, Wiley-Interscience, New York, 1982.

Rosen 1982. Stephen L. Rosen, *Fundamental Principles of Polymeric Materials,* Wiley-Interscience, New York, 1982.

Sauer 1975. J. A. Sauer and K. D. Pae, "Mechanical Properties of High Polymers," Chapter 7 in Herman S. Kaufman and Joseph J. Falcetta, eds., *Introduction to Polymer Science and Technology: An SPE Textbook,* Wiley-Interscience, New York, 1975.

Schultz 1974. Jerold Schultz, *Polymer Materials Science,* Prentice-Hall, New York, 1974.

Shen 1978. Mitchel Shen, "The Molecular and Phenomenological Basis of Rubberlike Elasticity," Chapter 4 in Frederick R. Eirich, ed., *Science and Technology of Rubber,* Academic Press, New York, 1978.

Smith 1977. Thor L. Smith, "Molecular Aspects of Rubber Elasticity," in J. M. Schultz, ed., *Treatise on Materials Science and Technology,* Vol. 10, Part A (Herbert Herman, series ed.), Academic Press, New York, 1977.

Sperling 1982. L. H. Sperling, "Molecular Motion in Polymers," *J. Chem. Educ.* **59,** 942–943 (1982).

Sternstein 1977. S. S. Sternstein, "Mechanical Properties of Glassy Polymers," in J. M. Schultz, ed., *Treatise on Materials Science and Technology,* Vol. 10, Part B (Herbert Herman, series ed.), Academic Press, New York, 1977.

Tager 1978. A. Tager, *Physical Chemistry of Polymers* (translated by David Sobolev and Nicholas Bobrov), Mir, Moscow, 1978 (Imported Publications, Chicago).

Tobolsky 1956. Arthur V. Tobolsky and Ephriam Catsiff, "Elastoviscous Properties of Polyisobutylene (and Other Amorphous Polymers) from Stress-Relaxation Studies. IX. A Summary of Results," *J. Polym. Sci.* **19,** 111–121 (1956).

Tobolsky 1960. Arthur V. Tobolsky, *Properties and Structure of Polymers,* John Wiley & Sons, New York, 1960.

Treloar 1975. L. R. G. Treloar, *The Physics of Rubber Elasticity,* 3rd ed., Clarendon Press, Oxford, 1975.

Van Wazer 1963. J. R. Van Wazer, J. W. Lyons, K. Y. Kim, and R. E. Colwell, *Viscosity and Flow Measurement, A Laboratory Handbook of Rheology,* Wiley-Interscience, New York, 1963.

Walters 1975. K. Walters, *Rheometry,* John Wiley & Sons, New York, 1975.

Ward 1971. I. M. Ward, *Mechanical Properties of Solid Polymers,* Wiley-Interscience, New York, 1971.

White 1978. James Lindsay White, "Rheological Behavior of Unvulcanized Rubber," Chapter 6 in Frederick R. Eirich, ed., *Science and Technology of Rubber,* Academic Press, New York, 1978.

Whorlow 1979. R. W. Whorlow, *Rheological Techniques,* John Wiley & Sons, New York, 1979.

Williams 1955. Malcolm L. Williams, Robert F. Landel, and John D. Ferry, "The Temperature Dependence of Relaxation Mechanisms in Amorphous Polymers and Other Glass-Forming Liquids," *J. Am. Chem. Soc.* **77,** 3701–3707 (1955).

Young 1981. Robert J. Young, *Introduction to Polymers,* Chapman and Hall, New York, 1981.

CHAPTER TWELVE

POLYMER STRUCTURE AND PHYSICAL PROPERTIES

Although interrelations between the molecular structure of polymers and their properties are mentioned throughout this book, they are emphasized in this chapter. First the structural features of polymers most directly responsible for determining their properties are classified, and in succeeding sections we describe how these structures influence various classes of physical properties. Finally, the property requirements associated with the uses of various polymers are reviewed.

Since the acceptance of the macromolecular hypothesis in the 1920's, it has been recognized that the unique properties of polymers—for example, the elasticity and abrasion resistance of rubbers, the strength and toughness of fibers, and the flexibility and clarity of films—must be attributed to their long-chain structure. In the examination of structure–property relationships it is advantageous to classify properties into those involving large and small deformations. The former class includes such properties as tensile strength and phenomena observed in the melt, while properties involving only small deformations include electrical and optical behavior, such mechanical properties as stiffness and yield point, and the glass and crystalline melting transitions.

Properties involving large deformations depend primarily on the long-chain nature of polymers and the gross configuration of their chains. Important factors for this group of properties include molecular weight and its distribution, chain branching and the related category of side-chain substitution, and crosslinking.

Physical properties associated with small deformations are influenced most by factors determining the manner in which chain atoms interact at small distances. The ability of polymers to crystallize, set by considerations of symmetry and steric effects, has major importance here, as do the flexibility of the chain bonds and the number, nature, and spacing of polar groups. To the extent that they influence the achievement of local order, gross configurational properties are also important. Similar considerations apply to amorphous polymers below the glass transition.

In crystalline polymers, the nature of the crystalline state introduces another set of variables influencing mechanical properties. These variables include nature of the crystal structure, degree of crystallinity, size and number of spherulites, and orientation. Some of these phenomena are in turn influenced by the conditions of fabrication of the polymer.

Finally, the properties of polymers can be varied importantly by the addition of other materials, such as plasticizers or reinforcing fillers. Properties involving both large and small deformations may be influenced in this way.

As described in Chapter 1, one of the most important determinants of polymer properties is the location in temperature of the major transitions, the glass transition and the crystalline melting point. It is appropriate therefore first to discuss the relations between molecular structure and these transition temperatures.

A. THE CRYSTALLINE MELTING POINT

As has been pointed out in Chapter 10D, crystalline melting in polymers is at least a pseudo-equilibrium process, and it is convenient here to describe it in thermodynamic terms, realizing that in a given situation an observed melting point T_m may not be the equilibrium value. Melting takes place when the free energy of the process is zero:

$$\Delta G = \Delta H_m - T_m \Delta S_m = 0$$

Thus T_m is set by the ratio $\Delta H_m/\Delta S_m$, and it is necessary to explore the effect of molecular structure upon both of these quantities to gain insight into the melting process.

The crystalline melting point is usually taken to be independent of molecular weight in the polymer range. It is assumed that ΔH_m and ΔS_m are made up of molecular-weight-independent terms H_0 and S_0 plus increments H_1 and S_1 for each chain unit. Thus, for degree of polymerization x,

$$T_m = \frac{\Delta H_m}{\Delta S_m} = \frac{H_0 + xH_1}{S_0 + xS_1} \longrightarrow \frac{H_1}{S_1} \qquad \text{as } x \longrightarrow \infty \qquad (12\text{-}1)$$

Melting Points of Homologous Series

The melting points of homologous series of various types of polymers are plotted in Fig. 12-1. The effect of the spacing of the polar groups on T_m is much as predicted by considering the polymers as derived from polyethylene by replacing methylenes with polar groups, but the overall level of T_m for each series remains to be considered.

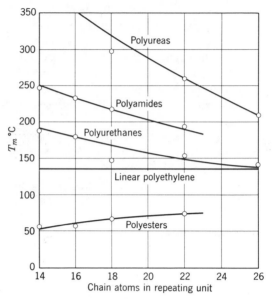

FIG. 12-1.   Trend of crystalline melting points in homologous series of aliphatic polymers (after Hill 1948).

Polyesters.   Although the low melting points of linear aliphatic polyesters were once attributed to unusual flexibility of the C—O chain bond (Bunn 1955), implying high entropies of fusion, both ΔS_m and ΔH_m are significantly lower in these polyesters than in polyethylene (Wunderlich 1958). Only recently have the reasons for these low values been elucidated in detail, as described later in this section.

Polyamides.   For polyamides it is observed that ΔH_m is lower than for polyethylene; therefore molar cohesion cannot account for high values of T_m. These result from low liquid-state entropies, leading to low values of ΔS_m, which arise from partial retention of hydrogen bonding in the melt and from chain stiffening due to the tendency for resonance of the type

within the amide group.

Similar considerations undoubtedly apply to the melting points of polyurethanes and polyureas, but values of ΔH_m and ΔS_m are not available.

As the spacing between the polar groups is increased, the melting points approach that of polyethylene. When the homologous series are examined in more detail, the melting points are found to vary in a more complex way with the spacing of the polar groups than is suggested in Fig. 12-1. An alternation of T_m with spacing is typical (Fig. 12-2). It results from differences in the crystal structure, which alternates in type for chain repeat units with odd and even numbers of carbon atoms.

Effect of Chain Flexibility and Other Steric Factors

Chain Flexibility.   The flexibility of chain molecules arises from rotation around saturated chain bonds. The potential energy barriers hindering this rotation range from 0.2 to 1.2 kJ/mole, the same order of magnitude as molecular cohesion forces. It is not surprising therefore that the flexibility of polymer chains is an important factor in determining their melting points. Thus polytetrafluoroethylene (T_m = 327°C) melts much higher than polyethylene because of its low entropy of fusion (Starkweather 1960), which results from the high stiffness of the polymer chains. Similarly, the high melting point of isotactic polypropylene (T_m = 165°C) is attributed to low entropy of fusion arising from stiffening of the chain in the melt because of the higher energy barrier for rotation about C—C bonds than in polyethylene. In neither case can a high heat of fusion account for the high value of T_m, since for both polytetrafluoroethylene and polypropylene ΔH_m is well below that of polyethylene as indicated in Table 10-1 (Dole 1959).

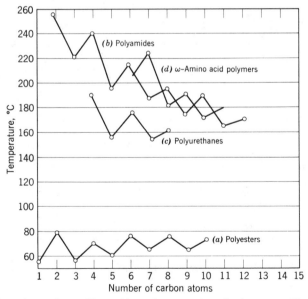

FIG. 12-2.   Dependence of crystalline melting point on spacing of polar groups. Number of carbon atoms refers to (a) acid for polyesters made with decamethylene glycol, (b) diamine for polyamides made with sebacic acid, (c) diamine for polyurethanes made with tetramethylene glycol, and (d) ω-amino acid polymers (Bannerman 1956).

The substitution of an inflexible group like the *p*-phenylene group,

for six chain CH_2 groups causes a marked rise in the melting point of the polymer (Edgar 1952). Some examples of this change are shown in Table 12-1. The rise is considerable when the *p*-phenylene group is connected to CH_2 groups, but greater still when it is connected to carbonyl,

since the group can resonate as a unit. Among other chain-stiffening groups are *p,p'*-diphenyl, 1,5- or 2,6-naphthyl, diketopiperazine,

TABLE 12-1. Effect of a *p*-Phenylene Group on the Melting Point of Condensation Polymers[a]

Repeating Unit	T_m (°C)
$-O(CH_2)_2OCO(CH_2)_6CO-$	45
$-O(CH_2)_2OCO\langle\ \rangle CO-$	265
$-NH(CH_2)_6NHCO(CH_2)_6CO-$	235
$-NH(CH_2)_6NHCO\langle\ \rangle CO-$	350[b]
$-O(CH_2)_8OCO(CH_2)_8CO-$	75[c]
$-OCH_2\langle\ \rangle CH_2OCOCH_2\langle\ \rangle CH_2CO-$	146
$-CH_2CH_2-$	135
$-CH_2\langle\ \rangle CH_2-$	380

[a]Edgar (1952).
[b]Decomposes
[c]Estimated.

triazole,

$$-C \underset{N-N}{\overset{NH}{\diamond}} C-$$

acetal, and thioketal.

Side-Chain Substitution.   In most cases, the substitution of nonpolar groups for hydrogens of a polymer chain leads to a reduction in T_m or possibly complete loss of crystallinity. If the substitution is random, as in branched polyethylene, the primary effect is a reduction in the size and perfection of the crystalline regions, usually accompanied by a decrease in the degree of crystallinity. The crystalline melting point of polyethylene is lowered 20–25°C on going from the linear to the branched material.

Replacement of an amide hydrogen with an alkyl group has a much larger effect, since hydrogen bonding is destroyed. In general, N-methyl nylons melt at least 100°C lower than their unsubstituted counterparts.

When an alkyl group is regularly substituted into a methylene chain, with the retention of stereoregularity so that crystallization is possible, two effects compete in setting T_m. As indicated for isotactic poly(α-olefins) in Table 12-2 (with poly-ethylene omitted because of its widely different crystal structure), an increase in the length of the side chain results in a looser crystal structure with an increasingly lower melting point. On the other hand, an increase in the bulkiness of the side chain increases T_m, since rotation in the side chain is hindered in the liquid state, with consequent decrease in ΔS_m.

TABLE 12-2.   Effect of Side-Chain Structure on the Crystalline Melting Point of Isotactic Poly(α-olefins)[a]

Side Chain	T_m (°C)
—CH$_3$	165
—CH$_2$CH$_3$	125
—CH$_2$CH$_2$CH$_3$	75
—CH$_2$CH$_2$CH$_2$CH$_3$	−55

—CH$_2$CHCH$_2$CH$_3$
     |
     CH$_3$
	196

CH$_3$
 |
—CH$_2$CCH$_2$CH$_3$
     |
     CH$_3$
	350

[a]Campbell (1959), Bawn (1960).

Entropy and Heat of Fusion

Entropy of Fusion. The effect of structure on the entropy of fusion can now be calculated, in good agreement with experiment, for many polymers. This had led, for example, to a satisfactory explanation (Hobbs 1970) of the low entropy of fusion of the linear aliphatic polyesters whose melting points are depicted in Fig. 12-1. Following Starkweather (1960), the entropy of fusion is considered to consist of independent contributions from volume change on melting and the change in the number of conformations that the chain can assume, from one in the crystal to a number determined by the type of chain bonds in the melt. The latter number can be calculated by an enumeration scheme, and is combined with knowledge of the potential barrier to rotation to compute the desired conformational entropy. It was concluded that, in comparison to polyethylene, the stiffening effect of the rigid ester bonds outweighs to a small degree the increased flexibility of the ester chains.

Heat of Fusion. The low melting points of the linear aliphatic polyesters must thus be attributed to their low heats of fusion. These are less easily explained in detail, but a very approximate analysis (Hobbs 1970) suggests that the cohesive forces to be overcome in fusion result almost entirely from methylene interactions (dispersion forces) between neighboring chains. Since there are fewer methylene groups per unit chain length in the polyesters, their heats of fusion are lower.

Dipole interactions from the ester carbonyl groups appear not to contribute to the heats of fusion of these polymers, supporting the inference that the dipole bonds in the crystal are almost entirely reformed in the melt. A similar effect was postulated for the polyamides by Dole (1959).

Effect of Copolymerization

When copolymers are made from monomers that form crystalline homopolymers, degree of crystallinity and crystalline melting point decrease as the second constituent is added to either homopolymer. The melting point depends on the mole fraction n of the crystallizing constituent by the relation of Eq. 10-4 (Flory 1949):

$$\frac{1}{T_m} - \frac{1}{T_m^0} = -\frac{R}{\Delta H_m} \ln n \qquad (12\text{-}2)$$

where T_m^0 is the melting point of the homopolymer and ΔH_m is its heat of fusion. A typical case is the copolymer of hexamethylene terephthalamide and hexamethylene sebacamide (Fig. 12-3).

If, however, the comonomers are isomorphous, that is, capable of replacing each other in the crystals, the melting point may vary smoothly over the composition range. An example is the copolymer of hexamethylene terephthalamide and hexamethylene adipamide, also shown in Fig. 12-3. Other variations may occur, including the formation of an alternating copolymer with a crystal structure and

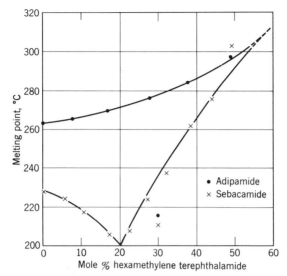

FIG. 12-3. Melting points of copolymers of hexamethylene adipamide and terephthalamide, and of hexamethylene sebacamide and terephthalamide (Edgar 1952).

melting point far different from those of either homopolymer. Block and graft copolymers may exhibit two crystalline melting points, one for each type of chain segment.

GENERAL REFERENCES

Alfrey 1967; Flory 1969; Sweeney 1969; Elias 1977, Chapter 10; Wunderlich 1980.

B. THE GLASS TRANSITION

Relation Between T_m and T_g

With few exceptions, polymer structure affects the glass transition T_g and the crystalline melting point T_m similarly. This is not unexpected, since similar considerations of cohesive energy and molecular packing apply to the amorphous and crystalline or paracrystalline regions, respectively, in accounting for the temperature levels at which the transitions occur. In consequence, T_m and T_g are rather simply related for many polymers (Beaman 1952; Boyer 1954, 1963): Depending on symmetry, T_g K is approximately one-half to two-thirds T_m K (Fig. 12-4). There are, however, some exceptions to this general rule. The most important appear (Pearce 1969) to be associated with crystallizability or hydrogen bonding in the amorphous

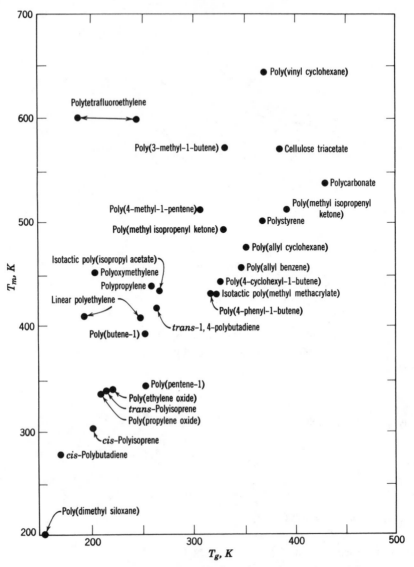

FIG. 12-4. Relation between T_m and T_g for various polymers (Boyer 1963).

regions of polymers. Within series of polymers where crystallizability is changed by control of tacticity (Reding 1962) or structure (Zimmerman 1968), ease of crystallization is associated with a low ratio of T_m to T_g, and difficulty with a high ratio. In polyamides, where T_m varies with ester group spacing as indicated in Section A, T_g is nearly constant at a level postulated (Woodward 1960, Komoto 1967) to be set by the energy required to break hydrogen bonds in the amorphous regions. Copolymers also have ratios of T_m to T_g differing from those in Fig. 12-4, as described below.

Effects of Molecular Weight and Diluents

In the polymer range, T_g is more dependent on molecular weight than is T_m (Section A), the relation having the form

$$T_g = T_g^\infty - \frac{k}{\bar{M}_n} \tag{12-3}$$

derived from temperature–volume considerations (Fox 1955), where T_g^∞ is the glass-transition temperature at infinite molecular weight, and k is about 2×10^5 for polystyrene (Fox 1950) and poly(methyl methacrylate) (Beevers 1960) and 3.5×10^5 for atactic poly(α-methyl styrene) (Cowie 1968).

From the considerations of Chapter 10, it is probable that chain ends, which lead to the terms H_0 and S_0 for low-molecular-weight substances, are usually associated with defects in the crystalline regions of polymers. Hence they are unlikely to make a significant contribution to T_m, which is the melting point of the most perfect crystalline regions. In amorphous polymers, however, the effect of chain ends on free volume (Chapter 11D) should retain importance in proportion to the concentration of ends; thus it is not surprising that T_g is found to vary with $\bar{M}_n$ in the polymer range.

While the effect of molecular-weight distribution on T_g is accounted for by the appearance of $\bar{M}_n$ in Eq. 12-3, the effect of low-molecular-weight diluents is worth noting. The primary example of this effect is plasticization, widely used to improve the flexibility of certain polymers and allow them to remain flexible well below T_g of the unplasticized resin (Chapters 14 and 17). Based on the weight of added constituent, plasticization is considerably more effective than copolymerization in lowering T_g.

Effect of Chemical Structure

The effects of the nature of the chain repeat units on T_g are closely related to intermolecular forces, chain stiffness, and symmetry. Probably the most important factor among these is hindrance to free rotation along the polymer chain resulting from the presence of stiff bonds or bulky side groups: Compare polybutadiene, $T_g = -85°C$; styrene–butadiene copolymer (25/75), $-55°C$; polystyrene, $+100°C$; poly(α-methyl styrene), $+150°C$; polyacenaphthalene, $+285°C$. The effect of intermolecular forces is quantified by the cohesive energy density or solubility parameter (Chapter 7A): Compare polypropylene, $\delta = 16.0$, $T_g = -20°C$; polyacrylonitrile, $\delta = 31.5$, $T_g = 90°C$. The effect of symmetry of the repeat unit is illustrated in Fig. 12-4: Compare poly(vinyl chloride), $T_g = 85°C$; poly(vinylidene chloride), $T_g = -17°C$.

Effect of Chain Topology

Copolymerization. The glass-transition temperatures of random copolymers usually fall between those of the corresponding homopolymers, T_g for the copolymer often being a weighted average given by

$$a_1 w_1 (T_g - T_{g1}) + a_2 w_2 (T_g - T_{g2}) = 0 \qquad (12\text{-}4)$$

where T_{g1} and T_{g2} refer to the homopolymers, w_1 and w_2 are the weight fractions of monomers 1 and 2 in the copolymer, and a_1 and a_2 depend on monomer type (Wood 1958). There are numerous deviations, both positive and negative, from this linear relationship, however.

The contrast between this behavior and the common depression of T_m by copolymerization is not surprising, since the changes at T_g do not require fitting a structure into a crystal lattice, and in consequence structural irregularity does not affect T_g as it does T_m.

While the above considerations for random copolymers lead to low ratios of T_m to T_g, it is sometimes found that block and graft copolymers may have long enough homogeneous chain segments to exhibit the properties of both homopolymers, rather than intermediate values. Thus a block copolymer in which one homopolymer has high, and the other low, softening and brittle temperatures may exhibit both a high softening point and a low brittleness temperature. Hence it may have values of T_m/T_g higher than those found for other polymer types.

Branching and Crosslinking.   The effects of chain branching and crosslinking on T_g can be explained in terms of free volume. The higher concentration of chain ends in a branched polymer increases the free volume and lowers T_g, whereas crosslinking lowers free volume and raises T_g. Roughly, the latter change may be accounted for in terms of the average molecular weight of the segment between crosslinks by an equation like Eq. 12-3. More complete treatments are cited in Nielsen (1969).

GENERAL REFERENCES

Alfrey 1967; Elias 1977, Chapter 10.

C. PROPERTIES INVOLVING LARGE DEFORMATIONS

Melt Properties

Melt Viscosity.   As discussed in Chapter 11A, the viscosity of a polymer melt is a strong function of weight-average molecular weight. Melt viscosity is also influenced by chain branching: In polyethylene and in silicone polymers melt viscosity decreases with increasing degree of long-chain branching at constant weight-average molecular weight, whereas in poly(vinyl acetate) melt viscosity increases under the same circumstances. The reason for this difference appears to lie in branch length: In experiments with poly(vinyl acetate) branched by graft polymerization (Long

1964), melt viscosity decreased at constant molecular weight when the added branches were shorter than the critical chain length Z_c at which the melt viscosity power law changes (Chapter 11A). Only when branches longer than Z_c were added did the viscosity increase.

Crosslinking has a pronounced effect on melt viscosity in that the latter becomes essentially infinite along with $\bar{M}_w$ at the onset of gelation. However, small, tightly crosslinked network particles may behave like rigid spheres and have little effect on melt viscosity.

The addition of low-molecular-weight species, as in plasticization, reduces melt viscosity by lowering average molecular weight. Bulky side groups may have a similar effect.

For many polymers an upper limit to $\bar{M}_w$ is set by fabrication requirements on the melt viscosity. At the same time, a lower limit on $\bar{M}_n$ may be set by requirements involving tensile strength, brittleness, or other mechanical properties. In such cases, which include polypropylene and probably other polyolefins, the best balance of properties is achieved when the distribution of molecular weights is made as narrow as possible.

However, the anticipated gain in ease of fabrication on decreasing $\bar{M}_w$ at constant $\bar{M}_n$ may not be fully realized, since fabricability depends upon melt viscosity at high shear stress. In contrast to low shear (Newtonian) viscosity, which depends upon $\bar{M}_w$, high-shear viscosity depends on a molecular-weight average between $\bar{M}_w$ and $\bar{M}_n$ (Rudd 1960).

Other Melt Properties. For polymers with very high melt viscosity such as polytetrafluoroethylene, the tensile strength of the melt becomes a property of importance. Like melt viscosity, it increases with increasing molecular weight. The viscoelastic or elastic properties of polymer melts decrease in magnitude with increasing molecular weight and with increasing chain branching.

Tensile Strength and Related Properties

Many polymer properties, including tensile strength, can be described by an equation of the type

$$\text{Property} = a - \frac{b}{\bar{M}_n} \tag{12-5}$$

This is the type of relation predicted for properties depending on the number of ends of polymer chains. Many such properties, including density and refractive index, attain constant values at molecular weights well below the polymer range. Tensile strength, however, varies significantly with molecular weight in the range of interest for polymers, although the variation may in fact be with an average between $\bar{M}_n$ and $\bar{M}_w$ (McCormick 1959). It has been shown (Flory 1945) that dependence on $\bar{M}_n$ implies that the tensile strength $\overline{TS}$ of a mixture of components with tensile strengths $(\overline{TS})_i$ is the weight average

$$\overline{TS} = \sum_i w_i (\overline{TS})_i \qquad (12\text{-}6)$$

If a polymer exhibits a yield point and then undergoes extensive elongation before tensile failure, its ultimate tensile strength increases with increasing molecular weight. Typical data for branched polyethylene are shown in Fig. 12-5. Important structure variables in this polymer were found to be degree of crystallinity and level of molecular weight; in Fig. 12-5 crystallinity is replaced by the equivalent property of density, and molecular weight by the logarithm of melt viscosity.

Although Fig. 12-5 shows that tensile strength is independent of degree of crystallinity for branched polyethylene, this independence does not carry over to linear polyethylenes, with densities in the range 0.94–0.96: the latter polymers have higher tensile strengths than branched polyethylenes of the same molecular weight (see also Chapter 11E). The difference is attributed to changes in the crystal morphology of the high-density polymers during tensile elongation (Chapter 10E).

Morphology is indeed important in determining the mechanical properties of crystalline polymers. Both tensile strength and the mechanism of failure are influenced by such factors as spherulite size and structure and the nature of interlamellar ties. Polymers with smaller, finer-textured spherulites tend to fail at high elongations after drawing, while those with large, coarse spherulites often fail by brittle fracture between spherulites at low elongations (Collier 1969).

Toughness in Rubber-Modified Glassy Plastics

One of the major developments leading to plastics with outstanding toughness has been the production of rubber-modified glassy plastics such as the ABS resins

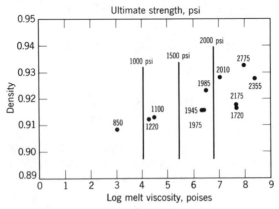

FIG. 12-5.   Dependence of ultimate tensile strength of polyethylene on molecular structure variables (Sperati 1953).

(Chapter 14), made preferably by polymerizing a continuous glassy matrix in the presence of small rubber particles. It is known that for optimum toughness the two-phase structure is essential, only a small amount (5–15%) of rubber is needed, the optimum size of the rubber particles is 1–10 μm, and the rubber–matrix interface should be well grafted.

The mechanism (Bucknall 1967) by which toughness is developed in these materials, in contrast to the brittle failure characteristic of the unmodified glassy polymer, is intimately related to the formation of crazes, regions of low-density material formed as a precursor to cracking in glassy polymers (Sternstein 1977). At low strains, the stress in the sample is largely borne by the matrix, and is concentrated at the rubber particles. As straining continues crazes are initiated here and grow with the absorption of energy as the matrix deforms. Ultimately the applied stress is distributed between the rubber and the crazed matrix. The rubber, now under tension, strengthens the crazed matter and fracture is thus delayed in favor of craze initiation elsewhere.

GENERAL REFERENCES

Alfrey 1967; Kargin 1968; Ward 1971; Boenig 1973, Chapter 7; Nielsen 1974; Elias 1977, Chapter 11.

D. PROPERTIES INVOLVING SMALL DEFORMATIONS

This section is concerned with structural determinants of a variety of properties involving small local deformations of polymers in contrast to the gross deformations discussed in Section C. Among these are such mechanical properties as stiffness, yield stress, elongation, and impact strength. Related to these properties are hardness, abrasion resistance, and flexural fatigue life, among others.

A second group of properties discussed includes solubility and related phenomena, such as swelling, cloud points, sorption of liquids, permeability to gases, and compatibility of plasticizers. The final class under discussion includes the effects of electromagnetic radiation, in such optical properties as refractive index and transparency and such electrical properties as dielectric constant, dielectric loss, and dielectric strength.

Effect of Crystallinity

Mechanical Properties.   The properties of crystalline polymers are emphasized in this section for two reasons: It is these polymers that are most widely utilized because of their mechanical properties, and structural features related to crystallinity may have profound effects on these properties.

The degree of crystallinity alone is effective in determining the stiffness and yield point for most crystalline plastics. As indicated for branched polyethylene in Fig. 12-6, these properties are independent of molecular weight. As a result, they

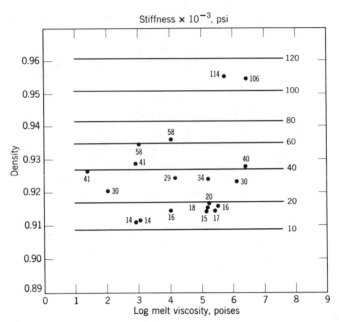

FIG. 12-6.   Dependence of stiffness of branched polyethylene on molecular structure variables (Sperati 1953). (See discussion for Fig. 12-5.)

can be expressed as single-valued functions of degree of crystallinity. As crystallinity decreases, both stiffness and yield stress decrease. As a result of the latter change, the chance of brittle failure is reduced.

Solubility and Related Properties.   As pointed out in Chapter 7, crystallinity decreases the solubility of polymers markedly, since the process of solution involves overcoming the heat and entropy factors associated with crystallization as well as those of the intermolecular interactions in the amorphous regions. Properties related to solubility, such as the cloud point of dilute solutions, are often functions of crystallinity relatively independent of molecular weight.

The solubility of liquids and gases in polymers is also strongly dependent on crystallinity, since solubility is usually confined to the amorphous regions. Permeability, the product of solubility and diffusivity, behaves similarly.

Plasticization is closely related to solubility, and the selection of an efficient and compatible plasticizer (see Chapter 17) involves considerations similar to those for the selection of solvents. Plasticization usually results in loss of crystallinity; however, if crystallinity is well developed it may not be possible to find a plasticizer sufficiently compatible with (soluble in) the polymer to have a significant effect on its properties.

Electrical and Optical Properties.   The primary effect of crystallinity on the electrical and optical properties is associated with the changes in dielectric constant and refractive index arising from the difference in density between the crystalline and amorphous regions. In the case of visible light, this difference leads to scattering, which may be large if the regions responsible (crystallites or lamellae, and spherulites) are significant in size compared to the wavelength of the light. Thus crystalline plastics usually appear translucent or opaque except in thin films, their transparency increasing with decreasing spherulite size.

Effect of Molecular Weight

Solubility.   Where the presence of a crystalline phase is not involved, solubility and related phenomena are inverse functions of molecular weight. This fact is reflected in the equations for the thermodynamic properties of polymer solutions, and forms the basis of fractionation methods, as discussed in Chapter 7.

Electrical and Optical Properties.   The interactions of electromagnetic radiation with polymers involve, at the most, the cooperative movement of small groups of atoms. The molecular weight dependence of these properties, obeying relations of the type of Eq. 12-1, vanishes at molecular weights far below the polymer range. Except as molecular weight influences some more direct structural determinant of these properties, they are independent of this variable.

Combined Effects of Crystallinity and Molecular Weight

Mechanical Properties.   A number of mechanical properties, including hardness, flexural fatigue resistance or flex life, softening temperature, elongation at tensile break (where plastic flow occurs), and sometimes impact strength, are influenced by both degree of crystallinity and molecular weight. Typical examples are the softening temperature of branched polyethylene as measured in the Vicat test (Chapter 9F), which increases with increasing molecular weight and increasing crystallinity (Fig. 12-7), and the flex life of polytetrafluoroethylene, which increases with increasing molecular weight and decreasing crystallinity (Fig. 12-8; ''standard specific gravity'' is a measure of molecular weight, as discussed in Chapter 14). Flex life is sometimes associated with brittleness, samples having low resistance to flexing being brittle. The chance of brittle failure is decreased by raising molecular weight, which increases brittle strength, and by reducing crystallinity as indicated previously. For amorphous polymers, impact strength is found to depend on the weight-average molecular weight.

As the degree of crystallinity decreases with temperature during the approach to T_m (Chapter 10D), stiffness and yield stress decrease correspondingly. These factors often set limits on the temperature at which a plastic is useful for mechanical purposes.

A major determinant of the behavior of a polymer on impact is the relation between the yield stress and the tensile strength in brittle failure, which may be designated the brittle strength. If the yield stress or strength is lower than the brittle strength, plastic flow begins (at the yield point in a tensile experiment) and the

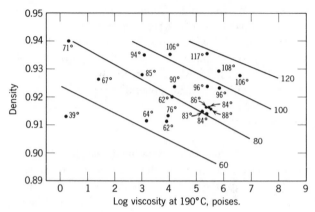

FIG. 12-7. Dependence of softening temperature, as measured by the Vicat test, of branched polyethylene on molecular structure parameters (Sperati 1953). (See discussion for Fig.
12-5.)

polymer is tough. If the brittle strength is lower, brittle failure takes place on
impact.

Effect of Polar Groups

Solubility and Related Properties.   In general, the introduction of polar groups
into polymers tends to decrease solubility, since strong polymer–polymer bonds
usually develop. The situation is complicated, however, by factors such as the
arrangement and bulkiness of the groups, which in turn influence crystallinity.

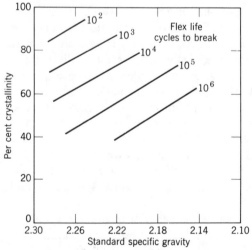

FIG. 12-8. Lines of equal flex life on a crystallinity–molecular weight "map" for polytetrafluoroethylene (Thomas 1956).

The solubility behavior of cellulose and its derivatives provides examples of the effect of polarity and substitution on solubility. Cellulose itself is insoluble because of strong hydrogen bonding and stiff chains which prevent hydrating molecules from penetrating its crystalline regions. Substitution of less polar groups for the hydroxyls on cellulose leads first to solubility in alkalis, which only swell the parent polymer, and then to water solubility when 0.5–1 hydroxyl per glucose unit is replaced. Crystallinity has usually disappeared at this stage. At higher degrees of substitution, water solubility is replaced by alcohol solubility, and then by solubility in organic solvents whose type depends on the nature of the substituent group. When all the substitutents are alike, complete substitution leads to crystallization and more limited solubility.

The permeability of polymers to gases and liquids decreases with increasing polarity, since a more polar polymer has a higher activation energy for diffusion. Even methyl groups contribute sufficiently to high cohesion and high activation energy in rubbers to impart low gas permeability. Thus butyl, nitrile, and chloroprene rubbers all have lower permeability than natural or butadiene-based rubbers.

Electrical Properties. The electrical properties of polymers depend more on an unbalance or asymmetry of dipoles than on the presence of polar groups *per se*. Thus both polyethylene and polytetrafluorethylene have low dielectric constant and dielectric loss (dissipation factor), but these quantities are much larger for polymers containing both hydrogen and fluorine or chlorine. Similarly, the dielectric loss of poly(2,5-dichlorostyrene) is much less than that of poly(3,4-dichlorostyrene) because of cancellation of the dipoles of the para chlorines.

Effect of Copolymerization

Mechanical Properties. The addition of a comonomer to a crystalline polymer usually causes a marked loss in crystallinity, unless the second monomer crystallizes isomorphous with the first (see Section *A*). Crystallinity typically decreases very rapidly, accompanied by reductions in stiffness, hardness, and softening point, as relatively small amounts (10–20 mol.%) of the second monomer are added. In many cases, a rigid, fiber-forming polymer is converted to a highly elastic, rubbery product by such minor modification.

The dependence of mechanical properties on copolymer composition in systems which do not crystallize results primarily from changes in intermolecular forces as measured by cohesive energy. Higher cohesive energy results in higher stiffness and hardness and generally improved mechanical properties.

The additional variable introduced by block or graft copolymerization can be used to alter the properties of such copolymers by changing their method of preparation. A block or graft copolymer consisting of long-chain segments of widely differing polarity can exist in solution with one or the other type of segment extended, the second relatively contracted, depending on solvent type. When isolated from solution, the copolymer has properties resembling those of the homopolymer corresponding to the extended segments (Merrett 1957). Thus natural rubber with chains of poly(methyl methacrylate) grafted to it was hard and stiff with a nontacky surface when isolated from a solution where the rubber chains were collapsed and the poly(methyl methacrylate) chains were extended. When prepared under the opposite conditions, it was limp, flabby, and self-adherent like rubber. A third form with intermediate properties was isolated from solvents in which both chain segments were relatively extended.

Solubility.   In essentially random copolymers of monomers whose homopolymers are noncrystalline, a property such as solubility varies more or less regularly from that of one homopolymer to that of the other as the relative proportions of the components are varied. The solubility of copolymers of this type is frequently low in solvents for either homopolymer, but high in mixtures of these solvents.

The solubility of graft and block copolymers is often unusually high, especially if the two components have widely different polarities. Block copolymers of polystyrene with poly(vinyl alcohol) or poly(acrylic acid), for example, are soluble in benzene, acetone, and water. In water the hydrophilic blocks are solubilized and extended, holding the tightly coiled hydrocarbon segments in solution much as a detergent solubilizes a hydrocarbon by micelle formation. In benzene the opposite situation occurs, but in an intermediate solvent such as acetone both blocks are relatively extended, as indicated by higher solution viscosity in intermediate solvents. These polymers act as efficient detergents and emulsifying and compatibilizing agents, but their virtually universal solubility makes it difficult to isolate or purify them.

Effect of Plasticization, Reinforcement, or Crosslinking

Plasticization.   The addition of a plasticizer usually reduces stiffness, hardness, and brittleness, and has a similar effect on other mechanical properties, since interchain forces are effectively reduced. These changes are accompanied by a reduction in T_g, as noted in Section B. As indicated in Chapter 17, plasticization is usually restricted to amorphous polymers or polymers with a low degree of crystallinity because of the limited compatibility of plasticizers with highly crystalline polymers.

Reinforcement and Crosslinking.   Whether carried out by chemical crosslinking in an unmodified amorphous polymer system or by the addition of a reinforcing filler such as carbon black in rubber (a process involving chemical bonding between polymer and filler), the addition of crosslinks leads to stiffer, stronger, tougher products, usually (in the case of rubbers) with enhanced tear and abrasion resistance as well. However, extensive crosslinking in a crystalline polymer may cause loss of crystallinity, with attendant deterioration of the mechanical properties depending on this factor. When this occurs, the initial trend of properties may be toward either enhancement or deterioration, depending on the degree of crystallinity of the unmodified polymer and the method of formation and location (crystalline or amorphous regions) of the crosslinks.

GENERAL REFERENCES

Kargin 1968; Ward 1971; Boenig 1973, Chapter 7; Nielsen 1974; Elias 1977, Chapters 11, 14, and 15.

E. PROPERTY REQUIREMENTS AND POLYMER UTILIZATION

The variables necessary to define the mechanical and physical properties of polymers have now been discussed (Fig. 12-9). The increase of T_m with molecular weight, leveling off as polymer molecular weights are reached, the related approximate behavior of T_g, and the continual increase of viscosity with molecular weight serve to define, in terms of the variables molecular weight and temperature, regions in which the properties of typical plastics, rubbers, viscous liquids, and so on, may be found.

Combinations of properties unique to polymers are evidenced in each of the major uses, including elastomers, fibers, and plastics, to which macromolecules are put. In this section the property requirements of these end uses are described and related to polymer structure.

Elastomers

It is a matter of experience that all substances exhibiting a high degree of rubberlike elasticity contain long-chain structures. The restoring force leading to elastic behavior results directly from the decrease in entropy associated with the distortion of a chain macromolecule from its most probable conformation. Two additional property requirements are imposed by the condition that there be sufficient freedom of molecular motion to allow the distortions to take place rapidly: First, the polymer must at its use temperature be above T_g and, second, it must be amorphous, at least in the undistorted state.

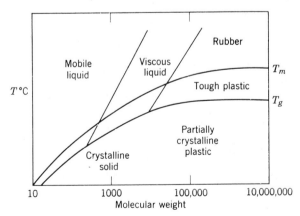

FIG. 12-9. Approximate relations among molecular weight, T_g, T_m, and polymer properties.

In contrast to the high local mobility of chain segments implied by these requirements, the gross mobility of chains in elastomers must be low. The motions of chains past one another must be restricted in order that the material can regain its original shape when the stress is released. This restriction of gross mobility is usually obtained by the introduction of a network of primary bond crosslinks in the material. (It cannot be obtained through secondary bond forces; these must be kept low in order to gain local segment mobility.) The crosslinks must be relatively few and widely separated, however, so that stretching to large extensions can take place without rupture of primary bonds.

The requirement of low cohesive energy limits the family of elastomers to polymers that are largely hydrocarbon (or fluorocarbon or silicone) in nature, with polar groups distributed at random and in not too great number, so that crystallinity is absent and T_g is sufficiently low, ideally in the range $-50 - -80°C$. The requirement of sites for crosslinking often leads to the choice of a diene as a monomer or comonomer for an elastomer.

In contrast to its equilibrium properties, a stretched elastomer should have the high tensile strength and modulus usually associated with crystalline plastics. Thus rubbers in which crystallinity can develop on stretching, such as natural rubber and its stereoregular synthetic counterparts, usually have more desirable properties than those with less regular structures. However, a reinforcing filler can sometimes impart to a rubber properties similar to those obtained on the development of crystallinity. Thus reinforced styrene–butadiene rubber has properties nearly equivalent to those of reinforced natural rubber, whereas its properties when crosslinked but not reinforced are much poorer.

Fibers

In contrast to elastomers, the requirements of high tensile strength and modulus characteristic of fibers are almost always obtained by utilizing the combination of molecular symmetry and high cohesive energy associated with a high degree of crystallinity. Usually the fiber is oriented to provide optimum properties in the direction of the fiber axis.

The end use requirements of fibers, particularly those involving textiles, lead to restrictions on several properties. The crystalline melting point T_m must be above a certain minimum, say 200°C, if the resulting fabric is to be subjected to ironing. On the other hand, spinning the polymer into a fiber requires either that T_m be below, say, 300°C and well below the temperature of decomposition of the polymer, or that the polymer be soluble in a solvent from which it can be spun. Other requirements limit solubility, for example, in solvents useful for dry cleaning.

The requirement of orientation in the fiber usually implies that T_g be not too high (since orientation by cold drawing and ironing are typically carried out at or above this temperature) or too low (since orientation and related characteristics, such as crease retention after ironing, must be maintained at room temperature).

Thus the selection of a polymer for use as a fiber involves a number of compromises, usually met by choosing a linear polymer with high symmetry and high intermolecular forces resulting from the presence of polar groups, high enough in molecular weight so that tensile strength and related properties are fully developed. Branching in the polymer chain is in general detrimental to fiber properties because branch points disrupt the crystalline lattice, lower the crystalline melting point, and decrease stiffness. Crosslinking, on the other hand, offers the possibility of obtaining strong interchain bonding. If crosslinks are formed after the polymer is spun into fiber, and are relatively few in number, improvement in fiber properties may result. Thus poly(vinyl alcohol), polyurethanes, and protein fibers may be crosslinked with formaldehyde to give higher melting point, lower solubility and moisture regain, and improved hand, while wool, a natural protein fiber, is crosslinked with cystine links.

General-Purpose and Specialty Plastics

The wide range of end uses of plastics requires a variety of property combinations; correspondingly, a wider variety of structures is important. In general, the properties of plastics are intermediate between those of fibers and elastomers, with much overlapping on either end. Thus typical plastics may have cohesive energies higher than those of elastomers but lower than those of fibers. However, a polymer useful as a fiber when oriented in this form may also be useful as a plastic, where orientation is not readily achieved in the massive pieces used; an example is nylon.

Optical Applications.   The requirement of good optical properties, especially in massive pieces, imposes severe limitations on the structure of polymers. In general, crystallinity must be absent, and most amorphous polymers exhibit softness and brittleness that exclude them from many applications. In thin films, structural requirements are not as severe, since crystallinity can often be tolerated if spherulite growth is inhibited and the material can be processed to give sufficiently smooth surfaces. Thus the normally incompatible requirements of clarity and toughness can both be met.

Electrical Applications.   To achieve low dielectric loss over a wide frequency range, the structure of a polymer must be selected on the basis of low polarity. All other requirements are of considerably less importance. It follows that polyethylene and polytetrafluoroethylene are the best materials for low-loss applications, particularly at high frequencies. At low frequencies, however, other plastics, such as poly(vinyl chloride), are useful.

Mechanical Applications.   Perhaps the most important property requirement for the use of polymers in mechanical applications is toughness. This property is usually achieved by selection of a polymer with a moderate, but not too high, degree of crystallinity (see the discussion of the mechanical properties of crystalline polymers, Chapter 11E). Often, a delicate balance of structural features is needed to achieve the desired combination of properties. Composite structures (as in glass-reinforced or rubber-modified plastics) often provide unique property combinations.

Economic Aspects. Though the preceding discussion could be extended considerably, it should be pointed out that a major factor in the selection of the appropriate plastic for a given use is economic. In many (some observers would say far too many) cases, a mass market is achieved by sacrificing properties for price. The plastics manufacturer must, therefore, consider what structures give the optimum combination of melt and solid-state properties. He would like, for example, to lower molecular weight to achieve low melt viscosity and more rapid fabrication; but he must maintain levels of molecular weight consistent with the development of good mechanical properties. The selection of raw materials is also important: The rise in prominence of olefin polymers was closely related to the low cost of the monomers. The cost of polymerization itself appears less important in most cases, but it is clear that monomers that are expensive to polymerize have less opportunity to yield large-volume plastics.

A "profile" of the important properties of common general-purpose and specialty plastics is given in Fig. 12-10.

Engineering Plastics

In contrast to general-purpose and specialty plastics, the term *engineering polymers* or plastics is applied to those materials that command a premium price, usually

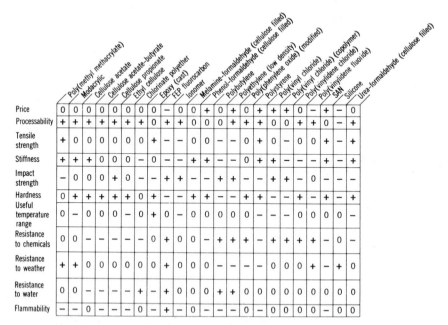

	Poly(methyl methacrylate)	Modacrylic	Cellulose acetate	Cellulose acetate-butyrate	Cellulose propionate	Ethyl cellulose	Chlorinate polyether	Epoxy (cast)	FEP fluorocarbon	Ionomer	Melamine-formaldehyde (cellulose filled)	Phenol-formaldehyde (cellulose filled)	Polybutylene	Polyethylene (low density)	Poly(phenylene oxide) (modified)	Polystyrene	Poly(vinyl chloride)	Poly(vinyl chloride) (copolymer)	Poly(vinylidene chloride)	Poly(vinylidene fluoride)	SAN	Silicone	Urea-formaldehyde (cellulose filled)
Price	0	0	0	0	0	0	0	0	−	0	0	+	0	+	0	+	+	+	0	−	+	−	0
Processability	+	+	+	+	+	+	+	0	+	+	0	0	0	+	+	+	0	0	+	+	0	−	+
Tensile strength	+	0	0	0	0	0	0	+	−	−	0	0	−	−	0	+	0	−	0	0	+	−	+
Stiffness	+	+	+	0	0	0	−	0	−	−	+	+	−	−	0	+	+	−	−	−	+	−	+
Impact strength	−	0	0	0	+	0	−	−	+	+	−	−	+	+	−	−	+	+	−	0	−	−	−
Hardness	0	+	+	+	+	+	0	+	−	−	+	+	−	−	+	+	−	−	+	−	+	−	+
Useful temperature range	0	−	0	0	0	−	0	+	0	−	0	0	0	0	0	−	−	−	0	0	0	0	−
Resistance to chemicals	0	0	−	−	−	−	−	0	+	0	0	−	+	+	+	−	+	+	+	+	−	0	−
Resistance to weather	+	+	0	0	0	0	0	0	0	+	0	0	0	−	−	−	−	−	0	+	−	+	0
Resistance to water	0	0	−	−	−	−	−	+	−	+	0	0	0	+	+	0	0	0	0	0	0	0	0
Flammability	−	−	0	−	−	−	0	−	+	−	0	−	−	−	0	−	0	0	0	0	−	0	0

FIG. 12-10. A "profile" of the properties of some general-purpose and specialty plastics (Billmeyer 1968): +, outstanding in the property indicated, among the best performers available; 0, acceptable performance in this property, still suitable in most cases; −, not recommended if this property is important to the intended use.

associated with relatively low production volume, because of their outstanding balance of properties which allows them to compete successfully with other materials (metals, ceramics) in engineering applications. They are strong, stiff, tough, abrasion-resistant materials capable of withstanding wide ranges of temperatures, and resistant to attack by weather, chemicals, and other hostile conditions. The value they contribute to the end product justifies their higher price per pound. Figure 12-11 compares engineering with general-purpose plastics for price and volume.

The outstanding properties of engineering plastics come primarily from their crystalline nature and strong intermolecular forces. Most of them have quite high melting points, ensuring retention of good physical properties to high temperatures, and good toughness over wide temperature ranges.

The necessary high melting points can be obtained in several ways, as discussed elsewhere. These include combining high degree of crystallinity with stiff polymer chains, or finding structural features substituting for crystallinity in imparting rigidity to the total material structure. This can be done in two ways: crosslinking and utilizing composite structures with an extremely rigid material, such as glass fibers.

So far, these alternatives have been utilized one at a time in almost all cases. That is, glass fibers are usually used to reinforce glassy plastics such as polyester–styrene copolymers, crosslinked polymers are not crystallizable, and so on. We are now beginning to see instances where two or more of these property-enhancing approaches are combined, with outstanding results—for example, the development of ladder polymers combining the features of crosslinking with crystallizability, and the use of high-performance inorganic fibers (such as boron) to reinforce crystalline plastics. Clearly, much effort is being expended toward progress in these directions.

A "profile" of the important properties of engineering plastics is given in Fig. 12-12.

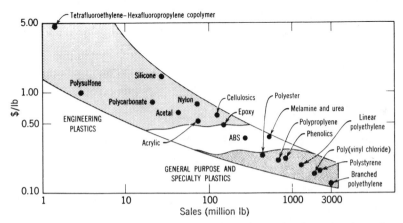

FIG. 12-11. Relationship between prices and production volumes of major engineering and general-purpose plastics (Gutoff 1969).

	ABS	Acetal	Polytetra-fluoroethylene	Polychlorotri-fluoroethylene	Nylon	Phenoxy	Polycarbonate	Polyimide	Poly(phenylene oxide)	Polyethylene (high density)	Polypropylene	Polysulfone
Price	0	0	−	−	−	−	0	−	−	+	+	−
Processability	0	+	−	+	+	0	0	−	0	+	+	+
Tensile strength	0		−	0	0	0	0	+	+	−	0	+
Stiffness	0		−	0	0	0	0	+	0	−	0	0
Impact strength	0	−	0	0	−	+	+	−	−	+	−	0
Hardness	0	+	−	0	0	+	+	+	+	−	0	+
Useful temperature range	−	0	+	0	0	−	0	+	0	0	0	0
Resistance to chemicals	0	0	+	+	0	0	0	+	0	+	+	+
Resistance to weather	0	0	+	+	−	0	0	+	0	−	−	0
Resistance to water	0	0	+	+	−	0	0	0	+	+	+	0
Flammability	−	−	+	+	0	0	0	+	0	+	+	0

FIG. 12-12.   "Profile" of the properties of some engineering plastics (Billmeyer 1968) (same key as for Fig. 12-10).

The Role of Plastics Among Materials

In the last two decades, the role of plastics in the overall materials market has changed drastically. Until the recession of 1980 put an end to the growth in the production of all materials, plastics and particularly engineering plastics had grown at a significantly higher annual rate than either metals or ceramics. By 1970 the annual production of plastics exeeded that of aluminum or copper on a weight basis, and it had been predicted (Gutoff 1969) that on a volume basis (consistent with the higher strength-to-weight ratios of the engineering plastics) production would exceed that of steel by the early 1980s. Until the drastic increase in the price of petroleum in the mid-1970s caused prices of plastics to rise disproportionally in comparison to those of other materials, the trend in plastics prices had for many years been downward in contrast to the trends for most metals. It is anticipated that when the world economy regains a semblance of normalcy, the expansion in the use of plastics seen for so many years in the past will continue.

GENERAL REFERENCES

Riley 1968; Bakker 1980; McQuiston 1980; National Research Council 1981; Platzer 1981; Thommes 1981; Kossoff 1982.

DISCUSSION QUESTIONS AND PROBLEMS

1.   Compare the crystalline melting points of the following classes of polymers at comparable numbers of chain atoms in the repeat unit, and explain in terms

of thermodynamic properties and their molecular origins: polyethylene, polyamides, polyesters, polyureas, polyurethanes.

2. Predict and explain the difference in crystalline melting point between the polymers with the repeat units —$(CH_2)_6NHCO(CH_2)_4CONH$— and —$(CH_2)_6N(CH_3)CO(CH_2)_4CON(CH_3)$—.

3. Explain why 6-nylon is soluble in some solvents at room temperature, whereas linear polyethylene is not.

4. Predict and explain which of the following have similar, and which different, properties at room temperature: polystyrene, poly(methyl methacrylate), poly(vinyl acetate), poly(vinyl alcohol).

5. What sort of comonomer could be used to raise the glass-transition temperature of poly(vinyl acetate)?

6. Why does polypropylene undergo a greater change in physical properties near T_g than does linear polyethylene?

7. Predict and explain the effect, if any, of varying molecular weight and degree of short-chain branching on each of the following properties of polyethylene: ultimate tensile strength, stiffness, T_m, sorption of organic liquids.

8. Explain the effect of molecular weight on the specific gravity of polytetrafluoroethylene.

BIBLIOGRAPHY

Alfrey 1967. Turner Alfrey, Jr., and Edward F. Gurnee, *Organic Polymers*, Prentice-Hall, Englewood Cliffs, New Jersey, 1967.

Bakker 1980. Marilyn Bakker, "Engineering Plastics," pp. 118–137 in Martin Grayson, ed., *Kirk–Othmer Encyclopedia of Chemical Technology*, 3rd ed., Vol. 9, Wiley-Interscience, New York, 1980.

Bannerman 1956. D. G. Bannerman and E. E. Magat, "Polyamides and Polyesters," Chapter 7 in C. E. Schildknecht, ed., *Polymer Processes*, Wiley-Interscience, New York, 1956.

Bawn 1960. C. E. H. Bawn, "High Polymers and Molecular Architecture," *Chem. Ind.* **1960**, 388–391 (1960).

Beaman 1952. Ralph G. Beaman, "Relation between (Apparent) Second-Order Transition Temperature and Melting Point," *J. Polym. Sci.* **9**, 470–472 (1952).

Beevers 1960. R. B. Beevers and E. F. T. White, "Physical Properties of Vinyl Polymers. Part I.—Dependence of the Glass-Transition Temperature of Polymethylmethacrylate on Molecular Weight," *Trans. Faraday Soc.* **56**, 744–752 (1960).

Billmeyer 1968. Fred W. Billmeyer, Jr., and Renée Ford, "The Anatomy of Plastics," *Sci. Technol.* **73**, 22–37, 81–82 (1968).

Boenig 1973. Herman V. Boenig, *Structure and Properties of Polymers*, John Wiley & Sons, New York, 1973.

Boyer 1954. R. F. Boyer, "Relationship of First-to-Second-Order Transition Temperatures for Crystalline High Polymers," *J. Appl. Phys.* **25**, 825–829 (1954).

Boyer 1963. Raymond F. Boyer, "The Relation of Transition Temperatures to Chemical Structure in High Polymers," *Rubber Chem. Technol.* **36**, 1303–1421 (1963).

Bucknall 1967. C. B. Bucknall, "The Relationship between the Structure and Mechanical Properties of Rubber-Modified Thermoplastics, Part One," *Br. Plastics* **40** (11), 118–122 (1967); "Part Two," **40** (12), 84–86 (1967).

Bunn 1955. C. W. Bunn, "The Melting Points of Chain Polymers," *J. Polym. Sci.* **16**, 323–343 (1955).

Campbell 1959. Tod W. Campbell and A. C. Haven, Jr., "The Relationship Between Structure and Properties of Crystalline, High-Melting Polyhydrocarbons," *J. Appl. Polym. Sci.* **1**, 73–83 (1959).

Collier 1969. John R. Collier, "Polymer Crystallization History and Resultant Properties," *Rubber Chem. Technol.* **42**, 769–779 (1969).

Cowie 1968. J. M. G. Cowie and P. M. Toporowski, "The Dependence of Glass Temperature on Molecular Weight for Poly(α-methyl styrene)," *Eur. Polym. J.* **4**, 621–625 (1968).

Dole 1959. M. Dole and B. Wunderlich, "Melting Points and Heats of Fusion of Polymers and Copolymers," *Makromol. Chem.* **34**, 29–49 (1959).

Edgar 1952. Owen B. Edgar and Rowland Hill, "The *p*-Phenylene Linkage in Linear High Polymers: Some Structure–Property Relationships," *J. Polym. Sci.* **8**, 1–22 (1952).

Elias 1977. Hans-Georg Elias, *Macromolecules ·1· Structure and Properties* (translated by John W. Stafford), Plenum Press, New York, 1977.

Flory 1945. Paul J. Flory, "Tensile Strength in Relation to Molecular Weights of High Polymers," *J. Am. Chem. Soc.* **67**, 2048–2050 (1945).

Flory 1949. Paul J. Flory, "Thermodynamics of Crystallization in High Polymers. IV. A Theory of Crystalline States and Fusion in Polymers, Copolymers, and Their Mixtures with Diluents," *J. Chem. Phys.* **17**, 223–240 (1949).

Flory 1969. Paul J. Flory, *Statistical Mechanics of Chain Molecules*, Wiley-Interscience, New York, 1969.

Fox 1950. Thomas G Fox, Jr., and Paul J. Flory, "Second-Order Transition Temperatures and Related Properties of Polystyrene. I. Influence of Molecular Weight," *J. Appl. Phys.* **21**, 581–591 (1950).

Fox 1955. T. G Fox and S. Loshaek, "Influence of Molecular Weight and Degree of Crosslinking on the Specific Volume and Glass Temperature of Polymers," *J. Polym. Sci.* **15**, 371–390 (1955).

Gutoff 1969. Reuben Gutoff, "Engineering Polymers: Risks and Rewards in the Decade Ahead," talk before the Commercial Chemical Development Association, New York, March 1969.

Hill 1948. R. Hill and E. E. Walker, "Polymer Constitution and Fiber Properties," *J. Polym. Sci.* **3**, 609–630 (1948).

Hobbs 1970. Stanley Y. Hobbs and Fred W. Billmeyer, Jr., "Heats and Entropies of Fusion of Linear Aliphatic Polyesters. II. Molecular Origins," *J. Polym. Sci. A-2* **8**, 1395–1409 (1970).

Kargin 1968. V. A. Kargin and G. L. Slonimsky, "Mechanical Properties," pp. 445–516 in Herman F. Mark, Norman G. Gaylord, and Norbert M. Bikales, eds., *Encyclopedia of Polymer Science and Technology*, Vol. 8, Wiley-Interscience, New York, 1968.

Komoto 1967. H. Komoto, "Physico-Chemical Studies of Polyamides. I. Polyamides having Long Methylene Chain Units," *Rev. Phys. Chem. Jpn.* **37**, 105–111 (1967).

Kossoff 1982. Richard M. Kossoff, "The Engineering Plastics Business," *Chemtech* **12**, 552–555 (1982).

Long 1964. V. C. Long, G. C. Berry, and L. M. Hobbs, "Solution and Bulk Properties of Branched Poly(vinyl Acetates). IV. Melt Viscosity," *Polymer* **5**, 517–524 (1964).

McCormick 1959. Herbert W. McCormick, Frank M. Brower and Leo Kin, "The Effect of Molecular Weight Distribution on the Physical Properties of Polystyrene," *J. Polym. Sci.* **39**, 87–100 (1959).

McQuiston 1980. Henry McQuiston, "Designing with Engineering Plastics," *Plastics Eng.* **36** (6), 18–25 (1980).

Merrett 1957. F. M. Merrett, "Graft Copolymers with Preset Molecular Configurations," *J. Polym. Sci.* **24**, 467–477 (1957).

National Research Council 1981. National Research Council, *ad hoc* Panel on Polymer Science and Engineering, *Polymer Science and Engineering: Challenges, Needs, and Opportunities,* National Academy Press, Washington, D.C., 1981.

Nielsen 1969. Lawrence E. Nielsen, "Cross-Linking—Effect on Physical Properties of Polymers," *J. Macromol. Sci. Rev. Macromol. Chem.* **C3,** 69–103 (1969).

Nielsen 1974. Lawrence E. Nielsen, *Mechanical Properties of Polymers and Composites,* Marcel Dekker, New York, 1974.

Pearce 1969. Eli M. Pearce, "Polymer Synthesis: Philosophy and Approaches," *Trans. N.Y. Acad. Sci.* **31,** 629–636 (1969).

Platzer 1981. Norbert Platzer, "Commodity and Engineering Plastics," *Chemtech* **11,** 90–94 (1981).

Reding 1962. F. P. Reding, E. R. Walter, and F. J. Welch, "Glass Transition and Melting Point of Poly(Vinyl Chloride)," *J. Polym. Sci.* **56,** 225–231 (1962).

Riley 1968. Malcolm W. Riley, "Materials, Selection," pp. 419–440 in Herman F. Mark, Norman G. Gaylord, and Norbert M. Bikales, eds., *Encyclopedia of Polymer Science and Technology,* Vol. 8, Wiley-Interscience, New York, 1968.

Rudd 1960. John F. Rudd, "The Effect of Molecular Weight Distribution on the Rheological Properties of Polystyrene," *J. Polym. Sci.* **44,** 459–474 (1960).

Sperati 1953. C. A. Sperati, W. A. Franta, and H. W. Starkweather, Jr., "The Molecular Structure of Polyethylene. V. The Effect of Chain Branching and Molecular Weight on Physical Properties," *J. Am. Chem. Soc.* **75,** 6127–6133 (1953).

Starkweather 1960. Howard W. Starkweather, Jr., and Richard H. Boyd, "The Entropy of Melting of Some Linear Polymers," *J. Phys. Chem.* **64,** 410–414 (1960).

Sternstein 1977. S. S. Sternstein, "Mechanical Properties of Glassy Polymers," in J. M. Schultz, ed., *Treatise on Materials Science and Technology,* Vol. 10, Part B (Herbert Herman, series ed.), Academic Press, New York, 1977.

Sweeney 1969. W. Sweeney and J. Zimmerman, "Polyamides," pp. 483–597 in Herman F. Mark, Norman G. Gaylord, and Norbert M. Bikales, eds., *Encyclopedia of Polymer Science and Technology,* Vol. 10, Wiley-Interscience, New York, 1969.

Thomas 1956. P. E. Thomas, J. F. Lontz, C. A. Sperati, and J. L. McPherson, "Effects of Fabrication on the Properties of Teflon Resin," *Soc. Plast. Eng. J.* **12** (6), 89–96 (1956).

Thommes 1981. G. Thommes, "Making THINGS out of Chemicals," *Chemtech* **11,** 285–287 (1981).

Ward 1971. I. M. Ward, *Mechanical Properties of Solid Polymers,* Wiley-Interscience, New York, 1971.

Wood 1958. Lawrence A. Wood, "Glass Transition Temperatures of Copolymers," *J. Polym. Sci.* **28,** 319–330 (1958).

Woodward 1960. A. E. Woodward, T. M. Crissman, and J. A. Sauer, "Investigations of the Dynamic Mechanical Properties of Some Polyamides," *J. Polym. Sci.* **44,** 23–34 (1960).

Wunderlich 1958. Bernhard Wunderlich and Malcolm Dole, "Specific Heat of Synthetic High Polymers, IX. Poly(ethylene Sebacate)," *J. Polym. Sci.* **32,** 125–130 (1958).

Wunderlich 1980. Bernhard Wunderlich, *Macromolecular Physics, Volume 3: Crystal Melting,* Academic Press, New York, 1980.

Zimmerman 1968. Joseph Zimmerman, "Melt Blend of Polyamides," U.S. Patent 3,393,252 (to E.I. du Pont de Nemours and Co.), 16 July 1968.

PART FIVE

PROPERTIES OF
COMMERCIAL POLYMERS

CHAPTER THIRTEEN

HYDROCARBON PLASTICS AND ELASTOMERS

In this and the following chapters of Part 5, we discuss the synthesis, structure, properties, and applications of the major commercial polymers in the early 1980's. Some material of historical interest is included, and there is implied a familiarity with the material covered in Parts 1–4 and 6.

In Parts 5 and 6, three reference series are extremely valuable and are cited extensively (and hence abbreviated) in the bibliography: *The Modern Plastics Encyclopedia* (*MPE*), cited with the date 1982 but published annually in October (readers should consult the latest volume for corresponding articles); the *Encyclopedia of Polymer Science and Technology* (*EPST*); and the *Kirk–Othmer Encyclopedia of Chemical Technology* (*ECT*).

A. POLYETHYLENE

Low-Density (Branched) Polyethylene

The first commercial ethylene polymer was branched polyethylene, commonly designated as low-density or high-pressure material to distinguish it from the essentially linear material described below. After a period of relatively slow growth in the 1940's, the production of branched polyethylene expanded rapidly; this was the first plastic with annual production exceeding 1 billion lb (in 1959). Record production volume of 7.9 billion lb/yr came in 1979, but by 1982 volume had dropped because of economic recession to around 6.3 billion lb (excluding an additional billion lb of LLDPE; see Section *C*), at prices around $0.30 per lb, about triple those of 1969.

Polyethylene was first produced in the laboratories of Imperial Chemical Industries, Ltd. (ICI), England, in a fortuitous experiment in which ethylene (and other chemicals that remained inert) was subjected to 1400 atm of pressure at 170°C. Traces of oxygen caused polymerization to take place. The phenomenon was first described by E. W. Fawcett in Staudinger 1936.

Polymerization. Ethylene (b.p., −104°C) is made from the thermal (steam) and catalytic cracking of a variety of hydrocarbons, ranging from ethane derived from natural gas to fuel oil. About 25 billion lb was produced in 1982, almost half of it used to produce ethylene polymers and copolymers.

High polymers of ethylene are made commercially at pressures between 1000 and 3000 atm (15,000–45,000 psi) or possibly higher, and temperatures as high as 250°C.

Traces of oxygen initiate the polymerization of ethylene readily. Rapid exothermic reactions can occur, and violent explosions have taken place. Many other possible impurities in the monomer, such as hydrogen and acetylene, act as chain transfer agents and must be carefully removed if high-molecular-weight products are to be obtained. Besides oxygen, peroxides (benzoyl, diethyl), hydroperoxides, and azo compounds have been used as initiators.

Ethylene polymerization can be carried out with benzene or chlorobenzene as solvent. At the temperatures and pressures used, both polymer and monomer dissolve in these compounds so that the reactions are true solution polymerizations. Water or other liquids may be added to dissipate the heat of reaction.

Batch polymerizations of ethylene cannot be carried out rapidly with reproducibility and good control. Long reaction times, consistent with good control, are not economical. In addition, chain branching becomes excessive at high conversion and results in poor physical properties of the product. As a result, balanced, continuous polymerization systems are preferred. Emulsion polymerization has had little success.

One continuous process utilizes tubular reactors, which may have diameters of less than 1 in. and lengths up to 100 ft. The stainless steel tube may be filled with water, and ethylene containing initiator and possibly benzene is introduced. Additional initiator and water or benzene can be injected at one or more points along the tube to keep the initiator concentration more nearly constant throughout the reactor. Ten or more percent of the ethylene is polymerized at the far end of the reactor. Here the gas and liquid phases are taken off continuously, the polymer is separated, and the ethylene is recycled after purification.

Another process utilizes bulk polymerization in a tower-type reactor. Ethylene containing trace amounts of oxygen is charged to the reactor at 1500 atm and 190°C. The reaction is kept essentially isothermal and carried to 10–15% conversion. The effluent from the reactor passes to a separatory vessel in which unconverted ethylene is removed for recycling. The molten polyethylene is chilled below its crystalline melting point and passed through the usual finishing steps.

Structure. Low-density polyethylene is a partially (50–60%) crystalline solid melting at about 115°C, with density in the range 0.91–0.94. It is soluble in many solvents at temperatures above 100°C, but only a few solvent mixtures provide borderline solubility at or near room temperature.

In 1940 infrared spectroscopy (Chapter 9B) revealed that low-density polyethylene contains branched chains. These branches are of two distinct types. Branching due to intermolecular chain transfer, arising from reactions of the type

$$R_1CH_2CH_2\cdot + R_2CH_2CH_2R_3 \xrightarrow[\text{hydrogen transfer}]{\text{intermolecular}} R_1CH_2CH_3 + R_2\overset{\cdot}{C}HCH_2R_3$$

| propagating chain | dead polymer molecule | | dead polymer molecule | propagating chain |

leads to branches which are, on the average, as long as the main polymer chain. This sort of branching has an observable effect on the solution viscosity of the polymer (Chapter 8E) and can be detected by comparing the viscosity of a branched polyethylene with that of a linear polymer of the same molecular weight.

The second branching mechanism in polyethylene is postulated to produce short-chain branching by intramolecular chain transfer:

$$RCH_2CH_2CH_2CH_2CH_2CH_2\cdot \xrightarrow[\text{ring formation}]{\substack{\text{transient} \\ \text{six-membered}}}$$

$$\xrightarrow[\text{hydrogen transfer}]{\text{intermolecular}} RCH_2\overset{\cdot}{C}HCH_2CH_2CH_2CH_3$$
propagating secondary free radical

The transient ring mechanism suggests four carbon atoms as the most probable length of the short branches. Infrared absorption studies and studies of degradation under bombardment with high-energy radiation of polyethylene and substituted polyethylenes of known branch structure suggest that both ethyl and butyl branches are present. A mechanism accounting for the ethyl branches assumes a transfer reaction of the type

after addition of one monomer unit to the radical resulting from a short-chain branching step by the previous mechanism.

Since the transient ring conformations are relatively probable during propagation, the short-chain branching mechanism accounts for the large majority of the chain ends observed in the infrared. A typical low-density polyethylene molecule may contain 50 short branches and less than one long branch on a number-average basis.

The molecular-weight distribution of typical polyethylenes has been found experimentally and theoretically to be very much broadened by the long-chain branching mechanism. Weight- to number-average molecular weight ratios $\bar{M}_w/\bar{M}_n$ of 20–50 are considered typical. The distribution of long branches among the molecules is also very broad: Even for highly branched polymer many molecules contain no long branches, while most of the branches are concentrated on a few very large molecules.

Infrared spectroscopy (Chapter 9B) provides much information about the chemical and physical structure of polyethylene. Structural features associated with the crystallinity of polyethylene are described in Chapter 10.

Properties. The physical properties of low-density polyethylene are functions of three independent structural variables: molecular weight, molecular-weight distribution or long-chain branching, and short-chain branching.

Short-chain branching has a predominant effect on the degree of crystallinity and therefore on the density of polyethylene. (Actually these properties are influenced by total chain branching, but the number of long-chain branch points per molecule in typical polyethylenes is so much less than the number of short-chain branch points that the former can be neglected.) Therefore, as discussed in Chapter 12, properties dependent on crystallinity, such as stiffness, tear strength, hardness, chemical resistance, softening temperature, and yield point, increase with increasing density or decreasing amount of short-chain branching in the polymer, whereas permeability to liquids and gases, toughness, and flex life decrease under the same conditions.

The effect of molecular weight is largely evidenced (Chapter 12C) in properties of the melt and properties involving large deformations of the solid. As molecular weight increases, so do tensile strength, tear strength, low-temperature toughness, softening temperature, impact strength, and resistance to environmental stress cracking, while melt fluidity, melt "drawability," and coefficient of friction (film) decrease. These properties are commonly compared on the basis of changes in melt index (Chapter 11A), which varies inversely with molecular weight.

The effect of long-chain branching on the properties of polyethylene is often evaluated in terms of the breadth of the molecular-weight distribution $\bar{M}_w/\bar{M}_n$. With other structural parameters held constant, a decrease in $\bar{M}_w/\bar{M}_n$ causes a decrease in ease of processing but an increase in tensile strength, toughness and impact strength, softening temperature, and resistance to environmental stress cracking.

The interrelation of density, melt index, and $\bar{M}_w/\bar{M}_n$ in producing physical properties desirable for specific end uses has been described (McGrew 1958). For film uses polymers producing tough and flexible films are needed. Injection-molding applications require polymers characterized by rigidity and good flow; pipe, by strength; and wire insulation, by good processing characteristics and resistance to stress cracking.

The mechanical properties of low-density polyethylene are between those of rigid materials like polystyrene and limp plasticized polymers like the vinyls. Polyethylene has good toughness and pliability over a wide temperature range. Its density falls off fairly rapidly above room temperature, and the resulting large dimensional changes cause difficulty in some fabrication methods. The relatively low crystalline melting point (about 115°C for typical materials) limits the temperature range of good mechanical properties.

The electrical properties of polyethylene are outstandingly good, probably ranking next to those of polytetrafluoroethylene for high-frequency uses. In thick sections polyethylene is translucent because of its crystallinity, but high transparency is obtained in thin films.

Polyethylene is very inert chemically. It does not dissolve in any solvent at room temperature, but is slightly swelled by liquids such as benzene and carbon tetrachloride which are solvents at higher temperatures. It has good resistance to acids and alkalis. At 100°C it is unaffected in 24 hr by sulfuric or hydrochloric acid but charred by concentrated nitric acid. It is often used in containers for acids, including hydrofluoric.

Polyethylene ages on exposure to light and oxygen, with loss of strength, elongation, and tear resistance. The probable point of attack is the tertiary hydrogens on the chain at branch points. Stabilizers retard the deterioration, but few are compatible enough with the polymer to do much good. The weathering of carbon-black-pigmented material is quite good. The polymer also undergoes some crosslinking when heated or worked at elevated temperatures. Few plasticizers or other additives are compatible with polyethylene in amounts larger than 1% or so.

Applications. Almost two-thirds of the low- and medium-density branched polyethylene produced has gone into film and sheeting uses for many years. Few competitive film materials have polyethylene's desirable combination of low density, flexibility without a plasticizer, resilience, high tear strength, moisture and chemical resistance, and little tendency for nicks or cuts to propagate. Blown film, produced by extruding a tube of polymer and expanding it by means of internal pressure of inert gas, thus drawing the polymer, is most widely produced. The tube may be slit to produce flat film or left as a seamless tube. Film thicknesses are usually 0.001–0.005 in. Over three-fourths of the polyethylene film produced goes into packaging applications, including bags and pouches and wrappings for produce, textile products, merchandise, frozen and perishable foods, and many other products. Other film uses include drapes and tablecloths, and extensive application in agriculture (greenhouses, ground cover, tank, pond, and canal liners, etc.) and construction (moisture barriers and utility covering material).

Extrusion coating for packaging materials is the second-largest market for low-density polyethylene, accounting for almost 10% of the use in the United States. The constructions, often laminates of foil, paper, and polyethylene, are used in milk-type cartons for a wide variety of foods and drinks.

Polyethylene has filled a long-standing need for a material that would effectively insulate electrical cables without introducing electrical losses at high frequencies. The nonpolar nature of the polymer makes it ideal for this purpose. Television, radar, and multicircuit long-distance telephone might well have been impossible without such insulating materials. Weathering introduces polar impurities such as carbonyl groups into the polymer and must be carefully guarded against. In addition to the high-frequency uses, polyethylene is being more generally utilized for me-

chanical protection of wire and cables, where its chemical inertness and light weight
are advantageous. About 5% of the branched polyethylene produced is used for
wire and cable insulation.

Other important uses for branched polyethylene are housewares, toys, and containers, lids, and closures produced by injection molding (about 10% of the production), rotational molding, powder coating, and pipe extrusion.

High-Density (Linear) Polyethylene

Linear polyethylene can be produced in several ways, including radical polymerization of ethylene at extremely high pressures, coordination polymerization of
ethylene, and polymerization of ethylene with supported metal-oxide catalysts.
Commercial production of linear polyethylene, using the second and third routes
named, began in 1957 and reached a volume of 5 billion lb/yr in 1979; its 1982
volume was about 4.8 billion lb, at list prices of $0.45–$0.50 per lb, heavily
discounted.

Polymerization. The coordination polymerization of ethylene (Chapter 4D) utilizes a catalyst prepared as a colloidal dispersion by reacting, typically, an aluminum
alkyl and $TiCl_4$ in a solvent such as heptane. Ethylene is added to the reaction
vessel under slight pressure, at a temperature of 50–75°C. Heat of polymerization
is removed by cooling. Polymer forms as a powder or granules, insoluble in the
reaction mixture. At the completion of the reaction, the catalyst is destroyed by
the admission of water or alcohol, and the polymer is filtered or centrifuged off,
washed, and dried.

Supported metal-oxide catalysts can be used in a variety of operating modes,
including fixed-bed, moving-bed, fluid-bed, or slurry processes. Ethylene is fed
with a paraffin or cycloparaffin diluent, at 60–200°C and around 500 psi pressure.
The polymer is recovered by cooling or by solvent evaporation.

Structure. Typical linear polyethylenes are highly (over 90%) crystalline polymers, containing less than one side chain per 200 carbon atoms in the main chain.
Melting point is above 127°C (typically about 135°C), and density is in the range
of 0.95–0.97. Infrared spectroscopy (Chapter 9B) gives detailed information on
the chemical and physical structure of the polymer. Structural features associated
with the crystallinity of linear polyethylene are discussed in Chapter 10.

Properties. Most of the differences in properties between branched and linear
polyethylenes can be attributed to the higher crystallinity of the latter polymers.
Linear polyethylenes are decidedly stiffer than the branched material (modulus of
100,000 versus 20,000 psi), and have a higher crystalline melting point and greater
tensile strength and hardness. The good chemical resistance of branched polyethylene is retained or enhanced, and such properties as low-temperature brittleness
and low permeability to gases and vapors are improved in the linear material.

Applications. The production of bottles and other containers by blow molding accounts for about 40% of the linear polyethylene made. The adjustment of structure variables to obtain high resistance to environmental stress cracking, allowing the material to be used in detergent bottles, produced a large expansion in this field.

About 25% of the linear polyethylene produced is used in the injection molding of crates, pails, tubs, caps and closures, and housewares. The higher stiffness and heat resistance of the linear material have led to its replacement of branched polyethylene in applications where these properties are important.

Other major uses of linear polyethylene include film and sheet, rotational molding, wire and cable insulation, extrusion coating, and pipe and conduit.

High and Ultrahigh Molecular-Weight Polyethylenes. Although most linear polyethylenes have weight-average molecular weights in the range 100,000–200,000, two higher-molecular-weight grades are widely used commercially:

"High-molecular-weight" (HMW) polyethylene has $\bar{M}_w$ between 300,000 and 500,000. Still processable by the usual techniques, it has improved environmental stress-crack resistance, impact and tensile strength, and long-term strength retention. HMW polyethylene is used for pipe, film, and large blow-molded containers where these properties are important.

"Ultrahigh-molecular-weight" (UHMW) polyethylene has $\bar{M}_w$ between 3,000,000 and 6,000,000. It has exceptional abrasion and impact resistance compared to other polyethylenes. Its uses depend on these properties coupled with those usually found for ethylene polymers, such as low coefficient of friction, and include bearings, sprockets, gaskets, valve seats, conveyor-belt parts, and other high-wear-resistance applications. Because of its extremely high molecular weight, UHMW polyethylene does not melt or flow in the normal thermoplastic manner, and most of its fabrication is based on modifications of the compression-molding technique. Fast molding methods combine screw ram injection molding and compression or transfer molding (Chapter 17).

Crosslinked Polyethylene

There has been considerable interest in converting polyethylene to a thermosetting material, in order to combine its low cost, easy processing, and good mechanical properties with the enhanced form stability at elevated temperatures, resistance to stress crack, and tensile strength expected in a crosslinked polymer.

Chemical Crosslinking. Incorporation of relatively stable peroxides, such as di-cumyl peroxide and di-*t*-butyl peroxide, provides a chemical means of crosslinking polyethylene. The peroxides are stable at normal processing temperatures but decompose to provide free radicals for crosslinking at higher temperatures in a post-processing vulcanization or curing reaction. Chemically crosslinked polyethylene is used in the wire and cable industry and is of interest for pipe, hose, and molded articles. Ethylene-propylene rubbers (Section *C*) are also vulcanized by peroxide curing systems.

Radiation Crosslinking   (Chapter 6*D*). The crosslinking of polyethylene by irradiation with high-energy electrons has been used in the commercial production
of films combining the properties typical of polyethylene with form stability up to
200°C and a significant increase in tensile strength. The film can be made heat
shrinkable by biaxial stretching. It is used for insulating (by wrapping) electrical
power cables, coils, transformers, and motors and generators.

GENERAL REFERENCES

Raff 1965–1966, 1967; Buckley 1968; Kresser 1969; Boysen 1981; Buchanan 1981;
    Hogan 1981; Paschke 1981; Short 1981*b*; Henkel 1982; Hug 1982; Snell 1982;
    Carol 1983.

B.   POLYPROPYLENE

With the commercial utilization of coordination polymerization (Chapter 4*D*) in
1957, the production of polypropylene became possible. In the intervening decades
this has become one of the world's major plastics, with U.S. production in 1982
of about 3.6 billion lb, down some 10% from a record 4.0 billion lb in 1981, at a
price of about $0.40 per lb.

Polymerization.   The polymerization of propylene, recovered from cracked gas
streams in olefin plants and oil refineries, is carried out with coordination catalysts
essentially as described in Section *A* for linear polyethylene. Ethylene, propylene,
and other α-olefins can be polymerized in the same equipment with very little
modification, leading to highly flexible operation. Catalysts and operating conditions must be selected with care to ensure that isotactic polypropylene (see the next
paragraph) is produced.

Structure.   Polypropylene can be made in isotactic, syndiotactic, or atactic form
(Chapter 10*A*). The crystallizability of isotactic polypropylene makes it the sole
form with properties of commercial interest. Isotactic polypropylene is an essentially
linear, highly crystalline polymer, with a melting point of 165°C. Its crystal structure
is described in Chapter 10*B*.

Properties.   Polypropylene is the lightest major plastic, with a density of 0.905.
Its high crystallinity imparts to it high tensile strength, stiffness, and hardness. The
resulting high strength-to-weight ratio is an advantage in many applications. Finished articles usually have good gloss and high resistance to marring. The high
melting point of polypropylene allows well-molded parts to be sterilized, and the
polymer retains high tensile strength at elevated temperatures.
    The low-temperature impact strength of polypropylene is somewhat sensitive to
fabrication and test conditions. This sensitivity results from the presence of a

dominating α-transition (Chapter 11D) in polypropylene at about 0°C, resulting in a marked loss in stiffness near this temperature. In high-density polyethylene, the dominant transition is the lower-temperature β-transition. Thus the restriction of molecular motion leading to brittle behavior takes place not far below room temperature in polypropylene, but at a much lower temperature in polyethylene.

To overcome brittleness, wide use is made of both random and block copolymers of propylene with ethylene. The block copolymers are the most impact resistant, and are used in injection molding applications. To retain transparency, random copolymers are used for film applications, while the homopolymer is used almost exclusively for filaments.

Polypropylene has excellent electrical properties and the chemical inertness and moisture resistance typical of hydrocarbon polymers. It is completely free from environmental stress cracking. However, it is inherently less stable than polyethylene to heat, light, and oxidative attack (presumably because of the presence of tertiary hydrogens) and must be stabilized with antioxidants and ultraviolet light absorbers for satisfactory processing and weathering. The resulting formulations are quite satisfactory, even for such applications as indoor–outdoor carpeting, but are more expensive.

Applications. Injection molding uses, including wide application in the automotive and appliance fields, account for almost half of the production of polypropylene. Another third is used as filament (rope, cordage, and webbing) and filament and staple for carpeting. Film uses run well behind.

GENERAL REFERENCES

Raff 1965–1966, 1967; Rebenfeld 1967; Frank 1968; Jezl 1969; Kresser 1969; Short 1980, 1981a; Buchanan 1981; Crespi 1981; Holtgrewe 1982.

C. OTHER OLEFIN-BASED POLYMERS AND COPOLYMERS

Olefin-Based Plastics

Polymers of α-olefins, including propylene, 1-butene, and higher homologues, and copolymers of these monomers with ethylene, can be prepared by coordination polymerization (Chapter 4D). Homopolymers of olefins higher in chain length than propylene that have been offered commercially include those of 1-butene and 4-methyl pentene-1. In addition, a number of ethylene copolymers are sold, including those with propylene, 1-butene, and isobutylene. Copolymers of ethylene with vinyl acetate, maleic anhydride, and ethyl acrylate are made commercially by radical polymerization. Copolymers with carboxyl-containing monomers are crosslinkable by metal–salt bridges. Some of these compositions are rubbery in nature, but they are listed here because they are primarily utilized as plastics.

Polymers of Higher Olefins. Isotactic poly(butene-1), known in the trade as polybutylene, is known in four crystalline modifications. A metastable tetragonal form results from crystallization from the melt, and changes irreversibly to the stable twinned hexagonal form in 5–7 days at room temperature. The remaining forms, orthorhombic and untwinned hexagonal, are formed only after crystallization from solution. Polybutylene's major use is as a pipe resin, based on its flexibility and resistance to creep, environmental stress cracking, chemicals, and abrasion. It has the highest hydrostatic design stress rating for any flexible thermoplastic. Blown-film applications are also important.

Copolymers of 4-methyl pentene-1 with undisclosed monomers, presumably used in small amounts to reduce degree of crystallinity, are known to the trade as poly-methylpentene. It is about 40% crystalline, with the extremely low density of 0.83. With very high transparency and optical clarity and a thermal expansion coefficient very near that of water, it is used in a variety of applications such as laboratory and medical ware.

"Linear Low-Density" (LLD) Polyethylene. LLD polyethylene (LLDPE), a co-polymer of ethylene with an alpha-olefin such as butene, hexene, or octene, is designed to simulate the short-chain branching and density of conventional branched polyethylene without the occurrence of long-chain branching. With unusually good melt-flow and physical properties, LLDPE has been one of the fastest growing new plastics in recent years, with about 1 billion lb produced in 1982. A major use is plastic trash bags.

Other Ethylene Copolymers. Copolymers of ethylene with ethyl acrylate, methyl acrylate, and vinyl acetate, are made by high-pressure radical polymerization. The main functions of the comonomer are to reduce crystallinity and introduce polarity, leading to more flexible and tougher products. The major uses of these materials are as films.

Ionomers. The word *ionomer* (Kinsey 1969) was coined as a generic term for a class of thermoplastics containing ionizable carboxyl groups which can create ionic crosslinks between chains. These substances are produced as copolymers of α-olefins with carboxylic acid monomers, such as methacrylic acid, followed by partial neutralization with a metal cation. Crosslinking thus occurs through metal "bridges." These crosslinks are labile at processing temperatures, allowing the ionomers to be extruded or molded in conventional equipment. The upper use temperature of the ionomers is limited, as might be expected, because the crosslinks begin to "melt out." The primary uses of this class of materials are centered around their combination of properties such as high transparency, toughness, flexibility, adhesion, and oil resistance. Food packaging, skin and blister packaging, and several shoe uses are examples.

Olefin-Based Rubbers

Chlorosulfonated Polyethylene. When polyethylene is treated with a mixture of chlorine and sulfur dioxide, some chlorine atoms are substituted on the chains and some sulfonyl chloride groups ($-SO_2Cl$) are formed. The chlorosulfonation can be carried out either on the solid material or in solution. There are two results of these modifications: (a) The chlorine atoms break up the regularity of the polyethylene chain structure so that crystallization is no longer possible, thus imparting an elastomeric character to the polymer, and (b) the sulfonyl chloride groups provide sites for crosslinking. A typical polymer contains 25–30% chlorine (one chlorine for every seven carbon atoms) and about 1.5% sulfur (one $-SO_2Cl$ for every 90 carbon atoms).

The elastomer can be crosslinked by a large variety of compounds, including many rubber accelerators. Metallic oxides are recommended for commercial cures. Fillers are not needed to obtain optimum strength properties.

Chlorosulfonated polyethylene is resistant to ozone, being better than neoprene and butyl rubber in this respect. Oxidative resistance and heat resistance are good. Chemical resistance is better than that of the common elastomers. The material is poor in "snap" and rebound and has low elongation and some permanent set. Its abrasion resistance, flex life, low-temperature brittleness, and resistance to crack growth are good.

Ethylene–Propylene Rubbers. Copolymerization of propylene with ethylene yields noncrystalline products that have rubbery behavior and are chemically inert because of their saturation. They must, however, be crosslinked by use of peroxides or radiation. To gain sites for crosslinking, a diene monomer is often added; the resulting terpolymers are known as *ethylene-propylene-diene monomer* (EDPM) elastomers. The major comonomers are 1,4-hexadiene, dicyclopentadiene,

and ethylidene norbornene,

(In the latter two, it is the double bond on the far left that is most favorably opened during polymerization.) These rubbers have several desirable properties, including resistance to ozone, oxygen, and heat. However, the additional cost resulting from the use of the expensive diene monomer and other sources has limited them to the special-purpose field to date, with 1982 production about 0.3 billion lb.

GENERAL REFERENCES

Raff 1965–1966; Buckley 1968; Rubin 1968; Kresser 1969; Brydson 1978, Chapter 12; Borg 1979; Johnson 1979*a*; Ofstead 1979; Holohan 1980; Kochhar 1981; Baker 1982; Bonotto 1982; Chatterjee 1982; Duncan 1982; Ohi 1982; Person 1982.

D. NATURAL RUBBER AND OTHER POLYISOPRENES

Natural Rubber

Natural rubber is a high-molecular-weight polymer of isoprene, in which essentially all the isoprenes have the *cis*-1,4 configuration (Chapter 10*A*). The natural polymer has a number-average degree of polymerization of about 5000 and a broad distribution of molecular weights.

Source.   Natural rubber can be obtained from nearly five hundred different species of plants. The outstanding source is the tree *Hevea brasiliensis,* from which comes the name *Hevea rubber.* Rubber is obtained from a latex that exudes from the bark of the *Hevea* tree when it is cut. Latex is an aqueous dispersion of rubber, containing 25–40% rubber hydrocarbon, stabilized by a small amount of protein material and fatty acids. The latex is gathered, coagulated, washed, and dried. Two different processes are used.

Crepe rubber results if a small amount of sodium bisulfite is added to bleach the rubber. The coagulum is rolled out into sheets about 1 mm thick and dried in air at about 50°C. If *smoked sheets* are to be made, the bleach is omitted and somewhat thicker sheets are rolled. These are dried in smokehouses at about 50°C in the smoke from burning wood or coconut shells.

Mastication.   It was discovered by Hancock in 1824 that rubber becomes a soft, gummy mass when subjected to severe mechanical working. This process is known as *mastication.* The addition of compounding ingredients is greatly facilitated by this treatment, which is usually carried out on roll mills or in internal mixers or plasticators (similar to extruders). Mastication is accompanied by a marked decrease in the molecular weight of the rubber. Oxidative degradation is an important factor in mastication, since the decrease in viscosity and the other property changes do not take place if the rubber is masticated in the absence of oxygen.

After mastication is complete, compounding ingredients are added, and the rubber mix is prepared for vulcanization. These processes are discussed in Chapter 19.

Properties.   The properties of vulcanized natural rubber form the model for the ideal elastomeric properties discussed in Chapter 11*B*, including rapid extensibility to great elongations, high stiffness and strength when stretched, and rapid and complete retraction on release of the external stress. The properties of natural rubber are discussed further in comparison with those of the major synthetic elastomers in Chapter 19.

Stereoregular Synthetic Polyisoprene

In 1955, production began of essentially *cis*-1,4-polyisoprenes with structures closely duplicating that of natural rubber. These polymers can be made by two processes that are almost identical except for the catalyst used: One is based on coordination polymerization (Chapter 4*D*), using a catalyst of titanium tetrachloride and an aluminum alkyl such as triisobutyl aluminum; the other is an anionic polymerization (Chapter 4*C*) with butyl lithium as the catalyst.

In a typical operation, isoprene (derived from petroleum) is mixed with a hydrocarbon solvent such as *n*-pentane. The catalyst is added, and the reaction allowed to take place at about 50°C and moderate pressures until a solids content of about 25% is reached. At this point the reaction mixture is a highly viscous "cement." A catalyst deactivator and an antioxidant are added, and solvent is removed in an extruder or a drum dryer. The polymer made by coordination polymerization has a molecular weight distribution similar to that of masticated natural rubber and requires no further processing before compounding, whereas that made with butyl lithium is higher in molecular weight and requires even longer mastication than natural rubber.

The properties of the *cis*-1,4-polyisoprenes are, commensurate with their structure, very nearly identical with those of natural rubber, and the synthetic polymers are not only complete replacements for the natural product, but are often preferred because of their greater cleanliness and uniformity. Production has, however, remained relatively small, dropping from around 175 million lb/year in the early 1970's to about 120 million lb in 1982.

GENERAL REFERENCES

Bean 1967; Kennedy 1968–1969; Barnard 1970; Brydson 1978, Chapters 2 and 5; McGrath 1979; Tucker 1979; St. Cyr 1982; Blackley 1983.

E. RUBBERS DERIVED FROM BUTADIENE

Although also made from alcohol during World War II, butadiene is now derived exclusively from petroleum. Fractionation of the products of cracking petroleum, either for producing olefins or for obtaining high-octane gasoline, yields a cut containing largely hydrocarbons of the butane and butene family. 1-Butene is separated and catalytically dehydrogenated in the vapor phase to butadiene.

Styrene–Butadiene Rubber (SBR)

Production of SBR (then known as GR-S) was begun in the United States during World War II. The product was designed to be similar to the German Buna-S (Chapter 19) but lower in molecular weight for easier processing. The rubber was made by emulsion polymerization using the so-called Mutual recipe (Table 6-2), at 50°C. After the war, product quality was improved by carrying out the polymerization at 5°C (41°F) with some being made at temperatures as low as − 10° or − 18°C. These changes were brought about by the use of more active initiators, such as cumene hydroperoxide and *p*-menthane hydroperoxide, and the addition of antifreeze components to the mixture. The product is known as *cold rubber*.

Anionic solution copolymerization of butadiene and styrene with alkyllithium catalysts is used to produce so-called *solution* SBR. This product has a narrower molecular-weight distribution, higher molecular weight, and higher *cis*-1,4-polybutadiene content than emulsion SBR. Tread wear and crack resistance are improved, as is economy because oil extension and carbon-black loading can be increased.

Consumption of SBR in the United States has remained for many years at about 50% of all rubber use. It reached almost 3 billion lb/year in 1979, but the reduction in the production and size of automobile tires, which account for about 75% of SBR use, reduced the production figure to around 2.2 billion lb in 1982. List prices were in the range $0.60–$0.80 per lb, but were heavily discounted.

Structure of SBR. By virtue of its free radical polymerization, SBR is a random copolymer. The butadiene units are found to be about 20% in the 1,2 configuration, 20% in the *cis*-1,4, and 60% in the *trans*-1,4 for polymer made at 50°C, with the percentage of *trans*-1,4 becoming higher for polymer made at lower temperatures. In consequence of its irregular structure, SBR does not crystallize.

Branching reactions due to chain transfer to polymer and to polymerization of both double bonds of a diene unit become extensive if conversion is allowed to become too high or a chain transfer agent is not used in SBR polymerization. However, SBR has been shown to have exactly one double bond per butadiene unit. Thus no extensive side reactions occur during its formation, at least up to about 75% conversion.

Processing of SBR. In general, the differences in mastication and vulcanization between SBR and natural rubber are minor. A reinforcing filler is essential to the achievement of good physical properties in SBR. However, some fillers other than carbon black reinforce it moderately well. SBR is compatible with the other major elastomers and can be used in blends. The techniques of oil extension and masterbatching are widely employed. These techniques, as well as the processing of SBR in general, are discussed further in Chapter 19.

Properties of SBR. Tire tread stocks made from regular SBR are inferior in tensile strength to those from natural rubber (3000 versus 4500 psi), whereas those from "cold rubber" are almost equivalent to *Hevea* (3800 psi). At elevated temperatures, however, regular and "cold" SBR lose almost two-thirds of their tensile strength whereas natural rubber loses only 25%. The ozone resistance of SBR is superior to that of natural rubber, but when cracks or cuts start in SBR they grow much more rapidly. Perhaps the most serious defect of both types of SBR for tire uses, however, is its poorer resilience and greater heat buildup. Tread wear of the synthetic material is at least as good as that of natural rubber. The weatherability of SBR is better than that of natural rubber.

"Cold" SBR is superior to the standard product because it contains less low-molecular-weight (nonreinforceable) rubber, less chain branching and crosslinking, and a higher proportion (70%) of the *trans*-1,4 configuration around the double bond.

Applications of SBR.   Originally, SBR was used in tires only of necessity, but "cold" SBR appears equal in most respects to natural rubber, especially for lighter-duty tire use. It is inferior to the natural product for truck tires. For many mechanical goods SBR is superior to natural rubber and is preferentially used because of its easier processing and good-quality end product. Such items include belting, hose, molded goods, unvulcanized sheet, gum, and flooring. Rubber shoe soles are made of SBR. Extruded goods and coated fabrics are other fields in which SBR offers advantages in processability. It is widely used for electrical insulation, although its properties are not as good as those of butyl rubber.

Nitrile Rubbers

The nitrile rubbers are polymers of butadiene and acrylonitrile, having ratios of the two monomers similar to the ratio of butadiene to styrene in SBR. They are noted for their oil resistance but are not suitable for tires. The first commercial nitrile rubbers were made in Germany.

The oil resistance of the nitrile rubbers varies greatly with their composition. Several grades are available commercially, ranging in acrylonitrile content between 18% (only fair oil resistance) and 40% (extremely oil resistant).

The nitrile rubbers are prepared in emulsion systems similar to those used for SBR. Because the monomer reactivity ratios are quite different, the compositions of the feed and the polymer differ markedly. This fact is usually taken into account by adjusting the monomer composition during the polymerization to achieve the desired polymer composition.

Nitrile rubber is used primarily for its oil resistance, by which is implied its low solubility, low swelling, and good tensile strength and abrasion resistance after immersion in gasoline or oils. Swelling of nitrile rubbers is greater in polar solvents than in nonpolar solvents, but they can be used in contact with water and antifreeze solutions. Resistance to ethylene glycol is good. The rubbers are inherently less resilient than natural rubber. Their heat resistance is good; and, if properly protected by antioxidants, they show satisfactory resistance to oxidative degradation as well.

Nitrile rubbers are extensively used for gasoline hoses, fuel tanks, creamery equipment, and the like. In addition they find wide application in adhesives and, in the form of latex, for impregnating paper, textiles, and leather. In all, about 165 million lb was used in the United States in 1982.

Stereoregular Polybutadienes

cis-1,4-Polybutadiene. This polymer is made by coordination or anionic polymerization in the same processes used for *cis*-1,4-polyisoprene (Section *D*) and has similar properties. It is utilized almost entirely in tires, blended with SBR and natural rubber. A small amount of natural rubber appears to be required to prevent the polybutadiene from crumbling during processing and to improve tack. Polybutadiene has high elasticity, low heat buildup, and good resistance to oxidation. It imparts outstanding abrasion resistance to truck and passenger tires, but cannot be used at levels higher than 40–50%, above which its major deficiency of poor skid resistance becomes too apparent. Use in 1982 was about 0.8 billion lb in the United States.

GENERAL REFERENCES

Saltman 1965; Kennedy 1968–1969; Rubin 1968; Brydson 1978, Chapters 6–9 and 14; Bauer 1979; Kuzma 1979; McGrath 1979; Robinson 1979; Hsieh 1981; Blackley 1983.

F. OTHER SYNTHETIC ELASTOMERS

Polyisobutylene and Butyl Rubber

Butyl rubbers are copolymers of isobutylene with a small amount of isoprene added in order to make them vulcanizable. Since the amount of comonomer is small, the methods of polymerization and the properties of the unvulcanized copolymer are similar to those of polyisobutylene itself.

Polymerization. The polymerization of isobutylene and its mixtures with diolefins typifies the industrial application of low-temperature cationic polymerization (Chapter 4*B*). Isobutylene polymerizes rapidly at $-80°C$ with Friedel–Crafts catalysts. In a bulk system at $-80°C$, rapid polymerization is induced by bubbling BF_3 gas through isobutylene. The heat of reaction can be absorbed by adding solid carbon dioxide to the monomer or by adding a low-boiling diluent such as pentane or ethylene which is refluxed. In a typical process, butyl rubber is manufactured by mixing isobutylene with 1.5–4.5% isoprene and methyl chloride as diluent. This mixture is fed to stirred reactors cooled to $-95°C$ by liquid ethylene. Catalyst solution, made by dissolving anhydrous aluminum chloride in methyl chloride, is added. Polymer forms at once as a finely divided product suspended in the reaction mixture. This slurry is pumped out of the reactor continuously as monomer and catalyst are added. The product mixture is passed into a large volume of hot agitated water in a tank where the volatile components are flashed off and recovered. An antioxidant and some zinc stearate to prevent agglomeration of the polymer particles are added at this point. The polymer is then filtered off, dried, and extruded.

Structure. Polyisobutylene and butyl rubber are amorphous under normal conditions, but crystallize on stretching. Most, if not all, of the isoprene units are present in the 1,4 structure. It is usually assumed that the polymers are linear, although the possibility of a small amount of branching has not been investigated in detail. The molecular weights of the polymers made by low-temperature ionic polymerization can be quite high; chain transfer agents such as diisobutylene are often added to control molecular weight at the 200,000–300,000 level.

Unstabilized polyisobutylenes are degraded by heat or light to sticky low-molecular-weight products. The usual rubber antioxidants or retarders of free radical reactions stabilize the polymers well.

Low-molecular-weight polyisobutylenes are liquids. As molecular weight increases, they change to balsamlike solids, and then to rubberlike polymers. Unless low-molecular-weight material is removed, even polyisobutylene of 100,000 molecular weight is sticky. Unlike natural rubber, polyisobutylene and butyl rubber do not crystallize on cooling and hence remain flexible to as low as $-50°C$.

The response of isobutylene polymers and copolymers to stress is quite different from that of natural rubber. Polyisobutylene is sluggish, showing large viscoelastic and viscous components in its response (Chapter 11C). The polymer has been widely used in the study of viscoelasticity.

The strong tendency toward cold flow in polyisobutylene prevents its direct application as an elastomer. It is used in adhesives, caulking compounds, pressure-sensitive tapes, and coatings for paper.

Properties. The properties of butyl vulcanizates are compatible with their structure. The very low residual unsaturation of the rubbers leads to outstanding chemical inertness. The closepacked linear paraffinic chains result in unusually low permeability to gases. Steric hindrance of the methyl groups on the chains causes high internal viscosity and viscoelastic response to stresses.

A property of great importance in elastomers is aging in the presence of oxygen. It has been found that the presence of a double bond in the skeletal structure of a polymer is very important in enhancing the rate of absorption of oxygen, and that the presence of methyl side groups is also significant but less so. Butyl rubber is therefore, as expected, less sensitive to oxidative aging than are most other elastomers except the silicones. Since methyl side groups appear to favor chain scission whereas double bonds favor crosslinking, butyl rubber becomes soft rather than brittle on oxidative degradation.

Butyl has much better ozone resistance than natural rubber. Its solvent resistance is typical of that of hydrocarbon elastomers. Its acid resistance is quite good.

The stress–strain properties of butyl rubber are similar to those of natural rubber. Both show the importance of crystallization in obtaining high tensile strength. Crystallization does not take place in butyl, however, until higher elongations are reached. The tear resistance of butyl is quite good and is retained well at high temperatures and for long times, in contrast to natural rubber. The electrical properties are quite good, as predicted from its nonpolar, saturated nature. The dynamic and elastic properties of butyl are marked by sluggishness over the temperature range $-30°$ to $+40°C$, characteristic of a polymer with high internal friction and high damping power. Rebound is slow and heat buildup is high.

Applications.   About 75% of the butyl rubber produced is used for inner tubes for tires. This usage was substantially reduced when tubeless passenger tires were introduced in 1953 (Herzegh 1981), but most nonpassenger tires still use tubes. Other uses for butyl are mainly in the area of mechanical goods.

Consumption of butyl rubber in the United States was about 290 million lb in 1982.

Polychloroprene (Neoprene)

The generic term neoprene denotes rubberlike polymers and copolymers of chloroprene, 2-chloro-1,3-butadiene. Neoprenes were the first synthetic rubbers developed in the United States. Although they are primarily known for their oil resistance, they are good general-purpose rubbers that can replace natural rubber in most of its uses and are satisfactory in a wide variety of applications. About 270 million lb was used in the United States in 1982.

Chloroprene is prepared by the catalytic addition of hydrogen chloride to vinylacetylene, which in turn is made by the catalytic dimerization of acetylene. This is an expensive process, however, and interest is turning to the production of chloroprene from butadiene. One process utilizes chlorination to 3,4-dichlorobutene-1, followed by dehydrochlorination.

The neoprenes are produced by emulsion polymerization. Some types are polymerized in the presence of sulfur, which introduces some crosslinking in the polymer. In these cases, the latex is allowed to age in the presence of an emulsion of tetraethylthiuram disulfide, which restores the plasticity of the polymer. The latex is then coagulated by acidification followed by freezing.

Polymerization appears to take place almost entirely in the *trans*-1,4 form. As a result the neoprenes are crystallizable elastomers.

The vulcanization of neoprene is different from that of other elastomers in that it can be vulcanized by heat alone. Zinc oxide and magnesium oxide are the preferred vulcanizing agents. The mechanism by which they cause crosslinking is not known. Sulfur vulcanizes neoprene very slowly, and the usual rubber accelerators are in general not effective—some, in fact, are potent retarders of the cure. A few chemicals are known to accelerate the vulcanization, however, among them antimony sulfide.

Unlike many other elastomers, neoprene vulcanizates have high tensile strength (3500–4000 psi) in the absence of carbon black. No reinforcing effect is found with any filler. Suitably protected neoprene vulcanizates are extremely resistant to oxidative degradation. Weathering resistance and ozone resistance are quite good. Neoprene is slightly inferior to nitrile rubber in oil resistance, but markedly better than natural rubber, butyl, or SBR. The dynamic properties of neoprene are superior to those of most other synthetics and only slightly inferior to those of natural rubber. They are less affected by elevated temperature than those of natural rubber. Neoprene has been shown to make excellent tires but cannot compete with other elastomers in price. Its major uses include wire and cable coatings, industrial hoses and belts, shoe heels, and solid tires. Gloves and coated fabrics are made from neoprene latex.

Thermoplastic Elastomers

The term *thermoplastic elastomers* is currently used to describe a wide variety of materials that have elastomeric properties at ambient temperatures, but process like thermoplastics, obviating the need for the vulcanization step to develop typical rubberlike elasticity. The best known and most widely used thermoplastic elastomers, to which the name originally applied, are block copolymers of styrene (S) with butadiene (B) or isoprene (I) with the block structure S–B–S or S–I–S. The morphology and properties of these materials are described in Chapter 17D. Other polymers commonly termed thermoplastic elastomers are polyurethanes and copolyesters (Chapter 15) and blends of ethylene-propylene copolymers (Section C) with polypropylene.

As a family, the thermoplastic elastomers are beginning to replace other specialty rubbers in a wide variety of uses, for example, replacing neoprene in adhesives and wire and cable insulation. Including all types, about 400 million lb was sold in 1982.

Other Elastomers

Several other elastomers have the status of specialty rubbers, at relatively low production volume and high price. Among these are chlorosulfonated polyethylene (Section C), the acrylate and various fluorocarbon elastomers (Chapter 14), polyurethane and polysulfide elastomers (Chapter 15), and silicone and epichlorhydrin rubbers (Chapter 16).

GENERAL REFERENCES

Buckley 1965; Friedlander 1965; Hargreaves 1965; Kennedy 1968–1969; Brydson 1978, Chapters 10, 11, and 17; West 1978; Baldwin 1979; Finelli 1979; Johnson 1979b; McGrath 1979; Bonk 1982; Ellerstein 1982; Blackley 1983.

DISCUSSION QUESTIONS AND PROBLEMS

1. Write the chemical reactions postulated for the formation of long- and short-chain branching in polyethylene.

2. Describe the relation between long-chain branching and molecular-weight distribution typical of low-density polyethylene.

3. Discuss the properties of LLDPE in relation to its branched-chain structure.

4. Compare the glass transitions in polyethylene and polypropylene and relate them to the impact strengths of the two polymers.

5. Account for the high transparency and optical clarity of polymethylpentene.

6. Describe the crosslink bonding in ionomers in terms of molecular structure, chemical sensitivity, and physical properties of the polymers.

7. Discuss the functions of the three chain constituents in (a) chlorosulfonated polyethylene and (b) EPDM rubber in producing elastomeric properties.

8. Write chemical formulas for typical chain structures of the three major types of EPDM rubber.

9. Name and show by chemical formulas all the ways in which isoprene can add to a growing chain during polymerization. Identify those most frequently found in commercially valuable polyisoprenes.

BIBLIOGRAPHY

Baker 1982. George Baker, "Ethylene–Methyl Acrylate," p. 81 in *MPE,*† 1982.

Baldwin 1979. F. B. Baldwin and R. H. Schatz, "Butyl Rubber," pp. 470–484 in *ECT,*‡ Vol. 8, 1979.

Barnard 1970. D. Barnard, J. I. Cunneen, P. B. Lindley, A. R. Payne, M. Porter, A. Schallamarch, W. A. Southorn, P. McL. Swift, and A. G. Thomas, "Rubber, Natural," pp. 178–256 in *EPST,*§ Vol. 12, 1970.

Bauer 1979. R. G. Bauer, "Styrene–Butadiene Rubber," pp. 608–625 in *ECT,* Vol. 8, 1979.

Bean 1967. Arthur R. Bean, Glenn R. Hines, Geoffrey Holden, Robert R. Houston, John A. Langlon, and Roger H. Mann, "Isoprene Polymers," pp. 782–855 in *EPST,* Vol. 7, 1967.

Blackley 1983. D. C. Blackley, *Synthetic Rubbers: Their Chemistry and Technology,* Elsevier, New York, 1983.

Bonk 1982. H. W. Bonk, "Thermoplastics Elastomers," pp. 120–122, 127 in *MPE,* 1982.

Bonotto 1982. Sergio Bonotto, "Ethylene–Ethyl Acrylate," pp. 80–81 in *MPE,* 1982.

† Joan Agranoff, ed., *Modern Plastics Encyclopedia 1982–1983,* Vol. 59, No. 10A, McGraw-Hill, New York, 1982.

‡ Martin Grayson, ed., *Kirk–Othmer Encyclopedia of Chemical Technology,* 3rd ed., Wiley-Interscience, New York.

§ Herman F. Mark, Norman G. Gaylord, and Norbert M. Bikales, eds., *Encyclopedia of Polymer Science and Technology,* Wiley-Interscience, New York.

Borg 1979. E. L. Borg, "Ethylene–Propylene Rubbers," pp. 492–500 in *ECT*, Vol. 8, 1979.

Boysen 1981. Robert L. Boysen, "High Pressure (Low and Intermediate Density) Polyethylene," pp. 402–420 in *ECT*, Vol. 16, 1981.

Brydson 1978. J. A. Brydson, *Rubber Chemistry,* Applied Science, London, 1978.

Buchanan 1981. P. R. Buchanan, "Olefin Fibers," pp. 357–385 in *ECT*, Vol. 16, 1981.

Buckley 1965. D. J. Buckley, "Butylene Polymers," pp. 754–795 in *EPST,* Vol. 2, 1965.

Buckley 1968. D. J. Buckley, B. S. Dyer, M. R. Day, and W. R. Bergenn, "Olefin Polymers," pp. 440–458 in *EPST,* Vol. 9, 1968.

Carol 1983. Frederick J. Carol, "The Polyethylene Revolution," *Chemtech* **13**, 222–228 (1983).

Chatterjee 1982. A. M. Chatterjee, "Polybutylene," pp. 58, 60 in *MPE,* 1982.

Crespi 1981. Giovanni Crespi and Luciano Luciani, "Polypropylene," pp. 453–469 in *ECT*, Vol. 16, 1981.

Duncan 1982. R. E. Duncan, "Ethylene–Vinyl Acetate," pp. 81–82 in *MPE,* 1982.

Ellerstein 1982. S. M. Ellerstein and E. R. Bertozzi, "Polysulfides," pp. 814–831 in *ECT*, Vol. 18, 1982.

Finelli 1979. A. F. Finelli, R. A. Marshall, and D. A. Chung, "Thermoplastic Elastomers," pp. 626–640 in *ECT*, Vol. 8, 1979.

Frank 1968. H. P. Frank, *Polypropylene,* Gordon and Breach, New York, 1968.

Friedlander 1965. Henry Z. Friedlander, "Chemically Resistant Polymers," pp. 665–683 in *EPST,* Vol. 3, 1965.

Hargreaves 1965. C. A. Hargreaves, II, and D. C. Thompson, "2-Chlorobutadiene Polymers," pp. 705–730 in *EPST,* Vol. 3, 1965.

Henkel 1982. R. N. Henkel, "High and Low Density Polyethylene," pp. 73–74, 76 in *MPE,* 1982.

Herzegh 1981. Frank Herzegh, "The Tubeless Tire," *Chemtech* **11**, 224–228 (1981).

Hogan 1981. J. Paul Hogan, "Linear (High Density) Polyethylene," pp. 421–433 in *ECT*, Vol. 16, 1981.

Holohan 1980. J. F. Holohan, Jr., J. Y. Penn, and W. A. Vredenburgh, "Hydrocarbon Resins," pp. 852–869 in *ECT*, Vol. 12, 1980.

Holtgrewe 1982. D. A. Holtgrewe, "Polypropylene," pp. 94,96,98 in *MPE,* 1982.

Hsieh 1981. Henry L. Hsieh, Ralph C. Farrar, and Kishore Udipi, "Anionic Polymerization: Some Commercial Applications," *Chemtech* **11**, 626–633 (1981).

Hug 1982. D. P. Hug, "UHMW Polyethylene," pp. 79–80 in *MPE,* 1982.

Jezl 1969. James L. Jezl and Earl M. Honeycutt, "Propylene Polymers," pp. 597–619 in *EPST,* Vol. 11, 1969.

Johnson 1979a. Paul R. Johnson, "Chlorosulfonated Polyethylene," pp. 484–491 in *ECT*, Vol. 8, 1979.

Johnson 1979b. Paul R. Johnson, "Neoprene," pp. 515–534 in *ECT*, Vol. 8, 1979.

Kennedy 1968–69. Joseph P. Kennedy and Erik Törnquist, eds., *Polymer Chemistry of Synthetic Elastomers,* Wiley-Interscience, New York; *Part I,* 1968; *Part II,* 1969.

Kinsey 1969. Roy H. Kinsey, "Ionomers, Chemistry and New Developments," *Appl. Polym. Symp.* **11**, 77–94 (1969).

Kochhar 1981. R. K. Kochhar, Yuri V. Kissin, and David L. Beach, "Polymers of Higher Olefins," pp. 470–479 in *ECT*, Vol. 16, 1981.

Kresser 1969. Theodore O. G. Kresser, *Polyolefin Plastics,* Van Nostrand Reinhold, New York, 1969.

Kuzma 1979. L. J. Kuzma and W. J. Kelly, "Polybutadiene," pp. 546–567 in *ECT*, Vol. 8, 1979.

McGrath 1979. James E. McGrath, "Elastomers, Synthetic—Survey," pp. 446–459 in *ECT*, Vol. 8, 1979.

McGrew 1958. Frank C. McGrew, "The Polyolefin Plastics Field—Present Technical Status," *Modern Plastics* **35** (7), 125,126,128,132–133 (1958).

Ofstead 1979. Eilert A. Ofstead, "Polypentenamers," pp. 592–607 in *ECT*, Vol. 8, 1979.

Ohi 1982. H. Ohi, "Polymethylpentene," p. 93 in *MPE*, 1982.

Paschke 1981. Eberhard Paschke, "Ziegler Process Polyethylene," pp. 433–452 in *ECT*, Vol. 16, 1981.

Person 1982. W. J. Person, "Linear Low Density Polyethylene," pp. 76,78 in *MPE*, 1982.

Raff 1965–1966. R. A. V. Raff and K. W. Doak, eds., *Crystalline Olefin Polymers,* Wiley-Interscience, New York; *Part I,* 1965; *Part II,* 1966.

Raff 1967. R. A. V. Raff, P. E. Campbell, R. V. Jones, E. D. Caldwell, Herbert N. Friedlander, and Peter J. Canterino, "Ethylene Polymers," pp. 275–454 in *EPST,* Vol. 6, 1967.

Rebenfeld 1967. Ludwig Rebenfeld, "Fibers," pp. 505–573 in *EPST,* Vol. 6, 1967.

Robinson 1979. H. W. Robinson, "Nitrile Rubber," pp. 534–546 in *ECT*, Vol. 8, 1979.

Rubin 1968. I. D. Rubin, *Poly(1-Butene): Its Preparation and Properties,* Gordon and Breach, New York, 1968.

St. Cyr 1982. David R. St. Cyr, "Rubber, Natural," pp. 468–491 in *ECT*, Vol. 20, 1982.

Saltman 1965. W. M. Saltman, "Butadiene Polymers," pp. 678–754 in *EPST,* Vol. 2, 1965.

Short 1980. James N. Short, "Processes for Polypropylene Commercialization," *Ind. Res. Dev.* **22,** (9) 109–112 (1980).

Short 1981a. James N. Short, "Polypropylene: Processes, Catalysis, Economics," *Chemtech* **11,** 238–243 (1981).

Short 1981b. James N. Short, "Low Pressure Linear (Low Density) Polyethylene," pp. 385–401 in *ECT*, Vol. 16, 1981.

Snell 1982. T. P. Snell, "HMW-High Density Polyethylene," p. 79 in *MPE*, 1982.

Staudinger 1936. H. Staudinger, "The Formation of High Polymers of Unsaturated Substances," *Trans. Faraday Soc.* **32,** 97–121 (1936).

Tucker 1979. Harold Tucker and S. E. Horne, Jr., "Polyisoprene," pp. 582–592 in *ECT*, Vol. 8, 1979.

West 1978. James C. West and Stuart L. Cooper, "Thermoplastic Elastomers," Chapter 13 in Frederick R. Eirich, ed., *Science and Technology of Rubber,* Academic Press, New York, 1978.

CHAPTER FOURTEEN

OTHER CARBON-CHAIN POLYMERS

In this chapter we extend the discussion of hydrocarbon polymers (Chapter 13) to the consideration of other macromolecules with a carbon-carbon backbone chain. Included here are polymers made from vinyl monomers, CH_2=CHX, and vinylidene monomers, CH_2=CY_2, and the fluorinated polymers. Discussed elsewhere are thermosetting resins made in part from vinyl monomers, such as "polyester" resins (Chapter 16C). It should be noted that in most of this chapter the term *vinyl* is used in the sense of monosubstituted ethylenic, while in the plastics industry (and in Section *D*) the term *vinyl resins* commonly refers to poly(vinyl chloride) and related plastics.

A. POLYSTYRENE AND RELATED POLYMERS

The family of styrene polymers includes polystyrene, copolymers of styrene with other vinyl monomers, polymers of derivatives of styrene, and mixtures of polystyrene and styrene-containing copolymers with elastomers. Total production of these polymers exceeded 3.2 billion lb as early as 1969, and reached about 4.6 billion lb in 1982, down some 20% from a peak in the late 1970s. Prices have about tripled since the early 1970's, however, to a level around $0.45 per lb in 1982. In comparison to the other major high-volume resins, the outlook for polystyrene seems relatively unpromising. The major application field for polystyrene is packaging, accounting for a third of its end markets. Most of the polymer is fabricated by injection molding or extrusion, with the foamed or expandable-bead product accounting for about 10% of the total and growing.

Polystyrene is a thermoplastic with many desirable properties. It is clear, transparent, easily colored, and easily fabricated. It has reasonably good mechanical and thermal properties, but is slightly brittle and softens below 100°C.

Monomer. Styrene (vinylbenzene),

$$\langle\hspace{-0.4em}\bigcirc\hspace{-0.4em}\rangle CH{=}CH_2$$

is made from benzene and ethylene. In one process, ethylene is passed into liquid benzene under pressure, in the presence of aluminum chloride catalyst:

$$\langle\hspace{-0.4em}\bigcirc\hspace{-0.4em}\rangle + CH_2{=}CH_2 \xrightarrow[90°C]{AlCl_3} \langle\hspace{-0.4em}\bigcirc\hspace{-0.4em}\rangle CH_2CH_3$$

The resulting ethylbenzene is dehydrogenated to styrene by passing it over an iron oxide or magnesium oxide catalyst at about 600°C. The styrene is then refined by distillation.

Polymerization. Although solution or emulsion polymerization may occasionally be use, most polystyrene is made either by suspension polymerization or by polymerization in bulk. All but the latter process are typically carried out as described in Chapter 6.

The bulk polymerization of styrene is begun in a "prepolymerizer," a stirred vessel in which inhibitor-free styrene is polymerized (usually with a peroxide initiator) until the reaction mixture is as concentrated in polymer as is consistent with efficient mixing and heat transfer. Normally, a solution containing about 30% polymer is as viscous as can be handled.

The syrupy mixture from the prepolymerizer then enters a cylindrical tower (about 40 ft long by 15 ft diameter) maintained essentially full of fluid. By cooling the upper part of the tower and heating the lower part, the polymerization is controlled to prevent runaways but to proceed to essentially pure molten polymer at the bottom. This melt is discharged through spinnerets or into an extruder producing small-diameter rod that is chopped, after cooling, into short lengths to provide the finished molding powder.

Structure and Properties. Polystyrene is a linear polymer, the commercial product being atactic and therefore amorphous. Isotactic polystyrene can be produced, but offers little advantage in properties except between the glass transition (about 80°C) and its crystalline melting point (about 240°C), where it is much like other crystalline plastics. Isotactic polystyrene is not of commercial interest because of increased brittleness and more difficult processing than the atactic product.

Like most polymers, polystyrene is relatively inert chemically. It is quite resistant to alkalis, halide acids, and oxidizing and reducing agents. It can be nitrated by fuming nitric acid, and sulfonated by concentrated sulfuric acid at 100°C to a water-soluble resin (see ion-exchange resins, below). Chlorine and bromine are substituted on both the ring and the chain at elevated temperatures. Polystyrene degrades at elevated temperatures to a mixture of low-molecular-weight compounds about half of which is styrene. The characteristic odor of the monomer serves as an identification for the polymer.

As made, polystyrene is outstandingly easy to process. Its stability and flow under injection-molding conditions make it an ideal polymer for this technique. Its optical properties—color, clarity, and the like—are excellent, and its high refractive index (1.60) makes it useful for plastic optical components. Polystyrene is a good electrical insulator and has a low dielectric loss factor at moderate frequencies. Its tensile strength reaches about 8000 psi.

On the other hand, polystyrene is readily attacked by a large variety of solvents, including dry-cleaning agents. Its stability to outdoor weathering is poor; it turns yellow and crazes on exposure. Two of its major defects in mechanical properties are its brittleness and its relatively low heat-deflection temperature of 82–88°C, which means that polystyrene articles cannot be sterilized.

Many of these defects can be overcome by proper formulating, or by copolymerization and blending as described below. For example, the addition of ultraviolet-light absorbers improves the light stability of polystyrene enough to make it useful in lighting fixtures such as fluorescent-light diffusers. Flame-retardant polystyrenes have been developed through the use of additives.

Copolymers of Styrene

Butadiene Copolymers. The most important of all copolymers of styrene in terms of volume are the styrene–butadiene synthetic rubbers discussed in Chapter 13*E*. Another group of styrene–butadiene copolymers is widely used in latex paints; these fall in the composition range 60 styrene:40 butadiene by weight. Styrene–butadiene block copolymers are also a major constituent of thermoplastic elastomers (Chapter 17*C*).

Heat- and Impact-Resistant Copolymers. A number of copolymers of styrene with a minor amount of comonomer have enhanced heat and impact resistance without loss of other desirable properties of polystyrene. Typical comonomers are those that increase intermolecular forces of attraction by introducing polar groups, or those that stiffen the chain and reduce rotational freedom through the steric hindrance of bulky side groups. In the first class are acrylonitrile (CH_2=CHCN), fumaronitrile (*trans*-NCCH=CHCN), and 2,5-dichlorostyrene. Comonomers with bulky side groups include *N*-vinylcarbazole and *N,N*-diphenylacrylamide.

The products of major commercial interest in this group contain, typically, 76% styrene and 24% acrylonitrile. Such copolymers have heat-deflection temperatures of 90–92°C and more resilience and impact resistance than polystyrene. Their color, however, is slightly yellow.

Ion-Exchange Resins.   Cationic type ion-exchange resins are produced by making a suspension polymerization of styrene with several per cent divinylbenzene. The product, in the form of uniform spheres, is sulfonated to the extent of about one —SO₃H group per benzene ring. Anionic ion-exchange resins are made by copolymerizing styrene with divinylbenzene and vinylethylbenzene. The polymer is treated with chloromethyl ether to put chloromethyl groups on the benzene rings. Quaternary ammonium salts are formed by reaction with tertiary amines such as trimethylamine, methyldiethanolamine, and dimethylpropanolamine.

Copolymers of styrene with methyl methacrylate are discussed in Section *B*.

Polymers of Styrene Derivatives

Many styrene derivatives have been synthesized, but only a few have found commercial importance. Poly(α-methyl styrene), with better heat resistance than polystyrene, has failed to survive because of brittleness and difficulty in fabrication due to a low ceiling temperature of 61°C. Poly(*p-tert*-butylstyrene) is used as a viscosity improver in motor oils, polychlorostyrene is valuable because of its self-extinguishing characteristics, and poly(sodium styrenesulfonate) is used as a water-soluble flocculating agent. It is predicted that poly(*p*-methyl styrene) may become of commercial interest because of a new monomer synthesis from toluene and ethylene that could result in lower prices; some key properties of the polymer are improved over those of polystyrene.

Rubber-Modified Polystyrene

Rubber is incorporated into polystyrene primarily to impart toughness. The resulting materials consist of a polystyrene matrix with small inclusions of the rubber (usually 5–10% polybutadiene or copolymer rubber). They are termed *impact polystyrene,* and account for over half of the polystyrene homopolymer produced. Grafting of the rubber to the polystyrene may occur if the rubber is present during the styrene polymerization; these materials are the most effective in enhanced impact strength, particularly if the rubber is slightly crosslinked; but the mechanical blends are also used.

ABS Resins

Like the rubber-modified polystyrenes, ABS resins are two-phase systems consisting of inclusions of rubber in a continuous glassy matrix. In this case the matrix is a styrene–acrylonitrile copolymer, and the rubber a styrene–butadiene copolymer, the name ABS deriving from the initials of the three monomers. Again, development of the best properties requires grafting between the glassy and rubbery phases. The ABS resins have higher temperature resistance and better solvent resistance than the high-impact polystyrenes and are true engineering plastics, particularly suitable for high-abuse applications. They can easily be decorated by painting, vacuum metalizing, and electroplating. In addition to fabrication by all the usual plastics techniques (Chapter 17), the ABS resins can be cold formed, a technique typical of metal fabrication.

About 700 million lb of ABS resins was sold in 1982.

Polystyrene Foams

Around 760 million lb of styrene homopolymer was used in 1982 to make a wide variety of foamed products. Most of these are based on foamed-in-place beads, made by suspension polymerization in the presence of a foaming agent such as pentane or hexane, liquid at polymerization temperature and pressure. Subsequent heating softens the resin and volatilizes the foaming agent.

GENERAL REFERENCES

Basdekis 1964; Abrams 1967; Boyer 1970; Platzer 1977; Radosta 1977; Morneau 1978; Evans 1982; Lantz 1982; Peppin 1982; Swett 1982; Platt 1983.

B. ACRYLIC POLYMERS

In this section are discussed the acrylate and methacrylate plastics, of which the most important is poly(methyl methacrylate), and the acrylic fibers, based largely on polyacrylonitrile.

Poly(methyl methacrylate)

Poly(methyl methacrylate) is a clear, colorless transparent plastic with a higher softening point, better impact strength, and better weatherability than polystyrene. It is available in molding and extrusion compositions, syrups, and cast sheets, rods, and tubes.

Monomers. The principal commercial processes for the production of acrylate esters are based on ethylene cyanohydrin (much like the acetone cyanohydrin route to methyl methacrylate described below), the carbonylation of acetylene, or the polymerization and subsequent depolymerization of β-propiolactone to give acrylic acid or (in the presence of an alcohol) an acrylate ester.

Methyl methacrylate is made by heating acetone cyanohydrin (from the addition of hydrocyanic acid to acetone) with sulfuric acid to form methacrylamide sulfate.

The latter is reacted (without separation) with water and methanol to give methyl methacrylate:

$$CH_3-\underset{\underset{CH_3}{|}}{\overset{\overset{OH}{|}}{C}}-CN \xrightarrow[125°C]{H_2SO_4} CH_2=\underset{\underset{O}{\|}}{C}CNH_2\cdot H_2SO_4 \xrightarrow[H_2O]{CH_3OH} CH_2=\underset{\underset{O}{\|}}{C}COCH_3$$

Polymerization. Poly(methyl methacrylate) for molding or extrusion is made by bulk or suspension polymerization, as described in Chapter 6. The production of cast sheets, rods, and tubes is carried out by bulk polymerization, starting in most cases with a syrup of partially polymerized methyl methacrylate with a convenient viscosity for handling. Shrinkage and heat evolution during polymerization are reduced by the use of a syrup.

Sheets of poly(methyl methacrylate) are commonly made by extrusion. Alternatively, they may be cast in cells made up two sheets of heat-resistant glass separated by a coated rubber gasket. The cell is filled with syrup and sealed, and polymerization is carried out at 60–70°C in an air oven or water bath, with a finishing treatment at 100°C. Peroxide or azo initiators may be used.

Properties. Poly(methyl methacrylate) is a linear thermoplastic, about 70–75% syndiotactic. Because of its lack of complete stereoregularity and its bulky side groups, it is amorphous. Both isotactic and syndiotactic poly(methyl methacrylate) have been prepared but have not been offered commercially. Poly(methyl methacrylate) is resistant to many aqueous inorganic reagents, including dilute alkalis and acids. It is quite resistant to alkaline saponification, in contrast to the polyacrylates. It undergoes pyrolysis almost completely to monomer by a chain reaction.

Perhaps the outstanding property of poly(methyl methacrylate) is its optical clarity and lack of color. Coupled with its unusually good outdoor weathering behavior, its optical properties make it highly useful in all applications where light transmission is important.

The mechanical and thermal properties of the polymer are good also. Tensile strength ranges as high as 10,000 psi. Impact strength is about equal to that of the impact-resistant styrene copolymers. Heat-deflection temperatures are above 90°C for the heat-resistant grades of poly(methyl methacrylate) molding powder. Electrical properties are good but not outstanding. Fabricability is quite good; only slightly higher temperatures are needed for molding poly(methyl methacrylate) than for polystyrene. Poly(methyl methacrylate) is less susceptible to crazing than is polystyrene.

A limitation to the optical uses of the material is its poor abrasion resistance compared to glass. Despite considerable effort, attempts to improve the scratch resistance or surface hardness of poly(methyl methacrylate) have so far been accompanied by deterioration in other properties, such as impact strength.

Applications. Injection-molded acrylic articles include automotive lenses, reflective devices, instrument and appliance covers, optical equipment, and home furnishings. Acrylic sheeting is used for signs, glazing (in particular, aircraft windows), furniture, partitions, and lighting-fixture diffusers. About 0.5 billion lb of acrylic plastics was sold in 1982.

Copolymers of methyl methacrylate, ethyl acrylate, and monomers containing reactive functional groups are widely used as thermosetting resins in baked enamel applications. The functionality can be derived from amides (acrylamide, methacrylamide), acids (acrylic acid, methacrylic acid, or others), hydroxyls (hydroxyalkyl acrylates or methacrylates), or oxiranes (glycidyl methacrylate).

Other Acrylic Plastics

Higher Methacrylates. The monomers of higher alkyl methacrylates are most conveniently prepared from methyl methacrylate by alcoholysis. The reaction is carried out at 150°C with an excess of the alcohol whose methacrylate ester is desired and a small amount of sulfuric acid catalyst. As described in Chapter 12*B*, the higher alkyl methacrylate polymers have glass transition temperatures lower than that of poly(methyl methacrylate). Poly(lauryl methacrylate) is widely used as a pour-point depressant and improver of viscosity–temperature characteristics for lubricating oils.

Polyacrylates. The rubbery, adhesive nature of the lower acrylates makes both suspension polymerization and casting less feasible than with the methacrylates. Consequently, solution and emulsion polymerizations are used commercially. Since the polyacrylates contain an easily removed tertiary hydrogen atom, they undergo some chain transfer to polymer when polymerized to high conversion. This leads to highly branched, less soluble materials.

Since the lower acrylate polymers have glass transition points below room temperature, they are typically soft and rubbery. As the size of the ester group decreases, the polymers become harder, tougher, and more rigid. The polyacrylates have been used in finishes, in textile sizes, and in the manufacture of pressure-sensitive adhesives.

Copolymers of ethyl acrylate with methyl methacrylate (to provide hardness and strength) and small amounts of hydroxyl, carboxyl, amine, or amide comonomers (to provide adhesion and thermosetting capability) are used to produce high-quality latex paints for wood, wallboard, and masonry in homes. Other copolymers are used for automotive and appliance coatings and a variety of specialty coatings, finishes, polishes, and adhesives.

Copolymers of ethyl acrylate with a few per cent of a chlorine-containing monomer such as 2-chloroethyl vinyl ether have elastomeric properties. Vulcanization reactions apparently involve the chlorine atom, ester group, and α-hydrogens on the chain. Numerous rubber vulcanization agents and accelerators vulcanize these polymers. They are of interest because of their heat resistance and their excellent resistance to oxidation, allowing their use to over 180°C.

Poly(acrylic acid) and Poly(methacrylic acid). Acrylic acid, CH_2=CHCOOH, can be prepared directly from ethylene cyanohydrin. Methacrylic acid can be prepared from acetone cyanohydrin. Salts of the acids can be polymerized directly, then acidified to give poly(acrylic acid) and poly(methacrylic acid). Alternatively, poly(methyl acrylate) or poly(methyl methacrylate) may be saponified to form the acids. Or polymethacrylonitrile may be treated with hydrogen chloride to give poly(methacrylic acid). The polymers are insoluble in their monomers and in most organic liquids, but are soluble in water and very soluble in dilute bases. Poly(acrylic acid) and poly(methacrylic acid) are too water sensitive to serve as plastics. They are brittle when dry, and on heating do not become thermoplastic but crosslink, char, and decompose. In solution they show typical polyelectrolyte behavior, including abnormally high viscosities. Because of this property they are useful as thickening agents for latices and for adhesives.

Cyanoacrylate Adhesives.   Methyl cyanoacrylate monomer is an extremely powerful adhesive. Adhesion occurs when the liquid monomer is spread in a thin layer between the surfaces to be bonded. Traces of bases (even as weak as alcohols or water) on the surfaces catalyze polymerization by an anionic mechanism. Adhesion arises in part from mechanical interlocking between polymer and surface, and in part from strong secondary bond forces.

Polyacrylamide.   This water-soluble polymer is used as a thickening agent and as a flocculant.

Acrylic Fibers

The *acrylic* fibers are polymers containing at least 85% acrylonitrile. Other monomers are often used in small amount to make the polymer amenable to dyeing with conventional textile dyes. Common comonomers are vinyl acetate, acrylic esters, and vinyl pyrrolidone. The last may be used as the homopolymer in a blend or grafted onto a copolymer backbone. A generic class of fibers closely related to the acrylics is the *modacrylic* fibers, containing 35–85% acrylonitrile and, usually, 20% or more vinyl chloride or vinylidene chloride.

Polyacrylonitrile.   Acrylonitrile can be made either by the direct catalytic addition of HCN to acetylene, or by the addition of HCN to ethylene oxide to give ethylene cyanohydrin, followed by dehydration. The monomer is soluble in water to the extent of about 7.5% at room temperature, and polymerization is usually carried out in aqueous solution by means of redox initiation. The polymer precipitates from the system as a fine powder.

During the time that other vinyl polymers were being developed commercially as plastics as well as fibers, polyacrylonitrile was considered a useless material because it could not be dissolved or plasticized. It softens only slightly below its decomposition temperature, and, because the polymer is insoluble in the monomer, it could not be polymerized into useful shapes by casting. This intractability was for a time attributed to crosslinking, but better knowledge of the requirements for solubility of polymers led to an extensive search (Houtz 1950) for molecules that might interact with the highly polar —C≡N groups and cause solution of the polymer. Several such solvents were discovered, among them dimethyl formamide and tetramethylene sulfone. Some concentrated aqueous solutions of salts, such as calcium thiocyanate, also dissolve polyacrylonitrile. With these solvents available, both wet- and dry-spinning techniques (Chapter 18) have been used for this polymer.

Properties of Acrylic Fibers.   The acrylic fibers exhibit the properties of high strength, stiffness, toughness, abrasion resistance, resilience, and flex life associated with the synthetic fibers as a class. They are relatively insensitive to moisture and have good resistance to stains, chemicals, insects, and fungi. Their weatherability is outstandingly good. In continuous filament form, they are considered to have superior "feel," while as crimped staple they are noted for bulkiness and a wool-like "hand."

In many respects the acrylic fibers are similar to the chlorine-containing fibers described in Section *D*, and to the condensation fibers discussed in Chapter 15. Further information on fibers and fiber technology is found in Chapters 15 and 18.

GENERAL REFERENCES

Bamford 1964; Davis 1964; Fram 1964; Luskin 1964; Miller 1964; Thomas 1964; Rebenfeld 1967; Beevers 1968; Kennedy 1968; Moncrieff 1975, Chapters 28, 32–37; Elias 1977, Chapter 25.8; Brydson 1978, Chapter 14; Cooper 1978; Hobson 1978; Kine 1978, 1981; Peng 1978; Morris 1978; Vial 1979; Krenz 1982.

C. POLY(VINYL ESTERS) AND DERIVED POLYMERS

The most widely used polymer of a vinyl ester is poly(vinyl acetate). It is utilized not only as a plastic, primarily in the form of emulsions, but also as the precursor for two polymers which cannot be prepared by direct polymerization, poly(vinyl alcohol) and the poly(vinyl acetals). None of these polymers is useful for molding or extrusion, but each is important in certain special applications.

Poly(vinyl acetate)

Monomer.   Vinyl acetate,

$$CH_2{=}CH{-}O{-}\overset{\displaystyle O}{\overset{\displaystyle \|}{C}}{-}CH_3$$

is prepared by the vapor-phase addition of acetic acid to acetylene:

$$CH{\equiv}CH + CH_3COOH \longrightarrow CH_3COOCH{=}CH_2$$

Almost 2 billion lb/yr was produced in 1979–1982, virtually all being used as polymers.

Polymerization. Bulk polymerization of vinyl acetate is difficult to control at high conversions, and in addition the properties of the polymer may deteriorate because of chain branching. Bulk or solution polymerization is usually stopped at low to medium conversion (20–50%), after which the monomer is distilled off. If a solvent (such as methanol, ethanol, ethyl acetate, or benzene) is present, the polymer may be precipitated out or further reacted in solution to one of the derived polymers. The polymerization may be done batchwise or continuously.

Poly(vinyl acetate) may also be made in suspension or emulsion systems. In suspension polymerizations, various additives must be used to coat the beads of polymer to prevent their sticking together during the drying operation. Most of the poly(vinyl acetate) emulsions commercially available are made by emulsion polymerization mechanisms (Chapter 6B).

Properties and Uses. This polymer is atactic and hence amorphous; stereoregular polymers have not been offered commercially. The glass transition temperature of poly(vinyl acetate) is only slightly above room temperature (29°C). As a result the polymer, while tough and form stable at room temperature, becomes sticky and undergoes severe cold flow at only slightly elevated temperatures. Lower-molecular-weight polymers are brittle but become gumlike when masticated and in fact are used in chewing gums. Poly(vinyl acetate) is water sensitive in certain physical properties, such as strength and adhesion, but does not hydrolyze in neutral systems.

Aside from the production of poly(vinyl alcohol), one major use of poly(vinyl acetate) is in the production of water-based emulsion paints. The low cost, stability, quick drying, and quick recoatability of emulsion paints have led to their wide acceptance. Poly(vinyl acetate) is often copolymerized with dibutyl fumarate, vinyl stearate, 2-ethylhexyl acrylate, or ethyl acrylate, or is plasticized, to obtain softer compositions for emulsion use.

Poly(vinyl acetate) is also widely used in adhesives, both of the emulsion type and of the hot-melt type.

Poly(vinyl alcohol)

Since vinyl alcohol is unstable with respect to isomerization to acetaldehyde, its polymer must be prepared by indirect methods.

Preparation. Poly(vinyl alcohol) is prepared by the alcoholysis of poly(vinyl acetate) (the less accurate terms *hydrolysis* and *saponification* are also used):

$$-(CH_2CH-)_x + xCH_3OH \longrightarrow -(CH_2CH-)_x + xCH_3COCH_3$$

with pendant groups: on the left monomer, O, $C=O$, CH_3; on the right monomer, OH; and on the CH_3COCH_3 product, O (carbonyl).

Ethanol or methanol can be used to effect the alcoholysis, with either acid or base as catalyst. Alkaline hydrolysis is much more rapid. The acid hydrolysis is more likely to give some ether linkages in the chain by a mechanism involving the loss of a molecule of water from two adjacent hydroxyl groups. This is an undesirable side reaction. The alcoholysis is usually carried out by dissolving the poly(vinyl acetate) in the alcohol, adding the catalyst, and heating. The poly(vinyl alcohol) precipitates from the solution.

Structure and Properties.   Although poly(vinyl alcohol) is amorphous when un-stretched, it can be drawn into a crystalline fiber, the hydroxyl groups being small enough to fit into a crystal lattice despite the atactic chain structure. Poly(vinyl alcohol) does not melt to a thermoplastic, but decomposes by loss of water from two adjacent hydroxyl groups at temperatures above 150°C. Double bonds are left in the chain, and, as more of these are formed in conjugate positions, severe discoloration takes place.

Poly(vinyl alcohol) is water soluble. It dissolves slowly in cold water, but more rapidly at elevated temperatures, and can usually be dissolved completely above 90°C. The aqueous solutions are not particularly stable, especially if traces of acid or base are present. The solutions can undergo a complex series of reversible and irreversible gelation reactions. For example, crosslinking can occur at ether linkages, resulting in increased viscosity through the formation of insoluble products.

Applications.   The major use of poly(vinyl alcohol) is as a textile size, particularly for cotton–polyester blends. Other used fall into two categories. In one type of application, use is made of the water solubility of the polymer. It serves as a thickening agent for various emulsion and suspension systems, and as a packaging film where water solubility is desired. A further outlet is wet-strength adhesives.

In the second type of end use, the final form of the polymer is insoluble in water as a result of chemical treatment. The use of poly(vinyl alcohol) as a textile fiber (*vinal* fiber) is the major example. The polymer is wet spun from warm water into a concentrated aqueous solution of sodium sulfate containing sulfuric acid and formaldehyde. The polymer is insolubilized by the formation of formal groups:

$$-CH_2-\underset{\underset{OH}{|}}{CH}-CH_2-\underset{\underset{OH}{|}}{CH}- \ + \ HCHO$$

$$\downarrow$$

$$-CH_2-\underset{\underset{O}{|}}{CH}-CH_2-\underset{\underset{O}{|}}{CH}- \ + \ H_2O$$
$$\underset{H_2}{C}$$

About one-third of the hydroxyl groups is reacted to insolubilize the fiber. Some interchain acetalization is desirable to reduce shrinkage of the fiber, but the amount must be carefully controlled.

Poly(vinyl alcohol) fibers have higher water absorption (about 30%) than other fibers. They can thus replace cotton in uses where the fiber is in contact with the body. The hand of the fabric can be varied from wool-like to linenlike. The fiber washes easily, dries quickly, and has good dimensional stability. Tenacity and abrasion resistance are good.

Poly(vinyl acetals)

An important use of poly(vinyl alcohol) is as the starting material for poly(vinyl acetals). By far the most important of these polymers is poly(vinyl butyral), used as the plastic interlayer for automotive and aircraft safety glazings. Lower poly(vinyl acetals), especially poly(vinyl formal), are utilized in enamels for coating electrical wire and in self-sealing gasoline tanks.

Poly(vinyl butyral). This polymer is made by condensing poly(vinyl alcohol) with butyraldehyde in the presence of an acid catalyst, usually sulfuric acid:

$$-(CH_2-CH-CH_2-CH-)_x + xC_3H_7CHO$$
$$\underset{OH}{|} \qquad \underset{OH}{|}$$

$$\downarrow$$

$$-(CH_2-CH-CH_2-CH-)_x + xH_2O$$
$$\underset{O}{\diagdown} \quad \underset{O}{\diagup}$$
$$\underset{\underset{\underset{C_3H_7}{|}}{C}}{\diagdown H \diagup}$$

The reaction can be carried out by starting with an aqueous solution of poly(vinyl alcohol) and adding the butyraldehyde and catalyst. The poly(vinyl butyral) precipitates as it is formed. Alternatively, one may start with a suspension of poly(vinyl alcohol) in a water–alcohol mixture, the poly(vinyl butyral) dissolving as it is formed. Poly(vinyl butyral) produced for safety-glass use requires some hydroxyl groups in order to have adequate strength and adhesion to glass. Consequently the condensation reaction is stopped when about one-quarter of the hydroxyl groups is left.

Safety Glass. The earliest safety-glass interlayer was cellulose nitrate. Its poor weathering qualities led to its general replacement in 1935 by cellulose acetate, and in 1940 by poly(vinyl butyral). Some of the advantages of the latter resin as an interlayer are its superior adhesion to glass, toughness, stability on exposure to sunlight, clarity, and insensitivity to moisture.

Poly(vinyl butyral) must be plasticized for safety-glass use. The usual plasticizers are high-molecular-weight esters such as dibutyl sebacate and triethyleneglycol di-2-ethylbutyrate. About 40–45 parts plasticizer per 100 parts resin by weight is used. The plasticizer may be incorporated in the resin in several ways. One common method is to mix the resin, plasticizer, and a solvent such as ethanol to form a plastic mass. This is extruded in sheet form into a saltwater bath where the solvent is leached out, leaving the plasticized sheet. Since the plasticized resin sticks to itself, it must be powdered with talc or sodium bicarbonate before being rolled up for storage.

Safety-glass laminates are made by washing and drying the poly(vinyl butyral) sheet, and placing it between two pieces of glass. These are subjected to mild heat and pressure to seal the assembly, and are then autoclaved at higher temperature and pressure to complete the lamination process. The usual automotive safety glass contains one plastic layer about 0.030 in. thick.

GENERAL REFERENCES

Pritchard 1970; Lindemann 1971a,b; Elias 1977, Chapter 25.5.

D. CHLORINE-CONTAINING POLYMERS

Polymers and copolymers of vinyl chloride and vinylidene chloride are widely used as plastics and, to a much smaller extent, fibers. In the trade, these polymers are known as "the vinyls" or "vinyl resins." Once the largest group of thermoplastic materials, the vinyl resins have long since been surpassed in volume by the olefin polymers. Production of the vinyls in 1982 was about 5.35 billion lb, down from a 1979 peak of 6.1 billion lb. Prices were in the range $0.32–$0.50 per lb.

Poly(vinyl chloride)

Monomer. Vinyl chloride is a gas boiling at $-14°C$. It is produced by the dehydrochlorination of ethylene dichloride, which is made by reacting ethylene with chlorine. Almost all the vinyl chloride produced is used in polymerization, with great care being taken throughout because of the carcinogenic nature of the monomer.

Polymerization. Suspension polymerization (Chapter 6B) is used for the production of well over 80% of poly(vinyl chloride). Small amounts are made by solution, emulsion, and bulk polymerization, despite difficulties in the latter process resulting from the insolubility of the polymer in its monomer.

Structure. Poly(vinyl chloride) is a partially syndiotactic material, with sufficient irregularity of structure that crystallinity is quite low. Its structural characterization is complicated by the possibility of chain branching and the tendency of the polymer to associate in solution.

Stability. Poly(vinyl chloride) is relatively unstable to heat and light. Thermal initiation involves loss of a chlorine atom adjacent to some structural abnormality which reduces the stability of the C—Cl bond, such as terminal unsaturation (Grassie 1964). The chlorine radical so formed abstracts a hydrogen to form HCl; the resulting chain radical then reacts to form chain unsaturation with regeneration of a chlorine radical. The reaction can also be initiated by ultraviolet light which is absorbed at unsaturated structures with liberation of an adjacent chlorine atom. In the presence of oxygen, both chain reactions are accelerated, and ketonic structures are formed in the chain.

Stabilizers are almost invariably added to improve the heat and light stability of the polymer. Metallic salts of lead, barium, tin, or cadmium are used. Oxides, hydroxides, or fatty-acid salts are most effective. Epoxy plasticizers (Chapter 16*D*) aid materially in stabilizing the resin. Free radical acceptance appears to be a prominent mechanism of stabilization. HCl acceptors have been used, but the degradation reactions are not autocatalytic as had previously been supposed.

Vinyl Resins

Rigid Compounds. The term *rigid vinyls* usually refers to unplasticized poly(vinyl chloride), or compositions with only a few percent of a plasticizer such as an epoxy resin (Chapter 16*D*). Often other polymers are mixed physically with the poly(vinyl chloride) to improve impact resistance (nitrile rubber, chlorinated polyethylene, ABS resins, or methyl methacrylate–butadiene–styrene terpolymer) or processability (styrene–acrylonitrile copolymer or poly(methyl methacrylate)).

Copolymers. The advantages in polymer properties resulting from the copolymerization of small amounts of vinyl acetate with vinyl chloride were discovered around 1928. The lower softening point and higher solubility of the copolymers make fabrication very much easier. Stability is improved, if anything, over that of the homopolymer. Color and clarity are also better. Polymerization methods are similar to those for vinyl chloride homopolymer except that emulsion polymerization has had less success. Commercial compositions containing 5–40% vinyl acetate are available.

Polymers containing around 10% vinylidene chloride have better tensile properties than pure poly(vinyl chloride). Copolymers containing 10–20% diethyl fumarate or diethyl maleate have improved workability and toughness and retain the high softening point of poly(vinyl chloride). Acrylic esters have also been used to impart improvements in solubility and workability.

Plasticization. Many properties of poly(vinyl chloride) and vinyl chloride–vinyl acetate copolymers are improved by plasticization (Chapter 17). The large majority of the commercial production of vinyl resins is in the form of plasticized compositions. The first important plasticizer for the vinyls was tricresyl phosphate, which has since been replaced by other esters because of its tendency to cause low-temperature brittleness in plasticized compounds. Dibutyl phthalate, dibutyl sebacate, and tributyl phosphate have also been used and in turn replaced. Dioctyl phthalate, trioctyl phosphate, dioctyl sebacate and adipate, and various low-molecular-weight polymers such as poly(propylene glycol) esters are now widely utilized as plasticizers for the vinyls. The plasticizers are usually added to the polymers on hot rolls or in a hot mixer such as a Banbury. The plasticizer content varies widely with the end use of the material, but typically may be around 30% by weight.

Plastisols and Organisols.   These liquid compositions are produced by spray-drying latices obtained from the emulsion polymerization of vinyl chloride. The latex particles are dispersed into plasticizers to make plastisols, or into mixtures of plasticizers and volatile organic liquids to make organisols. Other ingredients, such as stabilizers, fillers, colorants, surfactants, and possibly blowing agents or gelling agents, are also present. The polymer particles do not dissolve in the liquids, but remain dispersed until the mixture is heated. Fusion (plus loss of solvent from the organisols) then yields the final plastic object.

Applications.   About 55% of poly(vinyl chloride) is used as rigid resins, the remainder plasticized. The single largest use, in the rigid field, is pipe, accounting for 40% of production. It is used for water supply and distribution, agricultural irrigation, chemical processing, drain, waste, and vent pipe, sewer systems, and conduits for electrical and telephone cables.

The building construction market accounts for about another 30% of poly(vinyl chloride) production, including siding, window frames, gutters, and interior molding and trim (12%); flooring (5%); wire and cable insulation (7%); and wall coverings, upholstery, shower curtains, gaskets, and so on (5%). Automotive uses, meat and food packaging, bottles, footwear, outerwear, phonograph records, sporting goods, and toys are other significant markets. The major plastics application for poly(vinylidene chloride) is saran film.

Chlorine-Containing Fibers

Compositions.   The use of poly(vinyl chloride) as a fiber was patented as early as 1913, and the material was commercialized in 1931. It was soon withdrawn in favor of modified polymers with better solubility, including chlorinated poly(vinyl chloride) and copolymers of vinyl chloride with vinylidene chloride and with vinyl acetate. Currently, the materials of most interest in the class of chlorine-containing fibers are copolymers of about 60% vinyl chloride and 40% acrylonitrile, and of about 85% vinyl chloride and 15% vinyl acetate. Other comonomers are occasionally used. The generic classes of fibers in this group include *vinyon,* containing more than 85% vinyl chloride; *saran,* containing more than 80% vinylidene chloride; and *modacrylic,* defined in Section *B.*

Properties and Uses.   The chlorine-containing fibers are usually moderately crystalline and are drawn (typically while hot) 5- to 15-fold to increase strength and reduce extensibility. Like the acrylic (Section *B*) and condensation (Chapter 15) fibers, they are insensitive to moisture and have high wrinkle resistance and good resistance to chemicals, insects, fungi, and the like. They have poorer dimensional stability at high temperatures, and somewhat less strength, elasticity, and abrasion resistance than the condensation fibers.

GENERAL REFERENCES

Kaufman 1969; Koleski 1969; Sarvetnick 1969; Brighton 1971; Wessling 1971; Moncrieff 1975, Chapters 27, 29, and 30; Nass 1976–77; Burgess 1982; Brown 1982; Jeziorski 1982.

E.   FLUORINE-CONTAINING POLYMERS

Fluorine-containing polymers represent in many respects the extremes in polymer properties. Within this family are found materials of high thermal stability and concurrent usefulness at high temperatures (in some cases combined with high crystalline melting points and high melt viscosity), and extreme toughness and flexibility at very low temperatures. Many of the polymers are almost totally insoluble and chemically inert, some have extremely low dielectric loss and high dielectric strength, and most have unique nonadhesive and low friction properties.

In this section we distinguish between fully fluorinated *fluorocarbon plastics*, including only polytetrafluoroethylene and its fully fluorinated copolymers, and *fluoroplastics*, containing hydrogen or chlorine in addition to fluorine on the carbon–carbon backbone.

Polytetrafluoroethylene

Monomer.   Tetrafluoroethylene is a nontoxic gas boiling at $-76°C$. Its modern synthesis starts with the fluorination of chloroform by HF in a two-step process yielding $CHClF_2$. This is then dimerized by pyrolysis with loss of HCl to provide tetrafluoroethylene.

Polymerization.   The first polymer of tetrafluoroethylene was discovered by R. J. Plunkett (Garrett 1962) when an apparently empty cylinder of the gas was cut open to see why more material had not been obtained from it. The cylinder was partly filled with a waxy white powder shown to be the polymer.

Tetrafluoroethylene is usually polymerized with free radical initiators at elevated pressure in the presence of water. Redox initiation may be used; persulfates and hydrogen peroxide have been employed as initiators.

Structure (Sperati 1962). Polytetrafluoroethylene is a highly crystalline, orientable polymer. These facts indicate a regular structure, which implies the absence of any considerable amount of crosslinking. Branching is presumed to be absent, since branching mechanisms would involve breaking C—F bonds, which are estimated to be too strong to be ruptured easily. The chances are, therefore, that polytetrafluoroethylene consists of linear —CF_2—CF_2— chains.

The crystal structure and crystalline-phase transitions in polytetrafluoroethylene are discussed in Chapter 10*B*. These transitions, occurring near room temperature, involve a 1.3% volume change having an important effect on the mechanical properties of the polymer for some applications. The degree of crystallinity of the polymer as formed from the monomer is generally quite high, 93–98%. The crystalline melting point is 327°C.

The crystalline density of polytetrafluoroethylene is 2.30 g/cm³. Upon melting the polymer and subsequently cooling it, lower densities are obtained. Subsequent thermal treatment (annealing) generally increases the crystallinity of the solid. A standardized molding and annealing technique under conditions carefully controlled to prevent the formation of voids in the sample gives "standard" specific grav for various samples ranging between 2.15 and 2.28. This range of densities has been correlated with the molecular weight and thus the melt viscosity of the samples. Higher-molecular-weight polymers have more viscous melts and hence crystallize more slowly and reach lower crystallinities under standardized thermal treatment.

Despite the reputation of polytetrafluoroethylene as a totally insoluble polymer, solvents for it have been found at temperatures not far below its crystalline melting point. Solution viscosities have been measured in perfluorinated kerosene fractions at 300°C, the measurements giving clues as to the molecular weight of the polymer. Other information as to molecular weight is derived from the kinetics of polymerization of tetrafluoroethylene using radioactive initiators, and from the flow properties of the polymers (see the next paragraph). All the evidence points to unusually high number-average molecular weights of many millions for this polymer.

Polytetrafluoroethylene does not flow easily above its crystalline melting point. The viscosity of the polymer is very high because of restricted rotation about the chain bonds and high molecular weight. Upon the continued application of stress to the amorphous polymer above the crystalline melting point, the strength of the polymer is exceeded, causing fracture before the stresses necessary to induce rapid flow are reached. This unusual behavior necessitates unconventional fabrication techniques, as discussed below.

Polytetrafluoroethylene decomposes at elevated temperatures. In vacuum, monomer is the chief product. The vapor pressure of monomer in equilibrium with the polymer at 500°C is 1 mm Hg. At low temperatures (250–350°C) degradation appears to start at chain ends; at higher temperatures random cleavage becomes more important. In the presence of air the degradation is more complicated.

Properties. Polytetrafluoroethylene is extremely resistant to attack by corrosive reagents or solvents. Of many hundreds of reagents tested up to their boiling points, only alkali metals, either molten or dissolved in liquid ammonia, attack the polymer, presumably by removing fluorine atoms from the chain. Fluorine itself degrades the polymer on prolonged contact under pressure. For all practical purposes the polymer is completely unaffected by water. Its thermal stability is such that its electrical and mechanical properties do not change for long intervals (months) at temperatures as high as 250°C.

Molded polytetrafluoroethylene articles have high impact strength but are easily strained beyond the point of elastic recovery. In a tensile test molded articles begin to cold draw at 1500–2000 psi, elongating (with orientation) to 300–450% before breaking at a load of 2500–4500 psi. Highly oriented fibers have tensile strengths as high as 50,000 psi. Polytetrafluoroethylene is subject to cold flow under stress but exhibits some delayed elastic recovery. The polymer is not hard, but is slippery and waxy to the touch, and has a very low coefficient of friction on most substances. Its density is unusually high (2.1–2.3) and its refractive index unusually low (1.375).

Polytetrafluoroethylene has extremely good electrical properties. Its dielectric constant is low (2.0), and its loss factor for all frequencies tested, including those used for television and radar, is one of the lowest known for solids. The properties of polytetrafluoroethylene can be varied widely as a function of molecular structure and fabrication, as described in Chapter 12.

Fabrication.   Since polytetrafluoroethylene is almost completely insoluble and has impracticably low melt flow rates, most of the fabrication techniques ordinarily used with polymers (Chapter 17) are not suitable for this material. Several unusual techniques have been developed for putting polytetrafluoroethylene into usable shapes. Most of these techniques are variants of two important processing steps: (a) pressing the polymer cold into the desired shape, followed by (b) sintering at a temperature above the crystalline melting point (say, 380°C) to yield a dense, strong, homogeneous piece.

Techniques for molding polytetrafluoroethylene resemble those of powder metallurgy or ceramics processing more than those of polymer fabrication. Granular polytetrafluoroethylene powder is compressed at room temperature and 2000–10,000 psi to give a preform. Sintering of the preform follows.

Polytetrafluoroethylene may be extruded at slow rates in a ram or screw extruder. In these devices the ram or screw serves merely to compact the molding powder and feed it to the die, within which the material is heated above the crystalline melting point, shaped, and allowed to solidify.

The high degree of cohesion between cold pressed particles of polytetrafluoroethylene is utilized in a calendering process for making tape and coating wire. Molding powder is fed to calender rolls, which compress the particles into a solid structure. This passes into a sintering bath or oven.

Polytetrafluoroethylene is available in the form of aqueous dispersions of ultimate particles about 0.2 μm in diameter. They can be used for casting films or for dip coating or impregnating porous structures.

Mechanical properties of polytetrafluoroethylene, including resistance to wear and to deformation under load, stiffness, and compressive strength, can be enhanced by the use of fillers. Most of the desirable properties of the polymer (heat resistance, low friction, weatherability, etc.) are retained. Short glass fibers, graphite, carbon, and bronze are preferred fillers.

Polytetrafluoroethylene can also be provided in a microfibrous form, which can be spun from aqueous dispersions to provide fibers that can be sintered at around 385°C to develop low but useful strength.

Applications. The uses of polytetrafluoroethylene are largely those requiring excellent toughness, electrical properties, heat resistance, low frictional coefficient, or a combination of these. Among the electrical applications of the polymer are wire and cable insulation, insulation for motors, generators, transformers, coils, and capacitors, and high-frequency electronic uses. Chemical equipment such as gaskets, pump and valve packings, and pump and valve parts is made from the polymer. Low-friction and antistick applications include nonlubricated bearings, linings for trays in bakeries and other food-processing equipment, mold-release devices, and covers for heat-sealer plates of packaging machines. Low-molecular-weight polymers of tetrafluoroethylene dispersed as aerosols are effective dry lubricants. Uses for the fiber include gasketing, belting, pump and valve packing, filter cloths, and other industrial functions where essentially complete chemical and solvent resistance, combined with heat resistance to above 250°C, is required.

Fluorocarbon Copolymers

A copolymer of hexafluoropropylene and tetrafluoroethylene has a crystalline melting point near 265°C. It retains most of the properties of polytetrafluoroethylene, but has melt viscosity in the range allowing fabrication by conventional techniques. This copolymer is tough at liquid-air temperatures, yet retains adequate mechanical strength for continuous service at temperatures up to 200°C. Like polytetrafluoroethylene, it is chemically inert and has zero water absorption. It maintains a low dielectric constant and loss factor over the unusually wide range of 60 cycles to 60 megacycles. It has excellent weatherability, nonstick and low friction properties, and very low permeability to gases. The material can be processed by conventional extrusion, injection molding, compression, transfer, or blow molding, fluidized bed coating, or vacuum forming of sheets or film. It finds application as wire jacketing, as extruded film, rods, tubes, or complex shapes, and as an encapsulating resin. Important end uses, many of which are similar to those of polytetrafluoroethylene, include film for inner glazing in solar collectors.

Copolymers of perfluoroalkoxy monomers,

$$CF_2=CF \\ | \\ OR_f$$

with tetrafluoroethylene are very close in properties to polytetrafluoroethylene, with $T_m = 310°C$, but are melt processable by conventional molding and extrusion techniques. All the outstanding properties of the homopolymers are retained, including a recommended upper use temperature of 260°C; for properties related to deformation, the copolymer is superior.

Polychlorotrifluoroethylene

Monomer. Chlorotrifluoroethylene is made by dechlorination of trichlorotrifluoroethane. The monomer is less subject to spontaneous explosive polymerization than is tetrafluoroethylene. Unlike tetrafluoroethylene, however, chlorotrifluoroethylene is itself toxic.

Polymerization. As in the case of tetrafluoroethylene, the polymerization of chlorotrifluoroethylene is best carried out in an aqueous system using a redox initiator. The details of commercial polymerizations have not been disclosed.

Structure and Properties. Polychlorotrifluoroethylene is surpassed only by polytetrafluoroethylene and tetrafluoroethylene–hexafluoropropylene copolymers in chemical inertness and resistance to elevated temperatures. Differences in the properties of the polymers are a consequence of the lower symmetry of the chlorine-containing polymer. For example, the crystalline melting point of polychlorotrifluoroethylene is 218°C, and the polymer can be quenched to quite clear sheets in which crystallinity is absent. Slower cooling provides opportunity for growth of crystallites and spherulites and results in quite different optical and mechanical properties.

Polychlorotrifluoroethylene is soluble in a number of solvents above 100°C, and is swelled by several solvents at room temperature. The polymer is tough at −100°C and retains useful properties as high as 150°C. Although high compared to that of many other polymers, its melt viscosity is low enough that the usual fabrication techniques, such as molding and extrusion, are practicable. The electrical properties of polychlorotrifluoroethylene are inferior to those of polytetrafluoroethylene, especially for high-frequency applications, because of the more polar nature of the polymer. This material also has outstanding barrier properties to gases and is compatible with liquid oxygen. The combination of these unusual properties makes polychlorotrifluoroethylene irreplacable in several applications.

Poly(vinyl fluoride)

Poly(vinyl fluoride) is a highly crystalline plastic which is commercially available in the form of a tough but flexible film. The polymer has the outstanding chemical resistance of the fluoroplastics and excellent outdoor weatherability. It is extremely resistant to thermal degradation and maintains usable strength above 150°C while remaining tough at −180°C. It has low permeability to most gases and vapors and resists abrasion and staining.

The film has wide use as a protective coating in the building industry. In 0.001–0.002 in. thickness, bonded to wood, metal, or asphalt-based materials, it lasts many times longer than paints, enamels, or other surface coatings.

Poly(vinylidene fluoride)

Poly(vinylidene fluoride) is a crystalline plastic with a melting point of about 160°C. It has good strength properties and resists distortion and creep at both high and low temperatures. It has very good weatherability and chemical and solvent resistance.

This polymer can be processed by extrusion and by compression and injection molding. It is used as a coating, gasketing, and wire- and cable-jacketing material, and in piping, molded and lined tanks, pumps, and valves in the chemical and nuclear power industries.

Fluoroplastic Copolymers

Copolymers of ethylene with tetrafluoroethylene and with chlorotrifluoroethylene have essentially 1:1 alternating structures (Chapter 5A). They have high melting points (270 and 245°C, respectively), and retain much of the superior chemical resistance and good weatherability of the fully fluorinated materials. In other respects, however, their properties are more like those of the engineering thermoplastics, with good strength, toughness, and wear resistance over wide temperature ranges.

Copolymers of chlorotrifluoroethylene and vinylidene fluoride range from tough, flexible thermoplastics to elastomers, depending on composition. One of their outstanding properties is their resistance to the attack and penetration of powerful oxidizing agents such as propellant-grade red-fuming nitric acid and 90% hydrogen peroxide, typical of the fluorocarbons, coupled with the toughness and flexibility required for use as valves, ring seals, caulking compounds, and so on. They also find application in automobile transmissions and brake systems as gaskets and seals.

Copolymers of hexafluoropropylene and vinylidene fluoride are elastomers combining high resistance to heat and to fluids and chemicals with good mechanical properties. They are cured with amine-type systems. These materials can be compounded to remain serviceable for short periods as high as 300°C, and retain useful properties indefinitely at 200°C. Their resistance to the lubricants, fuels, and hydraulic fluids used in jet aircraft is unequalled by other elastomers. Their mechanical properties are respectable for any elastomer, and excellent among the oil-resistant types.

GENERAL REFERENCES

Brown 1967; Towler 1969; McCane 1970; Cohen 1971; Dohany 1971; Wall 1972; Moncrieff 1975, Chapter 40; Robertson 1976; Brydson 1978, Chapter 13, West 1979, 1980; Brasure 1980; Dohany 1980; Gangal 1980; Johnson 1980; Kim 1980; Sperati 1982.

DISCUSSION QUESTIONS AND PROBLEMS

1. Compare the behavior of polystyrene, poly(vinyl chloride), and polyacrylonitrile on thermal degradation, writing appropriate chemical equations.

2. Write chemical equations for (a) the polymerization of poly(vinyl acetate), (b) the formation of poly(vinyl alcohol), (c) the crosslinking of poly(vinyl alcohol) fibers, and (d) the formation of poly(vinyl butyral).

3. Postulate, and write equations for, long-chain branching in poly(vinyl acetate) arising at three different sites. Illustrate the fate of these branches as the polymer is hydrolyzed and reacetylated.

4. Discuss the relative physical properties of polystyrene, impact polystyrene, and ABS resins, and account for differences in terms of molecular structure.

5. Discuss (a) thermal stability and stabilization and (b) the effect of plasticization on thermal and physical properties, in vinyl resins.

6. Explain in terms of molecular structure the following properties of polytetrafluoroethylene: (a) melting point, (b) solubility, (c) melt flow behavior, (d) chemical inertness, and (e) coefficient of friction.

BIBLIOGRAPHY

Abrams 1967. Irving M. Abrams and Leo Benzera, "Ion-Exchange Polymers," pp. 692–742 in *EPST,* †
  Vol. 7, 1967.

Bamford 1964. C. H. Bamford, G. C. Eastmond, Gerald P. Ziemba, and A. Levovits, "Acrylonitrile
  Polymers," pp. 374–444 in *EPST,* Vol. 1, 1964.

Basdekis 1964. C. H. Basdekis, *ABS Plastics,* Reinhold, New York, 1964.

Beevers 1968. R. B. Beevers, "The Physical Properties of Polyacrylonitrile and Its Copolymers,"
  Macromol. Rev. **3,** 113–254 (1968).

Boyer 1970. Raymond F. Boyer, with K. E. Coulter, J. L. Duda, Edward F. Gurnee, Kon Sup Hyun,
  H. Kehde, Henno Keskkula, Alan E. Platt, and J. S. Vrentas, "Styrene Polymers" pp. 128–447
  in *EPST,* Vol. 13, 1970.

Brasure 1980. Donald E. Brasure, "Poly(Vinyl Fluoride)," pp. 57–64 in *ECT*‡ Vol. 11, 1980.

Brighton 1971. C. A. Brighton, with J. L. Benton, J. P. Dux, and G. C. Marks, "Vinyl Chloride
  Polymers," pp. 305–483 in *EPST,* Vol. 14, 1971.

Brown 1967. Henry C. Brown and Robert P. Bringer, "Fluorine-Containing Polymers," pp. 179–219
  in *EPST,* Vol. 7, 1967.

Brown 1982. W. E. Brown, "Vinylidene Chloride Polymers and Copolymers," pp. 113–114 in *MPE,*§
  1982.

† Herman F. Mark, Norman G. Gaylord, and Norbert M. Bikales, eds., *Encyclopedia of Polymer Science
and Technology,* Wiley-Interscience, New York.
‡ Martin Grayson, ed., *Kirk–Othmer Encyclopedia of Chemical Technology,* 3rd ed., Wiley-Interscience,
New York.
§ Joan Agranoff, ed., *Modern Plastics Encyclopedia 1982–1983,* Vol. 59, No. 10A, McGraw-Hill, New
York, 1982.

Brydson 1978. J. A. Brydson, *Rubber Chemistry,* Applied Science, London, 1978.

Burgess 1982. R. H. Burgess, ed., *Manufacture and Processing of PVC,* Macmillan, New York, 1981.

Cohen 1971. Frederic S. Cohen and Paul Kraft, "Vinyl Fluoride Polymers," pp. 552–540 in *EPST,* Vol. 14, 1971.

Coover 1978. H. W. Coover, Jr., and J. M. McIntyre, "2-Cyanoacrylic Ester Polymers," pp. 408–413 in *ECT,* Vol. 1, 1978.

Davis 1964. C. W. Davis and Paul Shapiro, "Acrylic Fibers," pp. 342–373 in *EPST,* Vol 1, 1964.

Dohany 1971. J. E. Dohany, A. A. Dukert, and S. S. Preston III, "Vinylidene Fluoride Polymers," pp. 600–610 in *EPST,* Vol. 14, 1971.

Dohany 1980. Julius E. Dohany, "Poly(Vinylidene Fluoride)," pp. 64–74 in *ECT,* Vol. 11, 1980.

Elias 1977. Hans-Georg Elias, *Macromolecules ·2· Synthesis and Materials* (translated by John W. Stafford), Plenum Press, New York, 1977.

Evans 1982. T. E. Evans, "Styrene–Acrylonitrile," pp. 116–117 in *MPE,* 1982.

Fram 1964. Paul Fram, "Acrylic Elastomers," pp. 226–246 in *EPST,* Vol. 1, 1964.

Gangal 1980. S. V. Gangal, "Polytetrafluoroethylene," pp. 1–24, and "Fluorinated Ethylene–Propylene Copolymers," pp. 24–35 in *ECT,* Vol. 11, 1980.

Garrett 1962. Alfred B. Garrett, "The Flash of Genius, 2. Teflon: Roy J. Plunkett," *J. Chem. Educ.* **39,** 288 (1962).

Grassie 1964. N. Grassie, "Thermal Degradation," Chapter 8*B* in E. M. Fettes, ed., *Chemical Reactions of Polymers,* Wiley-Interscience, New York, 1964.

Hobson 1978. P. H. Hobson and A. L. McPeters, "Acrylic and Modacrylic Fibers," pp. 355–386 in *ECT,* Vol. 1, 1978.

Houtz 1950. R. C. Houtz, " 'Orlon' Acrylic Fiber: Chemistry and Properties," *Textile Res. J.* **20,** 786–801 (1950).

Jeziorski 1982. R. J. Jeziorski and R. A. Wenzler, "Polyvinyl and Vinyl Copolymers," pp. 108, 110, 112 in *MPE,* 1982.

Johnson 1980. Richard L. Johnson, "Tetrafluoroethylene Copolymers with Ethylene," pp. 35–41 and "Tetrafluoroethylene Copolymers with Perfluorovinyl Ethers," pp. 42–49, in *ECT,* Vol. 11, 1980.

Kaufman 1969. Morris Kaufman, *The History of PVC,* Maclaren, London, 1969.

Kennedy 1968. R. K. Kennedy, "Modacrylic Fibers," pp. 812–839 in *EPST,* Vol. 8, 1968.

Kim 1980. Y. K. Kim, "Poly(Fluorosilicones)," pp. 74–81 in *ECT,* Vol. 11, 1980.

Kine 1978. Benjamin B. Kine and Ronald W. Novak, "Acrylic Ester Polymers," pp. 386–408 in *ECT,* Vol. 1, 1978.

Kine 1981. Benjamin B. Kine and R. W. Novak, "Methacrylic Polymers," pp. 377–398 in *ECT,* Vol. 15, 1981.

Koleski 1969. J. V. Koleski and L. H. Wartmen, *Poly(Vinyl Chloride),* Gordon and Breach, New York, 1969.

Krenz 1982. Heider Krenz, "Acrylic," pp. 18, 21, 22 in *MPE,* 1982.

Lantz 1982. J. M. Lantz, "ABS," pp. 6–7 in *MPE,* 1982.

Lindemann 1971a. Martin K. Lindemann, "Vinyl Alcohol Polymers," pp. 149–239 in *EPST,* Vol. 14, 1971.

Lindemann 1971b. Martin K. Lindemann, "Vinyl Ester Polymers," pp. 531–703 in *EPST,* Vol. 15, 1971.

Luskin 1964. Leo S. Luskin, Robert J. Myers, J. P. Bruise, H. W. Coover, Jr., and T. H. Wicker, Jr., "Acrylic Ester Polymers," pp. 246–342 in *EPST,* Vol. 1, 1964.

McCane 1970. Donald I. McCane, "Tetrafluoroethylene Polymers," pp. 523–670 in *EPST,* Vol. 13, 1970.

Miller 1964. M. L. Miller, "Acrylic Acid Polymers," pp. 197–226 in *EPST,* Vol. 1, 1964.

Moncrieff 1975. R. W. Moncrieff, *Man-Made Fibers,* Newnes-Butterworths, London, 1975.

Morneau 1978. G. A. Morneau, W. A. Pavelich, and L. G. Roettger, "ABS Resins," pp. 442–456 in *ECT*, Vol. 1, 1978.

Morris 1978. J. D. Morris and R. J. Penzenstadler, "Acrylamide Polymers," pp. 312–330 in *ECT*, Vol. 1, 1978.

Nass 1976–1977. Leonard I. Nass, ed., *Encyclopedia of PVC,* Dekker, New York, Vol. 1, 1976; Vols. 2 and 3, 1977.

Peng 1978. Fred M. Peng, "Acrylonitrile Polymers," pp. 427–442 in *ECT*, Vol. 1, 1978.

Peppin 1982. A. Peppin, "Styrene–Maleic Anhydride," pp. 117–118 in *MPE*, 1982.

Platt 1983. A. E. Platt and T. C. Wallace, "Styrene Plastics," pp. 801–847 in *ECT*, Vol. 21, 1983.

Platzer 1977. Norbert Platzer, "Elastomers for Toughening Styrene Polymers," *Chemtech* **7**, 634–641 (1977).

Pritchard 1970. J. G. Pritchard, *Poly(Vinyl Alcohol)—Basic Properties and Uses,* Gordon and Breach, New York, 1970.

Radosta 1977. Joseph A. Radosta, "Improving the Physicals of Impact Polystyrene," *Plastics Eng.* **33** (9), 28–30 (1977).

Rebenfeld 1967. Ludwig Rebenfeld, "Fibers," pp. 505–573 in *EPST*, Vol. 6, 1967.

Robertson 1976. A. B. Robertson and E. C. Lupton, Jr., "Fluorinated Plastics," pp. 260–287 in *EPST*, suppl. Vol. 1, 1976.

Sarvetnick 1969. Harold A. Sarvetnick, *Polyvinyl Chloride,* Van Nostrand Reinhold, New York, 1969.

Sperati 1962. C. A. Sperati and H. W. Starkweather, Jr., "Fluorine-Containing Polymers. II. Polytetrafluoroethylene," *Adv. Polym. Sci.* **2**, 465–495 (1962).

Sperati 1982. C. A. Sperati, "Fluoroplastics," pp. 35–39 in *MPE*, 1982.

Swett 1982. R. M. Swett, "Polystyrene," pp. 98, 100, 102 in *MPE*, 1982.

Thomas 1964. W. M. Thomas, "Acrylamide Polymers," pp. 177–197 in *EPST*, Vol. 1, 1964.

Vial 1979. T. M. Vial, "Acrylic Elastomers," pp. 458–469 in *ECT*, Vol. 8, 1979.

Wall 1972. Leo A. Wall, ed., *Fluoropolymers,* Wiley-Interscience, New York, 1972.

Wessling 1971. R. Wessling and F. G. Edwards, "Vinylidene Chloride Polymers," pp. 540–579 in *EPST*, Vol. 14, 1971.

West 1979. Arthur C. West and Allen G. Holcomb, "Fluorinated Elastomers," pp. 500–515 in *ECT*, Vol. 8, 1979.

West 1980. A. C. West, "Polychlorotrifluoroethylene," pp. 49–54 in *ECT*, Vol. 11, 1980.

CHAPTER FIFTEEN

HETEROCHAIN THERMOPLASTICS

This chapter includes discussion of those thermoplastic materials with backbone chains in which the regular sequence of carbon atoms, seen in the materials discussed in Chapters 13 and 14, is interrupted by the presence of other atoms, notably oxygen and nitrogen. Included are the polyamides and polypeptides characterized by the —CONH— group (Section *A*), a variety of polymers containing oxygen in the main chain (Section *B*), the plastics and fibers derived from cellulose (Section *C*), and a variety of other organic and inorganic polymers, mostly of interest for their superior high-temperature properties (Section *D*). Excluded are several types of polymers used exclusively, or nearly so, in crosslinked form. These materials, including the silicones and polyurethanes used as foams, are discussed in Chapter 16.

The polymers discussed in the present chapter have in common, for the most part, the high strength and toughness, stiffness and abrasion resistance, and retention of properties over wide temperature ranges associated with fibers and with engineering plastics. It is therefore appropriate that their description be contained in a single chapter.

A. POLYAMIDES AND POLYPEPTIDES

Polyamides

The coined word *nylon* has been accepted as a generic term for synthetic polyamides. The nylons are described by a numbering system that indicates the number of carbon atoms in the monomer chains. Amino acid polymers are designated by a single number, as 6-nylon for poly(ω-aminocaproic acid) (polycaprolactam). Nylons from diamines and dibasic acids are designated by two numbers, the first representing the diamine, as 66-nylon for the polymer of hexamethylenediamine and adipic acid, and 610-nylon for that of hexamethylenediamine and sebacic acid. Of greatest commercial importance are 6- and 66-nylon.

The production of nylon fibers in the United States was about 2.5 billion lb, whereas production for plastics use was only about 0.25 billion lb, in 1982. The drop in usage of both apparel and home furnishings in the early 1980s was reflected in a reduction in nylon production from a high of about 3.1 billion lb in 1979.

Development of the Nylons. The research leading to the production of synthetic fibers from polyamides began about 1928, when W. H. Carothers (Mark 1940) undertook a series of researches dealing with the fundamentals of polymerization processes. These studies were not aimed at the production of any specific product or process. In his fundamental researches, Carothers studied the condensation reactions of glycols and dibasic acids. He made a large number of polyesters ranging in molecular weight between 2500 and 5000. He also studied polymers of ω-amino acids, obtaining polyamides with a degree of polymerization of about 30. These were hard, insoluble, waxy materials.

Recognizing the equilibria involved in stepwise polymerization, Carothers and J. W. Hill introduced the use of a molecular still to remove the water liberated in the condensation process and to shift the equilibrium toward higher molecular weights. In this way they synthesized materials with molecular weights up to 25,000. These substances had properties so different from those previously studied that Carothers coined the term *superpolymers* to describe those having molecular weights above 10,000.

The superpolyesters obtained from glycols and dibasic acids were tough, opaque solids that melted at moderately elevated temperatures to clear, viscous liquids. Filaments could be pulled from the liquids by touching the surface with a rod and withdrawing it. When cool, these filaments could be drawn to several times their original length. They became tough, transparent, lustrous materials of high strength and elasticity, possessing as high tenacity when wet as dry. Examination by x-ray diffraction showed them to be crystalline and highly oriented.

The properties of the aliphatic superpolyesters made them unsuitable for textile fibers because of their relatively low melting points and high solubility. Mixed aliphatic polyester–polyamides had the same bad features as the polyesters. It was clear that success, if achieved at all, would involve the use of a polyamide fiber.

The next phase of the research was a concentrated effort to find polyamides which could be made into fibers. A large variety of amino acids, diamines, and dibasic acids was studied. Several of the products gave good fibers melting near 200°C and equal to silk in strength and pliability. In 1935 the polyamide of hexamethylenediamine and adipic acid was made. Like the others, it gave fibers by melt spinning or by dry spinning from phenol solutions. The fibers could be cold drawn to obtain high tenacity and elasticity. They melted at about 265°C and were insoluble in all common solvents.

This polymer was selected for commercial development because of its good balance of properties and the possibilities foreseen for the manufacture of the raw materials. At that time, adipic acid was made only in Germany, and hexamethylenediamine was little more than a laboratory curiosity. In Germany, the first commercial nylon was polycaprolactam, 6-nylon. For these historic reasons, the major fraction of nylon production in the United States has been 66-nylon, and that in Europe, 6-nylon. Despite some differences in upper use temperature resulting from the difference in crystalline melting points (265°C for 66-nylon, 225 for 6-nylon), the production, properties, and applications of the two can be considered together.

Production. The key chemical in the production of both 6- and 66-nylon is cyclohexane. Produced by catalytic hydrogenation of benzene, some 1.5 billion lb was made in 1982, severely down from 2.4 billion lb in 1979. Sixty percent was converted to adipic acid by oxidation, and 30% was converted to caprolactam. In turn, a portion of the adipic acid can be converted to hexamethylenediamine via the ammonium salt, adipamide, and adiponitrile, although other syntheses can be used.

The polymerization of caprolactam is carried out by adding water to open the rings and then removing the water again at elevated temperature, where linear polymer forms. An autoclave or a continuous reactor can be used. Polycaprolactam is in equilibrium with about 10% of the monomer, which must be removed by washing with water before the polymer can be spun. At the spinning temperature, more monomer is formed to restore the equilibrium, and this must again be removed to achieve good properties in the yarn.

Caprolactam can also be polymerized by ionic chain mechanisms, as described in Chapter 4E. The reaction can be carried out below the melting point of the nylon and at atmospheric pressure, making the technique very attractive for the production of large cast articles.

In the polymerization of 66-nylon, the achievement of the stoichiometric balance needed to obtain high-molecular-weight step-reaction polymers (Chapter 2) is simplified by the tendency of hexamethylenediamine and adipic acid to form a 1:1 salt that can be isolated because of its low solubility in methanol. This salt is dissolved in water and added to an autoclave with 0.5–1 mol.% acetic acid as viscosity stabilizer. As the temperature is raised, the steam generated purges the air from the vessel. Pressure is kept at 250 psi as the temperature is raised to 270–280°C. Pressure is then reduced, and a vacuum may be applied. After a total time of 3–4 hr, nitrogen pressure is used to extrude the nylon as a ribbon through a valve in the bottom of the autoclave. The ribbon is subsequently cut into cubes, the finished product for the plastics-grade material. For fiber use, the polymer (somewhat lower in molecular weight) is melt spun as described in Chapter 18B.

Properties. Both as plastics and as fibers, the nylons are characterized by a combination of high strength, elasticity, toughness, and abrasion resistance. Good mechanical properties are maintained up to 150°C, although a more conservative limit for plastics use is 125°. Both toughness and flexibility are retained well at low temperatures.

The solvent resistance of nylon is good; only phenols, cresols, and formic acid dissolve the polymer at room temperature. Strong acids degrade it somewhat. The polymer discolors in air at temperatures of about 130°C and is degraded by hydrolysis at elevated temperatures. Its outdoor weathering properties are only fair unless it is especially stabilized or pigmented with carbon black. Nylon has a moderately low specific gravity, 1.14. Its moisture resistance is fair; moisture acts as a plasticizer to increase flexibility and toughness. The electrical uses of nylon are restricted to low frequencies because of the polar groups in the polymer.

Fiber Applications.   About 60% of nylon production goes into housing-related outlets, especially carpets. Another 20% is used for apparel, and only about 10% for tire cord, down only slightly in volume over the last 20 years and accounting for nearly one-third of the market.

Plastics Applications.   The most important uses of the nylon plastics are as an engineering material—a substitute for metal in bearings, gears, cams, and so on. Nylon is well suited for such applications because of its high tensile and impact strength, form stability at high temperatures, good abrasion resistance, and self-lubricating bearing properties. The ability to be injection molded to close dimensional tolerances, augmented by its light weight, frequently gives the plastic a cost advantage in appliances. Rollers, slides, and door latches are often made of nylon. Bearings and thread guides in textile machinery may be made of nylon, which requires no lubricant that could damage yarn or cloth. Electrical wire is jacketed with nylon, which provides a tough, abrasion-resistant outer cover to protect the primary electrical insulation.

A number of specialty grades of the nylons are finding increased plastics use: higher-molecular-weight polymers, copolymers with olefin-based resins for increased impact resistance at low temperatures, and reinforced grades with increased strength, stiffness, and creep resistance. Fillers consist of either continuous mat or chopped glass, various minerals and glass, and minerals alone.

Other Polyamides.   Higher nylons (610, 612, 11, and 12) have significantly lower stiffness and heat resistance than 6- and 66-nylons, but have improved chemical resistance and lower moisture absorption. They are used as specialty plastics.

Nylon copolymers whose monomers include 6, 66, and 610 starting materials, terephthalic acid, and alkyl substituted diamines, are amorphous and therefore transparent though slightly yellow. These materials are also plastics.

The name *aramid* is applied to aromatic polyamides, two of which are firmly established in the specialty fiber markets. To use their Du Pont trade names, Nomex, is virtually nonflammable and achieved an early solid market because of this property; it is now progressing into other industrial applications where unusual heat resistance is required. Kevlar,

more readily crystallized and oriented as can be inferred from its para structure, is noted for its outstanding strength–weight ratio. It is stated (Layman 1982) that a Kevlar cable can match the strength of steel at the same diameter and 20% of the weight. The major present use of the material is in premium-quality tire cords, but its impact on that market is still virtually negligible.

A nylon made from dodecanedioic acid and bis-*para*-aminocyclohexyl methane,

(Du Pont's Quiana), spun in trilobal form, has a variety of good aesthetic qualities that lead to its use in luxury dress apparel applications.

Polypeptides

Polypeptides, the basis of natural proteins, are polyamides of α-amino acids. They are characterized by the repeat unit

where R varies widely among the 30 or so amino acids in proteins.

Although synthetic products can be made, the polypeptides of greatest interest are based on naturally occurring materials. They include wool and silk and a variety of fibers and plastics made from proteins extracted from nonfibrous natural products.

The importance of wool and silk is discussed in relation to other fibers in Chapter 18. Other protein-based fibers and plastics are at present of little more than historic interest.

Wool. Wool is a complex polypeptide, a polyamide made up of about 20 α-amino acids. The acids glycine, leucine and isoleucine, proline, cystine, and arginine, glutamic acid, and aspartic acid make up about two-thirds of its weight. About half of wool's mass is in the main polymer chain and half in the side chains. Since some 30% of the side chains have acid or amine groups along their length, wool is a huge polymeric "zwitterion" whose properties and solubility depend markedly upon pH.

Wool can exist in both the α- and β-keratin structures characteristic of the polypeptides: helical and extended sheet structures, respectively. The high reversible extensibility of wool (about 30% compared to 5% for cotton) derives from the fact that it can exist in either of these structures.

The microscopic appearance of wool is easily recognized by the presence of scales in the epidermal or surface layer of the fiber. These scales give wool the ability to felt or pack into dense stable mats.

Wool is noted for its high moisture absorption (up to 10–15%) and good insulating properties, its wrinkle resistance and elastic recovery, and its crimpiness. Its soil resistance, dyeability, and resistance to organic (dry-cleaning) solvents are good. On the other hand, it is attacked by moths and mildew, mild alkalis, and bleaches. It causes allergies. Its strength and stiffness are low compared to those of cotton. It has poor temperature resistance and weatherability in sunlight, and is adversely affected by hot water, with shrinkage and loss of luster and strength.

Despite these drawbacks, the virtues of wool are so great that it probably will continue to find extensive use, as it has to date, in suitings and other wearing apparel, blankets, yarns, carpets, felts, and upholstery. So far, no artificial fiber has equaled it in resilience, hand, insulating properties, dyeability, and flame resistance. The price of wool is notoriously unsteady, however, and its future may well depend upon its price in a competitive market. Production in the peak year of 1979 was only a little over 100 million lb in the United States.

Wool is crosslinked through cystine links. This structure gives swelling but not solubility in the natural product. On the addition of thioglycolic acid, however, the crosslinks are broken with the formation of sulfhydryl groups. The tensile strength of the fiber decreases, its stress–strain curve changes to resemble that of a typical elastomer, and it shrinks readily and becomes soluble in polar solvents.

Wool can be recrosslinked by a number of reagents, including alkyl dihalides. Since the thioethers resulting from the treatment are much more stable than the original disulfides, the new fiber has much better resistance to laundering and chemicals. Since moth damage usually begins with rupture of the disulfide links, the treated material has improved moth resistance. The scission and crosslinking can be made so mild that the appearance and hand of the material are not changed.

Silk.   Silk is a polypeptide made up of only four amino acids: glycine, H_2NCH_2COOH; alanine, $H_2NCH(CH)_3)COOH$; serine; and tyrosine. Glycine and alanine, present in a 2:1 ratio, account for over 75 mol.% of the material. Silk crystallizes in an antiparallel-chain pleated sheet β-keratin structure, in which only glycine and alanine residues occur. The other amino acids are present only in the amorphous regions of the fiber. There is no structure in silk equivalent to the α-keratin spiral in wool and other polypeptides. Thus the silk fiber has an elongation (20–25%) lower than that of wool but higher than that of cotton.

The silkworm spins a double fiber from which the two single fibers are easily separated as continuous filaments 400–700 yards long. Silk is thus more nearly analogous to continuous filament than to stable synthetic fibers.

Silk has been noted for thousands of years for its strength, toughness, and smooth, soft feel. It is resilient and shapes well. It is a poor conductor of heat and electricity. Its moisture regain is 11%; it withstands heat and hot water better than wool.

Some disadvantages of silk are those associated with its cultivation and gathering, unevenness of the fibers, loss of strength on dyeing, and poor launderability of the dyed fabrics. It lacks the strength of nylon for hosiery use but would otherwise be preferred for its warmer feel. There is relatively little effort in progress to improve the properties of silk.

GENERAL REFERENCES

Polyamides. Rebenfeld 1967; Schule 1969; Snider 1969; Sweeney 1969; Kohan 1973; Moncrieff 1975, Chapters 19–22; Elias 1977, Chapter 28.3; Jacobs 1977; Preston 1978; Fisher 1982; Galanty 1982; Putscher 1982; Reitano 1982; Saunders 1982; Welgos 1982.

Polypeptides. Elias 1977, Chapter 30.

B. POLYESTERS, POLYETHERS, AND RELATED POLYMERS

Polyesters

The properties of low melting point and high solubility caused Carothers (see Section A) to reject the linear aliphatic polyesters as fiber-forming candidates, and to this day no commercial product is based solely on these polymers. The stiffening action of the *p*-phenylene group in a polymer chain, however, leads to high melting points and good fiber-forming properties, as discussed in Chapter 12A. Commercially important polyesters are based on such polymers, of which poly(ethylene tere-phthalate) is the major product. These polymers are used for fibers, films, and blown bottles. About 4.3 billion lb was consumed as fibers in 1982, about 400 million lb as blown bottles, and about 240 million lb in extrusion and film uses.

Polymerization and Properties. The production of high-molecular-weight pol-yesters differs somewhat from that of similar polyamides. In the case of the nylons, the chemical equilibrium favors the polyamide under readily achieved polymeri-zation conditions. Stoichiometric equivalence is easily achieved by the use of salts, and amide interchange reactions are slow. In polyester formation, however, the equilibrium is much less favorable, and equivalence is more difficult to achieve, since the salts do not form. Furthermore, the dibasic aromatic acids are very difficult to purify because of low solubility and high melting point.

This situation has been met by taking advantage of the rapidity of ester-inter-change reactions. The acid, such as terephthalic acid, is converted to the dimethyl ester, which can easily be purified by distillation or crystallization. This is then allowed to react with the glycol by ester interchange. In practice a low-molecular-weight glycol is used and the reaction takes place in two steps. First, a low-molecular-weight polyester is made with an excess of glycol to ensure hydroxyl end groups. Then the temperature is raised and pressure lowered to effect condensation of these molecules by ester interchange with the loss of the glycol.

In the production of poly(ethylene terephthalate), terephthalic acid is made by the oxidation of p-xylene. The polymerization step is similar to that for the polyamides in so far as the equipment and conditions are concerned. The polymer coming from the autoclave is quenched from the molten state to below its glass transition point of about 80°C and is amorphous. Crystallinity develops on heating; the crystalline melting point is 265°C. The polymer is melt spun.

Because of its high crystalline melting point and glass transition temperature, poly(ethylene terephthalate) retains good mechanical properties at temperatures up to 150–175°C. Its chemical and solvent resistance is good, being similar to that of nylon.

Fiber Applications.   The properties of poly(ethylene terephthalate) fiber that influence its uses are its outstanding crease resistance and work recovery and its low moisture absorption. These properties arise from the stiff polymer chain and the resulting high modulus, and the fact that the interchain bonds are not susceptible to moisture. As a result, garments made from the polyester fibers are very resistant to wrinkling and can be washed repeatedly without subsequent ironing. The higher modulus of the polyester fibers and their wrinkle resistance are reminiscent of those of wool, but fully oriented polyester fibers are too stiff to have the crisp but soft hand of wool. It is possible, however, to impart this property to the polyesters by control of crystallinity and orientation so that their stress–strain curve is similar to that of wool. This is done in the production of staple polyester fiber.

The polyester fibers are seldom used alone, but are blended with cotton or wool to make summer- and medium-weight suiting and other goods. They have set new standards for the performance of wearing apparel through the "wash and wear" concept.

Polyester fibers are unique among the synthetics in their ability to form felts. These products are better than wool felts in resistance to temperature, to abrasion, and to further felting.

Polyester fiber is currently popular as a tire cord, with about 25% of the market for the last decade.

Plastics Applications.   The tensile strength of poly(ethylene terephthalate) film is about 25,000 psi, two to three times that of cellophane or cellulose acetate film (Section C). If the area of the specimen at the break point is considered, the tensile strength of this plastic is about twice that of aluminum and almost equal to that of mild steel.

The stiffness of poly(ethylene terephthalate) film is comparable to that of cellophane and other cellulosic films, but its resistance to failure on repeated flexing is unusually high. In one flex test it lasts for over 20,000 cycles compared to a few hundred for cellulosic films. Its tear strength is also better than that of the cellulosics. The impact strength of this material is three to four times that of any other plastic film. This toughness is a major advantage in typical applications, such as magnetic recording tape.

The major new plastics use for poly(ethylene terephthalate) in the last few years has been in the development of blown bottles for soft-drink use, with early emphasis on the 2-liter bottle size. Molding is usually done in two stages, with a preform injection molded, then blown to final bottle shape. Both are biaxially oriented to provide the necessary strength for the application, to withstand dropping (though an additional cap on the bottom of the bottle is required for extra protection) and the internal pressure of carbon dioxide.

Poly(butylene terephthalate) and various copolymers have minor plastics uses.

Ether and Acetal Polymers

Polyethers, by definition, contain the ether linkage —R—O—R—O— as part of their chain structure. By convention, the polymers in which R is a methylene group, —CH_2— (i.e., polyoxymethylenes), are described by the generic term *acetal resins,* the name polyethers being applied to polymers in which R is more complex. Epoxy resins, also polyethers, are discussed in Chapter 16C.

Polyoxymethylene.   Polyoxymethylene is a polymer of formaldehyde or of trioxane. Although polymeric products of formaldehyde have been known for over 100 years, and were studied in detail by Staudinger (1925), thermally stable polymers of formaldehyde were only recently prepared. The improved stability of these acetal resins allows them to be fabricated into useful articles. Copolymers, thought to contain small amounts of ethylene oxide as the comonomer, are also produced. About 80 million lb of acetal resins was produced in the United States in 1982.

Exceptionally pure formaldehyde (better than 99.9% CH_2O) is polymerized by an anionic mechanism in the presence of an inert solvent (e.g., hexane) at atmospheric pressure and a temperature, preferably between -50 and $+70°C$, where the solvent is liquid. A wide variety of anionic catalysts is suitable, including amines, cyclic nitrogen-containing compounds, arsines, stibines, and phosphines. The polymer is insoluble in the reaction mixture and is removed continuously as a slurry. Thermal stability is improved by acetylation of the hydroxyl end groups of the polymer using acetic anhydride:

$$—CH_2OCH_2OH + CH_3COO^- \longrightarrow$$
$$—CH_2OCH_2O^- + CH_3COOH \xrightarrow{(CH_3CO)_2O}$$
$$—CH_2OCH_2OCOCH_3 + CH_3COO^-$$

Acetal resins have properties characteristic of partially crystalline high-molecular-weight polymers. Typical specimens are about 75% crystalline, with a melting point of 180°C. The impact strength of the polymers is high, and both their stiffness

(450,000 psi) and yield stress (10,000 psi) are greater than those of other crystalline polymers. Moisture absorption is negligibly small, and the polymers are insoluble in all common solvents at room temperature. They can be processed by conventional molding and extrusion techniques.

Major uses for the acetal resins are as direct replacements for metals. Their stiffness, light weight, dimensional stability, and resistance to corrosion, to wear, and to abrasion have led to their replacing brass, cast iron, and zinc in many instances. Typical applications include automobile parts, such as instrument panels, door hardware, and pump housings and mechanisms; pipe, especially for oil field systems; and a wide variety of machine and instrument parts.

Polyglycols. Polymers made from ethylene oxide and its higher homologues

$$\underset{\displaystyle O}{CH_2CHR}$$

can be considered to be the condensation products of glycols, although they are in fact produced by ring-scission polymerization of the corresponding oxides (Chapter 4E). The product of greatest commercial interest is the polymer of ethylene oxide. This polymer combines the thermoplastic behavior and mechanical properties of a highly crystalline, high-molecular-weight polymer with complete water solubility.

Polyurethanes

Polyurethanes are polymers containing the group

$$\underset{\displaystyle -N-C-O-}{\overset{\displaystyle H \quad O}{| \quad ||}}$$

formed typically through the reaction of a diisocyanate and a glycol:

$$x\text{OCNRCNO} + x\text{HOR'OH} \longrightarrow -[\text{OCONHRNHCOOR'}-]_x$$

The polymers formed in this way are useful in four major types of product: foams, fibers, elastomers, and coatings. Polyurethane foams are invariably cross-linked; hence they are discussed with thermosetting resins in Chapter 16C.

Elastomers. Polyurethane elastomers are made in several steps. A *basic intermediate* is first prepared in the form of a low-molecular-weight (1000–2000) polymer with hydroxyl end groups. This may be a polyester, such as that made from ethylene glycol and adipic acid, a polyether, or a mixed polyester–polyamide.

The basic intermediate, which is here designated **B,** is then reacted with an aromatic diisocyanate to give a *prepolymer.* Typical isocyanates are 2,4-tolylenediisocyanate, 4,4-benzidenediisocyanate, and 1,5-naphthalenediisocyanate: Using the former as an example, a typical prepolymer can be represented as

$$H_3C \langle \rangle NHCOOBOCONH \langle \rangle CH_3 \qquad NCO$$
$$NCO \qquad NHCOOBOCONH \langle \rangle CH_3$$

The elastomer is vulcanized through the isocyanate groups by reaction with glycols, diamines, diacids, or amino alcohols. If water is used, carbon dioxide is eliminated during crosslinking, as in the production of urethane foams (Chapter 16C).

Polyurethane elastomers are noted for extremely good abrasion resistance and hardness, combined with good elasticity and resistance to greases, oils, and solvents. They make tire treads with unusually long life and are widely used in applications requiring outstanding abrasion resistance, such as heel lifts and small industrial wheels.

Fibers. Polyurethane fibers with unusually high elasticity are used for lightweight foundation garments and swimsuits. They have replaced rubber latex thread in this use.

Coatings. Coatings based on polyurethanes have very good resistance to abrasion and solvent attack plus good flexibility and impact resistance. They can be applied by dip, spray, or brush and adhere well to a wide variety of materials. They are suitable in applications for which unusual impact and abrasion resistance is required, such as gymnasium and dance floors and bowling pins, in magnet wire coatings, and in a variety of outdoor and marine uses because of their good weatherability.

Molding Compounds. With the development of methods for molding thermosetting materials (reaction injection molding), there has been a resurgence in the use of polyurethane molding compounds. These are discussed in Chapter 16C.

Polycarbonates

A polycarbonate plastic, characterized by the —OCOO— heterochain unit, can be made either from phosgene and bisphenol A (4,4'-dihydroxydiphenyl-2,2'-propane) or by ester exchange between bisphenol A and diphenyl carbonate. It has this structure:

$$-O \langle \rangle \overset{CH_3}{\underset{CH_3}{C}} \langle \rangle OC- \atop \overset{\|}{O}$$

Like the nylon, acetal, and polyether resins, this polymer is a crystalline thermoplastic of very good mechanical properties. It has unusually high impact strength, even at low temperatures, ascribed in part to a combination of relatively high ordering in the amorphous regions and considerable disordering in the crystalline regions. It has low moisture absorption, good heat resistance (useful to 140°C), and good thermal and oxidative stability in the melt. It is transparent and self-extinguishing. It can be processed by conventional injection molding and extrusion. Typical applications include telephone parts, business machine housings, machinery housings, automobile tail-light lenses, and unbreakable glazing applications. About 210 million lb of polycarbonate resins was produced in the United States in 1982.

Sulfur-Containing Polymers

Polysulfide elastomers are the reaction products of ethylene dihalides and alkali sulfides. It was discovered in 1920 that a rubberlike material could be made from these reagents. Commercial production began in 1930, and the products were quite successful despite some disadvantages (notably odor) because of their oil resistance.

The polysulfide rubbers are linear condensation polymers:

$$x\text{RCl}_2 + x\text{Na}_2\text{S}_y \longrightarrow -(\text{R}-\text{S}_y-)_x + 2x\text{NaCl}$$

The physical properties of the materials depend on the length of the aliphatic group and the number of sulfur atoms present. With four sulfurs per monomer all products are rubbery, whereas with only two sulfurs at least four methylene groups are needed in the dihalide to obtain elastomeric properties. The halides ordinarily used are ethylene dichloride, β,β'-dichloroethyl ether, and dichloroethyl formal. Cocondensation can be carried out.

The polysulfide elastomers are outstanding in oil and solvent resistance and in gas impermeability. They have good resistance to aging and ozone. On the other hand, they have disagreeable odors, poor heat resistance, poor abrasion resistance in tire treads, and low tensile strength (1500 psi). Most of their applications as rubbers depend on solvent resistance and low gas permeability. They include gasoline hoses and tanks, diaphragms and gaskets, and balloon fabrics.

Liquid polysulfide elastomers can be cured at room temperature and without shrinkage to tough, solvent-resistant rubbers. These liquid compounds are widely used as gasoline tank sealants and liners for aircraft and in a variety of other sealing and impregnating applications.

These liquid polymers, when combined with oxidizers, burn with great intensity and generate large volumes of gas. They form the basis of solid-fuel rocket propellants. Inorganic oxidizers, in finely ground form, are mixed with the liquid elastomeric binder and cast directly into the rocket motor, binding to the walls of the vessel for support and insulation of the motor case.

Other sulfur-containing polymers are engineering plastics with good property retention over wide temperature ranges. They include polyphenylene sulfides, discussed with the corresponding oxygen compounds in Section D; polysulfones,

and polyethersulfones,

GENERAL REFERENCES

Polyesters. Rebenfeld 1967; Farrow 1969; Goodman 1969; Hawthorne 1969; Moncrieff 1975, Chapter 24; Elias 1977, Chapter 26.4; Katz 1977; Davis 1982; Jaquiss 1982; Pengilly 1982; Scales 1982; Yarger 1982.

Polyethers. Akin 1962; Bevington 1964; Elias 1977, Chapter 26.2; Persak 1978, 1982; Vandenberg 1979; Braun 1982; Dreyfuss 1982; Newton 1982; Royal 1982; White 1982.

Polyurethanes. Saunders 1962, 1964; Ibrahim 1967; Blokland 1968; Bruins 1969; David 1969; Pigott 1969; Wright 1969; Moncrieff 1975, Chapter 26; Backus 1977; Brydson 1978, Chapter 16; Dominguez 1982.

Polycarbonates. Schnell 1964; Bottenbruch 1969; Fox 1982; Page 1982.

Sulfur-Containing Polymers. Berenbaum 1969; Eichhorn 1977; Elias 1977, Chapter 27.2; Ballintyne 1982; Boeke 1982; Hill 1982; Rigby 1982.

C. CELLULOSIC POLYMERS

The class of polymers based on cellulose includes the following: native cellulose, including wood and most plant matter, but of interest here primarily because of the widespread popularity of cotton as a fiber; regenerated cellulose, used as a fiber (viscose rayon) and a film (cellophane); chemical derivatives of cellulose, of which the organic esters, particularly the acetate, are most important; and minor polymers with structures similar to that of cellulose.

Cellulose

Cellulose is a polysaccharide made up of β-D(+)-glucose residues joined in linear chains having the structure

$$
\left[\begin{array}{c} \text{CH}_2\text{OH} \\ | \\ \text{C—O} \\ \diagup\text{H} \quad \diagdown \\ \text{—O—HC}_{\text{H} \quad \text{H}} \text{CH*—} \\ \text{C—C} \\ | \quad | \\ \text{HO} \quad \text{OH} \end{array} \right]_x
$$

With its three hydroxyl groups, cellulose has the opportunity of forming many hydrogen bonds. The resulting high intermolecular forces plus the regular structure of the polymer result in its having an unusually high degree of crystallinity. The crystalline melting point of cellulose is far above its decomposition temperature. The solubility of the polymer is very low; it is doubtful that solution ever takes place unless a chemical derivative is formed. Cellulose does swell, however, in hydrogen-bonding solvents, including water. The swelling is, of course, restricted to the amorphous regions of the structure. When native cellulose is dissolved via chemical reaction and then reprecipitated as pure cellulose, the product is known as *regenerated cellulose*.

Cotton

Cotton is about 95% cellulose, with small amounts of protein, pectin, and wax. The fiber can easily be recognized under the microscope because of its flattened, twisted shape.

Cotton fabrics launder well and show excellent resistance to alkalis. Since the strength of cotton is 25% greater wet than dry, it withstands repeated washings well, in contrast to most animal and synthetic fibers which lose strength when wet or moist. Cotton is less stiff than flax, hemp, or jute. Fabrics of cotton can be folded repeatedly without loss of strength. With proper construction and treatment, shrinkage and stretching can be controlled. Cotton is a good conductor of heat, better than silk or wool but not as good as linen. This property and its high moisture absorption give it comfort in wearing apparel. Cotton withstands heat better than the other natural fibers. It can be dyed readily, and the colors are good in lightfastness and washfastness. It is not affected by solvents and usually not attacked by moths.

About 5.4 billion lb of cotton was grown in the United States in 1978.

Mercerization. The chemical treatment of cotton known as *mercerization* to improve its luster has been practiced for many years. The treatment consists in swelling the cotton by concentrated alkali solution and then washing out the alkali.

The chemical changes during the process are similar to those in the formation of alkali cellulose. At the end of the treatment there is no molecular difference in the product, but it is somewhat more amorphous than native cotton and, unless the fibers were held under tension during the process, somewhat less oriented.

As a result of these changes, mercerized cotton is slightly lower in density than the untreated material and has increased water absorption, better dyeability, lower tensile strength, and higher extensibility.

Regenerated Cellulose

The term *regenerated cellulose* describes cellulose which has been dissolved by virtue of the production of a soluble chemical derivative, cellulose xanthate, and subsequently reprecipitated. When prepared as a fiber, regenerated cellulose is known as *viscose* or (viscose) *rayon*. Traditionally both viscose and cellulose acetate have been known as rayons, but the term is now restricted to viscose in order to avoid confusion. As a film, regenerated cellulose is known by the generic term *cellophane*.

Manufacture. The first step in the production of regenerated cellulose, starting with purified cellulose usually obtained from wood pulp, is the formation of *alkali* or *soda cellulose*. This is produced by the reaction of cellulose with aqueous NaOH at room temperature for 15 min to 2 hr. The excess alkali is then pressed out and the mass is shredded into "crumbs" and aged for several days at 25–30°C to promote oxidative degradation of the chains to the desired degree of polymerization.

The alkali cellulose is then converted to cellulose xanthate by the addition of carbon disulfide:

$$R\text{—OH} + CS_2 + NaOH \longrightarrow R\text{—O—}\overset{\overset{\displaystyle S}{\|}}{C}\text{—S—Na} + H_2O$$

where R is a cellulose residue. The average degree of xanthation is about one CS_2 for every two cellulose residues. Since the reaction takes place heterogeneously by the addition of liquid or gaseous CS_2 to the alkali cellulose crumbs, the xanthation in some regions may correspond to the dixanthate or higher.

The mass retains most of its physical form but becomes more sticky or gelatinous and takes on a deep orange color and a characteristic and highly unpleasant odor. It is held at 25–30°C for 1–3 hr with about 10% CS_2 by weight and then evacuated to remove excess CS_2. It is now dissolved in dilute NaOH, becoming completely soluble in this reagent for the first time. The solution is known as viscose. Its color and odor are due to products of side reactions such as Na_2CS_3, NaSH, H_2S, sulfides, and polysulfides.

A fresh viscose solution cannot easily be coagulated and must be allowed to "ripen" for a few days. The changes taking place during ripening are complex, and the reaction must be carefully controlled by time and temperature to avoid premature coagulation.

Cellulose xanthate is essentially unstable and decomposes gradually during the ripening process by hydrolysis and saponification. At the end of the ripening period, enough cellulose residues have been regenerated so that coagulation is imminent. The viscose is then spun into a bath containing sulfuric acid to effect the regeneration of the remaining cellulose residues and coagulate the polymer:

$$\underset{\displaystyle R-O-\overset{\displaystyle S}{\overset{\displaystyle \|}{C}}-S-Na} {} + H^+ \longrightarrow R-OH + CS_2 + Na^+$$

Viscose Rayon

Normal Tenacity.   Viscose rayon normally has a tenacity of 2–2.5 g/denier, somewhat lower than that of cotton (3–5 g/denier). Its wet tenacity is only about half the dry value but is still adequate for laundering. The moisture absorption of viscose is about twice that of cotton. The elasticity is not high (typically 15% elongation at break), and the fiber shows considerable viscoelastic character.

Viscose of normal tenacity is primarily an apparel fiber, but its popularity is declining. Several one-time major manufacturers have ceased producing the material, and the others are operating well below capacity.

High Tenacity.   In contrast to apparel rayon, a high-tenacity product (3–6 g/denier) is produced by stretching the fibers just short of their breaking point in the spinning bath. Considerable orientation (and crystallinity) is so introduced. High-tenacity rayon has lower elongation and moisture absorption than the apparel material and was used almost exclusively for tire cord. This market has, however, virtually disappeared with the increased use of nylon, polyester, and steel for that purpose.

Cellophane

In the manufacture of cellophane the viscose solution is extruded as film and then immersed in a bath of ammonium and sodium sulfate and dilute sulfuric acid, which removes the xanthate groups and precipitates the cellulose. The film is passed through various washing, bleaching, and desulfurizing baths, and finally through a bath of glycerol, glucose, or a polyhydric alcohol, which is imbibed and acts as a plasticizer. The film is then dried. For applications in which low moisture permeability is desired, it is coated with a mixture of nitrocellulose and various plasticizers and waxes.

Cellophane is a thin (0.001–0.002 in.) film of fair physical properties. Its tensile strength is good, but its tear strength, impact strength, and resistance to flexing are poor compared to those of newer film materials. The permeability of the uncoated film to water vapor and to water-soluble gases is extremely high. The coated material passes about 0.3 g of water vapor/in^2 per hour at 40°C. Cellophane is widely used as a wrapping and packaging material.

Cellulose sponges are made by stirring lumps of a salt such as sodium sulfate into viscose, coagulating, and washing out the salt to leave a porous product. Sponges may also be made by incorporating a blowing agent in the material.

Cellulose Acetate

Manufacture. As a polyhydric alcohol, cellulose can undergo esterification reactions. In the presence of a strong organic acid such as formic acid, equilibrium in the reaction

$$\text{acid} + \text{alcohol} \rightleftharpoons \text{ester} + \text{water}$$

lies partly to the right, and some ester is formed. With other organic acids, including acetic acid, equilibrium is well to the left and ester formation does not take place under normal circumstances. The easiest way to shift the equilibrium to the right is by the removal of water as it is formed in the reaction. To accomplish this, part of the acid is replaced by an acid chloride or anhydride. Sulfuric acid is also utilized in the reaction mixture. It acts as catalyst and also assists in removing water.

Purified cotton linters or wood pulp form the raw material for cellulose acetate. This cellulose is partially dried from its natural moisture content of 5–10% to reduce the water content of the subsequent reaction mixture. The end of the acetylation is indicated by complete solution of the cellulose. At this point complete acetylation to the triacetate has taken place. The mixture is now held at about 50°C until the desired chain degradation has been achieved, as indicated by the viscosity of the mixture.

To obtain a more easily spinnable composition, it is necessary to reverse the acetylation reaction to the point where the proper solubility relations are obtained. This corresponds to the acetone-soluble diacetate, or an acetyl content of about 30%. This stage of the reaction can be carried out in homogenous solution, thus preserving the maximum amount of chemical homogeneity in the final product.

At the end of the reaction to produce the diacetate the mixture is cooled and added to a large amount of 25–35% acetic acid. This precipitates the polymer in finely divided form. After the polymer is washed and dried, it is dissolved in acetone for the spinning operation.

Fiber Properties and Applications. The traditional acetate fiber (diacetate) has a tenacity of 1.2–1.5 g/denier and about the same moisture absorption as cotton. It tends to soil less and wash more easily than viscose. Since acetate is thermoplastic, pleats and creases can be permanently set. On the other hand, acetate is less susceptible to creasing and wrinkling in use than viscose because its resilience is greater.

With the advent of spinning processes using methylene chloride, cellulose triacetate became popular as an apparel fiber. Its properties are improved over those of the diacetate in moisture absorption, wrinkle resistance and creep retention, dyeability, and speed of drying.

About 1.3 billion lb of cellulose acetate and viscose rayon, combined, was produced in the United States in 1979.

Plastics Properties and Uses.   Cellulose acetate was introduced in sheet, rod, and tube form (as well as in lacquers and dopes) around the time of World War I. Its popularity as a substitute for cellulose nitrate was not high, and it was not widely used until injection molding became common in the 1930's. It has been for many years a very popular injection-molding material. The usual molding material has 50–55% acetyl content by weight. It is somewhere between the diacetate and the triacetate in composition. Higher acetyl content gives poorer flow but better moisture resistance; lower acetyl content gives better impact strength. A plasticizer is ordinarily used. Higher-softening grades are preferred for compression molding, and softer grades for extrusion. Cellulose acetate molded and extruded articles are used where extreme moisture resistance is not required.

Cellulose acetate is also utilized in sheet form. Thin films are cast from solution onto polished metal surfaces. The films are used in photography, for wrapping, and for making small envelopes, bags, and boxes for packaging.

The plastics uses of cellulose acetate plus the mixed organic esters described below amounted to about 100 million lb in 1982.

Mixed Organic Esters of Cellulose

Cellulose Acetate–Butyrate.   The mixed ester cellulose acetate–butyrate has several advantages in properties over cellulose acetate. These include lower moisture absorption, greater solubility and compatibility with plasticizers, higher impact strength, and excellent dimensional stability. The material used in plastics has about 13% acetyl and 37% butyryl content. It is an excellent injection-molding material. Moldings of cellulose acetate–butyrate are used in automobile hardware. Motion picture safety film is cellulose acetate–butyrate or cellulose propionate–butyrate. The outdoor weathering properties of cellulose acetate–butyrate are superior to those of the other cellulose esters, but not nearly as good, for example, as those of the acrylic resins.

Cellulose Propionate and Acetate-Propionate.   These materials are similar to the acetate–butyrate in properties and uses. The acetate-propionate has some application in surface coatings.

Cellulose Nitrate

Although cellulose nitrate accounts today for only a minute fraction of the volume of cellulose plastics, it has considerable historical interest (Chapter 1B). The major use of plastics-grade cellulose nitrate is in the coating field. The properties and applications of the material are dependent on its degree of nitration; the plastics and lacquer grades contain 10.5–12% nitrogen, corresponding roughly to the dinitrate. Explosives grades contain 12.5–13.5% nitrogen.

Manufacture. The esterification reaction of the hydroxyl group on cellulose with nitric acid is an equilibrium:

$$ROH + HONO_2 \rightleftharpoons RONO_2 + H_2O$$

which can be shifted to the ester side by the removal of water. The usual reagent for this purpose is sulfuric acid; the nitration mixture ordinarily used consists of sulfuric and nitric acids and a limited amount of water.

Fabrication. The temperature sensitivity of cellulose nitrate excludes it from all fabrication methods involving heat, such as molding and extrusion. The plastic is handled by so-called "block" methods. Cellulose nitrate, alcohol, and camphor, the plasticizer always used, are mixed in a dough mixer until homogeneous. Here for the first time the fibrous shape of the original cellulose is lost. The colloid is filtered under pressure and rolled on hot rolls (65–80°C) where much of the alcohol is evaporated. It is then sheeted into slabs that are pressed while hot into a homogeneous block about 6″ × 5′ × 2′ in size.

Sheets of any desired thickness are subsequently sliced from the block. They must be aged by storage at about 30°C until the last traces of solvent have diffused out. This may take from 2 hr to 6 months, depending on the thickness of the sheet. The final operation consists of finishing the surface of the sheet by pressing in contact with polished metal sheets or by buffing.

Rods, tubes, and other shapes can be formed by solvent extrusion of the plasticized cellulose nitrate, followed by removal of solvent as above.

Properties and Uses. Despite its disadvantages of flammability, instability, and poor weathering properties, cellulose nitrate is one of the cheapest and most highly impact-resistant plastics. It still has some applications as a plastic, but is now used mostly for lacquers.

Cellulose Ethers

The cellulose ethers are products of the reaction of an organic halide with cellulose swollen by contact with an aqueous base:

$$ROH + R'Cl + NaOH \longrightarrow ROR' + NaCl + H_2O$$

where ROH represents a cellulose residue and one of its hydroxyl groups, and R′Cl is the organic halide. Chlorides are preferred over bromides or iodides, despite their lower reactivity, because of their higher diffusion rate in the heterogeneous reaction system used.

Manufacture. Cellulose ethers are made by reacting cellulose, alkali, and the organic chloride at about 100°C. The ether content and the viscosity of the product are controlled by temperature, pressure, reaction time, and composition of the reacting mixture.

Ethyl Cellulose.   The most important of the cellulose ethers is ethyl cellulose. The commercial material has 2.4–2.5 ethoxy groups per glucose residue. It is a molding material which is heat stable and has low flammability and high impact strength. It is flexible and tough at temperatures as low as $-40°C$. Its electrical, mechanical, and weathering properties are good in comparison with those of other cellulosics, but not generally outstanding. Its softening point is low, and it has a high water absorption (but not as high as cellulose acetate) and some tendency to cold flow.

GENERAL REFERENCES

Ott 1954–1955; Bruxelles 1965; Haskell 1965; Hill 1965; Nissan 1965; Savage 1965; Rebenfeld 1967; Mitchell 1969; Bikales 1971; Elias 1977, Chapter 31.5; Andrews 1979; Bogan 1979; Greminger 1979; Serad 1979; Turbak 1979; Daul 1981; Brown 1982; Lundberg 1982; Rich 1982.

D.   HIGH-TEMPERATURE AND INORGANIC POLYMERS

The requirements of modern technology, including those related to the "space age," place increasing demands on the high-temperature behavior of all materials, including polymers. For the last few years there has been an increased effort to define and produce polymeric structures capable of demonstrating good mechanical properties for long periods of time at higher and higher temperatures.

Mark (1967) has pointed out several ways in which the melting or softening points of polymers can be raised, with concomitant improvement in high-temperature properties. These are: stiffening the chains by the addition of ring structures or other stiff elements, crosslinking the chains, and inducing crystallization. With our present capabilities, it has not been possible to devise structures in which all three of these approaches are combined; for example, crystallinity and crosslinking are still largely mutually exclusive. But the application of one or two of these principles has proved fruitful.

The results of adding aromatic rings to polymer chains have been described in Chapter 12A. The addition of other ring structures, usually heterocyclic, to form fused-ring groups even larger and stiffer than the phenylene moiety, is a logical extension of this idea. Beyond this, two approaches are suggested: One, attributed (Fraser 1969) to Marvel, is to produce ladder polymers whose structure is such that breaking one or a few bonds can not sever the chain; the other is to abandon polymers based on the carbon–carbon bond and go to other structures known collectively as inorganic polymers.

C. S. Marvel has stated (private communication) that the upper temperature limit for the stability of carbon-chain polymers in a nitrogen atmosphere is probably around 550–600°C, unless the structure is very similar to that of graphite itself (as "black orlon" may be—see below). If the polymer is completely free from hydrogen, the stability in air and in nitrogen should be essentially the same. These limits are conventionally determined by dynamic TGA (Chapter 4E); isothermal weight loss experiments at high temperatures sometimes yield other results.

The synthesis of ladder polymers, anticipated with enthusiasm a decade ago, has led to little more than laboratory curiosities (Overberger 1970). These polymers are defined as those consisting of an uninterrupted series of rings connected by links around which rotation cannot occur except by bond breaking. If their structure were perfect, the chain of a ladder polymer could be broken only if at least two bonds on the same ring were broken. In a degradation process in which all bonds are equally strong, this would be far less probable than a single rupture breaking a single-chain polymer. Thus ladder polymers should have far greater thermal stability. Such polymers may be formed by reactions of existing polymer chains; there are several well-known examples: One is the cyclization of poly(methyl vinyl ketone):

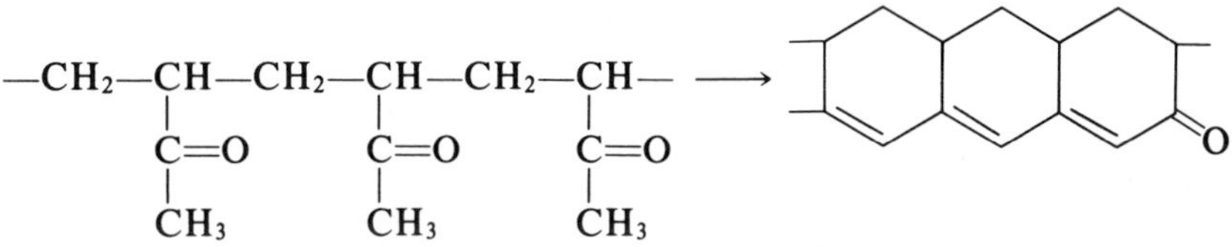

A second is the well-studied production of "black orlon" by the simultaneous cyclization and oxidation at 160–300°C of polyacrylonitrile, originally reported by Houtz (1950). The final material, whose structure may be

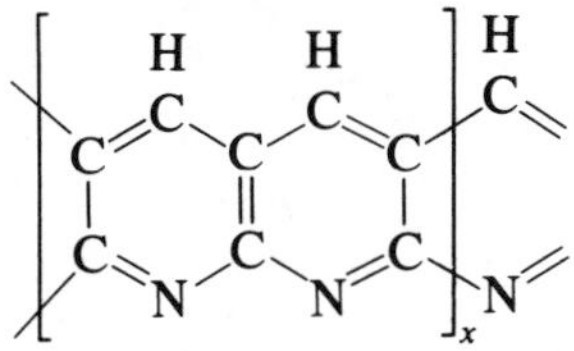

or may be an oxidized and aromatized reaction product of an intermediate of this structure, glows red but maintains its form in a blowtorch flame (as do some of the heterocyclic polymers described below), withstanding temperatures of 700–800°C without loss in properties (which are admittedly not too good). This material is, however, the precursor of the graphite fibers described in Chapter 17D.

High-Temperature Polymers

Polyimides. Polyimides with structures similar to

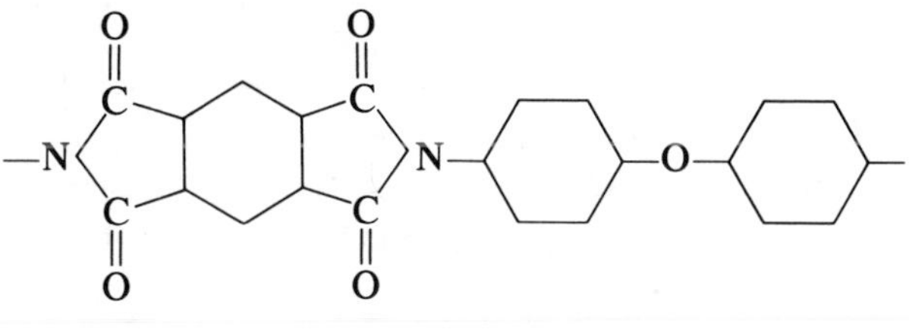

are made by the polycondensation of aromatic dianhydrides and aromatic amines. The first step is a soluble polyamic acid, which is converted to the polyimide by further condensation.

These materials are available as films and as fabricated solid parts. The polyamic acid is also sold as a coating solution. The polyimides retain usable properties at 300°C for months, and at 400°C for a few hours, and withstand exposures of a few minutes to temperatures well over 500°C. Depending on degree of condensation and structure, the polyimides can be used either as thermoplastic materials for compression molding, or thermosetting formulations.

Related commercial polymers include polyamide–imides,

and polyetherimides.

Poly(phenylene oxide).   The polymer

is made by oxidative coupling (Chapter 2*B*). It is most useful as blends with polystyrene, often filled, which have excellent dimensional stability at elevated temperatures, good electrical properties, and good resistance to aqueous environments. Water absorption is exceptionally low. As far as is disclosed, these are the most widely used of the high-temperature polymers described in this section, about 100 million lb being produced in 1982.

A related material, poly(phenylene sulfide), does not have the methyl substituents on the rings. Its use temperatures are higher than those of the modified phenylene oxide compositions.

Conductive Polymers.   Although no commercial products are yet known, there is considerable interest in polymers with a wide range of electrical conductive properties, produced for the most part by doping with inorganics such as arsenic pentafluoride or iodine, polymers such as polyacetylene,

$$-CH{=}CH{-}CH{=}CH{-}$$

poly-*p*-phenylene,

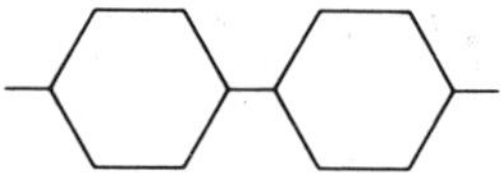

polypyrrole,

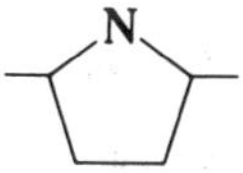

and the polyphenylene sulfide mentioned above (Mort 1980, Seymour 1981).

Miscellaneous High-Temperature Polymers.   Minor commercial polymers and others promoted for the purpose but not known to be commercially available are listed in Table 15-1.

TABLE 15-1.   Structures of Miscellaneous High-Temperature Polymers

Polybenzimidazole	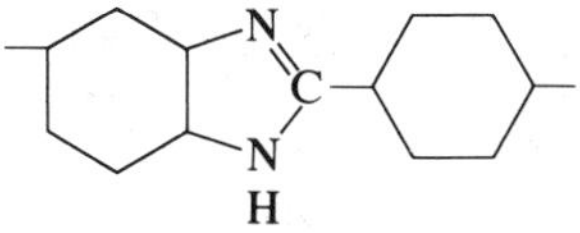
Polyarylate	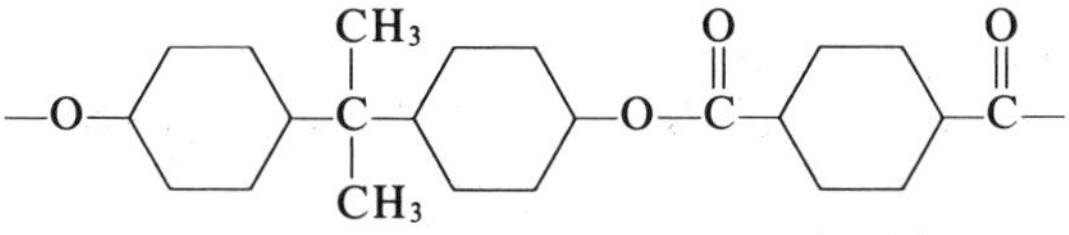 mixed isomers
Polyaryletherketone	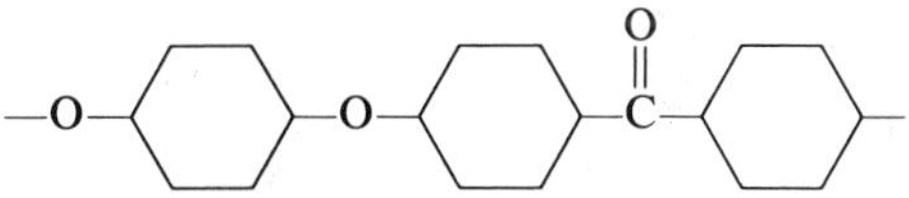
Polyarylethersulfone	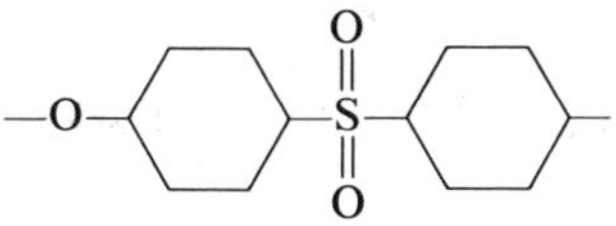
Polynorbornene	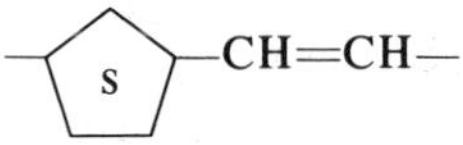

Inorganic Polymers

Fraser (1968) has described three major classes of inorganic polymers on which research has been concentrated.

1. Organic-inorganic polymers in which organic substituents are placed on inorganic chains. The outstanding example, and the only family of inorganic polymers to date, is the silicones, discussed in Chapter 16D. Research on other families of this type has centered on replacing the silicon in silicone-like structures with other elements, such as aluminum, tin, titanium, or boron. The objective is to improve the already good thermal stability of the silicones.

2. Metal chelate polymers, sometimes called coordination polymers (not to be confused with the entirely different conventional organic polymers produced by coordination polymerization—Chapter 4D). These materials can be prepared in several ways: by linking polydentate ligands by metal ions, as in metal acetonylacetonates; by polymer formation in the presence of metals, as in the production of polyphthalocyanines; by incorporation of metal ions in preformed polymers; by reacting chelates containing functional groups, as in the polymerization of basic beryllium carboxylates; and by the preparation of polymers containing ferrocene groups, such as biscyclopentadienyl iron.

3. Completely inorganic linear polymers based on silicon-nitrogen, phosphorous-nitrogen, or boron-nitrogen chains.

Two examples of the last-named class deserve mention. First, a wide variety of polymers has been prepared based on modifications of polydichlorophosphazene, —NCl_2—, (itself hydrolytically unstable) by nucleophilic substitution reactions. By way of example, the polymer

$$
\begin{array}{c}
OCH_2CF_3 \\
| \\
-N{=}P- \\
| \\
OCH_2CF_3
\end{array}
$$

is a stable, film-forming thermoplastic with $T_m = 242°C$ and $T_g = -66°C$. It is said to be more water-repellent than either the fluorocarbon or silicone polymers.

Second, boron nitride (BN) fibers can be produced by treatment with ammonia of a B_2O_3 precursor fiber. The BN fibers are stated to be stable and chemically inert at temperatures of 1800–2000°C.

GENERAL REFERENCES

High-Temperature Polymers. Fraser 1968, 1969; Hale 1969; Johnson 1969; Kovacic 1969; Levine 1969; Sroog 1969, 1976; Braunsteiner 1974; Marvel 1975; Elias 1977, Chapter 28.5; Hensel 1977; Schildknecht 1977; Cassidy 1980, 1982; Mort 1980; Ohm 1980; Preston 1980; Seymour 1981; Stening 1981; Boeke 1982;

Cekis 1982; Dickinson 1982; Feth 1982; Floryan 1982; Recchia 1982; Seanor 1982; Serfaty 1982; Sullo 1982.

Inorganic Polymers. Stone 1962; Hunter 1964; Andrianov 1965; Teach 1965; Block 1966; Allcock 1967, 1972, 1975, 1976, 1981, Chapter 7; Arledter 1967; Rebenfeld 1967; Venezky 1967; Neuse 1968; Sander 1969; Elias 1977, Chapter 33; Carraher 1978; Economy 1980; Peters 1981.

DISCUSSION QUESTIONS AND PROBLEMS

1. Write chemical equations for the polymerization, as practiced industrially, of (a) 66-nylon, (b) 6-nylon, (c) poly(ethylene terephthalate), (d) the polyurethane made from hexamethylene diisocyanate and 1,4-butanediol.

2. Describe and explain the relation between moisture content and toughness in 66-nylon.

3. Compare 66-, 610-, and 11-nylons with respect to water sensitivity.

4. What post-polymerization treatment is needed to develop the best properties in 6-nylon? Why?

5. In what respects does silk resemble the major synthetic fibers more than other natural fibers?

6. Write chemical formulas for (a) a repeat unit of cellulose, (b) the formation of cellulose acetate, (c) the reaction of cellulose with alkalis, (d) the formation of cellulose xanthate, (e) the regeneration of viscose, (f) the formation of ethyl cellulose.

7. Write the chemical formula for the presumed structure of the pyrolysis product of polyacrylonitrile.

8. Write chemical equations for the polymerization of (a) a polyimide, (b) poly(phenylene oxide), (c) a polycarbonate, (d) a polybenzimidazole, and (e) boron nitride fiber.

BIBLIOGRAPHY

Akin 1962. R. B. Akin, *Acetal Resins,* Reinhold, New York, 1962.

Allcock 1967. Harry R. Allcock, *Heteroatom Ring Systems and Polymers,* Academic Press, New York, 1967.

Allcock 1972. Harry R. Allcock, *Phosphorus–Nitrogen Compounds: Cyclic, Linear, and High-Polymeric Systems,* Academic Press, New York, 1972.

Allcock 1975. Harry R. Allcock, "Poly(organophosphazenes)," *Chemtech* **5,** 552–560 (1975).

Allcock 1976. Harry R. Allcock, "Polyphosphazenes: New Polymers with Inorganic Backbone Atoms," *Science* **193,** 1214–1219 (1976).

Allcock 1981. Harry R. Allcock and Frederick W. Lampe, *Contemporary Polymer Chemistry,* Prentice-Hall, Englewood Cliffs, New Jersey, 1981.

Andrews 1979. B. A. Kottes Andrews and Ines V. De Gruy, "Cotton," pp. 176–195 in *ECT,*† Vol. 7, 1979.

Andrianov 1965. K. A. Andrianov, *Metalorganic Polymers,* Wiley-Interscience, New York, 1965.

Arledter 1967. H. F. Arledter, F. L. Pundsack, and W. O. Jackson, "Fibers, Inorganic," pp. 610–690 in *EPST,*‡ Vol. 6, 1967.

Backus 1977. John K. Backus, "Polyurethanes," Chapter 17 in Calvin E. Schildknecht, ed., with Irving Skeist, *Polymerization Processes,* Wiley-Interscience, New York, 1977.

Ballintyne 1982. Nicolaas J. Ballintyne, "Polysulfone Resins," pp. 832–848 in *ECT,* Vol. 18, 1982.

Berenbaum 1969. M. B. Berenbaum, "Polysulfide Polymers," pp. 425–447 in *EPST,* Vol. 11, 1969.

Bevington 1964. J. C. Bevington and H. May, "Aldehyde Polymers," pp. 609–628 in *EPST,* Vol. 1, 1964.

Bikales 1971. N. M. Bikales and L. Segal, *Cellulose and Cellulose Derivatives,* John Wiley & Sons, New York, 1971.

Block 1966. B. B. Block, "Coordination Polymers," pp. 150–165 in *EPST,* Vol. 4, 1966.

Blokland 1968. R. Blokland, *Elasticity and Structure of Polyurethane Networks,* Gordon and Breach, New York, 1968.

Boeke 1982. P. J. Boeke, "Polyphenylene Sulfide," pp. 93–94 in *MPE,*§ 1982.

Bogan 1979. R. T. Bogan, C. M. Kuo, and R. J. Brewer, "Cellulose Derivatives, Esters," pp. 118–142 in *ECT,* Vol. 5, 1979.

Bottenbruch 1969. L. Bottenbruch, "Polycarbonates," pp. 710–764 in *EPST,* Vol. 10, 1969.

Braun 1982. David B. Braun and D. J. De Long, "Ethylene Oxide Polymers," pp. 616–632 in *ECT,* Vol. 18, 1982.

Braunsteiner 1974. E. E. Braunsteiner and H. F. Mark, "Aromatic Polymers," *J. Polym. Sci.* **D9,** 83–126 (1974).

Brown 1982. R. Malcolm Brown, Jr., ed., *Cellulose and Other Natural Polymer Systems: Biogenesis, Structure, & Degradation,* Plenum Press, New York, 1982.

Bruins 1969. Paul F. Bruins, ed., *Polyurethane Technology,* Wiley-Interscience, New York, 1969.

Bruxelles 1965. G. N. Bruxelles and V. R. Grassie, "Cellulose Esters, Inorganic," pp. 307–325 in *EPST,* Vol. 3, 1965.

Brydson 1978. J. A. Brydson, *Rubber Chemistry,* Applied Science, London, 1978.

Carraher 1978. Charles E. Carraher, Jr., John E. Sheats, and Charles U. Pittman, eds., *Organometallic Polymers,* Academic Press, New York, 1978.

Cassidy 1980. Patrick E. Cassidy, *Thermally Stable Polymers: Synthesis and Properties,* Marcel Dekker, New York, 1980.

Cassidy 1982. Patrick E. Cassidy, "Polyimides," pp. 704–719 in *ECT,* Vol. 18, 1982.

Cekis 1982. G. V. Cekis, "Polyamide–Imide," pp. 56, 58 in *MPE,* 1982.

Daul 1981. George C. Daul, "Rayon Revisited," *Chemtech* **11,** 83–87 (1981).

Davis 1982. Gerald W. Davis and Eric S. Hill, "Polyester Fibers," pp. 531–549 in *ECT,* Vol. 18, 1982.

†Martin Grayson, ed., *Kirk–Othmer Encyclopedia of Chemical Technology,* 3rd ed., Wiley-Interscience, New York.

‡Herman F. Mark, Norman G. Gaylord, and Norbert M. Bikales, eds., *Encyclopedia of Polymer Science and Technology,* Wiley-Interscience, New York.

§Joan Agranoff, ed., *Modern Plastics Encyclopedia 1982–1983,* Vol. 59, No. 10A, McGraw-Hill, New York, 1982.

Dickinson 1982. B. L. Dickinson, "Polyarylate," pp. 61–62 in *MPE*, 1982.

Dominguez 1982. R. J. G. Dominguez, "Polyurethane," pp. 102, 104, 106, 108 in *MPE*, 1982.

Dreyfuss 1982. P. Dreyfuss and M. P. Dreyfuss, "Tetrahydrofuran and Oxetane Polymers," pp. 645–670 in *ECT*, Vol. 18, 1982.

Economy 1980. James Economy, "Now That's an Interesting Way to Make a Fiber!," *Chemtech* **10**, 240–247 (1980).

Eichhorn 1977. Robert M. Eichhorn, "Polysulfone Coming on Strong for Electrical Applications," *Plastics Eng.* **33** (2), 53–56 (1977).

Elias 1977. Hans-Georg Elias, *Macromolecules ·2· Synthesis and Materials* (translated by John W. Stafford), Plenum Press, New York, 1977.

Farrow 1969. G. Farrow, E. S. Hill, and P. L. Weinle, "Polyester Fibers," pp. 1–41 in *EPST*, Vol. 11, 1969.

Feth 1982. George Feth, "Modified Phenylene Oxide," pp. 44, 49 in *MPE*, 1982.

Fisher 1982. William B. Fisher and Lamberto Crescentini, "(Polyamides) Caprolactam," pp. 425–436 in *ECT*, Vol. 18, 1982.

Floryan 1982. D. E. Floryan and I. W. Serfaty, "Polyetherimide: More Information on a New High-Performance Resin," *Modern Plastics* **59** (6), 146, 151 (1982).

Fox 1982. D. W. Fox, "Polycarbonates," pp. 479–494 in *ECT*, Vol. 18, 1982.

Frazer 1968. A. H. Frazer, *High Temperature Resistant Polymers*, Wiley-Interscience, New York, 1968.

Frazer 1969. A. H. Frazer, "High-Temperature Plastics," *Sci. Am.* **221** (1), 96–100, 103–105 (1969).

Galanty 1982. P. G. Galanty, E. C. Caughey, and P. P. Salatiello, "Nylon, Crystalline," pp. 49–50, 52 in *MPE*, 1982.

Goodman 1969. I. Goodman, "Polyesters," pp. 62–128 in *EPST*, Vol. 11, 1969.

Greminger 1979. G. K. Greminger, Jr., "Cellulose Derivatives, Ethers," pp. 143–163 in *ECT*, Vol. 5, 1979.

Hale 1969. Warren F. Hale, "Phenoxy Resins," pp. 111–122 in *EPST*, Vol. 10, 1969.

Haskell 1965. V. C. Haskell, "Cellophane," pp. 60–79 in *EPST*, Vol. 3, 1965.

Hawthorne 1969. J. M. Hawthorne and C. J. Heffelfinger, "Polyester Films," pp. 42–61 in *EPST*, Vol. 11, 1969.

Hensel 1977. Joseph D. Hensel, "When, Where, and How to Use Polyimides," *Plastics Eng.* **33** (10), 20–23 (1977).

Hill 1965. Roy O. Hill, Jr., B. P. Rouse, Jr., B. Sheldon Sprague, Lawrence I. Horner, and David J. Stanonis, "Cellulose Esters—Organic," pp. 325–454 in *EPST*, Vol. 3, 1965.

Hill 1982. H. Wayne Hill and D. G. Brady, "Poly(phenylene Sulfide)," pp. 793–814 in *ECT*, Vol. 18, 1982.

Houtz 1950. R. C. Houtz, " 'Orlon' Acrylic Fiber: Chemistry and Properties," *Textile Res. J.* **20**, 786–801 (1950).

Hunter 1964. D. N. Hunter, *Inorganic Polymers*, John Wiley & Sons, New York, 1964.

Ibrahim 1967. S. M. Ibrahim and A. J. Ultee, "Fibers, Elastomeric," pp. 573–592 in *EPST*, Vol. 6, 1967.

Jacobs 1977. Donald B. Jacobs and Joseph Zimmerman, "Preparation of 6,6-Nylon and Related Polyamides," Chapter 12 in Calvin E. Schildknecht, ed., with Irving Skeist, *Polymerization Processes*, Wiley-Interscience, New York, 1977.

Jaquiss 1982. Donald B. G. Jaquiss, W. F. H. Borman, and R. W. Campbell, "Polyesters, Thermoplastic," pp. 549–574 in *ECT*, Vol. 18, 1982.

Johnson 1969. R. N. Johnson, "Polysulfones," pp. 447–463 in *EPST*, Vol. 11, 1969.

Katz 1977. Manfred Katz, "Preparation of Linear Saturated Polyesters," Chapter 13 in Calvin E.

Schildknecht, ed., with Irving Skeist, *Polymerization Processes,* Wiley-Interscience, New York, 1977.

Kohan 1973. Melvin I. Kohan, ed., *Nylon Plastics,* Wiley-Interscience, New York, 1973.

Kovacic 1969. Peter Kovacic and Fred W. Koch, "Poly(phenylenes)," pp. 380–389 in *EPST,* Vol. 11, 1969.

Layman 1982. Patricia L. Layman, "Aramids, Unlike Other Fibers, Continue Strong," *Chem. Eng. News* **60** (6), 23 (1982).

Levine 1969. H. H. Levine, "Polybenzimidazoles," pp. 188–232 in *EPST,* Vol. 11, 1969.

Lundberg 1982. John Lundberg and Albin Turbak, "Rayon," pp. 855–880 in *ECT,* Vol. 19, 1982.

Mark 1940. H. Mark and G. Stafford Whitby, eds., *Collected Papers of Wallace Hume Carothers on High Polymeric Substances,* Interscience, New York, 1940.

Mark 1967. Herman F. Mark, "The Nature of Polymeric Materials," *Sci. Am.* **217** (3), 148–154, 156 (1967).

Marvel 1975. C. S. Marvel, "Trends in High-Temperature Stable Polymer Synthesis," *J. Macromol. Sci. Rev. Macromol Chem.* **C13,** 219–233 (1975).

Mitchell 1969. R. L. Mitchell and G. C. Daul, "Rayon," pp. 810–847 in *EPST,* Vol. 11, 1969.

Moncrieff 1975. R. W. Moncrieff, *Man-Made Fibers,* Newnes-Butterworths, London, 1975.

Mort 1980. J. Mort, "Conductive Polymers," *Science* **208,** 819–825 (1980).

Neuse 1968. Eberhard W. Neuse, "Metallocene Polymers," pp. 667–692 in *EPST,* Vol. 8, 1968.

Newton 1982. Robert A. Newton, "Propylene Oxide Polymers and Higher 1,2-Epoxide Polymers," pp. 633–645 in *ECT,* Vol. 18, 1982.

Nissan 1965. Alfred H. Nissan, Gunther K. Hunger, and S. S. Sternstein, "Cellulose," pp. 131–226 in *EPST,* Vol. 3, 1965.

Ohm 1980. Robert F. Ohm, "Polynorbornene, the Porous Polymer," *Chemtech* **10,** 183–187 (1980).

Ott 1954–55. Emil Ott, Harold M. Spurlin, and Mildred W. Grafflin, eds., *Cellulose and Cellulose Derivatives,* Interscience, New York, Parts I and II, 1954; Part III, 1955.

Overberger 1970. C. G. Overberger and J. A. Moore, "Ladder Polymers," *Adv. Polym. Sci.* **7,** 113–150 (1970).

Page 1982. S. L. Page, "Polycarbonate," p. 60 in *MPE,* 1982.

Pengilly 1982. B. W. Pengilly and J. W. Hill, "Thermoplastic Polyester: PET," pp. 64, 69, 70 in *MPE,* 1982.

Persak 1978. K. J. Persak and L. M. Blair, "Acetal Resins," pp. 112–123 in *ECT,* Vol. 1, 1978.

Persak 1982. K. J. Persak and W. G. Lofthouse, "Acetal Homopolymer," pp. 8, 17 in *MPE,* 1982.

Peters 1981. Edward N. Peters, "Inorganic High Polymers," pp. 398–413 in *ECT,* Vol. 13, 1981.

Pigott 1969. K. A. Pigott, "Polyurethanes," pp. 506–563 in *EPST,* Vol. 11, 1969.

Preston 1978. J. Preston, "Aramid Fibers," pp. 213–242 in *ECT,* Vol. 3, 1978.

Preston 1980. J. Preston, "Heat-Resistant Polymers," pp. 203–225 in *ECT,* Vol. 12, 1980.

Putscher 1982. Richard E. Putscher, "Polyamides, General," pp. 328–371 in *ECT,* Vol. 18, 1982.

Rebenfeld 1967. Ludwig Rebenfeld, "Fibers," pp. 505–573 in *EPST,* Vol. 6, 1967.

Recchia 1982. F. P. Recchia and W. J. Farrissey, "Thermoplastic Polyimide," pp. 82, 84 in *MPE,* 1982.

Reitano 1982. P. A. Reitano, "Nylon, Amorphous (Transparent)," pp. 52, 56 in *MPE,* 1982.

Rich 1982. R. P. Rich, "Cellulosic," pp. 29, 30, 32 in *MPE,* 1982.

Rigby 1982. R. B. Rigby and D. F. Dakin, "Sulfone Polymers," pp. 118–119 in *MPE,* 1982.

Royal 1982. C. A. Royal, "Acetal Copolymer," pp. 17–18 in *MPE,* 1982.

Sander 1969. Manfred Sander and H. R. Allcock, "Phosphorus-Containing Polymers," pp. 123–144 in *EPST,* Vol. 10, 1969.

Saunders 1962. J. H. Saunders and K. C. Frisch, *Polyurethanes: Chemistry and Technology. Part 1: Chemistry,* Wiley-Interscience, New York, 1962.

Saunders 1964. J. H. Saunders and K. C. Frisch, *Polyurethanes: Chemistry and Technology. Part 2: Technology,* Wiley-Interscience, New York, 1964.

Saunders 1982. J. H. Saunders, "Polyamide Fibers," pp. 372–405 in *ECT,* Vol. 18, 1982.

Savage 1965. A. B. Savage, E. D. Klug, Norbert M. Bikales, and David J. Stanonis, "Cellulose Ethers," pp. 459–549 in *EPST,* Vol. 3, 1965.

Scales 1982. R. E. Scales, "Thermoplastic Copolyester," p. 70 in *MPE,* 1982.

Schildknecht 1977. Calvin E. Schildknecht, "Synthesis of Some High-Temperature Polymers," Chapter 19 in Calvin E. Schildknecht, ed., with Irving Skeist, *Polymerization Processes,* Wiley-Interscience, New York, 1977.

Schnell 1964. Herman Schnell, *Chemistry and Physics of Polycarbonates,* Wiley-Interscience, New York, 1964.

Schule 1969. E. C. Schule, "Polyamide Plastics," pp. 460–482 in *EPST,* Vol. 10, 1969.

Seanor 1982. Donald A. Seanor, ed., *Electrical Properties of Polymers,* Academic Press, New York, 1982.

Serad 1979. George A. Serad and J. R. Sanders, "Cellulose Acetate and Triacetate Fibers," pp. 89–117 in *ECT,* Vol. 5, 1979.

Serfaty 1982. I. W. Serfaty and J. R. Bartolomucci, "Polyetherimide," pp. 72–73 in *MPE,* 1982.

Seymour 1981. Raymond B. Seymour, *Conductive Polymers,* Plenum Press, New York, 1981.

Snider 1969. O. E. Snider and R. J. Richardson, "Polyamide Fibers," pp. 347–460 in *EPST,* Vol. 10, 1969.

Sroog 1969. C. E. Sroog, "Polyimides," pp. 247–272 in *EPST,* Vol. 11, 1969.

Sroog 1976. C. E. Sroog, "Polyimides," *J. Polym. Sci.* **D11,** 161–208 (1976).

Staudinger 1925. H. Staudinger, "The Constitution of Polyoxymethylenes and Other High-Molecular Compounds" (in German), *Helv. Chim. Acta* **8,** 67–70 (1925).

Stening 1981. T. C. Stening, C. P. Smith, and P. J. Kimber, "Polyaryletherketone: High Performance in a New Thermoplastic," *Mod. Plastics* **58** (11), 86–87, 89 (1981).

Stone 1962. F. G. A. Stone and W. A. G. Graham, eds., *Inorganic Polymers,* Academic Press, New York, 1962.

Sullo 1982. N. A. Sullo, "Thermoset Polyimide," pp. 84, 93 in *MPE,* 1982.

Sweeny 1969. W. Sweeny and J. Zimmerman, "Polyamides," pp. 483–597 in *EPST,* Vol. 10, 1969.

Teach 1965. W. C. Teach and Joseph Green, "Boron-Containing Polymers," pp. 581–604 in *EPST,* Vol. 2, 1965.

Turbak 1979. Albin F. Turbak, Donald F. Durso, O. A. Battista, Henry I. Bolker, J. Ross Colvin, Nathan Eastman, Theodore N. Kleinert, Hans Krassig, and R. St. John Manley, "Cellulose," pp. 70–88 in *ECT,* Vol. 5, 1979.

Vandenberg 1979. E. J. Vandenberg, "Polyethers," pp. 568–582 in *ECT,* Vol. 8, 1979.

Venezky 1967. David L. Venezky, "Inorganic Polymers," pp. 664–691 in *EPST,* Vol. 7, 1967.

Welgos 1982. R. J. Welgos, "Polyamide Plastics," pp. 406–425 in *ECT,* Vol. 18, 1982.

White 1982. Dwain N. White and Glenn D. Cooper, "Aromatic Polyethers," pp. 594–615 in *ECT,* Vol. 18, 1982.

Wright 1969. P. Wright and A. P. C. Cumming, *Solid Polyurethane Elastomers,* Gordon and Breach, New York, 1969.

Yarger 1982. S. B. Yarger, "Thermoplastic Polyester: PBT," pp. 62, 64 in *MPE,* 1982.

CHAPTER SIXTEEN

THERMOSETTING RESINS

Thermosetting resins are those that change irreversibly under the influence of heat from a fusible and soluble material into one which is infusible and insoluble through the formation of a covalently crosslinked, thermally stable network. In contrast, *thermoplastic* polymers, discussed in Chapters 13–15, soften and flow when heat and pressure are applied, the changes being reversible.

In some of the systems considered in this chapter, network formation occurs with little or no heat required, as in the production of urethane foams and the use of unsaturated polyester resins. Furthermore, vulcanized rubbers consist of covalently crosslinked network polymers, usually formed by the application of heat. However, this network is generated in a separate postpolymerization step. With the exception of the silicone rubbers, thermosetting resins discussed in this chapter are those in which crosslinking occurs simultaneously with the final stages of polymerization, regardless of the amount of heat required in this step. The vulcanization of rubbers is discussed in Chapter 19A.

The most important thermosetting resins, both from a historic standpoint and in current commercial application, are formaldehyde condensation products with phenol (*phenolic resins*) or with urea or melamine (*amino resins*). Other thermosetting types include the epoxy resins, the unsaturated polyester resins, urethane foams, the alkyds widely used for surface coating, and minor types.

A. PHENOLIC AND AMINO RESINS

Phenolic Resins

Phenolic resins have been in commercial use longer than any other synthetic polymer except cellulose nitrate. In contrast to the latter, however, the sales of phenolic resins continued to rise at about 15% per year, reaching a peak of about 1.3 billion lb in 1979. Severely hit by the depression because of major uses in plywood and other construction uses, sales dropped 44% to about 0.97 billion lb† in 1982, at an average price of $0.45 per pound.

Reactions of Phenol and Formaldehyde.   Phenols react with aldehydes to give condensation products if there are free positions on the benzene ring ortho and para to the hydroxyl group. Formaldehyde is by far the most reactive aldehyde and is used almost exclusively in commercial production. The reaction is always catalyzed, either by acids or by bases. The nature of the product is greatly dependent on the type of catalyst and the mole ratio of the reactants.

The first step in the reaction is the formation of addition compounds known as methylol derivatives, the reaction taking place at the ortho or para position:

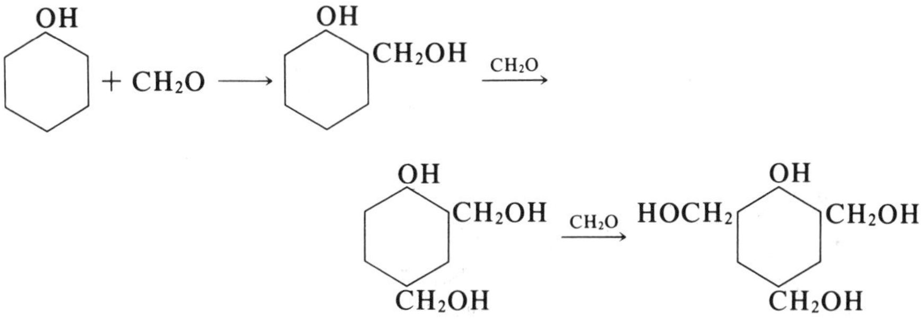

These products, which may be considered the monomers for subsequent polymerization, are formed most satisfactorily under neutral or alkaline conditions. In the presence of acid catalysts, and with the mole ratio of formaldehyde to phenol less than 1, the methylol derivatives condense with phenol to form, first, dihydroxy-diphenyl methane:

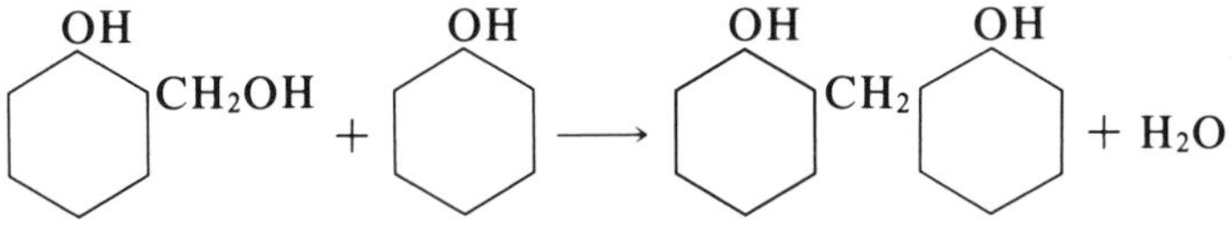

and, on further condensation and methylene bridge formation, fusible and soluble

†Beginning in 1981, phenolics production and sales volumes were reported as gross weights including solvents, roughly 85% higher than the figures cited.

linear low polymers called *novolacs* with the structure

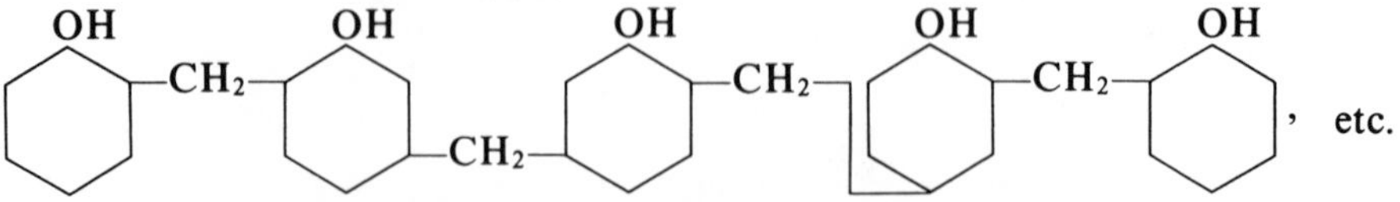

where ortho and para links occur at random. Molecular weights may range as high as 1000, corresponding to about ten phenyl residues. These materials do not themselves react further to give crosslinked resins, but must be reacted with more formaldehyde to raise its mole ratio to phenol above unity.

In the presence of alkaline catalysts and with more formaldehyde, the methylol phenols can condense either through methylene linkages or through ether linkages. In the latter case, subsequent loss of formaldehyde may occur with methylene bridge formation:

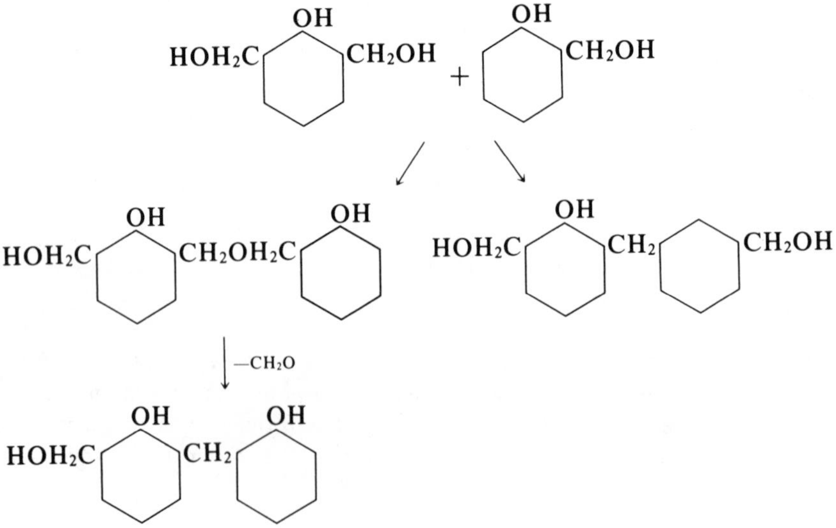

Products of this type, soluble and fusible but containing alcohol groups, are called *resoles*. If the reactions leading to their formation are carried further, large numbers of phenolic nuclei can condense to give network formation.

In summary, the four major reactions in phenolic resin chemistry are (a) addition to give methylol phenols, (b) condensation of a methylol phenol and a phenol to give a methylene bridge, (c) condensation of two methylol groups to give an ether bridge, and (d) decomposition of ether bridges to methylene bridges and formaldehyde, the latter reacting again by the first mode.

Production of Phenolic Resins. The formation of resoles and novolacs, respectively, leads to the production of phenolic resins by *one-stage* and *two-stage* processes. In the production of a one-stage phenolic resin, all the necessary reactants for the final polymer (phenol, formaldehyde, and catalyst) are charged into a resin kettle and reacted together. The ratio of formaldehyde to phenol is about $1.25:1$, and an alkaline catalyst is used. Two-stage resins are made with an acid catalyst, and only part of the necessary formaldehyde is added to the kettle, producing a mole ratio of about $0.8:1$. The rest is added later as hexamethylenetetramine, which decomposes in the final curing step, with heat and moisture present, to yield formaldehyde and ammonia which acts as the catalyst for curing.

The procedures for one- and two-stage resins are similar, and the same equipment is used for both. The reaction is exothermic and cooling is required. The formation of a resole or a novolac is evidenced by an increase in viscosity. Water is then driven off under vacuum, and a thermoplastic *A-stage* resin, soluble in organic solvents, remains. This material is dumped from the kettle, cooled, and ground to a fine powder.

At this point fillers, colorants, lubricants, and (if a two-stage resin) enough hexamethylenetetramine to give a final formaldehyde:phenol mole ratio of $1.5:1$ are added. The mixture is rolled on heated mixing rolls, where the reactions are carried further, to the point where the resin is in the *B stage,* nearly insoluble in organic solvents but still fusible under heat and pressure. The resin is then cooled and cut into final form. The *C stage*, the final, infusible, crosslinked polymer, is reached on subsequent fabrication—for example, by molding.

Properties and Uses. Especially when combined with suitable fillers, phenolic resins have good chemical and thermal resistance, dielectric strength, and dimensional stability. Products made with these resins are inherently low in flammability, are creep resistant, and have low moisture absorption.

By far the largest use of the phenolics (some 43% of production) is in heat-setting adhesives for plywood. Related uses in insulation and fibrous and granulated wood products account for almost another 25% of production.

For impregnating paper, wood, and other fillers, about 8% of the phenolic resins is produced as alcoholic solutions of one-stage resins (essentially varnishes), called laminating resins. These are used to produce decorative laminates for counter tops and wall coverings and industrial laminates for electrical parts, including printed circuits. The impregnated filler is dried in an air oven to remove volatiles and then is hot pressed between polished platens. Shaped products, including cutlery handles and toilet seats, are similarly made. A number of industrial applications are based on the excellent adhesive properties and bonding strength of the phenolics. These include the production of brake linings, abrasive wheels and sandpaper, and foundry molds (sand filled). Phenolics are also widely used in the production of ion-exchange resins, with amine, sulfonic acid, hydroxyl, or phosphoric acid functional groups.

Today less than 10% of the phenolics is used as injection-molded products, once their most popular outlet. These products are used primarily in the electrical and electronic industries, with other significant uses in appliances and housewares.

Amino Resins

The two important classes of amino resins are the condensation products of urea and of melamine with formaldehyde. They are considered together here because of the similarity in their production and applications. In general, the melamine resins have somewhat better properties but are higher in price. Production of amino resins in the United States in 1982 was about 1 billion lb.

Chemistry and Production.   Both melamine (I), a trimer of cyanamide, and urea

$$
\begin{array}{c}
NH_2 \\
| \\
N{=}\underset{\displaystyle \underset{\displaystyle}{}}{C}{-}N \\
H_2N{-}C \qquad C{-}NH_2 \\
N \\
I
\end{array}
$$

react with formaldehyde, first by addition to form methylol compounds and then by condensation in reactions much like those of phenol and formaldehyde. The methylol reaction takes the form

$$
O{=}C\!\!\begin{array}{c} \diagup NH_2 \\ \diagdown NH_2 \end{array} + CH_2O\ \longrightarrow\ \left[\begin{array}{c} O{=}C\!\!\begin{array}{c}\diagup HNCH_2OH \\ \diagdown NH_2\end{array} \\[2ex] O{=}C\!\!\begin{array}{c}\diagup HNCH_2OH \\ \diagdown HNCH_2OH\end{array}\end{array}\right.
$$

These may polymerize to crosslinked resins by loss of water. Other reactions that may enter into the condensation include the formation of methyleneurea,

$$
O{=}C\!\!\begin{array}{c} \diagup N{=}CH_2 \\ \diagdown NH_2 \end{array}
$$

and the following crosslinking reactions:

$$>N—CH_2—\overset{}{\boxed{OH + H}}—\overset{|}{N}—CH_2OH$$

$$\downarrow$$

$$>N—CH_2—\overset{|}{N}—CH_2OH + H_2O$$

$$>N—CH_2\boxed{OH + H}OCH_2—N<$$

$$\downarrow$$

$$>N—CH_2—O—CH_2—N< + H_2O$$

$$>N—\boxed{CH_2O\,H + HO}CH_2—N<$$

$$\downarrow$$

$$>N—CH_2—N< + H_2O + CH_2O$$

The production of the amino resins is similar to that of phenolic resins. Since the A-stage resin is water soluble, it is only partially dehydrated, the water solution being used to impregnate the filler. The molding resins are almost always filled with cellulose obtained from good-quality sulfite-bleached paper. Impregnation is carried out in a vacuum mixer, and the subsequent drying step carries the resin to the B stage. It is then ground to the desired particle size in ball mills.

Properties and Applications. A distinct advantage of the amino resins over the phenolics is the fact that they are clear and colorless, so that objects of light or pastel color can be produced. The tensile strength and hardness of the amino resins are better than those of the phenolics, but their impact strength and heat and moisture resistance are lower, although still characteristic of thermosetting resins. The melamine resins have better hardness, heat resistance, and moisture resistance than the ureas.

Almost three-quarters of the amino resins are used for adhesives, largely for plywood and furniture. The melamine resins give excellent, boil-resistant bonds, but for economy are usually blended with the ureas. The remaining quarter of production is divided roughly equally among uses as textile treatment and coatings resins, protective coatings, molding compounds, and paper treatment and coating resins.

Practically all urea molding compounds are cellulose filled, whereas the melamines, although predominantly cellulose filled, are also used with asbestos, glass or silica, and cotton fabric. Because of their poorer flow characteristics, the urea resins are usually compression molded, but injection molding is now common with both resins. Both resins can be preheated by high-frequency current because of their high polarity.

Because of their colorability, solvent and grease resistance, surface hardness, and mar resistance, the urea resins are widely used for cosmetic container closures, appliance housings, and stove hardware. The production of high-quality dinnerware from cellulose-filled compounds is the largest single molding use for the melamine resins.

Other Applications. The amino resins modify *textiles* such as cotton and rayon by imparting crease resistance, stiffness, shrinkage control, fire retardance, and water repellency. They also improve the wet strength, rub resistance, and bursting strength of *paper*. Alkylated resins, in which butyl- or amyl-substituted mono-methylol ureas or melamines are used, are combined with alkyd resins (Section *E*) to give *baking enamels*. The urea-based enamels are used for refrigerator and kitchen appliances, and the melamine formulations in automotive finishes.

GENERAL REFERENCES

Widmer 1965; Kentgen 1969; Elias 1977, Chapters 26.3 and 28.2; Updegraff 1977, 1978; Turro 1978; Brode 1982; Keegan 1982; Lichtenberg 1982.

B. UNSATURATED POLYESTER RESINS

Polyesters of several diverse types are useful as polymers. This section is concerned only with those in which the dibasic acid or the glycol, or both, contain double-bonded carbon atoms. It is further restricted to cases in which radical chain polymerization involving these double bonds and a vinyl monomer, usually styrene, is made to take place in the presence of a fibrous filler, generally glass. Other polyester systems are the fiber-forming saturated polyesters (Chapter 15*B*), the polyester intermediates in the production of urethane elastomers (Chapter 15*B*) or foams (Section *C*), and the alkyd paint and molding resins (Section *E*). The polyester systems discussed here have been known from time to time as *low-pressure laminating resins, contact resins, polymerizable polyesters,* and *styrenated polyesters.* The term *reinforced plastics,* however, is more general and includes the use of thermosetting resins other than the unsaturated polyesters, among them the phenolic, epoxy, melamine, diallyl phthalate, and alkyd resins, all discussed elsewhere in this chapter.

Badly depressed by the recession in marine products, construction, and transportation, the sales of unsaturated polyester formulations was about 860 million lb in 1982, down 12% from 1981 and at the lowest level in 6 years; this is less than half of estimated production capacity. Prices were about $0.60 per lb.

Composition of Reinforced Polyester Systems

Much of the versatility of the reinforced polyester systems lies in the wide variation in resin composition and fabrication methods possible, allowing the properties of the product to be tailored to the requirements of the application. This variability is manifested in the large number of resin components used. Most of the material produced is, however, based on a few key components.

The unsaturation in the polyester is usually supplied by the inclusion of maleic anhydride as one component. In addition, a saturated acid or anhydride is often used, such as phthalic anhydride. A higher proportion of unsaturated acid gives a more reactive resin, with improved stiffness at high temperatures, while more of the saturated components give less exothermic cures and less stiff resins. Propylene glycol is the most widely used dihydroxy component of the polyester, and styrene is the most popular by far of the crosslinking components. A typical formulation of a general-purpose polyester may contain 43 lb of phthalic anhydride, 19 lb of maleic anhydride, and 38 lb of propylene glycol. To these 100 lb the formulator will add about 70 lb of styrene for crosslinking. The amount of glass-fiber filler added to this may vary between 100 and 300 lb, depending upon the end use.

Formulation

The acid and glycol components of the polyester resin are mixed in a resin kettle and polymerized by step reaction to a molecular weight of 1000–5000, which is in the highly viscous liquid range. After cooling, the mixture is thinned down to a pourable liquid by the addition of the monomer. An inhibitor such as hydroquinone is then added to prevent premature polymerization. When kept cool, the mixture is stable for months to years.

Cure is begun by adding an initiator, usually an organic peroxide, such as benzoyl peroxide, or a hydroperoxide. Typically, promoters or accelerators are used to promote the decomposition of the initiator at room temperature and, thus, rapid low-temperature curing. Common accelerators are cobalt naphthenate or alkyl mercaptans. Cure takes place in two stages: The initial formation of a soft gel is followed by rapid polymerization. The heat evolved can lead to quite high temperatures in large masses of resin.

Fabrication

The fabrication of reinforced plastic articles is traditionally divided into laminating and molding processes, but this classification has been modified as newer techniques have been developed. The following discussion applies to the processing of a variety of reinforced plastics and composites in addition to the polyesters.

Laminating. Laminating processes are those in which separate plies of reinforcing material are coated or impregnated with resin and pressed together until cured to a single reinforced structure. The operation can be done batchwise, but most processing is now done continuously on an in-line conveyor on which the reinforcements are impregnated with resin, cured, and trimmed.

Molding. The simplest molding process for reinforced plastics is *contact molding* or open or hand lay-up molding, a process closely resembling laminating. A single mold is used; the reinforcing material is placed on the mold, impregnated with resin, and allowed to cure in air. Variations include *bag molding,* where a bag or blanket is used to apply low pressure to the open surface of the material, and *spray-up* techniques, in which the resin and sometimes chopped glass are sprayed onto the mold surface.

Matched-die molding includes processes involving two dies that, together, match or conform to all the dimensions of the finished piece. The material can be placed in the die in the form of mats or preforms of reinforcing material with the resin separately added; "pre-pregs," where the preform has been impregnated; or premixes, where the resin, reinforcement, and other extenders have been premixed.

Conventional molding techniques, described in Chapter 17*A*, can also be used with reinforced plastics if the reinforcement consists of short lengths of fiber, for example. In injection and transfer molding, some damage to the reinforcement may occur, but this is minimized in compression molding. *Stamping* is a variation in which the heated impregnated reinforcement is placed in an open mold, which quickly closes or strikes the materials, holds for 5–15 sec for cooling, then opens again for part removal and reloading. Very fast cycles are obtained.

Unsaturated polyester resins also form the basis of *sheet-molding compounds,* described in Chapter 17*C*.

Filament Winding. In this process, continuous filaments (fibers) are wound onto or in a form. Usually either the form or the supply of filament revolves in the process. The resin is used to bind or encapsulate the fibers, resulting in a solid structure. Filament winding is used for pipes for gases and chemicals, tanks, drive shafts, and bearings, and also for nonsymmetrical objects such as automobile leaf springs and helicopter blades.

Pultrusion. Originally devised for making glass fiber-reinforced fishing rods, this process resembles extrusion (Chapter 17*B*) except that the materials are pulled rather than pushed through a die. Continuous filament bundles, saturated with resin, are pulled through a curing die that shapes the profile while the resin is polymerizing.

Properties and Applications

The most important properties of the unsaturated polyester systems include ease of handling, rapid curing with no volatiles evolved, light color, dimensional stability, and generally good physical and electrical properties.

The major applications of glass-reinforced polyester resins fall in the following categories: boat hulls, whose popularity has risen spectacularly; transportation, including passenger car parts and bodies and truck cabs; consumer products, including such diverse items as luggage, chairs, and fishing rods; trays, pipe, and ducts; electrical applications; appliances; construction applications, largely sheet and paneling; and missile and radome uses.

GENERAL REFERENCES

Boenig 1969; Parker 1977; Clavadetscher 1982; Ewald 1982; Jensen 1982; McLarty 1982; Menzer 1982; Raymer 1982.

C. EPOXY RESINS AND POLYURETHANES

These thermosetting resins are considered together because of their similar chemistry, resulting from the reaction of a glycol with a reagent other than a diacid, as seen in the resins just discussed.

Epoxy Resins

The epoxy resins are fundamentally polyethers, but retain their name on the basis of their starting material and the presence of epoxide groups in the polymer before crosslinking. About 310 million lb of epoxy resins was sold in the United States in 1982, at an average price of $1.25 per lb.

Chemistry of Preparation and Curing. The epoxy resin used most widely is made by condensing epichlorohydrin with bisphenol A, diphenylol propane. An excess of epichlorohydrin is used, to leave epoxy groups on each end of the low-molecular-weight (900–3000) polymer:

Depending on molecular weight, the polymer is a viscous liquid or a brittle high-melting solid.

Other hydroxyl-containing compounds, including resorcinol, hydroquinone, glycols, and glycerol, can replace bisphenol A. No epoxides other than epichlorohydrin are available at attractive prices, however.

The epoxy resins are cured by many types of materials, including polyamines, polyamides, polysulfides, urea– and phenol–formaldehyde, and acids or acid anhydrides, through coupling or condensation reactions. The reaction with amines involves opening the epoxide ring to give a β-hydroxyamino linkage:

$$\overset{\displaystyle O}{-CH_2\overset{\diagup\diagdown}{C}HCH_2} + RNH_2 \longrightarrow CH_2\overset{\displaystyle \overset{OH}{|}}{C}HCH_2NHR$$

Acids and acid anhydrides react through esterification of the secondary hydroxyl groups on the epoxy resin as well as with the epoxide groups. The phenolic and amino resins may react in several ways, including condensation of methylol groups with the secondary hydroxyls of the epoxy, and reaction of the epoxide groups with phenolic hydroxyls and amino groups.

Epoxy resins can also be cured by cationic polymerization, using Lewis acid catalysts such as BF_3 and its complexes, which form polyethers from the epoxide groups.

Properties and Applications.   The major use of the epoxy resins is as surface-coating materials, which combine toughness, flexibility, adhesion, and chemical resistance to a nearly unparalleled degree. In addition to curing systems of the types described above, the epoxies can be esterified with drying or nondrying oil fatty acids and then cured by either air drying or baking.

Epoxy resins can be used in both molding and laminating techniques to make glass fiber-reinforced articles with better mechanical strength, chemical resistance, and electrical insulating properties than those obtained with unsaturated polyesters. Only the higher price of the epoxies prevents their wider use in this field.

Casting, potting, encapsulation, and embedment are widely practiced with the epoxy resins in the electrical and tooling industries. Liquid resins are often utilized, while hot-melt solids have some application.

Other important uses include industrial flooring, adhesives and solders, foams, highway surfacing and patching materials, and stabilizers for vinyl resins.

Polyurethanes

As indicated in Chapter 15B, urethane polymers contain the group —NHCOO— and are formed through the reaction of a diisocyanate and a glycol. In the production of urethane foams, excess isocyanate groups in the polymer react with water or carboxylic acids to produce carbon dioxide, blowing the foam, at the same time that crosslinking is effected.

Urethane foams can be made in either flexible or rigid form, depending on the nature of the polymer and the type of crosslinking produced. The flexible foam is more widely used, however, accounting for about 65% of the 1.5 billion lb produced in the United States in 1982.

Thermosetting polyurethane molding compounds fabricated by reaction injection molding (RIM) provide a new use for this polymer type, fabricated by what may become a major new processing technique.

Urethane Chemistry   Like polyurethane elastomers, urethane foams are made in several steps. A *basic intermediate* of molecular weight around 1000 is a polyether made from poly(1,4-butylene glycol), sorbitol polyethers, or others. The basic intermediate is bifunctional if flexible foams are desired and polyfunctional if rigid foams are to be made.

As in elastomer production, the intermediate is reacted with an aromatic diisocyanate, usually tolylenediisocyanate, to give a *prepolymer* of structure similar to that illustrated in Chapter 15B. Catalysts based on tertiary amines or on stannous soaps are added to achieve rapid production of foam. Crosslinking takes place through the formation of urea linkages:

$$RNCO + H_2O \longrightarrow RNH_2 + CO_2$$
$$R'NCO + RNH_2 \longrightarrow RNHCONHR'$$

The use of low-boiling inert liquids, in particular fluorocarbons, to augment or replace the chemical blowing action described above has led to certain property advantages in the final foams, such as low thermal conductivity characteristic of the entrapped fluorocarbon gas. The ingredients of the foam may be partly expanded (frothed) with an inert gas in a preliminary step, and subsequently expanded to form the final object in a separate second step.

Flexible Foams.   The use of flexible urethane foams for cushions for furniture and automobiles has displaced rubber foam in these applications because of improved strength, lower density, and easier fabrication.

The "one-shot" process, wherein polyether intermediate, tolylenediisocyanate, and catalysts are mixed just before foaming, is widely used for the production of flexible urethane foams. Most of the material is made in the form of slab stock in a continuous process, the foam being up to 8 ft wide and 3–4 ft high. It is cut into 10–60 ft lengths and, after curing 10–24 hr, is further cut up for sale to fabricators.

The foam can also be produced in molding processes, with or without external heating.

Rigid Foams.   Rigid urethane foams are resistant to compression and may be used to reinforce hollow structural units with a minimum of weight. In addition, they consist of closed cells and so have low rates of heat transmission. They develop excellent adhesion when they are formed in voids or between sheets of material.

Finally, they are resistant to oils and gasoline and do not absorb appreciable amounts of water. These properties make the rigid foams valuable for sandwich structures used in prefabrication in the building industry, for thermal insulation in refrigerators, portable insulated chests, and so on, and for imparting buoyancy to boats.

RIM Urethanes. Reaction injection molding (Chapter 17A) is a process that allows polymerization and crosslinking to take place simultaneous with forming of a part into its final shape. Because of the rapid curing of polyurethanes, compatible with the fast cycle times of RIM, these polymers seem exceptionally well suited to RIM processing. The major application to date has been in producing microcellular elastomers. Fillers can be used to produce quite stiff products. Most of the technology is currently used in the automotive industry, to produce bumper covers, with larger exterior parts such as fenders under development.

GENERAL REFERENCES

Saunders 1962–1964; Lee 1967; Bruins 1968, 1969; David 1969; Pigott 1969; Backus 1977; Elias 1977, Chapter 26.2.3; Salva 1977; Bauer 1980; Sherman 1980; Breitigam 1982; Dominguez 1982.

D. SILICONE POLYMERS

Like carbon, silicon has the capability of forming covalent compounds. Silicon hydrides (silanes) up to Si_6H_{14} are known. The Si–Si chain becomes thermally unstable at about this length, however, so that polymeric silanes are unknown. The siloxane link

$$-\overset{|}{\underset{|}{Si}}-O-\overset{|}{\underset{|}{Si}}-$$

is more stable, and is the one found in commercial silicone polymers. Unlike carbon, silicon does not form double or triple bonds. Thus silicone polymers are usually formed only by condensation-type reactions.

Silicone polymers became available commercially during World War II. They are particularly noted for their stability at temperatures as high as 150°C. The variety of products available ranges from liquids (lubricants, water repellents, release agents, defoamers) through greases and waxes to resins and rubbers.

Chemistry of the Silicones

The study of the chemistry of silicon and its compounds began with the discovery of the element in 1824. Soon afterwards $SiCl_4$ was prepared by reacting silicon and chlorine, and ethyl silicate was first made in 1844 by the reaction of the tetrachloride with ethanol. The compounds of silicon were studied intensively from that time on.

Silicone polymers are made from organosilicon intermediates prepared in various ways from elemental silicon, which is produced by reducing quartz in an electric furnace.

The intermediates ("monomers") are compounds of the type SiR_nX_{4-n}, where R is an alkyl or aryl group and X is a group which can be hydrolyzed to —SiOH, such as chlorine or alkoxy. The intermediates are made by a direct synthesis in which the R and X groups are attached simultaneously to the silicon by a high-temperature reaction of a halide with silicon in the presence of a metal catalyst. The chief reaction is, for example,

$$2CH_3Cl + Si \longrightarrow Si(CH_3)_2Cl_2$$

but a number of side reactions occur.

Polymerization

Silicone polymers are produced by intermolecular condensation of silanols, which are formed from the halide or alkoxy intermediates by hydrolysis:

$$-\overset{|}{\underset{|}{Si}}Cl + H_2O \longrightarrow -\overset{|}{\underset{|}{Si}}OH + HCl$$

$$-\overset{|}{\underset{|}{Si}}OH + HO\overset{|}{\underset{|}{Si}}- \longrightarrow -\overset{|}{\underset{|}{Si}}-O-\overset{|}{\underset{|}{Si}}- + H_2O$$

The desired siloxane structure is obtained by using silanols of different functionality, the alkyl (R) groups in the intermediate being unreactive.

Silicone Fluids

The silicone fluids are low polymers produced by the hydrolysis reaction mentioned above, in which a predetermined mixture of chlorosilanes is fed into water with agitation. In many cases, the cyclic tetramer predominates in the resulting mixture.

These compounds, not polymers in the sense of this book, are used as cooling and dielectric fluids, in polishes and waxes, as release and antifoam agents, and for paper and textile treatment.

Silicon Elastomers

Silicone elastomers are high-molecular-weight linear polymers, usually polydimethylsiloxanes. They can be cured in several ways:

a. By free-radical crosslinking with, for example, benzoyl peroxide, through the formation of ethylenic bridges between chains.

b. By crosslinking of vinyl or allyl groups attached to silicon through reaction with silylhydride groups:

$$\underset{\underset{R'}{|}}{\overset{\overset{R}{|}}{Si}}-CH{=}CH_2 + H-\underset{\underset{R'''}{|}}{\overset{\overset{R''}{|}}{Si}}- \longrightarrow -\underset{\underset{R'}{|}}{\overset{\overset{R}{|}}{Si}}-CH_2-CH_2-\underset{\underset{R'''}{|}}{\overset{\overset{R''}{|}}{Si}}-$$

c. By crosslinking linear or slightly branched siloxane chains having reactive end groups such as silanols. In contrast to the above reactions, this yields Si—O—Si crosslinks.

The latter mechanism forms the basis of the curing of room-temperature vulcanizing (RTV) silicone elastomers. These are available as two-part mixtures in which all three essential ingredients for the cure (silanol-terminated polymer, crosslinking agent such as ethyl silicate, and catalyst such as a tin soap) are combined at the time the two components are mixed, and as one-part materials using a hydrolyzable polyfunctional silane or siloxane as crosslinker, activated by atmospheric moisture.

Silicone elastomers must be reinforced by a finely divided material such as silica if useful properties are to be obtained. These materials are outstanding in low-temperature flexibility (to $-80°C$), stability at high temperatures (to $250°C$), and resistance to weathering and to lubricating oils. They are used as gaskets and seals, wire and cable insulation, and hot gas and liquid conduits. They are valuable in surgical and prosthetic devices. The RTV elastomers are very convenient for caulking, sealing, and encapsulating.

Silicone Resins

In contrast to the silicone fluids and elastomers, silicone resins contain Si atoms with no or only one organic substituent. They are therefore crosslinkable to harder and stiffer compounds than the elastomers, but many must be handled in solution to prevent premature cure. They are, in fact, usually made by hydrolysis of the desired chlorosilane blend in the presence of a solvent such as mineral spirits, butyl acetate, toluene, or xylene. These materials are usually cured with metal soaps or amines.

The silicone resins are used primarily as insulating varnishes, impregnating and encapsulating agents, and in industrial paints. A part to be coated is typically dipped into the resin solution, and drained or scraped free of excess resin. The solvent is allowed to evaporate, and the resin is cured in an oven.

GENERAL REFERENCES

Lichtenwalner 1970; Brydson 1978, Chap. 15.6; Owen 1981; Kookootsedes 1982; Hardman 1982.

E. MISCELLANEOUS THERMOSETTING RESINS

Alkyd Resins

Alkyd resins (the name deriving from *al*cohol + a*cid*) are polyesters used primarily in organic paints, with some molding applications. About 480 million lb was produced for paints, varnishes, and lacquers in 1982.

Among many possible compositions for alkyd resins, perhaps the most common is based on phthalic anhydride and glycerol (''glyptal''). Other polyhydric alcohols commonly used are glycols, pentaerythritol, and sorbitol; other acids include maleic anhydride, isophthalic, and terephthalic.

Many alkyd resins are modified by the addition of fatty acids derived from mineral and vegetable oils. If the acids are unsaturated, the resulting resins are the air-drying type which cure by oxidation of the (saponified) acids (*drying oils*). Baking type alkyds cure by heat alone, or by cocondensation with alkylated amino resins.

Surface coatings may be classified as *lacquer* types, in which drying involves only the evaporation of solvent, and *varnish* types, in which chemical reactions also take place on drying. These reactions may involve free-radical crosslinking of drying oils, the triglycerides of unsaturated acids such as oleic, linoleic, and ricinoleic.

Alkyd resins are applied in surface coating uses as follows.

Plasticizing Resins. These alkyds are of the glycerol–phthalic acid or glycerol–sebacic acid type and are used in lacquers along with natural resins such as shellac to impart flexibility—a plasticizing action.

Drying Resins. Drying alkyds contain, in addition to the glycerol phthalate components, some drying oils or acids of drying oils. They are, of course, used in varnish-type coatings. Here phenol–aldehyde or, better, urea– or melamine–aldehyde resins may be added to improve hardness.

Hard Resins. The hard resins are usually made from maleic anhydride combined with glycerol and rosin. Their function is to improve film hardness and gloss in both lacquer and varnish coatings.

Allyl Resins

Although allylic monomers are not used directly in chain-reaction or addition polymerization because of the stability and consequent low reactivity of the allyl radical, diallyl esters can be crosslinked by polymerization through their double bonds to give thermosetting resins. Two major types are of commercial interest.

Diallyl Phthalate.   Prepolymers (i.e., partially polymerized but still thermoplastic resins) of diallyl phthalate and diallyl isophthalate are used as molding compounds and in the production of glass fiber-reinforced laminates. They are cured by peroxide catalysts to heat- and chemical-resistant products with good dimensional stability and electrical insulating properties.

Allyl Diglycol Carbonate.   The ester diethylene glycol bisallyl carbonate, $(CH_2{=}CHCH_2OCOOCH_2CH_2)_2O$, is used directly for casting clear glass-like products similar to cast poly(methyl methacrylate) but much harder. The castings are cured with peroxide-type catalysts. They are used for special glazing applications and as spectacle lenses and for other optical purposes.

Furane Resins

Furane resins are based on furfuraldehyde,

$$\begin{array}{c} HC{\rule{1em}{0.4pt}}CH \\ \| \qquad \| \\ HC_{\diagdown O \diagup} C{-}CH \\ \| \\ O \end{array}$$

which is derived from waste vegetable matter. It is used in a thermosetting resin in combination with phenol, or converted to furfuryl alcohol, which itself can be thermoset by acids, or reacted with aldehydes or ketones to give polymerizable products. The resins are all dark in color but are strong, have good chemical resistance, and penetrate porous surfaces well.

GENERAL REFERENCES

Patton 1962; Mraz 1964; Lanson 1978; Schildknecht 1978; Harrington 1982; Mckillip 1982; Sare 1982.

DISCUSSION QUESTIONS AND PROBLEMS

1. Describe the molecular species present in each of the following stages in phenol–formaldehyde polymerization, using structural formulas, giving approximate degrees of polymerization, and so forth: (a) resole or A-stage resin, (b) novolac, (c) B-stage resin (alkaline catalyst), and (d) final product (C-stage).

2. Suggest two methods of fabrication for (a) an amino resin, (b) an unsaturated polyester resin, (c) an epoxy resin, (d) a polyurethane foam, and (e) a silicone rubber.

3. Write chemical formulas for each type of repeat unit found in a typical polymer of each kind mentioned in Question 2.

4. State one advantage and one disadvantage of amino resins in comparison to phenolic resins.

BIBLIOGRAPHY

Backus 1977. John K. Backus, "Polyurethanes," Chapter 17 in Calvin E. Schildknecht, ed., with Irving Skeist, *Polymerization Processes*, Wiley-Interscience, New York, 1977.

Bauer 1980. Ronald S. Bauer, "The Versatile Epoxies," *Chemtech* **10**, 692–700 (1980).

Boenig 1969. H. V. Boenig, "Polyesters, Unsaturated," pp. 129–168 in *EPST*,† Vol. 11, 1969.

Breitigam 1982. W. V. Breitigam, "Epoxy," pp. 32, 34–35 in *MPE*,‡ 1982.

Brode 1982. George L. Brode, "Phenolic Resins," pp. 384–416 in *ECT*,§ Vol. 17, 1982.

Bruins 1968. Paul F. Bruins, ed., *Epoxy Resins Technology*, Wiley-Interscience, New York, 1968.

Bruins 1969. Paul F. Bruins, ed., *Polyurethane Technology*, Wiley-Interscience, New York, 1969.

Brydson 1978. J. A. Brydson, *Rubber Chemistry*, Applied Science, London, 1978.

Clavadetscher 1982. Dave Clavadetscher, "Closed Mold Processing," pp. 357–358, 360–361 in *MPE*, 1982.

David 1969. D. J. David and H. B. Staley, *Analytical Chemistry of the Polyurethanes*, Wiley-Interscience, New York, 1969.

Dominguez 1982. R. J. D. Dominguez, "Polyurethane," pp. 102, 104, 106, 108 in *MPE*, 1982.

Elias 1977. Hans-Georg Elias, *Macromolecules ·2· Synthesis and Materials* (translated by John W. Stafford), Plenum Press, New York, 1977.

Ewald 1982. G. W. Ewald, "Pultrusion," pp. 365–367 in *MPE*, 1982.

Hardman 1982. Bruce B. Hardman and Arnold Torkelson, "Silicones," pp. 922–962 in *ECT*, Vol. 20, 1982.

Harrington 1982. H. J. Harrington, "Alkyd Polyester," p. 61 in *MPE*, 1982.

Jensen 1982. J. C. Jensen and J. F. Keeney, "Unsaturated Polyesters," pp. 71, 72 in *MPE*, 1982.

Keegan 1982. J. F. Keegan, "Phenolic," pp. 42–44 in *MPE*, 1982.

Kentgen 1969. W. A. Kentgen, "Phenolic Resins," pp. 1–73 in *EPST*, Vol. 10, 1969.

Kookootsedes 1982. G. J. Kookootsedes, "Silicone," pp. 114, 116 in *MPE*, 1982.

Lanson 1978. H. J. Lanson, "Alkyd Resins," pp. 18–50 in *ECT*, Vol. 2, 1978.

Lee 1967. H. Lee and K. Neville, "Epoxy Resins," pp. 209–271 in *EPST*, Vol. 6, 1967.

Lichtenberg 1982. D. W. Lichtenberg, "Amino," pp. 23–24, 29 in *MPE*, 1982.

†Herman F. Mark, Norman G. Gaylord, and Norbert M. Bikales, eds., *Encyclopedia of Polymer Science and Technology*, Wiley-Interscience, New York.

‡Joan Agranoff, ed., *Modern Plastics Encyclopedia 1982–1983*, Vol. 59, No. 10A, McGraw-Hill, New York, 1982.

§Martin Grayson, ed., *Kirk–Othmer Encyclopedia of Chemical Technology*, 3rd ed., Wiley-Interscience, New York.

Lichtenwalner 1970. H. Lichtenwalner and M. V. Sprung, "Silicones," pp. 464–569 in *EPST,* Vol. 12, 1970.

Mckillip 1982. W. J. Mckillip, "Furan," p. 39 in *MPE,* 1982.

McLarty 1982. J. L. McLarty, "Filament Winding," p. 363 in *MPE,* 1982.

Menzer 1982. A. B. Menzer, "Continuous RP Laminating," p. 362 in *MPE,* 1982.

Mraz 1964. Richard G. Mraz and Raymond P. Silver, "Alkyd Resins," pp. 663–734 in *EPST,* Vol. 1, 1964.

Owen 1981. M. T. Owen, "Why Silicones Behave Funny," *Chemtech* **11,** 288–292 (1981).

Parker 1977. Earl E. Parker and John R. Peffer, "Unsaturated Polyester Resins," Chapter 3 in Calvin E. Schildknecht, ed., with Irving Skeist, *Polymerization Processes,* Wiley-Interscience, New York, 1977.

Patton 1962. T. C. Patton, *Alkyd Resin Technology,* Wiley-Interscience, New York, 1962.

Pigott 1969. K. A. Pigott, "Polyurethanes," pp. 506–563 in *EPST,* Vol. 11, 1969.

Raymer 1982. J. F. Raymer, "Open Mold Processing," pp. 364–365 in *MPE,* 1982.

Salva 1977. Manilal Salva and Irving Skeist, "Epoxy Resins," Chapter 16 in Calvin E. Schildknecht, ed., with Irving Skeist, *Polymerization Processes,* Wiley-Interscience, New York, 1977.

Sare 1982. E. J. Sare, "Allyl," pp. 22–23 in *MPE,* 1982.

Saunders 1962–1964. J. H. Saunders and K. C. Frisch, *Polyurethanes: Chemistry and Technology. Part 1: Chemistry* (1962) and *Part 2: Technology* (1964), Wiley-Interscience, New York, 1962, 1964.

Schildknecht 1978. C. E. Schildknecht, "Allyl Monomers and Polymers," pp. 109–129 in *ECT,* Vol. 2, 1978.

Sherman 1980. Stanley Sherman, John Gannon, Gordon Buchi, and W. R. Howell, "Epoxy Resins," pp. 267–290 in *ECT,* Vol. 9, 1980.

Turro 1978. N. J. Turro, *Urea–Formaldehyde Resins,* Benjamin Cummings, Menlo Park, California, 1978.

Updegraff 1977. Ivor H. Updegraff and T. J. Sven, "Condensations with Formaldehyde," Chapter 14 in Calvin E. Schildknecht, ed., with Irving Skeist, *Polymerization Processes,* Wiley-Interscience, New York, 1977.

Updegraff 1978. Ivor H. Updegraff, Sewell T. Moore, William F. Herbes, and Philip B. Roth, "Amino Resins and Plastics," pp. 440–469 in *ECT,* Vol. 2, 1978.

Widmer 1965. Gustave Widmer, "Amino Resins," pp. 1–94 in *EPST,* Vol. 2, 1965.

PART SIX

POLYMER PROCESSING

CHAPTER SEVENTEEN

PLASTICS TECHNOLOGY

As the earlier sections of this book have suggested, the unique physical and mechanical properties of polymers are responsible in large part for their important place in our modern life. But these properties can be developed and utilized only by fabricating the polymer into useful articles or shapes. The methods by which this fabrication is carried out are described in this chapter.

Fabrication methods are largely determined by the rheological properties of the polymer in question. A primary consideration is whether the material is *thermoplastic,* that is, retains the ability to flow at elevated temperatures for relatively long times, or *thermosetting,* that is, subject to (controlled) crosslinking reactions at the temperatures necessary to induce flow, so that the ability to flow is rather quickly lost in favor of form stability. Other considerations of importance in selecting fabrication methods are softening temperature, stability, and, of course, the size and shape of the end product.

Tadmor (1979) points out that methods of processing polymers, including plastics, fibers, and elastomers, have several elementary steps in common: handling of particulate solids, melting, pressurization and pumping, mixing, and often a final step of devolatilization and stripping off undesired components. The only way in which the various processing methods differ is in the step of forming, in which the material is given its final shape.

In this chapter fabrication methods pertinent to plastics as distinguished from fibers or rubbers are considered. Techniques specific to the latter classes of materials are discussed in Chapters 18 and 19. The first three sections of this chapter describe various forming methods. Section *D* discussed compositions involving more than one polymer type, or one polymer and a reinforcing material that is an integral part of the structure of the composite, while Section *E* is devoted to other additives, including inert fillers and plasticizers, and to some of the other steps mentioned by Tadmor. Finally, Section *F* provides an abridged table of polymer properties.

457

A. MOLDING

Molding processes are those in which a finely divided plastic is forced by the application of heat and pressure to flow into, fill, and conform to the shape of a cavity (mold). One of the oldest methods of polymer processing, molding can be carried out in many different ways. We start with the simplest and oldest of these, compression molding.

Compression Molding

In *compression molding,* the polymer is put between stationary and movable members of a mold (Fig. 17-1). The mold is closed, and heat and pressure are applied so that the material becomes plastic, flows to fill the mold, and becomes a homogeneous mass. The necessary pressure and temperature vary considerably depending upon the thermal and rheological properties of the polymer. For a typical compression-molding material they may be near 150°C and 1000–3000 psi. A slight excess of material is usually placed in the mold to insure its being completely filled. The rest of the polymer is squeezed out between the mating surfaces of the mold in a thin, easily removed film known as *flash.*

Injection Molding

Most thermoplastic materials are molded by the process of *injection molding.* Here the polymer is preheated in a cylindrical chamber to a temperature at which it will flow and then is forced into a relatively cold, closed mold cavity by means of quite high pressures applied hydraulically, traditionally through a plunger or ram, but today almost invariably by means of a reciprocating screw that serves the dual

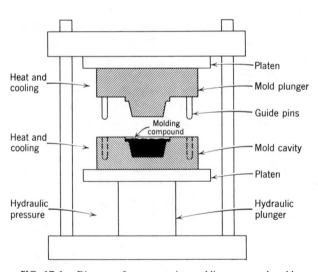

FIG. 17-1.   Diagram of a compression-molding press and mold.

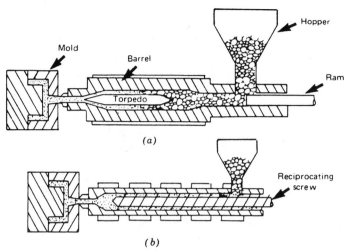

FIG. 17-2.   Sketch of (a) a ram-fed and (b) a screw-fed injection-molding machine. (Reprinted with
permission from Holmes-Walker 1975. Copyright 1975 by John Wiley & Sons.)

purposes of providing the molten polymer mass (as it does in extrusion, Section
B) and forcing it into the mold. With reference to Fig. 17-2, the screw rotates to
pick up the particulate polymer, compact and melt it, mix the melt, and deliver it
to the entrance to the mold. The screw then moves forward (to the left in the figure),
to force a fixed volume of the molten polymer into the closed mold. The melt
temperature may be considerably higher than in compression molding, and pressures
of hundreds to thousands of tons are common. After the polymer melt has solidified
in the cool mold, the screw rotates and moves backward to ready the charge of
polymer for the next cycle. Meanwhile the mold is opened and the molded article
is removed.

An outstanding feature of injection molding is the speed with which finished
articles can be produced. Cycle times of 10–30 sec are common, as are multicavity
molds allowing the production of many parts per cycle. Articles weighing up to
many kilograms can be produced.

Other Molding Techniques

Blow Molding.   This operation can be carried out either on an extruder (Section
B) or a reciprocating-screw injection machine. A section of molten polymer tubing
(*parison*) is extruded into an open mold (Fig. 17-3). By means of compressed air
or steam the plastic is then blown into the configuration of the mold. This technique
is widely used for the manufacture of bottles and similar articles. In the case of
large articles, such as 2-liter beverage bottles, the parison may previously have
been injection molded and oriented to provide additional strength to the final blown
piece.

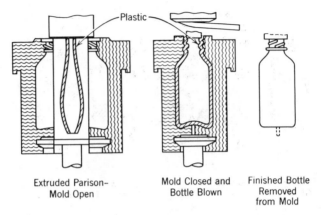

FIG. 17-3.   Sketch of blow-molding process (courtesy of the Society of the Plastics Industry, Inc.).

Reaction–Injection Molding (RIM).   This relatively new process (Fig. 17-4) starts with the unreacted components that lead to a polymer product. They are pumped in predetermined amounts into a mixing head, where they are thoroughly mixed and injected into a warm (~65°C) mold under relatively low pressures (50–100 psi). Polymerization takes place in the mold. Meanwhile, the mixing head is cleaned, with unused monomers being recycled. To date, RIM has been used almost entirely for the molding of polyurethanes (Chapter 16C), but commercialization of nylon and epoxy RIM is anticipated.

Rotational Molding.   In this technique powdered polymer is loaded into a relatively inexpensive closed mold, which is intensively heated while being rotated biaxially. The polymer coats the inner walls of the mold to a uniform thickness and is fused there. The method has advantages for producing large hollow parts, and can be used to produce multiwall constructions by successive steps.

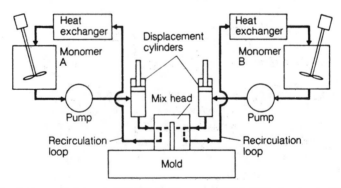

FIG. 17-4.   Basic components of a RIM processing system (Hall 1982. Reprinted with permission from *Modern Plastics Encyclopedia 1982–1983*. Copyright 1982 by McGraw-Hill, Inc.).

Thermoset Molding. The injection molding of thermosetting resins differs only slightly from the injection molding process described above for thermoplastics. First, the ingredients (B-stage resin, Chapter 16*A*) are heated just enough to make them fluid but not to initiate cure. In this stage their viscosity is considerably lower than that of molding-grade thermoplastics. Molds are usually heated, and this plus the frictional heat generated as the resin is injected into the mold raise its temperature enough to initiate cure. The part may be removed from the mold when it is form stable but before the cure is completed; stored heat in the part allows the cure to be completed in a minute or so, before the part can cool.

Transfer Molding. In this variation of compression molding, which evolved into the forerunner of the injection molding of thermosetting resins just described, the resin is placed in a separate chamber, called the pot, preheated to just below cure temperature. It is then injected or transferred into the heated mold where cure takes place.

GENERAL REFERENCES

McKelvey 1962; Hull 1968; Rubin 1973, 1977; Han 1976; Middleman 1977; Tulley 1977; Dym 1979; Tadmor 1979; Olmsted 1982; Richardson 1982; Rosen 1982; and specifically the following:

Blow Molding. Fredrickson 1982; Irwin 1982; Rainville 1982.
RIM. Becker 1979; Kubiak 1980; Schneider 1981; Hall 1982; Sneller 1982*a*.
Rotational Molding. Ramazzotti 1979; Fair 1982.
Thermoset Molding. Kiesau 1980; Hull 1982.
Transfer Molding. O'Brien 1982.

B. EXTRUSION

In the extrusion process, polymer is propelled continuously along a screw through regions of high temperature and pressure where it is melted and compacted, and finally forced through a die shaped to give the final object (Fig. 17-5). A wide variety of shapes can be made by extrusion, including rods, channels and other structural shapes, tubing and hose, sheeting up to several feet wide and one-quarter in. or more thick, and film of similar width down to a few thousandths of an inch in thickness.

The screw of an extruder is divided into several sections, each with a specific purpose (Chung 1977). The feed section picks up the finely divided polymer from a hopper and propels it into the main part of the extruder. In the compression section, the loosely packed feed is compacted, melted, and formed into a continuous stream of molten plastic. Some external heat must be applied, but much is generated

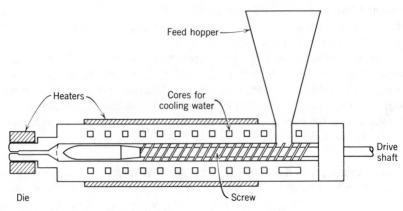

FIG. 17-5. Diagram of a plastics extruder.

by friction. The metering section contributes to uniform flow rate, required to produce uniform dimensions in the finished product, and builds up sufficient pressure in the polymer melt to force the plastic through the rest of the extruder and out of the die. Since viscous polymer melts can be mixed only by the application of shearing forces (their viscosity is too high to allow turbulence or diffusion to contribute appreciably to mixing), an additional working section may be needed before the die.

Modern trends in extruder usage include the *twin-screw* or *multiple-screw* extruder, in which two screws turn side by side in opposite directions, providing more working of the melt, and the *vented* extruder, having an opening or vent at some point along the screw that can be opened or led to vacuum to extract volatiles from the polymer melt.

Coextrusion

Films or sheets consisting of layers of two or more different polymers can be produced by mixing the molten streams from a like number of extruders in a multimanifold die. This process can be used to combine materials to provide combinations of properties that cannot be obtained in a single polymer. For example, a film for packaging food may consist of three layers imparting, respectively, high strength, low oxygen permeability, and heat sealability.

Film Extrusion

The extrusion of film can be carried out either by casting as in sheet extrusion (see below) or by the blown-film process. In the latter a tubular die is used, from which a hollow tube of product is extruded vertically upward toward a film tower, as shown in Fig. 17-6. The tube is blown into a thin cylindrical film by air introduced through the die and trapped in the film bubble. At the top of the tower, the bubble (by now cool) is collapsed, and subsequently slit into a flat film.

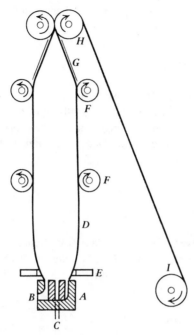

FIG. 17-6. Schematic view of film blowing: (A) film die, (B) die inlet from extruder, (C) air inlet and valve, (D) plastic tube (bubble), (E) air ring for cooling, (F) guide rolls, (G) collapsing frame, (H) pull rolls, and (I) windup roll. (Reprinted with permission from Richardson 1974. Copyright 1974 by John Wiley & Sons.)

Pultrusion

In this variation of extrusion, material is pulled rather than pushed through a die or mold. The process is used to produce continuous lengths of fiber-reinforced thermosetting resin (Fig. 17-7). The reinforcing materials are continuously pulled through a bath of liquid resin that saturates each individual fiber. Excess resin is

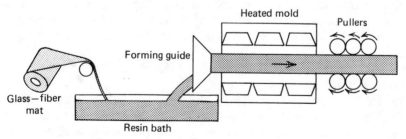

FIG. 17-7. Schematic diagram of the pultrusion process. (Reprinted from Martin 1979, by courtesy of *Plastics Engineering.*)

removed by a forming guide, and the mix is then pulled through a heated mold that shapes and cures it.

Other Extrusion Processes

In *sheet extrusion,* the product from a sheet die is passed between rolls that control its thickness and apply a desired surface finish by polishing and embossing. *Wire and cable extrusion* employs dies through which bare copper or aluminum wire is passed, and in which it is coated with plastic insulation. Extrusion rates can be as high as 8000 ft/min.

GENERAL REFERENCES

McKelvey 1962; Westover 1968; Brydson 1973; Richardson 1974; Fisher 1976; Han 1976; Batiuk 1977; Klein 1977; Middleman 1977; Tadmor 1978, 1979; Levy 1981; Haisser 1982; Richardson 1982; Rosen 1982; and specifically the following:

Coextrusion.   Brown 1981; Sneller 1982*b*.
Pultrusion.   Martin 1979; Anderson 1981.

C. OTHER PROCESSING METHODS

Calendering

Calendering is a process used for the continuous manufacture of sheet or film. Granular resin, or thick sheet, is passed between pairs of highly polished heated rolls under high pressure (Fig. 17-8). For the production of thin film, a series of pairs is used, with a gradual reduction in roll separation as the stock progresses through the unit. Proper calendering requires precise control of roll temperature, pressure, and speed of rotation. By maintaining a slight speed differential between a roll pair, it is often possible to impart an exceedingly high gloss to the film or sheet surface. An embossed design can be produced on the surface by means of a calender roll, appropriately engraved. By calendering a mixture of granular resin chips of varying color, it is possible to produce unusual decorative effects (e.g., marblization) in the product; this technique is widely employed in the manufacture of flooring compositions.

A large fraction of the resin used in calendering is poly(vinyl chloride) (Chapter 14*D*). The products range from thin films to thick, semirigid sheets. It should be noted that, as with virtually all processing techniques, much more than just the calender itself is required to transform the polymer from raw material to finished resin (Fig. 17-9).

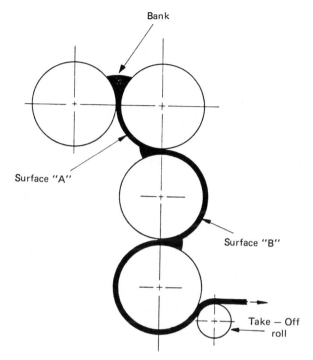

FIG. 17-8. Schematic diagram of four-roll inverted-L calender. (Reprinted from Perlberg 1977, p. 1378, by courtesy of Marcel Dekker, Inc.)

Casting

In casting processes, a liquid material is poured into a mold and solidified by physical (e.g., cooling) or chemical (e.g., polymerization) means, and the solid object is removed from the mold. Casting utilizes low-cost equipment (molds can be made out of soft, inexpensive materials such as rubber and plaster) but is a relatively slow process.

Casting of Thermosetting Resins. Thermosetting resins are cast by stopping the polymerization at the A stage, where the resin is still fusible and fluid. After the mold is filled, the resin is cured in an oven.

Casting of Vinyl Polymers. Vinyl polymers, primarily the acrylic resins, are cast by preparing a syrup of monomer and polymer, and polymerizing the monomer in the mold. Sheets, rods, and tubes are prepared as described in Chapter 14*B*.

Film Casting. Films, including photographic film and cellophane, are made by flowing a solution of the polymer onto an extremely smooth surface in the form of a large polished wheel or, occasionally, a metal belt or band. After the solvent has

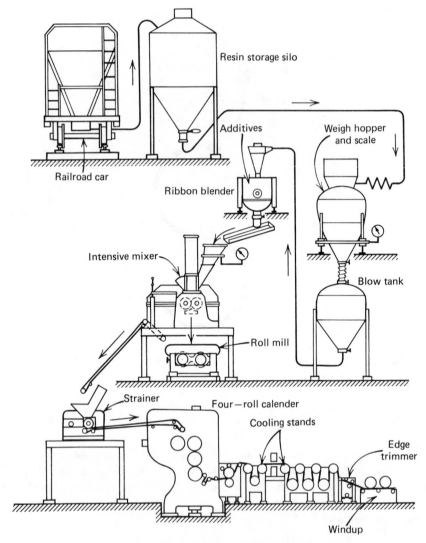

FIG. 17-9. Schematic diagram of typical poly(vinyl chloride) calendering system showing the principal elements that are required to convert powdered resin into a finished sheet. (Reprinted from Watkins 1976, by courtesy of *Plastics Engineering.*)

evaporated (or, in the case of cellophane, the polymer has coagulated) the film is stripped from the casting surface.

Epoxy Resins. Cast epoxies are widely used for dies because of their good dimensional stability and high impact strength.

Polyesters. Clear (unreinforced) polyester resins (Chapter 16*B*) are cast in open molds of ceramic, wood, plastic, or metal for both industrial and hobby craft work.

Nylon. Casting from the monomer is used for the anionic polymerization oı caprolactam to thick-walled parts of nylon-6.

Urethane Elastomers. Liquid urethane prepolymers (Chapter 15*B*) can be used in several casting methods, including hand batch, centrifugal or rotational, solvent, or spray techniques.

Coating

The technology of coating fabrics and paper, a major outlet for some plastics, is a subject whose magnitude and complexity place it outside the scope of this book. The plastic may be utilized as a melt, solution, latex, paste, or enamel or lacquer. It may be applied to the substrate by spreading with a knife, brushing, using a roller, calendering, casting, or extrusion. Coating processes include *dipping,* in which a form (such as for a rubber glove) is dipped into a suspension or latex of polymer and then into a bath of coagulating agent. After several dips to build up the desired thickness, the film is stripped from the form and subjected to heat treatments to cure or crosslink the resin. In *"slush" molding,* a hollow mold is filled with a more viscous latex or "slush" of partly plasticized material such as a vinyl plastisol (Chapter 14*D*). The excess is poured out, leaving a film which is heat treated and removed. The processes are very flexible and involve inexpensive equipment.

Foaming

The production of plastic foams is accomplished by generating a gas in a fluid polymer, usually at an elevated temperature. The chemical production of a gas during polymerization to form urethane foams is described in Chapter 16*C*. Thermoplastics are foamed by incorporating either a blowing agent, which decomposes to a gas at elevated temperature, or an inert gas. The references cited at the end of this section provide comprehensive reviews.

Forming

Postforming Thermosetting Resins. Laminated sheets of thermosetting resins are formed into various shapes by a process resembling that used for sheet metal. The sheet is heated before the final thermosetting reaction, shaped quickly in a mold or around a form, and held in place with light pressure until it sets up. This type of processing is widely practiced with *sheet-molding compound* (SMC), an inexpensive mix of about one-third unsaturated polyester resin (Chapter 16*B*), one-third filler (usually $CaCO_3$), and one-third chopped glass fiber. The liquid resin and filler, together with thickeners (MgO or CaO, that form weak ionic crosslinks with the acid groups on the polyester), and small amounts of curing agent (usually an organic peroxide), are mixed and spread onto layers of polyethylene film as shown in Fig. 17-10. Chopped glass fibers, about 1 in. long, are added, and the

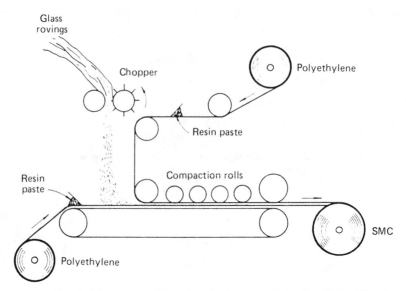

FIG. 17-10.   Sheet-molding compound line. (Reprinted with permission from Rosen 1982. Copyright 1982 by John Wiley & Sons.)

sheet is compacted to wet out the glass. The polyethylene is subsequently removed from the hardened (but not polymerized) sheet, which is formed and heated to effect the cure. SMC is widely used for automobile body parts such as front ends and fender extensions, and for other structural parts.

Forming Thermoplastic Sheets.   Vacuum forming is used widely for the manipulation of cellulose acetate and acrylic resin sheeting. A sheet of the plastic is warmed and laid across a hollow mold cavity, and a vacuum is drawn on the cavity. Atmospheric pressure forces the sheet to conform with the mold; after cooling of the sheet, the vacuum is released, and the formed object removed.

Many modifications of this basic process are in use. In one, an inflatable rubber bag forces the sheet to conform with the mold contour. In another variation the sheeting is first dished by vacuum to a curvature greater than ultimately desired, a *convex* mold is placed into the mold cavity, and the vacuum is released while the sheet is still warm. The sheet recovers sufficiently to conform to the shape of the convex mold and is frozen in that shape by cooling.

Laminating and Low-Pressure Molding

Both lamination, a high-pressure process, and low-pressure molding involve (a) the impregnation of sheets (wood, paper, fabric) with a liquid or dissolved thermosetting resin which acts as an adhesive, (b) assembly of the individual sheets, and (c) compression and curing.

In low-pressure molding or laminating, the sheets of impregnated material are laid over a mold and held in place by a rubber mattress or bag which is inflated with steam to provide heat and pressure to hold the laminate in place and effect the cure. Many variations of the process are practiced.

Lamination differs from low-pressure molding in that standard shapes are produced in sufficient quantity to allow economical application of hot-press methods. Resins requiring higher temperatures for curing can also be used.

GENERAL REFERENCES

McKelvey 1962; McDonagh 1977; Middleman 1977; Tadmor 1979; Crump 1980; Richardson 1982; Rosen 1982; and specifically the following:

Calendering.   Meinecke 1965; Perlberg 1977; Eighmy 1982.

Casting.   Wallis 1965; Metz 1977; Schaer 1977; Luskin 1982.

Coating.   Higgins 1965, 1967; Metz 1977; Werner 1977; Kiernan 1982.

Foaming.   Skotchdopole 1965; Benning 1969; Suh 1980; Cargile 1982; Gordon 1982; Herbert 1982; Schrafft 1982.

Forming.   Monroe 1965; Foster 1982; Rosen 1982, Chapter 20.4; Teal 1982.

Laminating.   Power 1968; Pasquale 1982.

D. MULTIPOLYMER SYSTEMS AND COMPOSITES

It has been stated in this book and elsewhere that advances in producing high-performance polymer-based compositions will in the future more likely result from the utilization of new mixtures of polymers and mixtures of polymers with reinforcing components, rather than from new polymer compositions. There is already much evidence to support this view, and here we provide a few final remarks on the two types of compositions just described, which for convenience we denote multipolymer systems and composites, respectively. (Terminology in this area is not fully established, as will be seen.)

Multipolymer Systems

It was pointed out in Chapter 7 that mixtures of two (or more) polymers do not in general form miscible systems. All but a small minority form two-phase systems, the mechanical properties of which are very poor except in special circumstances. Alfrey (1980) suggests that the circumstance leading to good, indeed often superior, properties in immiscible systems is the development of strong interfacial adhesion between the two phases. If this does not occur naturally, it may be possible to bring it about by the addition of a third polymer that adheres well to both phases, or by the use of appropriate block or graft copolymers. Some examples of useful multipolymer systems, several of which are discussed elsewhere in the text, are given below.

Multilayer Films. The production of multilayer films by coextrusion was described briefly in Section *B*. If there is good adhesion between the layers, the mechanical properties of these laminates are generally additive, but there are some striking exceptions. For example, the presence of a high-elongation layer may prevent tensile failure due to transverse crack formation in a hard, brittle layer, but occasionally the opposite effect can be seen.

Polymer Blends. Both miscible and immiscible blends are known and utilized commercially. The best known commercial example of a miscible blend is that of polystyrene and poly(phenylene oxide) (PPO). The latter is a tough, temperature-resistant engineering plastic, but difficult to process. The blend is both easier to process than PPO and higher in heat resistance and toughness than polystyrene.

Probably the best-known examples of immiscible polymer blends (sometimes called alloys, but the exact meanings and differentiation of the terms is not clear) are high-impact polystyrene and the ABS resins (Chapter 14*A*). The former is produced by dissolving rubber in styrene monomer, which is then polymerized. Phase separation takes place, and the structure is controlled to have a continuous polystyrene phase with dispersed rubber particles which themselves have inclusions of polystyrene. In ABS resins, inclusions of butadiene-based rubber exist in a continuous phase of styrene–acrylonitrile copolymer; the rubber molecules are graft copolymerized to styrene–acrylonitrile copolymer in order to enhance the interfacial adhesion. Similar considerations dictate the block copolymer structure of the thermoplastic elastomers described in Chapter 13*F*.

Interpenetrating Polymer Networks. A new class of multipolymer systems has recently been synthesized in the form of two independent interpenetrating polymer networks. One can, for example, take a lightly crosslinked polymer A, swell it with a second monomer B, plus a crosslinking agent, and polymerize B, or one can select the A and B systems to polymerize together by different mechanisms, for example, a styrene–divinyl benzene mixture polymerizing by radical chain mechanism together with a polyurethane network forming by a step-reaction mechanism. Phase separation will in general take place, but the structure greatly restricts the sizes of the domains of each phase. Novel synergistic properties have been observed in some cases of interpentrating polymer networks. No commercial applications are yet known.

Composites

The term *composites* is generally applied to fiber-reinforced engineering structural materials, in which the fibers are continuous or long enough that they can be oriented to produce enhanced strength properties in one direction. Several examples of such systems have been mentioned in earlier chapters, for example the polyester resins reinforced by continuous glass fibers by filament winding (Chapter 16*B*) and by pultrusion (Section *B* of this chapter). Other fibers providing outstanding strength in this application (albeit still at very high costs) are graphite and aramid (aromatic polyamide, Chapter 15A). The epoxy resins (Chapter 16C) are favored as the matrix resin for best properties.

The graphite fibers used in high-performance composites are produced by further treatment of the ladder polymer made by pyrolyzing polyacrylonitrile (Chapter 15D) (although other precursors can be used, such as rayon). If the polyacrylonitrile is carried through the first oxidation stage (at 250–400°C) under tension, and then pyrolyzed at 1500–2500°C to drive off all atoms but carbon, fibers with a graphite structure to a greater or lesser degree are obtained. In an epoxy matrix, continuous graphite fibers yield composites with nearly the tensile strength and stiffness of steel, but of course with a much lower maximum use temperature. The weight of the plastic composite is about a quarter that of the metal structure with comparable properties.

Particularly if higher production results in lower prices, an important application for these composites is anticipated in the automobile industry, where, for example, they can be used to produce leaf springs for both cars and trucks, drive shafts (by filament winding) and many other parts, and in the aircraft industry.

GENERAL REFERENCES

Multipolymer Systems. Alfrey 1980.

Polymer Blends. Manson 1976; Paul 1978; Fenelon 1981; Jalbert 1982; Kamal 1982; Olabisi 1982.

Interpenetrating Polymer Networks. Sperling 1981.

Composites. Agarwal 1980; Beardmore 1980; Piggott 1980; Tsai 1980; Delmonte 1981; Hancox 1981; Hull 1981; McGarry 1981; Hearons 1982; Lubin 1982; Miner 1982; Peterson 1982; Sheldon 1982.

E. ADDITIVES AND COMPOUNDING

Fillers

Many plastics are virtually useless alone but are converted into highly serviceable products by combining them with particulate or fibrous solids. Phenolic and amino resins are almost always filled with substances like wood flour, pure short-fiber cellulose, or powdered mica. These materials greatly enhance dimensional stability, impact resistance, tensile and compressive strength, abrasion resistance, and thermal stability. The use of glass fiber as a reinforcing filler for polyester resins is another important illustration. Soft thermoplastics like coumarone-indene and hydrocarbon resins are usually blended with very large amounts (over 80% by weight) of mineral solids such as crushed quartz, limestone, or clay. In compounds of this type, the resin functions as an interparticle adhesive; these products often have poor tensile properties but excellent compressive strength, abrasion resistance, and dimensional stability. A final and extremely important example of the beneficial effect of fillers is the reinforcement of rubber, discussed in Chapter 19B.

The principal fillers used in plastics can be divided into two types: particulate and fibrous. Among the particulate fillers are silica products, including sand, quartz, and diatomaceous earth; silicates, including clay, mica, talc, asbestos, and some synthetic silicates; glass, including granules, flakes, and solid and hollow spheres (the latter in syntactic foams); inorganic compounds, including chalk, limestone, alumina, magnesia, and zinc oxide, barytes, silicon carbide, and others; metal powders; and finely divided cellulosics (wood flour) and synthetic polymers (fluorocarbons and others).

Fibrous fillers, some quite old (cellulosics) and some new "space-age" products (metallic fibers, whiskers) include cellulosic fibers, such as alpha-cellulose and cotton flock; synthetic fibers, including nylon, polyester, acrylic, and poly(vinyl alcohol); carbon fibers made by pyrolizing materials such as rayon; boron filaments made by depositing boron from a BCl_3–H_2 mixture onto tungsten wire, and having tensile strengths approaching one-half million psi; and single-crystal fibers of aluminum or beryllium oxide, silicon or boron carbide, or others.

In recent years the use of mineral fillers has increased markedly, with emphasis on calcium carbonate, which dominates the market. Custom compounding companies now offer "filler concentrates," much like color concentrates, to provide an easy way for small processors to incorporate fillers into their products.

Plasticizers

Plasticizers are added to plastics to improve flow and, therefore, processability, and to reduce the brittleness of the product. This is achieved by lowering the glass transition temperature below room temperature, thus achieving a change in properties from those of a hard, brittle, glasslike solid to those of a soft, flexible, tough material (Chapter 12D). An example is the plasticization of poly(vinyl chloride) and vinyl chloride–acetate copolymers (Chapter 14D). Similar changes in properties can, of course, be brought about by altering the molecular structure of the polymer (e.g., by copolymerization, sometimes called *internal plasticization*).

The basic requirements that must be met by a plasticizer are compatibility and permanence. The plastizer must be miscible with the polymer. This implies a similarity in the intermolecular forces active in the two components, and explains why compatibility is difficult to achieve with a nonpolar polymer such as polyethylene. Permanence requirements demand low vapor pressure and a low diffusion rate of the plasticizer within the polymer, both of which are obtained by the use of high-molecular-weight plasticizers.

The efficiency of the plasticizer in bringing about the desired changes in properties is important in determining the proportion in which the plasticizer must be added to the resin. Plasticizer efficiency may be evaluated by a number of different semiempirical tests. Some of these measure the amount of nonsolvent needed to cause phase separation when added to polymer–plasticizer solutions (dilution ratio), the viscosity of dilute solutions of the polymer in the plasticizer, polymer–solvent interaction constants measured on these solutions, the depression of the glass transition temperature, the melt viscosity of the plasticized polymer, the electrical or mechanical properties of the plasticized polymer, or the molecular size and shape or viscosity of the plasticizer itself. Not all these tests, needless to say, rate plasticizer candidates in the same order. The selection of a particular plasticizer still depends to a large extent upon empirical results rather than theoretical predictions.

The following types of plasticizers are in common use:

a. Phthalate esters, accounting for over half of the total volume of plasticizers used. Di(2-ethylhexyl) phthalate (dioctyl phthalate, DOP) is the dominant plasticizer for poly(vinyl chloride).

b. Phosphate esters, chiefly tricresyl phosphate, valued primarily for their flameproofing characteristics.

c. Adipates, azelates, oleates, and sebacates, used chiefly in vinyl resins for improving low-temperature flexibility.

d. Epoxy plasticizers, produced by reacting hydrogen peroxide with unsaturated vegetable oils and fatty acids.

e. Fatty acid esters from natural sources, used primarily as extenders to reduce cost (*secondary plasticizers*).

f. Glycol derivatives, employed mainly as lubricants and mold-release agents, and as plasticizers for poly(vinyl alcohol).

g. Sulfonamides, used to plasticize cellulose esters, phenolic and amino resins, and amide and protein plastics.

h. Hydrocarbons and hydrocarbon derivatives, serving as secondary plasticizers.

Other Additives

Antioxidants. The role of antioxidant in preventing or inhibiting the oxidation of polymers is usually filled by a substance which itself is readily oxidized, although in some cases the antioxidant may act by combining with the oxidizing polymer to form a stable product. Common antioxidants fall in the classes of phenols, aromatic amines, and salts and condensation products of amines and aminophenols with aldehydes, ketones, and thio compounds.

Colorants. Colorants for plastics include a wide variety of inorganic and organic materials. A few are molecularly dispersed (oil-soluble *dyes*) or have small particle size and a refractive index near that of the plastic (*organic pigments* such as the phthalocyanines) and lead to transparent colored products when incorporated into transparent plastics. Others, including *inorganic pigments,* impart opacity to the plastic. Common colorants for plastics include, among many others, titanium dioxide and barium sulfate (white), phthalocyanine blues and greens, ultramarine blues, chrome greens, quinacridone reds and magentas, molybdate oranges, cadmium reds and yellows, iron oxide and chrome yellows, carbon black, flake aluminum for a silver metallic effect, and lead carbonate or mica for pearlescence. The coloring of plastics is normally carried out by adding the colorants to the powdered plastic, tumbling, and compounding on hot rolls or in an extruder. Other techniques are occasionally used: For example, colored castings of poly(methyl methacrylate) are produced by dissolving or dispersing the colorants in the syrup before polymerizing. Compounding companies provide color concentrates, consisting of up to 70% pigment in resin, that can be "let down" with up to 200 times as much uncolored resin by the plastics processor. Available in a wide variety of pigments and of resins, they provide an easy (but somewhat more expensive) means for the small plastic processor to obtain a well-dispersed pigment mixture for satisfactory coloring.

Ultraviolet-light absorbers are not colorants in the above sense, but have similar light-modifying effects.

Flame Retardants.   The most useful material imparting flame retardance to plastics is antimony trioxide. It must be used with a source of available chlorine to be effective; it is presume that antimony oxychloride is the active flame-retarding agent. Aside from this compound and other antimony derivatives, the phosphate ester plasticizers are widely used for reducing flammability, especially in vinyl resins.

The major factors in reducing the flammability of materials appear to be (a) elimination of volatile fuel, as by cooling; (b) production of a thermal barrier, as by charring, thus eliminating fuel by reducing heat transfer; and (c) quenching the chain reactions in the flame, as by adding suitable radical scavengers.

Stabilizers.   In addition to antioxidants and flame retardants, certain polymers require stabilizers to achieve and maintain utility. An important example is the vinyl resins, whose stabilization is discussed in Chapter 14D. Other applications of stabilizers include the use of carbon black to prevent photochemical degradation (by excluding light), as in polyethylene, and of ultraviolet light absorbers, such as hydroxybenzophenones, to improve the light stability of both plastics and their colorants.

Compounding

The term *compounding* is applied both to the selection of additives to modify the properties of a polymer, and to their incorporation with the polymer to give a homogeneous mixture, in a form suitable for efficient use in the subsequent processing or fabrication step. It is in the latter sense that compounding is here considered.

The traditional compounding device in the plastics industry is the two-roll mill, which looks and operates much like the top half of a calender (Fig. 17-8). By proper selection of temperatures and speeds of rotation, the plastic is made to adhere to the front roll, except as it is cut off by the operator. The compounding ingredients are added to the plastic mass as it passes between the rolls.

The roll mill has been supplanted in many operations by the compound or mixer extruder, an extruder in which the function of the mixing section of the screw is emphasized. The use of the extruder for compounding has many advantages: contamination is reduced, inert atmospheres or vacuum may be utilized, continuous processes are more readily achieved, etc.

Other compounding devices in common use are internal mixers such as kneaders, masticators, and paddle blenders, tumblers, and blenders.

GENERAL REFERENCES

Fillers. Frissel 1967; Katz 1978; Blumberg 1980; Stayner 1982; Tapper 1982; Solomon 1983.

Plasticizers. Buttery 1960; Darby 1969; Krauskopf 1976; Sears 1982*a,b*; Beeler 1982.

Additives. Mascia 1974.

Antioxidants. Paolino 1982.

Colorants. Hopmeir 1969; Webber 1979; Owen 1982; Faulks 1982.

Flame Retardants. Blake 1982.

Stabilizers. Brilliant 1982; Chasar 1982.

Compounding. Bajaj 1982; Eise 1982; Mathews 1982; Reising 1982.

F. TABLES OF PLASTICS PROPERTIES

Tables 17-1 and 17-2 list comparative properties for typical examples of the major commercial thermoplastic and thermosetting resins, respectively. It should be noted that the properties of a given plastic can vary widely, depending on compounding, fabrication, thermal history, and many other variables. The numbers in the tables may represent only a fraction of the total range of values attainable and should be used only for comparative purposes.

DISCUSSION QUESTIONS AND PROBLEMS

1. Compare compression and injection molding for speed, cost per part, investment cost, and flexibility in types of material that can be handled.

2. Describe the relative advantages and disadvantages of transfer molding, compression molding, and injection molding of thermosetting materials.

3. Under what circumstances can reaction injection molding be used in place of one of the methods listed in Question 2, and what are its advantages when it can be used?

4. Describe the blow molding and the blown film extrusion processes, indicating where orientation takes place.

TABLE 17-1. Typical Properties of Commercial Thermoplastic Polymers[a]

Property	ABS	Acetal Resin	Acrylic		Cellulose Acetate	Fluoroplastics		
			Poly(methyl methacrylate)	Impact Acrylic		Polytetra-fluoroethylene	Fluorinated Ethylene–Propylene Copolymer	Poly(vinylidene fluoride)
Specific gravity (g/cm^3)	1.02–1.04	1.42	1.17–1.20	1.08–1.18	1.22–1.34	2.14–2.20	2.12–2.17	1.75–1.78
Refractive index (n_D^{25})	—	1.48	1.49	—	1.46–1.50	1.35	1.34	1.42
Tensile strength (psi)	3500–6200	10,000	7000–11,000	5000–9000	1900–9000	2000–5000	2700–3100	5500–7400
Elongation (%)	5–60	25–75	2–10	>15–50	6–70	200–400	250–330	25–500
Tensile modulus (10^5 psi)	2–3.5	5.2	4.5	2–4	0.6–4.0	0.58	0.50	1.20
Impact strength (ft-lb/in. of notch)	3–8	1.4–2.3	0.3–0.5	0.5–4.5	0.4–5.2	3.0	No break	3.6–4.0
Heat-deflection temperature (°F, 264 psi)	200–218	255	155–210	165–215	111–195	250[b]	—	176–194
Dielectric constant (1000 cycles)	2.4–4.5	3.7	3.0–3.6	2.5–3.5	3.4–7.0	<2.1	2.10	7.46
Dielectric loss (1000 cycles)	0.004–0.007	0.0048	0.03–0.05	0.02–0.035	0.01–0.07	<0.0002	0.0001	0.019
Water absorption, (one-eighth in. bar, 24 hr, %)	0.2–0.45	0.25	0.3–0.4	0.2–0.4	1.7–6.5	0.00	0.00	0.04–0.06
Burning rate	Slow	Slow	Slow	Slow	Slow to self-extinguishing	None	None	None
Effect of sunlight	Yellows	Chalks	None	Slight	Slight	Note	None	None
Effect of strong acids or bases	Attacked, acids	Attacked	Attacked	Attacked, acids	Decomposes	Very resistant	Very resistant	Very resistant
Effect of organic solvents	Soluble	Resistant	Soluble	Soluble	Soluble	Very resistant	Very resistant	Very resistant
Clarity	Opaque	Opaque	Transparent	Opaque	Transparent	Opaque	Transparent	Transparent

476

Property	Ionomer	Polyamide (66-Nylon)	Poly(phenylene oxide)–Polystyrene blend	Polybutene	Polycarbonate
Specific gravity (g/cm^3)	0.93–0.96	1.13–1.15	1.06	0.910–0.915	1.2
Refractive index (n_D^{25})	1.51	1.53	—	1.50	1.586
Tensile strength (psi)	3500–5500	9000–12,000	9600	3800–4400	8000–9500
Elongation (%)	350–450	60–300	60	300–380	100–130
Tensile modulus (10^5 psi)	0.2–0.6	1.8–4.2	3.55	0.26	3.5
Impact strength (ft-lb/in. of notch)	6–15	1.0–2.0	5	No break	12–17.5
Heat-deflection temperature (°F, 264 psi)	100–120	150–220	265	130–140	265–285
Dielectric constant (1000 cycles)	2.4–2.5	3.9–4.5	2.64–3.14	2.25	3.02
Dielectric loss (1000 cycles)	0.0015	0.02–0.04	0.0004–0.0018	0.005	0.0021
Water absorption (one-eighth in. bar, 24 hr, %)	0.1–1.4	1.5	0.07	<0.01–0.026	0.15
Burning rate	Very slow	Self-extinguishing	Burns	Burns	Self-extinguishing
Effect of sunlight	Requires protection	Discolors	Requires protection	Crazes	Slight
Effect of strong acids or bases	Attacked, acids	Attacked, acids	Very resistant	Attacked, acids	Attacked
Effect of organic solvents	Resistant	Resistant	Soluble	Resistant	Soluble
Clarity	Transparent	Opaque	Opaque	Opaque	Transparent

TABLE 17-1 (continued)

Property	Polyethylene			Poly(ethylene terephthalate)	Polyimide	Polypropylene	Polysulfone
	Low Density (Branched)	"Linear" Low Density	High Density (Linear)				
Specific gravity (g/cm^3)	0.910–0.925	0.918–0.935	0.941–0.965	1.34–1.39	1.43	0.902–0.906	1.24
Refractive index (n_D^{25})	1.51	1.51	1.54	1.64	—	1.49	1.633
Tensile strength (psi)	600–2300	1900–4000	3100–5500	8500–10,500	10,500	4300–5500	10,200^c
Elongation (%)	90–800	100–950	20–1000	50–300	5.0–7.0	200–700	50–100
Tensile modulus (10^5 psi)	0.14–0.38	0.38–0.75	0.6–1.8	4.0–6.0	4.5	1.6–2.3	3.6
Impact strength (ft-lb/in. of notch)	>16	1.0–9.0	0.5–2.0	0.25–0.65	1.1	0.5–2.0	1.3^d
Heat-deflection temperature (°F, 264 psi)	90–105	—	110–130	100–106	650	125–140	345
Dielectric constant (1000 cycles)	2.25–2.35	2.25–2.35	2.30–2.35	3.46–4.5	3.4	2.2–2.6	3.13
Dielectric loss (1000 cycles)	<0.0005	<0.0005	<0.0005	0.002–0.03	0.002	<0.0005–0.0018	0.001
Water absorption (one-eighth in. bar, 24 hr, %)	<0.015	—	<0.01	0.1–0.2	0.32	<0.01	0.22
Burning rate	Very slow	Very slow	Very slow	Burns	None	Slow	Self-extinguishing
Effect of sunlight	Requires protection	Requires protection	Requires protection	Slight	None	Requires protection	Slight
Effect of strong acids or bases	Resistant	Resistant	Resistant	Resistant	Attacked, alkalis	Resistant	Resistant
Effect of organic solvents	Resistant below 80°C	Resistant	Resistant below 80°C	Resistant	Very resistant	Resistant below 80°C	Soluble
Clarity	Opaque	Opaque	Opaque	Transparent	Opaque	Opaque	Transparent

Property	Polystyrenes			Poly(vinyl chloride)		Poly(vinylidene chloride) (Film)
	Polystyrene	Impact Polystyrene	Styrene–Butadiene Thermoplastic Elastomer	Rigid	Plasticized	
Specific gravity (g/cm^3)	1.04–1.09	1.04–1.10	0.93–1.10	1.35–1.45	1.16–1.35	1.65–1.70
Refractive index (n_D^{25})	1.59–1.60	—	1.52–1.55	1.52–1.55	—	—
Tensile strength (psi)	5000–12,000	1500–7000	600–3000	5000–9000	1500–3500	2800–3500
Elongation (%)	1.0–2.5	2–80	300–1000	2.0–40.0	200–450	250–400
Tensile modulus (10^5 psi)	4–6	1.5–5	0.008–0.5	3.5–6	—	0.5–0.8
Impact strength (ft-lb/in. of notch)	0.25–0.4^d	0.5–11	No break	0.4–20	Varies	0.3–1.0
Heat-deflection temperature (°F, 264 psi)	220	210	−150	130–175	—	130–150
Dielectric constant (1000 cycles)	2.4–2.65	2.4–4.5	2.5–3.4	3.0–3.3	4.0–8.0	3.4
Dielectric loss (1000 cycles)	0.0001–0.0003	0.0004–0.002	0.001–0.003	0.009–0.017	0.07–0.16	0.015
Water absorption (one-eighth in. bar, 24 hr, %)	0.03–0.10	0.05–0.6	0.19–0.39	0.07–0.4	0.15–0.75	0.1
Burning rate	Slow	Slow	Slow	Self-extinguishing	Slow to self-extinguishing	None
Effect of sunlight	Yellows	Loses strength	Slight	Slight	Slight	None
Effect of strong acids or bases	Attacked, acids	Attacked, acids	Attacked	Resistant	Resistant	Resistant
Effect of organic solvents	Soluble	Soluble	Soluble	Soluble	Soluble	Soluble
Clarity	Transparent	Opaque	Transparent	Transparent	Transparent	Transparent

^{a}Modern Plastics (1979–1982).
^{b}66 psi.
^{c}At yield.
^{d}One-quarter in. bar.

TABLE 17-2. Typical Properties of Commercial Thermosetting Resins[a]

		Phenol–Formaldehyde		Polyester		Silicone Cast/RTV None	Sheet-Molding Compound	Polyurethane
Property	Epoxy, Cast None	None	Wood or Cotton	Cast None	Glass Cloth			
Specific gravity (g/cm^3)	1.11–1.40	1.25–1.30	1.34–1.45	1.10–1.46	1.50–2.10	0.99–1.50	1.65–2.6	1.05
Refractive index (n_D^{25})	1.55–1.61	—	—	1.52–1.57	—	1.43	—	—
Tensile strength (psi)	4000–13,000	7000–8000	5000–9000	6000–13,000	30,000–50,000	350–1000	8000–25,000	10,000–11,000
Elongation (%)	3.0–6.0	1.0–1.5	0.4–0.8	<5.0	0.5–2	100–300	3	3–6
Tensile modulus (10^5 psi)	3.5	7.5–10	8–17	3.0–6.4	15–45	0.09	14–25	—
Impact strength (ft-lb/in. of notch)	0.2–1.0	0.20–0.36	0.24–0.60	0.2–0.4	5–30	—	7–22	0.4
Heat-deflection temperature (°F, 264 psi)	115–550	240–260	260–340	140–400	—	—	375–500	190–200
Dielectric constant (1000 cycles)	3.5–4.5	4.5–6.0	4.4–9.0	2.8–5.2	4.2–6.0	2.7	4.15–5.17	3.43–4.22
Dielectric loss (1000 cycles)	0.002–0.02	0.03–0.08	0.04–0.20	0.005–0.025	0.01–0.06	0.01–0.002	0.008–0.017	0.005–0.0006
Water absorption (one-eighth in. bar, 24 hr, %)	0.08–0.15	0.1–0.2	0.3–1.2	0.15–0.60	0.05–0.50	0.02	0.1–0.25	1.5
Burning rate	Slow	Very slow	Very slow	Burns	Burns	Self-extinguishing	None	Burns
Effect of sunlight	None	Darkens	Darkens	Yellows	Slight	None	Slight	Slight
Effect of strong acids or bases	Attacked	Attacked	Attacked	Attacked	Attacked	Attacked	Attacked, acids	Resistant
Effect of organic solvents	Resistant	Resistant	Resistant	Attacked	Attacked	Attacked	Resistant	Resistant
Clarity	Transparent	Transparent	Opaque	Transparent	Translucent	Transparent	Opaque	Transparent

[a]*Modern Plastics* (1979–1982).

5. Describe the various sections of an extruder screw, indicating the purpose of each one.

6. Why is it feasible to produce multilayered films or sheets by coextrusion; why are blends not formed instead?

7. Discuss the conditions under which immiscible polymer blends can have useful properties.

8. Compare the structures of impact polystyrene and ABS resins.

9. Define and describe processes for making interpenetrating polymer networks.

10. Discuss the use of (a) fillers and (b) plasticizers to improve and tailor polymer properties.

BIBLIOGRAPHY

Agarwal 1980. D. B. Agarwal and Lawrence J. Broutman, *Analysis & Performance of Fiber Composites,* Wiley-Interscience, New York, 1980.

Alfrey 1980. Turner Alfrey, Jr., and Walter J. Schrenk, "Mulipolymer Systems," *Science* **208**, 813–818 (1980).

Anderson 1981. Roger A. Anderson, "New Structural Pultrusions are Stronger, More Corrosion-Resistant," *Plastics Eng.* **37** (10), 21–22 (1981).

Bajaj 1982. J. K. L. Bajaj, "Liquid and Paste Mixers," pp. 337, 340, 343, 344 in *MPE,*† 1982.

Batiuk 1977. Martin Batiuk, "Polyvinyl Chloride Extrusion," Chapter 23 in Leonard I. Nass, ed., *Encyclopedia of PVC,* Vol. 3, Marcel Dekker, New York, 1977.

Beardmore 1980. P. Beardmore, J. J. Harwood, K. R. Kisman, and R. E. Robertson, "Fiber-Reinforced Composites: Engineering Structural Materials," *Science* **208**, 833–840 (1980).

Becker 1979. Walter E. Becker, ed., *Reaction Injection Molding,* Van Nostrand Reinhold, New York, 1979.

Beeler 1982. A. D. Beeler and D. C. Finney, "Plasticizers," pp. 193–194, 198 in *MPE,* 1982.

Benning 1969. Calvin James Benning, *Plastic Foams,* John Wiley & Sons, New York, 1969.

Blake 1982. W. P. Blake, "Flame Retardants," pp. 171, 173 in *MPE,* 1982.

Blumberg 1980. John G. Blumberg, James S. Falcone, Jr., Leonard H. Smiley, and David I. Netting, "Fillers," pp. 198–215 in *ECT,*‡ Vol. 10, 1980.

Brilliant 1982. S. D. Brilliant, "Heat Stabilizers," pp. 180–182 in *MPE,* 1982.

Brown 1981. Randy J. Brown and James W. Summers, "Designing Profile Dies for Coextrusion," *Plastics Eng.* **37** (9), 25–29 (1981).

Brydson 1973. J. A. Brydson and D. G. Peacock, *Principles of Plastics Extrusion,* Applied Science, London, 1973.

Buttery 1960. D. N. Buttery, *Plasticizers,* 2nd ed., Franklin, Palisades, New Jersey, 1960.

Cargile 1982. H. M. Cargile and J. M. Tower, "Melt-Processible Structural Foam Molding," pp. 275–276, 278 in *MPE,* 1982.

†Joan Agranoff, ed., *Modern Plastics Encyclopedia 1982–1983,* Vol. 59, No. 10A, McGraw-Hill, New York, 1982.

‡Martin Grayson, ed., *Kirk–Othmer Encyclopedia of Chemical Technology,* 3rd ed., Wiley-Interscience, New York.

Chasar 1982. D. W. Chasar, J. T. Lai, and P. N. Son, "Ultraviolet Stabilizers," pp. 209–210 in *MPE*, 1982.

Chung 1977. Chan I. Chung, "A Guide to Better Extruder Screw Design," *Plastics Eng.* **33** (2), 34–37 (1977).

Crump 1980. E. Lea Crump, "Film and Sheeting Materials," pp. 216–246 in *ECT*, Vol. 10, 1980.

Darby 1969. J. R. Darby and J. K. Sears, "Plasticizers," pp. 228–306 in *EPST,*† Vol. 10, 1969.

Delmonte 1981. John Delmonte, *Technology of Carbon & Graphite Fiber Composites,* Van Nostrand Reinhold, New York, 1981.

Dym 1979. Joseph B. Dym, *Injection Molds & Molding: A Practical Manual,* Van Nostrand Reinhold, New York, 1979.

Eighmy 1982. G. W. Eighmy, Jr., "Calendering," pp. 220, 222 in *MPE*, 1982.

Eise 1982. Kurt Eise, "Fluxed Melt Mixers," pp. 335, 337–338 in *MPE*, 1982.

Fair 1982. R. L. Fair, "Rotational Molding," pp. 367–368 in *MPE*, 1982.

Faulks 1982. B. F. Faulks, "Color Concentrates," pp. 143–144 in *MPE*, 1982.

Fenelon 1981. Paul J. Fenelon, "Multicomponent Polymer Systems Could be Industry Bellwether," *Plastics Eng.* **37** (9), 37–41 (1981).

Fisher 1976. E. G. Fisher, *Extrusion of Plastics,* 3rd ed., John Wiley & Sons, New York, 1976.

Foster 1982. Joe Foster, "Solid Phase Pressure Forming," p. 376 in *MPE*, 1982.

Fredrickson 1982. R. B. Fredrickson, "Stretch-Blow Molding," pp. 219, 221 in *MPE*, 1982.

Frissell 1967. W. J. Frissell, "Fillers," pp. 740–763 in *EPST,* Vol 6, 1967.

Gordon 1982. J. B. Gordon, "Expandable Polystyrene Molding," pp. 272–273 in *MPE*, 1982.

Haisser 1982. A. S. Haisser, "Extrusion," pp. 246, 250, 254, 258, 260, 262, 264, 267, 268 in *MPE*, 1982.

Hall 1982. C. M. Hall, "Reaction Injection Molding," pp. 355–357 in *MPE*, 1982.

Han 1976. Chang Dae Han, *Rheology in Polymer Processing,* Academic Press, New York, 1976.

Hancox 1981. N. L. Hancox, ed., *Fibre Composite Hybrid Materials,* Macmillan, New York, 1981.

Hearons 1982. J. S. Hearons, "(Fibrous Reinforcements) Glass," pp. 153–154 in *MPE*, 1982.

Herbert 1982. Victor Herbert, "Multicomponent Liquid Foam Processing," pp. 278, 289 in *MPE*, 1982.

Higgins 1965. D. G. Higgins and Arthur H. Landrock, "Coating Methods," pp. 765–830 in *EPST,* Vol. 3, 1965.

Higgins 1967. D. G. Higgins, "Fabrics, Coated," pp. 467–489 in *EPST*, Vol. 6, 1967.

Holmes-Walker 1975. W. A. Holmes-Walker, *Polymer Conversion,* Halsted Press, John Wiley & Sons, New York, 1975.

Hopmeir 1969. A. P. Hopmeir, L. M. Greenstein, and Anthony J. Petro, "Pigments," pp. 157–219 in *EPST*, Vol. 10, 1969.

Hull 1968. John L. Hull, Lee J. Zukor, G. E. Pickering, Richard E. Duncan, David R. Ellis, Robert A. McCord, and A. B. Hitchcock, "Molding," pp. 1–157 in *EPST*, Vol. 9, 1968.

Hull 1981. Derek Hull, *Introduction to Composite Materials,* Cambridge University Press, Cambridge, 1981.

Hull 1982. J. L. Hull, "Injection Molding Thermosets," pp. 314, 316 in *MPE*, 1982.

Irwin 1982. Christopher Irwin, "Extrusion—Blow Molding," pp. 212–214, 217 in *MPE*, 1982.

Jalbert 1982. R. L. Jalbert, "Alloys," pp. 127–128 in *MPE*, 1982.

Kamal 1982. Musa R. Kamal, "Upgrading Plastics Performance—Part 2," *Plastics Eng.* **38** (12), 31–35 (1982).

Katz 1978. Harry S. Katz and John V. Milewski, *Handbook of Fillers and Reinforcements for Plastics,* Van Nostrand Reinhold, New York, 1978.

Kiernan 1982. E. F. Kiernan, "Solvent Casting of PVC Film," p. 226 in *MPE*, 1982.

Kiesau 1980. Milfried Kiesau, "Thermoset Molding: Injection, Transfer or Compression—Which is Best for Your Job?," *Plastics Eng.* **36** (8), 29–32 (1980).

Klein 1977. Imrich Klein and Jules W. Lindau, "Extrusion," Chapter 9 in Herman S. Kaufman and Joseph J. Falcetta, eds., *Introduction to Polymer Science and Technology—An SPE Textbook,* John Wiley & Sons, New York, 1977.

Krauskopf 1976. Leonard G. Krauskopf, "Plasticizers," Chapter 11 in Leonard I. Nass, ed., *Encyclopedia of PVC,* Vol. 1, Marcel Dekker, New York, 1976.

Kubiak 1980. Richard S. Kubiak, "Taking RIM Beyond the Urethanes," *Plastics Eng.* **36** (3), 55–61 (1980).

Levy 1981. Sidney Levy, *Plastics Extrusion Technology Handbook,* Industrial Press, New York, 1981.

Lubin 1982. George Lubin, *Handbook of Composites,* Van Nostrand Reinhold, New York, 1982.

Luskin 1982. L. S. Luskin, "Casting of Acrylic," pp. 222–224 in MPE, 1982.

McDonagh 1977. John M. McDonagh, "Polymer Fabrication Processes," Chapter 11 in Herman S. Kaufman and Joseph J. Falcetta, eds., *Introduction to Polymer Science and Technology—An SPE Textbook,* John Wiley & Sons, New York, 1977.

McGarry 1981. Frederick J. McGarry, "Laminated and Reinforced Plastics," pp. 968–978 in *ECT*, Vol. 13, 1981.

McKelvey 1962. James M. McKelvey, *Plastics Processing,* John Wiley & Sons, New York, 1962.

Manson 1976. J. A. Manson and L. H. Sperling, *Polymer Blends and Composites,* Plenum Press, New York, 1976.

Martin 1979. Jeffrey D. Martin, "Pultrusion: The Other Process," *Plastics Eng.* **35** (3), 53–57 (1979).

Mascia 1974. L. Mascia, *The Role of Additives in Plastics,* John Wiley & Sons, New York, 1974.

Mathews 1982. George Mathews, *Polymer Mixing Technology,* Elsevier, New York, 1982.

Meinecke 1965. Eberhard Meinecke, "Calendering," pp. 802–819 in *EPST*, Vol. 2, 1965.

Metz 1977. Hyman Metz and Carl Pettigrew, "Miscellaneous Molding, or Casting, and Coating Operations using Vinyl Powders and Liquids," Chapter 27 in Leonard I. Nass, ed., *Encyclopedia of PVC,* Vol. 3, Marcel Dekker, New York, 1977.

Middleman 1977. Stanley Middleman, *Fundamentals of Polymer Processing,* McGraw-Hill, New York, 1977.

Miner 1982. L. H. Miner, "(Fibrous Reinforcements) Aramid," p. 152, "Aramid/Carbon," p. 154, "Aramid/Glass," p. 155, "Aramid/Carbon/Glass," p. 155 in *MPE*, 1982.

Monroe 1965. Sam Monroe, "Bag Molding," pp. 300–316 in *EPST,* Vol. 2, 1965.

O'Brien 1982. J. C. O'Brien, "Transfer Molding," pp. 388–389 in *MPE*, 1982.

Olabisi 1982. Olagoke Olabisi, "Polyblends," pp. 443–478 in *ECT*, Vol. 18, 1982.

Olmsted 1982. B. A. Olmsted, "Injection Molding Thermoplastics," pp. 296, 298, 302, 306, 311, 314 in *MPE*, 1982.

Owen 1982. J. E. Owen, R. A. Charvat, and F. A. Waksmunski, "Colorants," pp. 135, 136, 138, 140, 142, 143 in *MPE*, 1982.

Paolino 1982. R. Rankin Paolino, "Antioxidants," p. 130, 132 in *MPE*, 1982.

Pasquale 1982. J. A. Pasquale III, "Laminating of Film," pp. 318–320 in *MPE*, 1982.

Paul 1978. D. R. Paul and Seymour Newman, eds., *Polymer Blends,* Academic Press, New York, 1978.

Perlberg 1977. S. Everett Perlberg and Peter R. A. Burnett, "Calendering and Calender Laminating,"

Chapter 25 in Leonard I. Nass, ed., *Encyclopedia of PVC*, Vol. 3, Marcel Dekker, New York, 1977.

Peterson 1982. H. L. Peterson, "(Fibrous Reinforcements) Carbon," pp. 152–153 in *MPE*, 1982.

Piggott 1980. Michael R. Piggott, *Load-Bearing Composite Materials*, Pergamon Press, New York, 1980.

Power 1968. G. E. Power, "Laminates," pp. 121–163 in *EPST*, Vol. 8, 1968.

Rainville 1982. Dewey Rainville, "Injection-Blow Molding," pp. 217, 219 in *MPE*, 1982.

Ramazzotti 1979. D. J. Ramazzotti, "Rotational Molding: A Process Whose Time Has Come," *Plastics Eng.* **35** (10), 47–49 (1979).

Reising 1982. Tom Reising, "Dry Solids Mixers," pp. 333–335 in *MPE*, 1982.

Richardson 1974. Paul N. Richardson, *Introduction to Extrusion*, John Wiley & Sons, New York, 1974.

Richardson 1982. Paul N. Richardson, "Plastics Processing," pp. 184–206 in *ECT*, Vol. 18, 1982.

Rosen 1982. Stephen L. Rosen, *Fundamental Principles of Polymeric Materials*, Wiley-Interscience, New York, 1982.

Rubin 1973. Irvin I. Rubin, *Injection Molding Theory & Practice*, Wiley-Interscience, New York, 1973.

Rubin 1977. Irvin I. Rubin, "Injection Molding," Chapter 10 in Herman S. Kaufman and Joseph J. Falcetta, eds., *Introduction to Polymer Science and Technology—An SPE Textbook*, John Wiley & Sons, New York, 1977.

Schaer 1977. Leonard S. Schaer, "Processing Plastics—Spin Casting Thermoset Parts," *Plastics Eng.* **33** (2), 48–49 (1977).

Schneider 1981. Fritz W. Schneider, "Step up to RIM," *Plastics Eng.* **37** (3), 89–93 (1981).

Schrafft 1982. Fred Schrafft, "Extruding Thermoplastic Foams," pp. 274–275 in *MPE*, 1982.

Sears 1982a. J. K. Sears and N. W. Touchette, "Plasticizers," pp. 111–183 in *ECT*, Vol. 18, 1982.

Sears 1982b. J. Kern Sears and Joseph R. Darby, *Technology of Plasticizers*, Wiley-Interscience, New York, 1982.

Sheldon 1982. R. P. Sheldon, *Composite Polymeric Materials*, Elsevier, New York, 1982.

Skotchdopole 1965. R. E. Skotchdopole, "Cellular Materials," pp. 80–130 in *EPST*, Vol. 3, 1965.

Sneller 1982a. Joseph Sneller, "Nylon RIM Surges. So does RIM Epoxy. The Surprise? They're Boosting Urethanes," *Mod. Plastics* **59** (5), 46–49 (1982).

Sneller 1982b. Joseph Sneller, "A Sure Bright Spot in '83: Sheet and Film Barrier Coextrusions," *Mod. Plastics* **59** (12), 58–61 (1982).

Solomon 1983. D. H. Solomon and D. G. Hawthorne, *Chemistry of Pigments and Fillers*, John Wiley & Sons, New York, 1983.

Sperling 1981. L. H. Sperling, *Interpenetrating Polymer Networks and Related Materials*, Plenum Press, New York, 1981.

Stayner 1982. Vance Stayner, "Nonfibrous Property Enhancers," pp. 156–158, 160, 163 in *MPE*, 1982.

Suh 1980. K. W. Suh and R. E. Skotchdopole, "Foamed Plastics," pp. 83–126 in *ECT*, Vol. 11, 1980.

Tadmor 1978. Zehev Tadmor and Imrich Klein, *Engineering Principles of Plasticating Extrusion*, Krieger, Melbourne, 1978.

Tadmor 1979. Zehev Tadmor and Costa G. Gogos, *Principles of Polymer Processing*, Wiley-Interscience, New York, 1979.

Tapper 1982. Michael Tapper, "Fillers/Extenders," pp. 163, 166, 168, 171 in *MPE*, 1982.

Teal 1982. Kevin Teal, "Thermoforming," pp. 371, 374, 376 in *MPE*, 1982.

Tsai 1980. Stephen W. Tsai and H. Thomas Hahn, *Introduction to Composite Materials*, Technomic, Westport, Connecticut, 1980.

Tulley 1977. F. T. Tulley and B. C. Harris, "Injection Molding of Polyvinyl Chloride," Chapter 24 in Leonard I. Nass, ed., *Encyclopedia of PVC*, Vol. 3, Marcel Dekker, New York, 1977.

Wallis 1965. Benedict L. Wallis, "Casting," pp. 1–20 in *EPST*, Vol. 3, 1965.

Watkins 1976. W. D. Watkins, "Calendering," *Plastics Eng.* **32** (6), 23–25 (1976).

Webber 1979. Thomas G. Webber, ed., *Coloring of Plastics*, Wiley-Interscience, New York, 1979.

Werner 1977. Arnold C. Werner, "Spread Coating," Chapter 26 in Leonard I. Nass, ed., *Encyclopedia of PVC*, Vol. 3., Marcel Dekker, New York, 1977.

Westover 1968. R. F. Westover, "Melt Extrusion," pp. 533–587 in *EPST*, Vol. 8, 1968.

CHAPTER EIGHTEEN

FIBER TECHNOLOGY

The wide range of materials classified as fibers includes natural and synthetic, organic and inorganic products. Some polymers used as fibers, such as nylon and cellulose acetate, serve equally well as plastics. In the final analysis, the classification of a substance as a fiber depends more on its shape than on any other property. One common definition of a fiber requires that its length be at least 100 times its diameter. Artificial fibers can usually be made in any desired ratio of length to diameter. Among the natural fibers, cotton, wool, and flax are often found with lengths 1000–3000 times their diameter; coarser fibers such as jute, ramie, and hemp have lengths 100–1000 times their diameter.

To be useful as a textile material, a synthetic polymer must have suitable characteristics with respect to several physical properties. These include a high softening point to allow ironing, adequate tensile strength over a fairly wide temperature range, solubility or meltability for spinning, a high modulus or stiffness, and good textile qualities such as those defined in Section A. In addition to these primary requirements, many other properties of the material are important if it is to be suitable for textile applications. Some of these are listed in Table 18-1. The desirable values of these properties and the molecular structure needed to obtain them are discussed in the following sections of this chapter.

The textile industry is quite complex, and many of the technologies that comprise it are intricate and obscure. A new fiber must conform to a set of desirable properties that have been largely decided upon in advance. Manipulation of molecular structure and synthesis variables does not allow freedom to achieve all the desired properties, so compromise is inevitable. Even a successful compromise in physical properties does not ensure success for a new fiber. Other factors that must be favorable include all the processes leading to some knitted or woven cloth ready for wearing trials, a sound economic situation and competitive price structure, a good supply of raw materials, and, eventually, a good record of customer acceptance.

TABLE 18-1. Fiber Properties Important in Textile Uses

Chemical	Physical	Biological	Fabric Properties
Stability toward	Mechanical	Toxicological	Appearance
Acids	Tenacity	Dermatological	Drape
Bases	Elongation	Resistance to	Hand
Bleaches	Stiffness	Bacteria	Luster
Solvents	Flex life	Molds	Comfort
Heat	Abrasion resistance	Insects	Warmth
Sunlight	Work recovery		Water sorption
Aging	Tensile recovery		Moisture retention
Flammability	Thermal		Wicking
Dyeability	Melting point		Stability
	Softening point		Shape
	Glass transition		Shrinkage
	temperature		Felting
	Decomposition		Pilling
	temperature		Crease resistance
	Electrical		Crease retention
	Surface resistivity		

United States consumption of fibers in 1982 was very close to 10 billion lb, down some 25% from the peak year of 1979. About 75% of the 1982 total was synthetic fibers, with virtually all of the remaining 25% being one natural fiber, cotton. Figures from earlier chapters show that polyester is by far the most widely used synthetic (3.1 billion lb in the United States in 1982), with the nylons (1.9 billion lb) and the acrylics (0.65 billion lb) also major, followed by the rayons and polypropylene. All others, including wool, play only a minor role.

A. TEXTILE AND FABRIC PROPERTIES

Definitions of Textile Terms

Types of Fibers. Most artificial fibers may be obtained either as a very long *continuous filament,* or as *staple,* made by cutting continuous filament into relatively short lengths. Natural fibers, with the exception of silk, are obtained only in the form of staple.

Denier. The *denier* of a fiber, a measure of its size, is defined as the weight in grams of 9000 m of the fiber. It is thus proportional to the density of the fiber and to its cross-sectional area. At least ten other measures of the size of fibers are in use, but the denier is the most widely accepted.

Tenacity. The tensile strength of a fiber is usually expressed in terms of *tenacity.* Tenacity is defined as the strength per unit size number, such as the denier, where the size number is expressed as a weight per unit length. Tenacity is thus a function of the density of the fiber as well as its tensile strength. Two fibers having the same denier and the same breaking strength have the same tenacity, usually expressed in grams per denier; if they have the same density, they have the same tensile strength as well. Tensile strength (pounds per square inch) = tenacity (grams per denier) × density × 12,791.

Moisture Content and Moisture Regain.   Most textile fibers absorb some moisture from their surroundings. If the amount of moisture present at equilibrium under standard conditions (65% relative humidity and 70°F) is expressed as a percentage of the total weight of the (moist) fiber, it is known as the *moisture content*; if it is expressed as a percentage of the oven-dry weight (110°C) of the fiber, it is known as the *moisture regain.*

Crimp.   *Crimp* is the waviness of a fiber, a measure of the difference between the length of the unstraightened and that of the straightened fiber. Some naturally occurring fibers, notably certain wools, have a natural crimp. Crimp can be artificially produced in fibers by suitable heat treatment or by rolling them between heated, fluted rolls.

Fabric Property Terms.   The esthetic qualities of fabrics are defined in terms such as *appearance, hand* (or *handle*), and *drape.* Although the qualitative meanings of these terms are obvious, their quantitative definitions and interpretations in terms of fiber properties are quite difficult. Hand and drape are largely determined by the tensile and elastic behavior of the fiber.

Properties of Textile Fibers

Electrical Properties.   Since fibers are not normally used in electrical applications, their only electrical property of great interest is their resistivity. Too high resistivity leads to the development of static electrical charges, which cause the fabric to cling unpleasantly and to be difficult to clean. Some of the synthetic fibers, including the nylons, polyesters, and acrylics, are poor in this respect.

Mechanical Properties.   The mechanical properties of fibers are quite complex and have been the subject of much experimental work. A stressed textile fiber is a complicated viscoelastic system, in which a number of irreversible processes can take place. Some typical fiber stress–strain curves are shown in Figs. 18-1 and 18-2. They may be divided roughly into two groups: the silklike curves, featuring almost constant high modulus to the break point, and the wool-like curves, featuring a sharp drop in modulus at low stress followed by a long period of elongation at almost constant stress. Intermediate characteristics are shown by some of the synthetics such as nylon: The first region of reversible elasticity is followed by an irreversible region in which drawing takes place. When orientation is complete, another reversible elastic region is found. This may end at the point of failure, or plastic flow may take place.

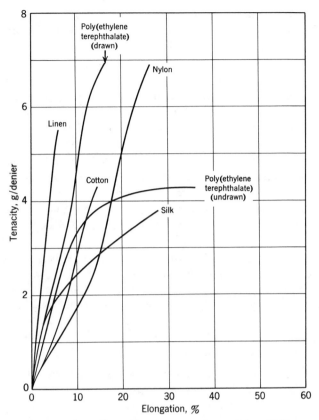

FIG. 18-1. Stress–strain curves of silklike fibers (Heckert 1953).

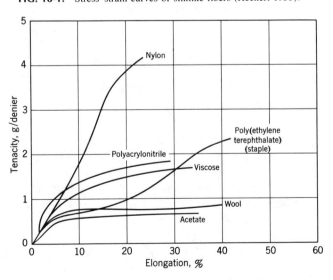

FIG. 18-2. Stress–strain curves of wool-like fibers (Heckert 1953).

The tenacity at break of typical fibers ranges from about 1 g/denier for the rayons up to 10 for highly oriented polyester and over 20 for the strongest aramid and carbon fibers. The range of 1–5 g/denier is considered suitable for textile applications.

Moisture Regain.   The moisture regain of the synthetic polyesters, nylons, and acrylics is so much lower than that of any of the natural fibers that the synthetics may be classed as hydrophobic materials. This is an advantage for rapid drying.

Dyeability.   Hydrophobic fibers are in general difficult to dye. The acrylics are particularly bad in this respect, and modification of the polymer structure plus the development of new classes of dyes was resorted to in order to provide a wide range of pleasing colors in this material. Nylon and polyester are intermediate in dyeability, while cellulose and the cellulosic fibers are eminently dyeable.

Chemical Stability.   All textile fibers must be stable to water, dry-cleaning solvents, and dilute acids, alkalis, and bleaches, and all are reasonably adequate in this respect. Silk and nylon are the least satisfactory fibers from the standpoint of weatherability, whereas the acrylics are outstanding in this respect.

Fabric Properties

Esthetic Factors.   (See preceding definitions of textile terms.) Silk is the outstanding example of a fiber with good esthetic properties. Filament acetate, which possibly ranks next, is replacing filament viscose rayon in many applications for which luster, hand, and drape are more important than good mechanical properties. Among the staple fibers, the polyesters appear to be most nearly wool-like and most pleasing.

Comfort.   The comfort of a fabric is a highly important but little understood property. It is related to the structure of the fabric in determining ventilation and heat-insulating characteristics and wicking ability. Comfort is not, however, dependent upon high moisture absorption in the fiber.

Crease Resistance and Crease Retention.   These two properties are related in that a fabric that can be creased only with difficulty tends to retain its crease. The thermoplastic fibers such as nylon, polyesters, and acrylics are very good in both respects. Creasing in wear becomes more important at high moisture contents. Cellulosics are naturally poor in crease resistance, although they may be modified by resin finishes to give good crease resistance.

Fabric Stability.   The stability of shape and dimensions of the synthetic fibers is outstandingly good. Wool retains its shape well in garments, but felts and shrinks under wet treatment.

The phenomenon of "pilling" has become important in some of the synthetics. A small nodule or pill can be formed on some fabrics by gentle rubbing, as the surface fibers are raised and tangled. In wool these pills are not important because they wear off rapidly, but in nylon and the polyesters the pills do not break away and may become quite unsightly.

Wear Resistance.   There is no laboratory test for abrasion in a fabric that is correlated with actual use. There is evidence that two or more independent parameters are required to define wear in a fabric. In general, acetate rayon is most easily abraded, with viscose, cotton, and wool following as a group (except that viscose when wet is not much better than acetate), while the polyesters, acrylics, and nylon are progressively better.

GENERAL REFERENCES

Mark 1967–1968; McIntyre 1968; Goswami 1977; McGovern 1980; Peters 1980; Roberts 1980.

B.   SPINNING

The conversion of bulk polymer to fiber form is accomplished by *spinning*. In most cases, spinning processes require solution or melting of the polymer (an exception is the spinning of fibers from an aqueous dispersion, as in the case of polytetrafluoroethylene).

If a polymer can be melted under reasonable conditions, the production of a fiber by *melt spinning* is preferred over solution processes. When melt spinning cannot be carried out, a distinction as to type of process is made depending upon whether the solvent is removed by evaporation (*dry spinning*) or by leaching out into another liquid which is miscible with the spinning solvent but is not itself a solvent for the polymer (*wet spinning*). Dry spinning has some advantages over wet spinning. The three processes have many features in common.

The conversion of the spun polymer melt or solution to a solid fiber involves cooling, solvent evaporation, or coagulation, depending on the type of spinning used. The rates of these processes decrease in the order listed. Cooling of a fine filament is normally very rapid and can be controlled within relatively narrow limits. Solvent evaporation involves simultaneous outward mass transfer and inward heat transfer, the rate-controlling step invariably being outward diffusion of solvent. Coagulation involves two-way mass transfer, the coagulating agent (e.g., acid) diffusing inward, and the products of coagulation (e.g., salts, H_2S) diffusing out. As a consequence of these facts, it is very easy to obtain a melt-spun fiber that possesses uniform properties throughout its cross section, but almost impossible to do so with solvent-spun or coagulated fibers. In addition, the rapid cooling of melt-spun fibers tends to produce an almost circular cross section, whereas with solvent-

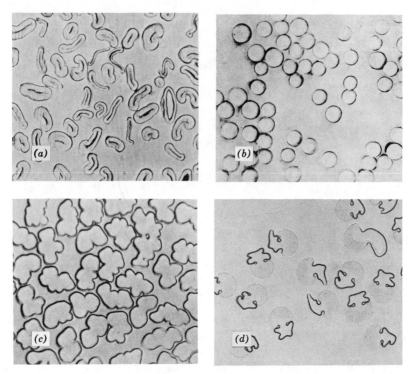

FIG. 18-3.   Photomicrographs (× 250) showing cross-sectional shapes of fibers: (a) cotton, (b) melt-spun polyester, (c) dry-spun acetate, (d) wet-spun conjugate viscose fiber [(a)–(c) Riley 1956, (d) Hicks 1967].

spun fibers the cross sections are usually elliptical, and with coagulated fibers (because of the absence of strong surface-tension effects) the cross sections are ordinarily highly convoluted, as shown in Fig. 18-3.

It is found, however, that convoluted fibers have desirable esthetic properties, particularly hand and luster, so that an effort is made to produce the effect at will. This can be done in melt spinning by the use of noncircular orifices in the spinneret (see below), or in several types of spinning by extruding two different polymers through the same orifice to produce a single *conjugate* fibers (Hicks 1967), as also depicted in Fig. 18-3. The production of such bicomponent and biconstituent fibers is becoming increasingly important, especially since they provide a means for producing the textured products described in Section *C*.

Melt Spinning

The process of melt spinning is inherently simple. Molten polymer is pumped at a constant rate under high pressure through a plate called a *spinneret* containing a large number of small holes. The liquid polymer streams emerge downward from the face of the spinneret, usually into air. They solidify and are brought together

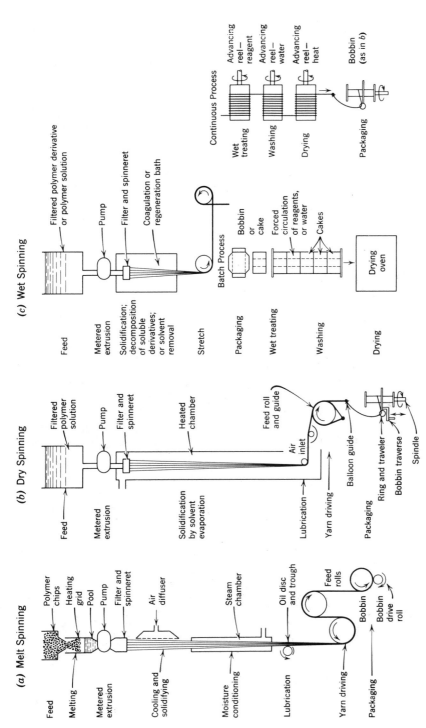

FIG. 18-4. Schematic diagrams of the three principal methods of spinning fibers (Riley 1956).

to form a thread and wound up on bobbins. A subsequent drawing step is necessary to orient the fibers.

The polymer is melted by contacting a hot grid in the form of steel tubing, which is heated by electric current or by hot vapors (Fig. 18-4a). It is usually necessary to protect the polymer melt from oxygen by blanketing it with steam or an inert gas such as carbon dioxide or nitrogen. If the viscosity of the molten polymer is low, it may pass directly to the metering (constant-rate) pump. For melts of higher viscosity a booster pump may be used. Other methods of melting have been proposed, including the use of an extrusion-type screw, one section of which can serve as its own metering pump. Methods in which the metering pump is replaced, for example, by a source of gas pressure or other device, do not appear to offer sufficiently precise control to hold the denier of fine yarns constant.

The spinneret may consist of a 2- to 3-in. diameter steel disk about one-quarter in. thick with 50–60 countersunk holes 0.010 in. or less in diameter. The denier of the filament is determined not by the diameter of the holes but by the rate at which polymer is pumped through the spinneret and the rate at which the filaments are wound up.

The filaments emerge from the spinneret face into air and begin to cool. An air blast may be used to speed up the cooling process. After the filaments have traveled far enough to become solid (about 2 ft) they are brought together and wound up. Speeds of about 2500 ft/min are usually employed.

The filaments as spun are almost completely unoriented. Most of the stretching that occurs between the spinneret and windup does so while the filament is still molten, and there is sufficient time for molecular orientation to relax before the fiber cools and crystallizes. Consequently a separate drawing step is necessary to produce the orientation of the crystallites necessary for optimum physical properties. In the drawing step somewhat lower speeds were traditionally required than can be achieved in spinning, so the two steps were usually carried out separately; now they are often done in a single continuous operation. The drawing step utilizes two sets of rolls (Fig. 18-5), one to feed the undrawn yarn from a supply package at velocity v_1 and the other, moving about four times as fast, to collect the drawn yarn at velocity v_2. The filaments may pass over a metal pin between the two sets

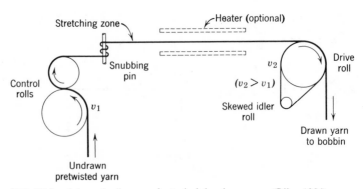

FIG. 18-5. Schematic diagram of a typical drawing process (Riley 1956).

of rolls; drawing is localized in the neighborhood of the pin. The yarn is then collected on a strong metal bobbin. Freshly drawn yarn has a tendency to contract somewhat in length. Heat is evolved during the drawing process, since work is done on the polymer.

Melt spinning is used for polyester and nylon fibers, among several others, and thus accounts for the vast majority of synthetic fiber production.

Dry Spinning

In the dry-spinning process, the filament is formed by the evaporation of solvent from the polymer solution into air or an inert gas atmosphere. The solvent must, of course, be volatile. Dry spinning has been used for many years for spinning cellulose acetate from acetone solution, and is now also employed for spinning polyacrylonitrile and the aramid fibers from solutions in dimethyl formamide or dimethyl acetamide.

So far no process is used in which the polymer is made in the same solvent from which it is dry spun. Therefore the first step in dry spinning is always the preparation of the polymer solution. The choice of solvent for spinning is based on considerations of solvent power, boiling point, heat of evaporation, stability, toxicity, ease of recovery, and so on. Nonpolar solvents are preferred because of convenient boiling points and nonhygroscopic nature, but may cause hazardous buildup of static charge. Low-boiling solvents with high heats of evaporation may cause polymer precipitation or coagulation on the surface of the fiber, in part due to condensation of moisture as the fiber cools when solvent evaporates. This leads to loss of luster and strength. High solution concentrations are desired, and elevated temperatures are often used to keep viscosities from becoming too high. Cellulose acetate is spun from 20–45% solutions with 400–1000 poise viscosity at 40°C. About 22% of the acrylic fibers is produced by spinning polyacrylonitrile from solution in dimethyl formamide. Because of the higher boiling point of this solvent, up to 50% residual solvent can remain in the fiber after it is spun; this must be washed out in a subsequent step. As in other types of spinning, polymer solutions are usually filtered just before dry spinning.

Spinning is done in a vertical tubular cell (Fig. 18-4b), jacketed for temperature control, in which air, steam, or inert gas may be passed through either concurrently or countercurrently as required. Downward spinning is preferred for small-denier fibers and upward spinning for high deniers, to control draw better by eliminating the influence of gravity. Dry spinning is carried out at rates as high as 2500–3000 ft/min, about the same as melt spinning.

Wet Spinning

In the wet-spinning process, a solution of a polymer or polymer derivative is spun through a spinneret into a liquid that can coagulate the polymer or derivative (Fig. 18-4c). The chemical reaction to convert a derivative to the final polymer can take place simultaneously or later. Wet spinning is commonly used for viscose, cellulose, and some synthetic fibers such as polyacrylonitrile spun from salt solutions.

The essential feature in wet spinning is the transfer of the mass of the solvent from the polymer to the coagulating bath. This transfer is not accompanied by heat of solidification of the polymer, as in melt spinning, or heat of evaporation of the solvent, as in dry spinning. A heat of chemical reaction may be present but is not an essential part of the process.

As in other types of spinning, the surface or interfacial tension forces around the filament are quite strong so that it is difficult to produce at the spinneret a filament with a cross section other than circular. These forces also tend to make the filament break up into drops. This tendency is opposed by the viscosity of the polymer; hence more viscous solutions are easier to spin.

It is also advantageous to work with as high concentration of polymer as possible, for obvious reasons. This again leads to high viscosity, and the practical limit may be reached above which the polymer cannot be filtered, pumped, or extruded. For this reason some polymers are wet spun at elevated temperatures where viscosity is lower. Higher temperatures usually promote crystallization and lead to denser, stronger fibers. Cellulose fibers may be spun at 50°C, acrylonitrile fibers as high as 160–185°C, from either 20–28% solutions in dimethyl acetamide or 10–15% solutions in 45–55% aqueous sodium thiocyanate. In each case the fiber is spun into a more dilute aqueous solution of the solvent.

During coagulation several processes take place simultaneously, including diffusion, osmosis, and salting out. Because of the interplay of these processes, coagulation occurs in rather different ways for different fiber–solvent systems. Usually coagulation is rapid and the fibers cannot be stretched greatly during this step. The skin of the fiber sets to a gel, and later the volume of material in the core is reduced as the rest of the solvent is removed. The skin then has to fold to accommodate the reduced volume. This gives a characteristic wrinkled cross section to some fibers, such as viscose rayon. It is difficult to produce by wet spinning fibers whose final cross section is circular, despite their initial production at the spinneret in this shape. As a result of the shrinkage, the skin of the fibers is oriented much more than the core.

In other cases, coagulation is much slower and the filaments may be stretched to as much as 30 times their original length before coagulation (as in the spinning of cellulose from cuprammonium solutions into water).

The need for sufficient time for coagulation and other treatments depending on diffusion, and the considerable viscous drag of the coagulating baths, limit the rates of wet spinning to 150–300 ft/min in the usual cases. These speeds are low enough that drawing can be done immediately after spinning in a continuous operation.

Wet-spun yarn may be collected on a bobbin or as a loose cake, which is subjected to chemical treatments, washing, and drying. Alternatively, these steps may be carried out continuously as indicated in Fig. 18-4c.

Other Spinning Methods

Insoluble and effectively nonfusible polymers such as polytetrafluoroethylene can be spun by the following technique or others achieving the same purpose: To a solution of ripened cellulose xanthate, an aqueous dispersion of polytetrafluoroethylene is added until the polymeric constituents are about 95% fluoro polymer and 5% cellulosic. This mixture is wet spun. The cellulose is then decomposed completely, and the polytetrafluoroethylene sintered into a continuous fiber, by contact with a metal roll heated to around 390°C.

Elastomeric fibers based on the polyurethanes can be reaction spun by the following method: A hydroxy-terminated polyester or polyether of molecular weight 1000–2000 is reacted with a diisocyanate in a 1:2 mole ratio to form an isocyanate-terminated prepolymer. This is spun into an aqueous solution of a diamine, with polymer formation and fiber formation taking place simultaneously. More often, however, these polymers are synthesized, then either wet or dry spun from solutions in dimethyl formamide or dimethyl acetamide.

GENERAL REFERENCES

Corbière 1967; Mark 1967; Siclari 1967; Smith 1967; Ziabicki 1967; McIntyre 1968; Middleman 1977; Hobson 1978; Preston 1978; Peters 1980.

C. FIBER AFTER-TREATMENTS

Although it is beyond the scope of this book to treat in detail the production of fabrics from fibers, there are certain intermediate steps between spinning and weaving which are ultimately dependent on the physical, chemical, and molecular structure of the fiber.

All natural and some synthetic fibers must be washed or scoured to free them of natural oils, dirt, chemicals, and other foreign impurities. They must be lubricated or sized or both for proper processing into cloth. They will very likely be dyed for pleasing appearance, and they may have any of a variety of treatments applied to impart or control crease resistance, softening, water repellency, slipping, dimensional stability, shrinkage, or many other properties.

After some or all of these treatments, the fiber is ready to be transformed into a fabric. The most important method of doing this is weaving, in which a set of yarns running lengthwise (warp) is interlaced with a second set at right angles (filling). Other methods of producing fabrics include knitting, in which a series of yarns is looped together, and the manufacture of nonwoven fabrics, papers, and felts, whereby the fibers are bonded together into flat sheets by heat, pressure, and, possibly, bonding agents. Variations of the weaving process lead to pile fabrics, laces, braids, and so on. The properties of the fabric depend on the fiber properties, the yarn construction (lengths and diameters of the fibers, size of yarns, amount of twist), the fabric construction (number of yarns per unit area and patterns in which they are combined), and the finishes applied to the fibers and the fabric.

Scouring

The removal of impurities from textile materials, called *scouring,* is carried out by the use of surface-active agents such as soaps and synthetic detergents. These materials have the property of reducing the surface tension of water, and their main function in scouring is to reduce the interfacial tension of their solution toward fats and oils. Most textile fibers contain such fats, oils, or waxes, either naturally or inadvertently or purposely added during various operations. The principal task of scouring is to remove these substances, since most of the other ingredients of the soil are embedded in the fatty material. On repeated laundering, however, some fine soil may be redeposited on the fat-free surface of the fiber, find its way into cavities on the surface, and become extremely resistant to removal. In contrast to initial scouring, one important function of detergents in household and laundry use is to prevent this redeposition of soil. Soil release finishes are also used, particularly with permanent-press garments (see below).

Lubrication

Lubrication of fibers is necessary to reduce their friction against themselves and against elements of the processing machinery. In the lubrication of yarns, it is necessary to preserve a high static coefficient of friction to keep the yarn in place on its spool or cone, while reducing the dynamic coefficient so as to obtain high speed of movement of the yarn without the generation of heat.

Lubricants may be vegetable or mineral oils or suitably refined petroleum products. Vegetable oils have been preferred because of easy and complete removal by saponification. The best lubricants are water soluble, such as the poly(alkylene glycols).

Another function of the lubricant, which is especially important in the case of the moisture-resistant synthetic fibers, is to reduce the static electric charge on the fibers by lowering their surface resistivity.

Sizing

A *size* is a surface coating used to protect the yarn during weaving. It makes the yarn smooth by binding protruding fibers onto the core of the yarn. Since the fabric is usually dyed after weaving, the size must be easily removed.

Starch is used almost exclusively to size cotton and, to some extent, rayon and wool. It is unsuitable for continuous-filament yarns of rayon, acetate, or the synthetics because of poor adhesion. Gelatin, poly(vinyl alcohol), or other polymeric materials are used to size these yarns.

Dyeing

Dyeing consists in placing fiber in an (aqueous) solution of a dye and leaving it there until an equilibrium is established in which most of the dye is adsorbed in the fiber and only a small part remains in the dye bath. When equilibrium is reached, the dye concentration is uniform throughout the cross section of the fiber. To attain these conditions, the dye must be preferentially adsorbed in the fiber by means of some type of intermolecular bond such as a van der Waals or hydrogen bond. It is clear also that because of the size of the dye molecules they can penetrate only the amorphous regions of the polymer. Two requirements for successful dyeing are sites to which dye molecules can bond and amorphous regions in the fiber.

Natural cellulosic and protein fibers were known for centuries to dye well with acid dyes which bonded to the hydroxyl or amino groups in the fibers. When cellulose acetate was introduced, a new problem arose which was solved by the use of so-called dispersed dyes containing terminal amine or hydroxyl groups. They are water insoluble and are used as dispersions, hence the name. It seems probable that these dyes undergo hydrogen bonding to the carbonyl oxygen of the acetate groups.

Synthetic fibers have often been very difficult to dye. Polyamide, polyester, and acrylic fibers contain sites for bonding dye molecules, but have such compact structures that dye absorption is extremely slow. Rate of dyeing has been increased by going to higher temperatures, or by utilizing swelling agents for the polymers. The dye sites can be modified, as, for example, in the treatment of acrylic fibers with copper salts so as to form sites having affinity for anionic dyes. New classes of dyes have been synthesized for several of the synthetics, greatly alleviating this problem. In addition, many fibers are colored by the incorporation of pigments, often organic pigments, prior to spinning, leading to so-called spun-dyed products. Aspland (1983) has reviewed methods for dyeing textile fibers.

Titanium dioxide is often added to the polymer before spinning as a delustering agent for the resulting fiber.

Finishing

Finishes That Affect Hand. Under the heading of finishes which affect the feel or hand of fabrics come those which increase or decrease the natural friction between the fibers. Thus lubricants make the fabrics feel softer or have increased pliability. On the other hand, antislip finishes such as rosin, carboxymethyl cellulose, and hydroxyethyl cellulose impart a harsh feel to the fabric.

Conditioners. Ethylene and propylene glycols and their low polymers, with the ability to absorb and retain moisture, act as softening, plasticizing, and antistatic agents for the hydrophilic textile fibers, and can counteract to some extent the harshness or stiffness which accompanies the use of antislip agents.

Water-Repellent Finishes. Treatment with waxes such as paraffin emulsions imparts water repellency to cellulosic fabrics, but the effect is not permanent. Permanence can be enhanced by subsequent treatment of the fabric with aluminum or zirconium salts in solution, but the best water repellents at present are those which use stearoylamide pyridinium chloride. This molecule contains the stearoyl radical to give the desired water repellency, the amidomethyl radical which can form a semiacetal with a hydroxyl group of cellulose, and the pyridinium group to solubilize the molecule in water so as to bring it into intimate contact with the fiber. Baking the fiber causes the molecule to split off pyridine hydrochloride, leaving the stearoyl amidomethyl group attached to the cellulose.

Silicones also impart water repellency to fabrics.

Permanent-Press Finishes.   One of the major advances in textile technology in recent years has been the introduction of resin finishes which impart the wrinkle- and crease-resistant properties associated with "wash and wear" fabrics. Resin finishes are currently applied to the majority of shirts, blouses, dresses, trousers, and retail piece goods produced.

The resin finishes consists of aqueous solutions of urea–formaldehyde or melamine–formaldehyde precondensates, or of cyclic ureas such as dimethylol ethylene urea. It is thought that wrinkle resistance is imparted through the crosslinking of adjacent molecular chains in the fiber, rather than from resin formation in or between the fibers. The deposition of the crosslinked polymer between fibers is specifically avoided, since it leads to stiffness and increased body of the fabric.

The fabric is treated with an aqueous solution of the monomers containing an acid or acid-generating catalyst. Excess liquid is removed, and the fabric is dried and cured by heat treatment. Crosslinking is demonstrated by the reduced swelling of the fibers. The treatments function by reducing the irreversible elongation of the fiber: A greater force is needed to stretch the fiber, but a larger percentage of the elongation is recovered.

These thermosetting permanent-press systems work very well with polyester fabrics, but not as well with cotton, the properties of which are otherwise much in demand as an outstanding textile fiber. Polyester–cotton blends were developed to provide a material combining the best features of both polymers, and have become by far the most popular fiber blend for permanent-press goods.

Textured Yarns

The ability to stabilize the physical form of the thermoplastic fibers by heat setting has been utilized in the production of a number of "textured" yarns which have unusually high elasticity. These yarns are made in a number of ways: The original method of knitting a fabric, heat setting it, and then de-knitting it has long since been superseded by other methods in which the fibers are crimped by passing them over a hot knife edge or between hot gear teeth, stuffed into a tightly enclosed space or twisted tightly and then heat treated, or looped randomly by a turbulent air blast.

Nonwoven Fabrics

A nonwoven fabric is a web or continuous sheet of staple-length fibers, laid down mechanically. The fibers may be deposited in a random manner or preferentially oriented in one direction. The sheet is then bonded together. Polyamides, polyesters, polypropylene, and polyethylene can be used, with 66-nylon most widely utilized.

The spun fibers, which may be drawn, are laid down directly onto a porous belt, often with the aid of an electrostatic charge. They are then bonded in an oven or a calender, using either melt-bonding or a chemical adhesive. Other methods reminiscent of paper making can be used to form the sheet.

The major uses of these spun-bonded nonwoven fabrics are for carpet backing, in bedding and home furnishings, as disposable apparel including diapers and sanitary goods, durable paper, coated fabrics, and in ''hidden'' fabric uses such as furniture components and garment interlinings.

GENERAL REFERENCES

Bell 1965; Valko 1965, Ellis 1966; Lewis 1966; Steele 1966; Bikales 1968; Hearle 1969; Wray 1970; Goswami 1977; Drehlich 1981; Porter 1981.

D. TABLE OF FIBER PROPERTIES

Table 18-2 lists comparative properties for typical examples of the major textile fibers. It should be noted that the properties of fibers of the same chemical type can vary widely depending on precise composition, heat treatment, yarn structure, and many other variables. The values in Table 18-2 should be used only for comparative purposes.

DISCUSSION QUESTIONS AND PROBLEMS

1. Define (a) fiber, (b) staple, (c) denier, and (d) tenacity.

2. What are typical values of the crystalline melting point and the glass-transition temperature for textile fibers, and why?

3. What is the approximate denier of a fiber 0.02 mm in diameter if the specific gravity of the polymer is 1.2?

4. Sketch typical stress–strain curves for silklike and wool-like fibers.

5. How do the stress–strain properties of typical fibers change with draw ratio?

6. Name the technique(s) for spinning each of the following fibers, giving typical temperatures and naming any other substances present: (a) polyester, (b) acrylic, (c) nylon, (d) aramid, (e) viscose rayon, (f) acetate rayon, (g) polytetrafluoroethylene.

7. Will the following changes from a linear structure be likely to improve fiber properties? (a) To a slightly crosslinked structure and (b), to a branched-chain structure. Give examples to support your conclusions.

TABLE 18.2. Properties of Fibers[a]

| Property | Polyester | | | | Amide | | Cellulosic | | |
| | Continuous | | | | | | | | |
	Regular Tenacity	High Tenacity	Staple	Carbon	6-Nylon	66-Nylon	Viscose Rayon	Acetate Rayon	Triacetate
Tenacity, dry (g/denier)	2.8–5.6	6.8–9.5	2.4–7.0	23	4.0–7.2	2.3–6.0	0.7–3.2	1.2–1.4	1.1–1.3
Tenacity, wet (g/denier)	2.8–5.6	6.8–9.5	2.4–7.0	23	3.7–6.2	2.0–5.5	0.7–1.8	0.8–1.0	0.8–1.0
Tensile strength (thousands psi)	50–90	106–168	39–106	515	73–100	40–106	28–47	20–24	18–22
Elongation (%, at break)	24–42	12–25	12–55	1.5	17–45	25–65	15–30	25–45	26–35
Elastic recovery (%, from elongation)	76/3	88/3	81/3	100	98/10	88/3	82/2	65/4	65/5
Stiffness (g/denier)	10–30	30	12/17	1500	18–23	5–24	6–17	3.5–5.5	5.2
Toughness (g-cm)	0.4–1.1	0.5–0.7	0.2–1.1	—	0.7–0.9	0.8–1.3	0.2	0.2–0.3	0.15
Specific gravity (g/cm^3)	1.38	1.39	1.38	1.77	1.14	1.14	1.5	1.32	1.3
Moisture regain (%, 70°F, 65% RH)	0.4	0.4	0.4	—	2.8–5	4.0–4.5	11–13	6.3–6.5	3.2
Melt temperature (°C)[b]	250	250	250	320	220	250	180d	260	300
Chemical resistance	Good	Good	Good	Good	Fair	Fair	Fair	Fair	Fair
Bleach resistance	Resistant	Resistant	Resistant	Poor	Bleaches	Bleaches	Resistant	Bleaches	Resistant
Solvent resistance	Excel.	Excel.	Excel.	Excel.	Good	Good	Excel.	Fair	Poor
Sunlight resistance	Good	Good	Good	Excel.	Fair	Fair	Good	Good	Poor
Abrasion resistance	Excel.	Excel.	Excel.	Poor	Excel.	Excel.	Good	Fair	Fair

Property	Acrylic	Polyethylene	Polypropylene	Polyurethane	Aramid		Poly-tetrafluoro-ethylene	Cotton	Silk	Wool
					Kevlar	Nomex				
Tenacity, dry (g/denier)	2.2–2.6	1.0–3.0	3.0–4.0	0.7–0.9	22	4.0–5.3	0.9–2.0	2.1–6.3	2.8–5.2	1.0–1.7
Tenacity, wet (g/denier)	1.8–2.1	1.0–3.0	3.0–4.0	—	22	3.0–4.1	0.9–2.0	2.5–7.6	2.4–4.4	0.9–1.5
Tensile strength (thousands psi)	32–39	11–35	35–47	11–14	400	90	25–31	42–125	45–83	17–28
Elongation (%, at break)	20–28	20–80	80–100	400–625	2.5	22–32	19–140	3–10	13–31	20–50
Elastic recovery (%, from elongation)	73/3	95/5	96/5	97/50	100/2	—	—	45/5	33/20	63/20
Stiffness (g/denier)	10	2–12	20–30	0.1–0.2	975	—	1.2–8.8	42–82	76–117	24–34
Toughness (g-cm)	0.40	1–3	1–3	2.0	—	0.85	0.15	—	—	—
Specific gravity (g/cm³)	1.16	0.92	0.91	1.21	1.44	1.38	2.1	1.50	1.25	1.30
Moisture regain (%, 70°F, 65% RH)	1.5	~0	0.01	1.3	4.5	6.5	0	8.5	11	17
Melt temperature (°C)[b]	235	110–120	160–180	230	500d	370d	290	—	—	—
Chemical resistance	Good	Excel.	Excel.	Good	Excel.	Good	Excel.	Fair	Good	Fair
Bleach resistance	Bleaches	Resistant	Excel.	Poor	Poor	Resistant	Resistant	Bleaches	Bleaches	Bleaches
Solvent resistance	Excel.	Excel.	Excel.	Good	Excel.	Excel.	Excel.	Excel.	Good	Good
Sunlight resistance	Good	Fair	Fair	Fair	Fair	Fair	Excel.	Good	Fair	Good
Abrasion resistance	Good	Good	Good	Good	Good	Good	Good	Good	Good	Fair

[a]*Textile World* **138** (8) (1982).
[b]d = decomposes.

8. Describe briefly the functions of the following after-treatments: (a) scouring, (b) lubrication, (c) sizing.

9. Name two treatments or reagents used to produce each of the following finishes on fibers: (a) soft hand, (b) antislip, (c) antistatic, (d) water repellent, (e) permanent press.

10. Discuss briefly the production, composition, structure, and uses of nonwoven fabrics.

BIBLIOGRAPHY

Aspland 1983. J. R. Aspland, "Textile Color Application Processes," *Color Res. Appl.* **8**, 205–214 (1983).

Bell 1965. T. E. Bell and H. de V. Partridge, "Bleaching," pp. 438–484 in *EPST,*† Vol. 2, 1965.

Bikales 1968. Norbert M. Bikales, "Nonwoven Fabrics," pp. 345–355 in *EPST,* Vol. 9, 1968.

Corbière 1967. J. Corbière, "Fundamental Aspects of Solution Dry-Spinning," pp. 133–167 in H. F. Mark, S. M. Atlas, and E. Cernia, eds., *Man-Made Fibers, Science and Technology,* Vol. 1, Wiley-Interscience, New York, 1967.

Drehlich 1981. Arthur Drehlich, "Nonwoven Textile Fabrics (Staple Fibers)," pp. 109–124 in *ECT,*‡ Vol. 16, 1981.

Ellis 1966. J. R. Ellis and G. M. Gantz, eds., and Emery I. Valko, "Dyeing," pp. 235–375 in *EPST,* Vol. 5, 1966.

Goswami 1977. B. C. Goswami, J. G. Martindale, and F. L. Scardino, *Textile Yarns: Technology, Structure and Application,* John Wiley & Sons, New York, 1977.

Hearle 1969. John W. S. Hearle, Percy Grosberg, and Stanley Backer, *Structural Mechanics of Fibers, Yarns and Fabrics,* Wiley-Interscience, New York, 1969.

Heckert 1953. W. W. Heckert, "Synthetic Fibers from Condensation Polymers," Chapter 5 in Samuel B. McFarlane, ed., *Technology of Synthetic Fibers,* Fairchild, New York, 1953.

Hicks 1967. E. M. Hicks, E. A. Tippetts, J. V. Hewett, and R. H. Brand, "Conjugate Fibers," pp. 375–408 in H. F. Mark, S. M. Atlas, and E. Cernia, eds., *Man-Made Fibers, Science and Technology,* Vol. 1, Wiley-Interscience, New York, 1967.

Hobson 1978. P. H. Hobson and A. L. McPeters, "Acrylic and Modacrylic Fibers," pp. 355–386 in *ECT,* Vol. 1, 1978.

Lewis 1966. Charles E. Lewis, "Dyes," pp. 376–405 in *EPST,* Vol. 5, 1966.

McGovern 1980. John N. McGovern, "Fibers, Vegetable," pp. 182–197 in *ECT,* Vol. 10, 1980.

McIntyre 1968. J. E. McIntyre, "Man-Made Fibers, Manufacture," pp. 374–404 in *EPST,* Vol. 8, 1968.

Mark 1967. H. F. Mark and S. M. Atlas, "Principles of Spinning in Emulsion and Suspension," pp. 237–240 in H. F. Mark, S. M. Atlas, and E. Cernia, eds., *Man-Made Fibers, Science and Technology,* Vol. 1, Wiley-Interscience, New York, 1967.

Mark 1967–1968. H. F. Mark, S. M. Atlas, and E. Cernia, eds., *Man-Made Fibers, Science and Technology,* Wiley-Interscience, New York; Vol. 1, 1967; Vols. 2 and 3, 1968.

†Herman F. Mark, Norman G. Gaylord, and Norbert M. Bikales, eds., *Encyclopedia of Polymer Science and Technology,* Wiley-Interscience, New York.

‡Martin Grayson, ed., *Kirk–Othmer Encyclopedia of Chemical Technology,* 3rd ed., Wiley-Interscience, New York.

Middleman 1977. Stanley Middleman, *Fundamentals of Polymer Processing,* McGraw-Hill, New York, 1977.

Peters 1980. Timothy V. Peters, "Fibers, Elastomeric," pp. 166–182 in *ECT,* Vol. 10, 1980.

Porter 1981. K. Porter, "Nonwoven Textile Fabrics (Spunbonded)," pp. 72–104 in *ECT,* Vol. 16, 1981.

Preston 1978. J. Preston, "Aramid Fibers," pp. 213–242 in *ECT,* Vol. 3, 1978.

Riley 1956. J. L. Riley, "Spinning and Drawing Fibers," Chapter 18 in Calvin E. Schildknecht, ed., *Polymer Processes,* Interscience, New York, 1956.

Roberts 1980. William J. Roberts, "Fibers, Chemical," pp. 148–166 in *ECT,* Vol. 10, 1980.

Siclari 1967. F. Siclari, "Fundamental Aspects of Wet-Spinning Solutions," pp. 95–132 in H. F. Mark, S. M. Atlas, and E. Cernia, eds., *Man-Made Fibers, Science and Technology,* Vol. 1, Wiley-Interscience, New York, 1967.

Smith 1967. Arthur L. Smith, "Fibers, Identification," pp. 593–609 in *EPST,* Vol. 6, 1967.

Steele 1966. Richard Steele and D. D. Gagliardi, "Crease Resistance," pp. 307–330 in *EPST,* Vol. 4, 1966.

Valko 1965. Emery I. Valko and Guiliana C. Teroso, "Antistatic Agents," pp. 204–229 in *EPST,* Vol 2, 1965.

Wray 1970. G. R. Wray, "Textile Processing," pp. 692–727 in *EPST,* Vol. 13, 1970.

Ziabicki 1967. Andrzej Ziabicki, "Principles of Melt-Spinning," pp. 169–236 in H. F. Mark, S. M. Atlas, and E. Cernia, eds., *Man-Made Fibers, Science and Technology,* Vol. 1, Wiley-Interscience, New York, 1967.

CHAPTER NINETEEN

ELASTOMER TECHNOLOGY

The final class of high polymers to be considered in Part 6 is elastomers. Like fibers, elastomers are considered apart from other polymeric materials because of their special properties. Unlike fibers, elastomers do not in general lend themselves to plastics uses; elastomers must be amorphous when unstretched and must be above their glass transition temperature to be elastic, whereas plastics must be crystalline or must be used below this temperature to preserve dimensional stability.

About 5.8 billion lb of rubber was used in the United States in 1982. Synthetics accounted for 72% of the total consumed, with SBR in largest volume by a wide margin, as indicated in Table 19-1. Automobile tires remain the largest single end use of rubber.

History of Synthetic Rubber. The first attempts to produce synthetic rubbers centered around the homopolymerization of dienes, particularly isoprene because it was known to be the monomer for natural rubber. It was found in the late nineteenth century that rubberlike products could be made from isoprene by treating it with hydrogen chloride or allowing it to polymerize spontaneously on storage. These materials could be vulcanized with sulfur, becoming more elastic, tougher, and more heat resistant.

Around 1900 it was discovered that other dienes such as butadiene and 2,3-dimethylbutadiene could be polymerized to rubberlike materials spontaneously by alkali metals or by free radicals. Application was made of these facts during World War I in Germany, where 2,3-dimethylbutadiene was polymerized spontaneously.

After World War I this research on rubberlike products continued, with the emphasis shifting to butadiene because of its more ready availability, at that time from acetaldehyde via the aldol synthesis. Alkali metals were used as initiators, and the products were called *buna rubbers* from the first letters of butadiene and the symbol Na for sodium.

506

TABLE 19-1. Approximate United States Consumption of Rubber in 1982[a]

Type	Consumption (million lb)
Natural rubber	1600
Synthetics	
SBR	2200
Polybutadiene	800
EPDM	300
Butyl	290
Neoprene	270
Nitrile	170
Polyisoprene	120
Other	50
Total synthetics	4200
Total	5800

[a]Estimates from previous chapters and *Rubber World* **187** (6), 8 (1983).

Improvements in processing and properties were made in two ways: Copolymers of butadiene with vinyl monomers, notably styrene, were introduced, and emulsion polymerization was adopted. The significance of initiators, reduction activators, and oxygen became known. The importance of maintaining low conversion or utilizing modifiers such as CCl_4 and long-chain mercaptans was discovered. By the beginning of World War II, acceptable polymers (buna-S) were being produced in Germany containing 68–70% butadiene and 30–32% styrene.

In review of earlier chapters, it may be recalled that the unique properties of elastomers include their ability to stretch and retract rapidly, exhibit high strength and modulus while stretched, and recover fully on release of the stress. To obtain these properties, certain requirements are placed upon the molecular structure of the compounds: They must be high polymers, be above their glass transition temperatures, be amorphous in the unstretched state (but preferably develop crystallinity on stretching), and contain a network of crosslinks to restrain gross mobility of the chains.

A. VULCANIZATION

The process by which a network of crosslinks is introduced into an elastomer is called *vulcanization*. The chemistry of vulcanization is complex and has not been well understood throughout the century of practice of the process since its discovery by Goodyear in 1839. The profound effects of vulcanization, however, are clear: it transforms an elastomer from a weak thermoplastic mass without useful mechanical properties into a strong, elastic, tough rubber (Table 19-2). The tensile strength, stiffness, and hysteresis (representing loss of energy as heat) of natural rubber before and after vulcanization are shown in Fig. 19-1, and the effects of

TABLE 19-2.   Properties Typical of Raw, Vulcanized, and Reinforced Natural Rubber

Property	Raw Rubber	Vulcanized Rubber[a]		Reinforced Rubber
Tensile strength, psi	300	3000		4500
Elongation at break (%)	1200	800		600
Modulus (psi)[b]	—	400		2500
Permanent set	Large		Small	
Rapidity of retraction (snap)	Good		Very good	
Water absorption	Large		Small	
Solvent resistance (hydrocarbons)	Soluble		Swells only	

[a]Not reinforced
[b]Tensile stress at 400% elongation, a measure of stiffness.

extent of vulcanization on these and other properties of elastomers are shown schematically in Fig. 19-2.

For many decades following Goodyear's first experiments in heating rubber with small amounts of sulfur, this process provided the best and most practical method for bringing about the drastic property changes described by the term *vulcanization*, not only in natural rubber but also in the diene synthetic elastomers such as SBR, butyl, and nitrile rubbers. It has been found since, however, that neither heat nor

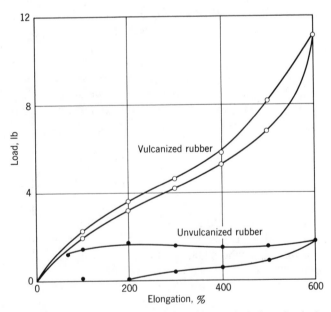

FIG. 19-1.   Stress–strain curves to 600% elongation and back, typical of unvulcanized and vulcanized natural rubber.

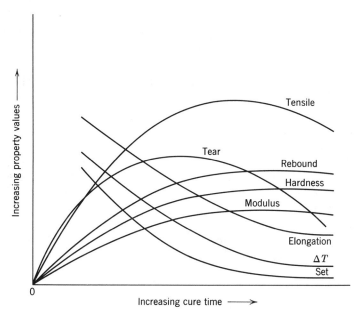

FIG. 19-2. Schematic representation of the effect of extent of vulcanization (cure time) on various physical properties of elastomers (after Garvey 1959). ΔT is heat buildup in a compression-flex test.

sulfur is essential to the vulcanization process. Rubber can be vulcanized or *cured* without heat by the action of sulfur chloride, for example. A large number of compounds that do not contain sulfur can vulcanize rubber; these fall generally into three groups: oxidizing agents (selenium, tellurium, organic peroxides, nitro compounds), generators of free radicals (organic peroxides, azo compounds, many accelerators, etc.), and phenolic resins. Thus it is clear that there is not a single method or chemical reaction of vulcanization.

Since the chemical reactions associated with vulcanization are varied and involve only a few atoms in each polymer molecule, a definition of vulcanization in terms of the physical properties of the rubber is necessary. In this sense, vulcanization may be defined as any treatment that decreases the flow of an elastomer, increases its tensile strength and modulus, but preserves its extensibility. There is little doubt that these changes are due primarily to chemical crosslinking reactions between polymer molecules. As might be expected, such properties as tensile strength are relatively insensitive to the onset of these crosslinking reactions; tensile strength does indeed change tenfold during curing, but this is evidence of the profound alteration of polymer properties by the process. Tests based on melt flow are more sensitive to initial crosslinking reactions and are widely used in the rubber industry.

Although vulcanization takes place by heat in the presence of sulfur alone, the process is relatively slow. It can be speeded many fold by the addition of small amounts of organic or inorganic compounds known as *accelerators*. Many accelerators require the presence of still other chemicals known as *activators* or *promoters* before their full effects are realized. These activators are usually metallic oxides, such as zinc oxide. They function best in the presence of a rubber-soluble metallic soap, which may be formed during the curing reaction from the activator and a fatty acid. The most efficient combination of chemicals for sulfur vulcanization includes sulfur, an organic accelerator, a metallic oxide, and a soap.

Chemistry of Vulcanization

Sulfur Vulcanization.   The curing of rubber with sulfur alone is quite slow and no longer commercially practiced. Even today the exact mechanisms of the curing reactions are not fully understood. Both free-radical and ionic intermediates have been postulated.

Accelerated Sulfur Vulcanization.   Typical accelerators for sulfur vulcanization include thiazoles, for example, 2-mercaptobenzothiazole,

dithiocarbamates, for example, tetramethyl thiuram disulfide,

and amines, for example, diphenyl guanidine,

In addition, the retarder *N*-(cyclohexylthio)phthalimide,

has proved very effective in inhibiting premature vulcanization. These chemicals are usually used with zinc oxide and a fatty acid, a typical recipe being 0.5–2 parts accelerator, 0.1–1 part retarder, 2–10 parts zinc oxide, 1–4 parts fatty acid, and 0.5–4 parts sulfur per 100 parts rubber.

Again, mechanisms are not known in detail, but it is thought that the accelerator reacts with sulfur to give sulfides of the type Ac—S_x—Ac, where Ac is a free radical derived from the accelerator. These polysulfides then interact with the rubber to give intermediates of the type rubber—S_x—Ac. These react further, with the final crosslinks being of the general type

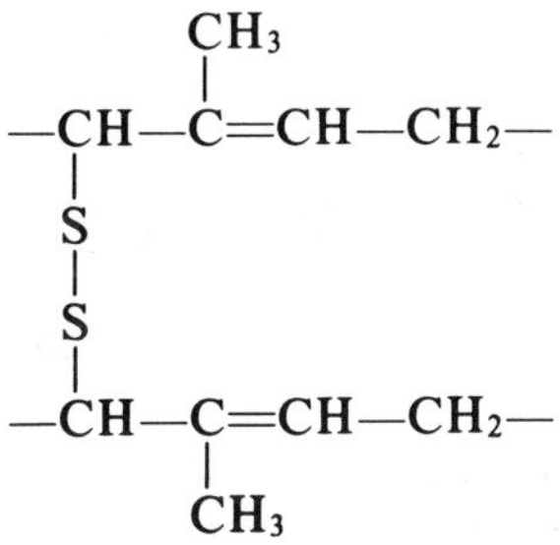

$$
\begin{array}{c}
\mathrm{CH_3} \\
| \\
\mathrm{-CH-C=CH-CH_2-} \\
| \\
\mathrm{S} \\
| \\
\mathrm{S} \\
| \\
\mathrm{-CH-C=CH-CH_2-} \\
| \\
\mathrm{CH_3}
\end{array}
$$

Many other possibilities exist. The role of the zinc oxide, which produces zinc ions in the presence of the fatty acid, appears to be to form chelates with the accelerator and sulfides, retarding and thus smoothing the course of the vulcanization. The double bonds in the elastomer are not saturated in this mechanism, but contribute by enhancing the activity of the allylic hydrogens, at which the crosslinking takes place.

Nonsulfur Vulcanization. The same activation effect of the elastomer double bonds is postulated to play a role in nonsulfur vulcanization, according to the following scheme:

a. A free radical R· is formed by the decomposition or oxidation of the curing agent, or as a step in the oxidative degradation of the rubber.

b. This free radical initiates vulcanization by abstracting a hydrogen atom from one of the α-methylene groups (in natural rubber the methyl side group directs the attack to the methylene group nearest it):

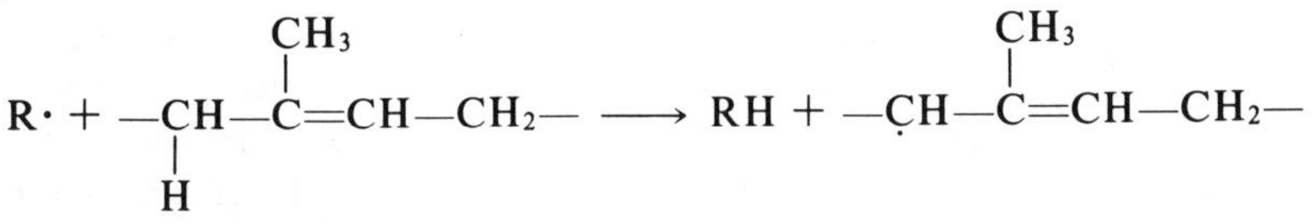

$$
\mathrm{R\cdot} + \begin{array}{c} \mathrm{CH_3} \\ | \\ \mathrm{-CH-C=CH-CH_2-} \\ | \\ \mathrm{H} \end{array} \longrightarrow \mathrm{RH} + \begin{array}{c} \mathrm{CH_3} \\ | \\ \mathrm{-CH-C=CH-CH_2-} \end{array}
$$

c. The rubber free radical then attacks a double bond in an adjacent polymer chain. This results in the formation of a crosslink and the regeneration of a free radical in a reaction analogous to propagation in an addition poly-merization (Chapter 3):

$$\underset{\bullet}{-CH}-\overset{\overset{\displaystyle CH_3}{|}}{C}=CH-CH_2- \;+\; -CH_2-\overset{\overset{\displaystyle CH_3}{|}}{C}=CH-CH_2-$$

$$\downarrow$$

$$-CH-\overset{\overset{\displaystyle CH_3}{|}}{C}=CH-CH_2-$$

$$-CH_2-\overset{\overset{\displaystyle CH_3}{|}}{\underset{\bullet}{C}}\!-\!\!-\!\!-CH-CH_2-$$

Vulcanization may continue by several such propagation steps.

Chain transfer may also occur. Termination probably occurs by reaction of the rubber free radical with a free radical fragment of the curing agent. In contrast to addition polymerization in a fluid system, the termination reaction between two rubber free radicals is considered unlikely because of the low probability of two such radicals coming into position to react, owing to the high viscosity of the medium.

Since free-radical intermediates also exist in accelerated sulfur vulcanization, similar reactions no doubt take place in that case also. Some double bonds are saturated in this way, some by addition of hydrogen sulfide (a known product of sulfur in vulcanization), and some by still other mechanisms.

Physical Aspects of Vulcanization

Vulcanization with sulfur present takes place when 0.5–5 parts (by weight) of sulfur is combined with 100 parts of rubber. If the reactions are allowed to continue until considerably more sulfur has combined (say 30–50 parts per 100 parts rubber), a rigid, nonelastomeric plastic known as *hard rubber* or *ebonite* is formed. Tensile strength reaches a maximum, with elongation remaining high, in the early *soft cure* region; then both fall to much lower values with increasing amount of combined sulfur.

The region of maximum tensile properties in the neighborhood of 5–10 parts combined sulfur is known as the region of *optimum cure* for this *mix*. Beyond this point the *stock* (i.e., mixture) is said to be *overcured* and is likely to be leathery, that is, stiffer and harder than at the optimum cure but weaker and less extensible. At high proportions of combined sulfur the strength of the material increases again, and the elongation becomes very low as the hard rubber region is reached.

Many properties of soft vulcanizates, including tensile and tear strength, stiffness, and hardness, go through a shallow maximum as the cure time is increased. Others, such as elongation and permanent set, drop continuously. The effect of increasing temperature is, as expected, to move the curves to shorter times. Experiments in which the composition of the mix is varied show that the initial rise in tensile strength is accompanied by the rapid incorporation of sulfur in the rubber, and that, as the rate of sulfur addition decreases, the rate of increase of tensile strength begins to decline. It is also known that the total amount of combined sulfur does not measure the extent of change in physical properties during cure, and that the ratio of sulfur combined to double bonds used up varies widely from one mix to another.

These facts suggest that processes of crosslinking and degradation occur simultaneously during vulcanization. Additional evidence for this view arises from the fact that badly overcured stocks may undergo *reversion,* that is, become soft and tacky with notable loss in strength. Such stocks resemble oxidatively degraded rubber. Reversion differs from overcuring in that it usually takes place on prolonged heating when not enough sulfur is present to get into the region of leathery cures.

GENERAL REFERENCES

Wolfe 1971; Coran 1978, 1983; Morton 1981.

B. REINFORCEMENT

For many uses, even vulcanized rubbers do not exhibit satisfactory tensile strength, stiffness, abrasion resistance, and tear resistance. Fortunately, these properties can be enhanced by the addition of certain fillers to the rubber before vulcanization. Fillers for rubber can be divided into two classes: *inert fillers,* such as clay, whiting, and barytes, which make the rubber mixture easier to handle before vulcanization but have little effect on its physical properties; and *reinforcing fillers,* which do improve the above-named unsatisfactory properties of the vulcanized rubber. Carbon black is the outstanding reinforcing filler for both natural and synthetic rubbers. The added effects of reinforcement with carbon black on the properties of rubbers over the effects of vulcanization alone are illustrated in Table 19-2. Although the nature of reinforcement is not completely understood, it appears to add a network of many relatively weak "fix points" to the more diffuse network of strong primary bond crosslinks introduced by vulcanization. Vulcanization restrains the long-range movements of the polymer molecules but leaves their local segmental mobility high; reinforcement stiffens the mass and improves its toughness by restricting this local freedom of movement.

Nonreinforced vulcanized rubbers, sometimes called pure-gum vulcanizates, are relatively soft, pliable, and extensible, and are most useful for such items as rubber bands, tubing, and gloves. Rubber would be of little value in modern industry were it not for the reinforcing effects of carbon black.

Types of Fillers

Only a few fillers are reinforcing for even one or two of the important mechanical properties of natural rubber, while some weaken the vulcanizates in one or more important respects. The latter are known as *inert fillers*. In general it is found that the reinforcing action of a filler depends upon its nature, the type of elastomer with which it is used, and the amount of filler present.

Carbon black is the only important reinforcing filler for most elastomers, although silica and silicate fillers can be used in some cases, as with the silicone elastomers (Chapter 16D). Important nonreinforcing fillers include talc, zinc oxide, and magnesium carbonate.

Nature of the Filler.   The chemical composition of the filler is a primary determinant of its reinforcing action. Carbon blacks are always more efficient reinforcing agents than, for example, talc, even though particle size, surface condition, and other properties can be varied widely for both. Within a single chemical class, particle size and surface condition are important variables.

Nature of the Elastomer.   Although a good reinforcing filler such as carbon black increases the tear and abrasion resistance of all the current major elastomers, it has no effect in enhancing the tensile strength of neoprene or butyl rubber, in contrast to the marked tensile reinforcement it causes in natural rubber, SBR, nitrile rubber, and even the polysulfide elastomers. These differences are ascribed to a conflict between two effects of the filler. (a) A reinforcing filler stiffens and strengthens the structure by introducing a network of many relatively weak fix points, and (b) simultaneously, it may interfere with the ability of the polymer to crystallize at high elongations, simply by its bulkiness in the system. Since stiffness and tensile strength are greatly enhanced by crystallization, the specific effect of the filler depends upon which of these actions predominates.

Amount of Filler.   Increasing amounts of reinforcing filler cause continuing improvement in properties until a maximum is reached, representing the optimum loading for this composition. Beyond this point, additional filler merely acts as diluent and the properties of the vulcanizate again deteriorate.

Carbon Black

Various colloidal forms of carbon constitute the most important reinforcing filler for rubber. Carbon blacks are the only inexpensive materials that reinforce all three of the important properties—tensile strength, tear resistance, and abrasion resistance. Many different carbon blacks are used, differing mainly in particle size, surface condition, and degree of agglomeration. Roughly, the degree of reinforcement increases with decreasing particle size of the black down to the practical lower limit at which blacks can be made, about 100 Å in diameter.

Almost all carbon blacks used for reinforcing elastomers are made by burning gas or oil in a furnace in limited air and removing the carbon from the off-gases by a centrifugal and electrostatic precipitation. They range from 300 to 800 Å in particle diameter, and are often designated by their effects on the vulcanizates, as high-abrasion, high-modulus, semireinforcing, or conductive furnace blacks.

Electron microscopy shows that carbon blacks consist of irregular aggregates of approximately spherical subunits, sometimes called nodules, fused together in branched, chainlike structures. The nodules are thought to be paracrystalline domains with a crystal structure similar to that of graphite except that the sheets of hexagonally close-packed carbon atoms are arranged randomly above one another. This suggests that unsatisfied bonds are present which may be important in reinforcement. The surfaces of the particles contain much adsorbed material, including hydrogenated and oxygenated structures. As a result the pH of a water slurry of the black may range from 2.5 to 11.

Effects of Carbon Black Structure on Reinforcement. The three parameters—particle size, pH, and *structure index*, that is, extent of agglomeration or chain structure—appear to be predominant in determining the reinforcing behavior of blacks. Tensile strength, abrasion and tear resistance, hardness, and toughness increase with decreasing particle size, whereas rebound and ease of processing become poorer. Blacks with a high structure index are difficult to disperse and give high stiffness and hardness but low tensile strength, toughness, and electrical resistivity. The pH influences the rate of vulcanization, depending upon the acidic or basic nature of the accelerator.

The stiffness of a reinforced vulcanizate rises steeply with the loading of carbon black and is nearly independent of the particle size of the black. These two features are unique for this property. Other properties, such as tensile strength, go through a maximum as loading is increased. They also depend upon particle size, as mentioned above.

The tensile strength of natural rubber can be increased about 40% by reinforcement. In SBR, however, the properties of the pure-gum vulcanizates are very poor, since the elastomer does not crystallize. The tensile strength of SBR can be raised about tenfold by reinforcement, making it equivalent to natural rubber for fully reinforced stocks. The tear resistance of SBR is also poor for unloaded vulcanizates, but is equivalent to that of natural rubber after reinforcement.

Abrasion resistance is highly important for heavy-duty rubber such as tire tread stock. The abrasion resistance of both natural rubber and SBR can be improved at least fivefold by proper reinforcement. Unfortunately, the resilience of the rubber decreases with increasing loading of filler. As a consequence hysteresis loss and heat buildup increase. Thus the use of a reinforcing filler represents a compromise between adequate abrasion and tear resistance and abnormal heat buildup.

Origin of Reinforcement. Only two conditions must be met for significant reinforcement of a rubber by carbon black. First, the particle size of the black must be small, usually between 200 and 500 Å. This ensures a large surface area and thus a large filler–rubber interface. Second, the rubber must "wet" the carbon black. This not only assures dispersion of the black in the rubber, but implies that the rubber–black adhesion approaches in strength the cohesion of the rubber to itself. This is necessary if the rubber–black bonds are to survive the large strains associated with high elongations. This adhesion is produced in part by secondary bond forces, which alone are enough to account for reinforcement, and in part by chemical grafting of rubber on to the surface of the black particle, almost certainly as the result of radical processes. While primary chemical bonding between the rubber and a filler is not essential for reinforcement, it does lead to the unique combination of properties that result when the filler is carbon black.

GENERAL REFERENCES

Burgess 1965; Garvey 1970; Kraus 1970, 1978; Morton 1981.

C. ELASTOMER PROPERTIES AND COMPOUNDING

Mechanical Properties of Elastomers

The stress–strain behavior of pure gum (i.e., nonreinforced) vulcanizates of the various elastomers depends markedly on the molecular structure, polarity, and crystallizability of the polymers. Nitrile rubber and SBR, being copolymers of complex and irregular structure, do not crystallize at all, but natural rubber and butyl rubber crystallize on stretching at room temperature. In these elastomers the crystallites have a stiffening effect like that of a reinforcing filler with the result that their stress–strain curves turn up markedly at higher elongations. The similarity of these stress–strain curves to those predicted from the kinetic theory of rubber elasticity (Chapter 11B) should be noted.

Unvulcanized rubber undergoes a very large amount of mechanical conditioning when first exposed to tensile stress. Figure 19-1 shows the first cycle of loading and unloading of a tensile specimen of unvulcanized natural rubber. As successive cycles take place, the changes in resistance to stretching, tensile strength, energy absorption, and permanent set become smaller. This mechanical conditioning is not accompanied by molecular weight degradation. Such degradation does occur, however, in the process of milling raw rubber to reduce its melt viscosity to the desired range before the incorporation of fillers, and so on. Here the process is complicated by the presence of oxygen. Vulcanized rubber undergoes a similar but much less extensive change in stress–strain properties on successive cycles of test.

The mechanical properties of elastomers depend greatly upon the rate of testing. Since service conditions for automobile tires (by far the most important end use of elastomers) involve rapid cyclic stresses, dynamic test methods are essential if the results are to correlate with the performance of the finished article.

Oxidative Aging of Elastomers

Rubbers that retain double bonds in their vulcanized structure, such as natural rubber, SBR, and nitrile rubber, are sensitive to heat, light, and particularly oxygen. Unless protected with antioxidants (of which the phenyl α- and β-naphthylamines are the most common) the rubbers age by an autocatalytic process accompanied by an increase in oxygen content. The results of the aging may be either softening or embrittlement, suggesting competing processes of degradation and crosslinking of the polymer chains. The first step in the process is presumed to be the same as that in the oxidation of any olefin, that is, the formation by free radical attack of a peroxide at a carbon atom next to a double bond:

$$-CH_2CH=CH- \longrightarrow -\overset{\cdot}{C}H-CH=CH- + H\cdot$$

$$H\cdot + -\overset{\cdot}{C}H-CH=CH- + O_2 \longrightarrow -\underset{\underset{OH}{\overset{|}{O}}}{\overset{|}{C}}H-CH=CH-$$

Subsequent steps may include crosslinking or chain cleavage or involvement of the double bond, probably initially through an epoxide group. Many of the steps in the reaction resemble closely those of certain types of vulcanization.

Many rubbers are sensitive also to attack by ozone, requiring protection by antiozonants. These are often derivatives of *p*-phenylene diamine, and are thought either to react with the ozone before it can undergo reaction with the rubber surface, or to aid in reuniting chains severed by ozone.

Natural rubber is far more sensitive than most of the other elastomers to both oxygen and ozone attack.

Compounding

The term *compounding* describes the selection of additives and their incorporation into a polymer so as to give a homogeneous mixture ready for subsequent processing steps. Except for differences in the nature of the compounding ingredients, especially as required for vulcanization and reinforcement and described under those headings, the compounding of rubbers is similar to that of most plastics. Typical equipment described in Chapter 17 is used for rubber as well as plastics compounding.

In the case of natural rubber, SBR, and the other synthetic rubbers produced by emulsion polymerization, the polymer is first available in the form of a latex. The normal procedure in compounding is to coagulate and dry the latex, and then to masticate and compound the rubber on mills or in other equipment. With SBR, however, two important compounding steps, *oil extending* and *masterbatching,* are commonly carried out before coagulation of the latex.

Oil Extending.   The use of hydrocarbon oils to dilute, or extend, rubbers has been known for many years. Since 1950, this practice has been applied to SBR polymerized to high molecular weight. The oil serves as a plasticizer and softener, reducing the melt viscosity of the rubber to that normally required in compounding. The process, which leads to a less expensive final product, is widely used: About 90% of the tire tread stock produced in this country is oil extended to some extent. The oil is added as an emulsion to the latex before coagulation.

Masterbatching.   A convenient method of mixing rubber and carbon black is the coprecipitation of a mixture of the rubber latex and an aqueous slurry of the black. This method gives adequate mixing when the particle size of the black is similar to that of the latex particle and the number of each type of particle per cubic centimeter before the coagulation can be made nearly the same. This situation holds for SBR and the most useful blacks.

GENERAL REFERENCES

Cox 1965; Maassen 1965; Cooper 1966; Stagg 1968; Barnard 1970; Garvey 1970; Studebaker 1978; Morton 1981.

D.   TABLE OF ELASTOMER PROPERTIES

Table 19-3 lists values of some physical and chemical properties of major commercial elastomers. In comparison with the corresponding tables in Chapters 17 and 18 it should be noted that the properties of elastomers depend in very large extent on the details of vulcanization, reinforcement, and compounding. It is, consequently, almost futile to assign "typical" values of properties according to polymer type. Therefore most of the data in Table 19-3 are of no more than qualitative significance, and comparisons should be made with extreme caution.

DISCUSSION QUESTIONS AND PROBLEMS

1.   Write chemical equations for the following processes in natural rubber: (a) the initial attack in oxidative degradation, (b) propagation in a vulcanization occurring by a free-radical mechanism, and (c) a typical crosslinking step in accelerated sulfur vulcanization.

2.   Identify and write chemical structures for (a) two organic accelerators of different types, (b) a common activator system, (c) a common rubber antioxidant, (d) the bonds thought to be responsible for reinforcement, (e) an elastomer that does not require reinforcement, (f) an elastomer for which reinforcement is essential for the development of good properties.

TABLE 19-3.   Typical Properties of Commercial Elastomers[a]

Property	Natural Rubber	SBR	Acrylate	Butyl	Chlorosulfonated Polyethylene	EPDM	Epichloro-hydrin	Fluorinated Rubbers
Tensile strength (psi)	4000	3500	2200	3000	2800	3000	2500	2400
Elongation (%)	700	700	400	700	500	300	400	400
Modulus (psi, 300–400% elongation)	2500	2500	—	1000	—	—	—	250
Dynamic properties	Excel.	Good	Good	Poor	—	—	—	Poor
Permanent set	Low	Low	Low	Moder.	—	—	—	High
Tear resistance	Excel.	Good	Fair	Good	Good	Good	Good	Fair
Abrasion resistance	Fair	Good	Good	Good	—	—	—	—
Adhesion	Excel.	Excel.	Good	Good	Excel.	Excel.	Fair	Good
Electrical properties	Excel.	Fair	Poor	Excel.	Fair	Excel.	Fair	Excel.
Gas permeability	High	High	Low	Low	—	—	—	—
Upper use temperature (°C)	80	110	150	100	120	150	120	230
Lower use temperature (°C)	−50	−50	−20	−50	−50	−40	−45	−40
Weather resistance	Fair	Fair	Excel.	Excel.	Excel.	Excel.	Good	Excel.
Ozone resistance	Poor	Fair	Excel.	Excel.	Excel.	Excel.	Good	Excel.
Oil resistance	Poor	Poor	Excel.	Poor	Good	Poor	Good	Excel.
Gasoline resistance	Poor	Poor	Fair	Poor	Fair	Poor	Good	Excel.
Water swelling	Excel.	Good	Poor	Excel.	Excel.	Excel.	Good	Excel.
Adhesion to metal	Excel.	Excel.	Good	Good	Excel.	Excel.	Excel.	Excel.

TABLE 19-3 (Continued)

Property	Neoprene	Nitrile	Polybutadiene (cis-1,4)	Polyisoprene (cis-1,4)	Polysulfide	Silicone	Urethane
Tensile strength (psi)	4000	4000	3000	4000	1200	1500	8000
Elongation (%)	700	600	700	750	400	800	700
Modulus (psi, 300–400% elongation)	1000	1500	—	2500	1400	—	1200
Dynamic properties	Fair	Poor	—	Excel.	Poor	Poor	Good
Permanent set	Moder.	Moder.	—	Low	High	High	Moder.
Tear resistance	Good	Good	Excel.	Excel.	Good	Poor	Excel.
Abrasion resistance	Good	—	—	Fair	Poor	Poor	Excel.
Adhesion	Excel.	Excel.	Excel.	Excel.	Poor	Poor	Excel.
Electrical properties	Good	Fair	Good	Good	Excel.	Excel.	Excel.
Gas permeability	Moder.	Moder.	—	High	Low	—	Fair
Upper use temperature (°C)	100	120	100	80	80	230	100
Lower use temperature (°C)	−50	−50	−60	−50	−50	−80	−50
Weather resistance	Excel.	Poor	Fair	Fair	Excel.	Excel.	Excel.
Ozone resistance	Good	Fair	Poor	Poor	Good	Excel.	Excel.
Oil resistance	Fair	Excel.	Poor	Poor	Excel.	Fair	Excel.
Gasoline resistance	Good	Good	Poor	Poor	Excel.	Poor	Excel.
Water swelling	Good	Excel.	Excel.	Excel.	Excel.	Excel.	Fair
Adhesion to metal	Excel.	Excel.	Excel.	Excel.	Excel.	Excel.	Excel.

[a]Studebaker (1978) and Morton (1981).

3. Sketch graphs showing the following (for natural rubber except as noted): (a) stress–strain curves for unvulcanized, vulcanized, and reinforced stocks; (b) stress–strain curves for vulcanized and reinforced stocks of SBR and natural rubber; (c) tensile strength, tear strength, and permanent set versus extent of vulcanization; (d) tensile strength versus carbon-black loading; (e) modulus versus surface area of carbon black for two values of the structure index.

4. Define the following terms: gum, loaded stock, mastication, masterbatching, oil extension, overcure, reinforcement, reversion, scorch, vulcanization.

5. List (a) four physical properties whose change or lack of change in vulcanization is significant and (b) four physical properties that improve on reinforcement.

6. List the ingredients in a typical vulcanization mix, identifying each one as to purpose, chemical type or structure, and approximate amount used.

BIBLIOGRAPHY

Barnard 1970. D. Barnard, J. I. Cunneen, P. B. Lindley, A. R. Payne, M. Porter, A. Schallamarch, W. A. Southorn, P. McL. Swift, and A. G. Thomas, "Rubber, Natural," pp. 178–256 in *EPST,*† Vol. 12, 1970.

Burgess 1965. K. A. Burgess, F. Lyon, and W. S. Stoy, "Carbon," pp. 820–836 in *EPST*, Vol. 2, 1965.

Cooper 1966. W. Cooper, "Elastomers, Synthetic," pp. 406–482 in *EPST*, Vol. 5, 1966.

Coran 1978. A. Y. Coran, "Vulcanization," Chapter 7 in Frederick R. Eirich, ed., *Science and Technology of Rubber,* Academic Press, New York, 1978.

Coran 1983. A. Y. Coran, "The Art of Sulfur Vulcanization," *Chemtech* **13,** 106–116 (1983).

Cox 1965. William L. Cox, "Antiozonants," pp. 197–203 in *EPST*, Vol. 2, 1965.

Garvey 1959. B. S. Garvey, Jr., "History and Summary of Rubber Technology," Chapter 1 in Maurice Morton, ed., *Introduction to Rubber Technology,* Reinhold, New York, 1959.

Garvey 1970. B. S. Garvey, Jr., "Rubber Compounding and Processing," pp. 280–304 in *EPST*, Vol. 12, 1970.

Kraus 1970. G. Kraus, "Reinforcement," pp. 42–57 in *EPST*, Vol. 12, 1970.

Kraus 1978. Gerard Kraus, "Reinforcement of Elastomers by Particulate Fillers," Chapter 8 in Frederick R. Eirich, ed., *Science and Technology of Rubber,* Academic Press, New York, 1978.

Maassen 1965. G. C. Maassen, R. J. Fawcett, and W. R. Connell, "Antioxidants," pp. 171–197 in *EPST*, Vol. 2, 1965.

Morton 1981. Maurice Morton, ed., *Rubber Technology,* 2nd ed., Van Nostrand Reinhold, New York, 1973; reprinted by Krieger, New York, 1981.

Stagg 1968. Richard Stagg, "Latexes," pp. 164–195 in *EPST*, Vol. 8, 1968.

Studebaker 1978. M. L. Studebaker and J. R. Beatty, "The Rubber Compound and Its Composition," Chapter 9 in Frederick R. Eirich, ed., *Science and Technology of Rubber,* Academic Press, New York, 1978.

Wolfe 1971. James R. Wolfe, Jr., "Vulcanization," pp. 740–757 in *EPST*, Vol. 14, 1971.

†Herman F. Mark, Norman G. Gaylord, and Norbert M. Bikales, eds., *Encyclopedia of Polymer Science and Technology,* Wiley-Interscience, New York.

APPENDIX ONE

LIST OF SYMBOLS

Symbols in dimensional formulas have the following significance:

M = mass	T = time	Q = electric charge
L = length	θ = temperature	

Symbol	Dimension	Common Unit	Definition	Chapter Where First Defined
Å	L	$\cdots$	Angstrom unit	
A	ML^2T^{-2}	kJ	Helmholz free energy, work content	$1C$
A	T^{-1}	sec^{-1}	Collision frequency factor (first-order reaction)	$3C$
A	$M^{-1}L^3T^{-1}$	liter/mole-sec	Collision frequency factor (second-order reaction)	$3C$
A	$ML^{-1}T^{-1}$	poise	Frequency factor for viscous flow	$11A$
A_1	$\cdots$	mole/g	First virial coefficient	$8B$
A_2	$M^{-1}L^3$	ml-mole/g^2	Second virial coefficient	$7C$
A_3	$M^{-2}L^6$	ml^2-mole/g^3	Third virial coefficient	$8B$
a	$\cdots$	$\cdots$	Arbitrary constant	
a	$\cdots$	$\cdots$	Exponent in modified Staudinger equation	$8E$

Symbol	Dimension	Common Unit	Definition	Chapter Where First Defined
a	L	Å	Length of crystal unit cell axis	10*B*
a_T	$\cdots$	$\cdots$	"Shift factor" for time–temperature superposition	11*B*
b	$\cdots$	$\cdots$	Arbitrary constant	
b	L	Å	Length of crystal unit cell axis	10*B*
b	L	cm or Å	Size parameter of Gaussian distribution	11*B*
C	$\cdots$	$\cdots$	Arbitrary constant	
C	$\cdots$	$\cdots$	Transfer constant in chain polymerization	3*B*
C_m	$\cdots$	$mole^{1/2}/g^{1/2}$	Thermodynamic constant of the Flory–Krigbaum dilute solution theory	7*C*
c	ML^{-3}	g/cm^3	Concentration	1*D*
c	ML^{-3}	g/dl	Concentration	8*E*
c	L	Å	Length of crystal unit cell axis	10*B*
D	L^2T^{-1}	cm^2/sec	Diffusion constant	8*D*
d	$\cdots$	$\cdots$	Signifies the total derivative	
d	L	cm	Diameter of spherical particle	8*C*
E	ML^2T^{-2}	kJ	Energy content	1*C*
E	$M^{-1/2}LT^{-1}$	$(J\text{-}cm^3)^{1/2}/$ mole	Molar attraction constant	7*A*
E	ML^2T^{-2}	kJ	Activation energy	3*C*
e	$\cdots$	$\cdots$	Base of natural logarithms	
e	$\cdots$	$\cdots$	Polarity factor in the Alfrey–Price equation	5*C*
F_1, F_2	$\cdots$	$\cdots$	Mole fractions in copolymer	5*A*
f	$\cdots$	$\cdots$	Denotes a functional relationship	

Symbol	Dimension	Common Unit	Definition	Chapter Where First Defined
f	$\cdots$	$\cdots$	Functionality	2C
f	$\cdots$	$\cdots$	Initiator efficiency	3B
f	$\cdots$	$\cdots$	Fraction of polymer in given phase	7D
f	MT^{-1}	dyne-sec/cm	Frictional coefficient	8D
f	MLT^{-2}	dyne	Force	11B
f	$\cdots$	$\cdots$	Volume fraction	11D
f_1, f_2	$\cdots$	$\cdots$	Mole fractions in monomer feed	5A
G	ML^2T^{-2}	kJ	Gibbs free energy	1C
G	$ML^{-1}T^{-2}$	psi or dyne/ cm^2	Modulus of elasticity	11B
g	$\cdots$	$\cdots$	Ratio of sizes of branched and unbranched molecules	7B
g	$\cdots$	$\cdots$	Coefficient of Γ^2c^2 in expansion of osmotic pressure	8B
g	$\cdots$	$\cdots$	Symmetry tensor in EPR analysis	9B
H	ML^2T^{-2}	kJ	Heat content or enthalpy	1C
H	$M^{-1}L^2$	mole-cm^2/g^2	Light-scattering calibration constant	8C
H	$L^{-1}T^{-1}Q^{-1}$	gauss	Magnetic field strength	9B
h	ML^2T^{-1}	erg-sec	Planck's constant	9B
I	ML^2T^{-2}	W	Transmitted light flux	8C
I_0	ML^2T^{-2}	W	Incident light flux	8C
i	$\cdots$	$\cdots$	Summation index	
J	$M^{-1}LT^2$	cm^2/dyne	Elastic compliance	11C
j	$\cdots$	$\cdots$	Summation index	
K	$\cdots$	$\cdots$	Equilibrium constant	2C
K	$M^{-1}L^3T^{-1}$	liter/mole-sec	Rate constant for initiation (second order)	4B
K	$M^{-1}L^2$	mole-cm^2/g^2	Light-scattering calibration constant	8C
K	$M^{-1}L^3$	dl-mole$^{1/2}$/g$^{3/2}$	Constant in Flory viscosity equation	8C

Symbol	Dimension	Common Unit	Definition	Chapter Where First Defined
K', K''	$\cdots$	$\cdots$	Empirical constants in modified Staudinger viscosity equations	8E
k, k'	$\cdots$	$\cdots$	Arbitrary constants	
k	$M^{-1}L^3T^{-1}$	liter/mole-sec	Rate constant (second order)	2C
k	$ML^2T^{-2}\theta^{-1}$	erg/deg-molecule	Boltzmann's constant	7C
k_d	T^{-1}	sec^{-1}	Rate constant for decomposition of initiator (first order)	3B
k', k''	$\cdots$	$\cdots$	Coefficients in Huggins viscosity equation	8E
l	L	cm or Å	Length	7B
ln	$\cdots$	$\cdots$	Abbreviation for natural logarithm	
log	$\cdots$	$\cdots$	Abbreviation for logarithm to the base 10	
M	$\cdots$	g/mole	Molecular weight	1D
$\bar{M}_n$	$\cdots$	g/mole	Number-average molecular weight	1D
$\bar{M}_v$	$\cdots$	g/mole	Viscosity-average molecular weight	8E
$\bar{M}_w$	$\cdots$	g/mole	Weight-average molecular weight	1D
$\bar{M}_z$	$\cdots$	g/mole	z-Average molecular weight	8D
m	M	g	Mass of a molecule	1D
m	$\cdots$	$\cdots$	Denotes meso symmetry	10A
m	L^{-2}	cm^{-2}	Number of elastic chains per unit area	6B
$\bar{m}_n$	M	g	Number-average molecular mass	1D
N	$\cdots$	$\cdots$	Number of molecules	1D
N_0	M^{-1}	$mole^{-1}$	Avogadro's number	8C
n	$\cdots$	$\cdots$	Mole fraction	7C
n	$\cdots$	$\cdots$	Refractive index	8C

Symbol	Dimension	Common Unit	Definition	Chapter Where First Defined
n	$\cdots$	$\cdots$	Exponent in Avrami equation	11D
P	$ML^{-1}T^{-2}$	psi or dyne/ cm^2	Pressure	3D
P	$\cdots$	$\cdots$	Reactivity of radical in Alfrey–Price equation	5C
P	$\cdots$	$\cdots$	Probability of an event	10A
$P(\theta)$	$\cdots$	$\cdots$	Angular light-scattering function	8C
p	$\cdots$	$\cdots$	Extent of reaction	2C
$\tilde{p}$	$\cdots$	$\cdots$	Reduced pressure	7C
Q	$\cdots$	$\cdots$	General reactivity factor in Alfrey–Price equation	5C
q	$\cdots$	$\cdots$	Fraction of crosslinkable monomers on a chain	6D
R	$ML^2T^{-2}\theta^{-1}$	cm^3-atm/mole deg, erg/mole deg, or J/ mole deg	Gas constant	3C
R	$\cdots$	$\cdots$	Ratio of volumes of dilute and precipitated phases	7D
R_θ	L^{-1}	cm^{-1} sr	Rayleigh ratio	8C
r	$\cdots$	$\cdots$	Ratio of number of A and B groups present in stepwise polymerization	2C
r	$\cdots$	$\cdots$	Denotes racemic symmetry	10A
r	L	cm or Å	Distance from origin	7B
$r,(\overline{r^2})^{1/2}$	L	cm or Å	Root-mean-square end-to-end distance	7B
r_1,r_2	$\cdots$	$\cdots$	Monomer reactivity ratios	5A
S	$ML^2T^{-2}\theta^{-1}$	kJ/deg	Entropy	1C

Symbol	Dimension	Common Unit	Definition	Chapter Where First Defined
$s,(\overline{s^2})^{1/2}$	L	cm or Å	Radius of gyration or root-mean-square distance of chain segments from center of gravity	7B
s	T	sec	Sedimentation constant	8D
s	$ML^{-1}T^{-2}$	psi or dyne/cm²	Stress or applied force per unit area	11A
T	θ	K or °C	Temperature	1C
T_g	θ	K or °C	Glass-transition temperature	11D
T_m	θ	K or °C	Crystalline melting point	10D
$\tilde{T}$	$\cdots$	$\cdots$	Reduced temperature	7C
t	T	sec	Time, efflux time	2C
u	$\cdots$	$\cdots$	Size parameter for light scattering of spheres	8C
V	L^3	cm³	Volume, specific volume, molar volume	3C
$\tilde{V}$	$\cdots$	$\cdots$	Reduced volume	7C
υ	$ML^{-3}T^{-1}$	mole/liter-sec	Rate of reaction	2B
υ	$\cdots$	$\cdots$	Volume fraction	7A
υ	$\cdots$	$\cdots$	Size parameter for light scattering of random coils	8C
υ	LT^{-1}	cm/sec	Velocity	11A
$\bar{\upsilon}$	$M^{-1}L^3$	cm³/mole	Partial molar or specific volume	7C
W	$\cdots$	$\cdots$	Probability function	7C
W	ML^2T^{-2}	kJ	Work	11B
w	$\cdots$	$\cdots$	Weight, weight fraction	1D
x	L	cm or Å	Distance, space coordinate	
x	$\cdots$	$\cdots$	Chain length, number of segments in chain, degree of polymerization	2C
x	$\cdots$	$\cdots$	Size parameter for light scattering of rods (Fig. 8-7 only)	8C

Symbol	Dimension	Common Unit	Definition	Chapter Where First Defined
$\bar{x}_n$	...	...	Number-average degree of polymerization	2C
$\bar{x}_w$	...	...	Weight-average degree of polymerization	2C
y	...	...	Space coordinate	
y	...	...	Parameter of distribution functions	3C
y	...	...	Size parameter for light scattering of discs (Fig. 8-7 only)	8C
Z	...	...	Number of atoms in polymer chain	11A
z	...	...	Space coordinate	
z	...	...	z-Average (see $\bar{M}_z$)	
z	...	...	Parameter of distribution functions	3C
α	...	...	Denotes first in a series	
α	...	...	Branching coefficient	2D
α	...	...	Expansion factor of dissolved polymer chain	7B
α	$L^3\theta^{-1}$	cm^3/deg	Volume expansion coefficient	11D
β	...	...	Denotes second in a series	
β	$ML^3T^{-1}Q^{-1}$	Bohr magneton	Magnetic moment of electron spin	9B
β	(angle)	degree	Angle of crystal unit cell	10B
Γ	...	...	Gamma function	3C
Γ	$M^{-1}L^3$	cm^3/g	Coefficient of c in expansion of osmotic pressure	8B
γ	...	...	Strain or deformation in response to applied stress	11B
$\dot{\gamma}$	T^{-1}	sec^{-1}	Shear rate	11A
Δ	...	...	Signifies a change in or difference between the values of a thermodynamic function for two states	1C

Symbol	Dimension	Common Unit	Definition	Chapter Where First Defined
δ	$\cdots$	$\cdots$	Ratio of termination to propagation rate constants in copolymerization	5A
δ	$M^{1/2}L^{-1/2}T^{-1}$	$(J/cm^3)^{1/2}$	Solubility parameter	7A
δ	$\cdots$	$\cdots$	Chemical shift (NMR)	9B
δ	(angle)	deg	Phase difference due to energy absorption	9G
δ^2	$ML^{-1}T^{-2}$	J/cm^3	Cohesive energy density	7A
ϵ	$\cdots$	$\cdots$	Function of reactivity ratios	5B
ζ	$\cdots$	$\cdots$	Size parameter of crystallization theory	10D
η	$ML^{-1}T^{-1}$	dyne sec/cm², poise	Viscosity	8E
η_{inh}	$M^{-1}L^3$	dl/g	Inherent viscosity	8E
η_r	$\cdots$	$\cdots$	Relative visocity	8E
η_{red}	$M^{-1}L^3$	dl/g	Reduced viscosity	8E
η_{sp}	$\cdots$	$\cdots$	Specific viscosity	8E
$[\eta]$	$M^{-1}L^3$	dl/g	Intrinsic viscosity	8E
Θ	θ	K	Flory temperature, where polymer–solvent interactions are zero	7B
θ	$\cdots$	$\cdots$	Fraction of surface sites occupied	4D
θ	(angle)	deg	Angle	8C
κ	$\cdots$	$\cdots$	Thermodynamic constant of Flory–Krigbaum dilute solution theory	7C
λ	L	cm or Å	Wavelength of light in air	8C
λ_s	L	cm or Å	Wavelength of light in solution = λ/n	8C
μ	$\cdots$	$\cdots$	Parameter of distribution functions	3C
μ	$ML^3T^{-1}Q^{-1}$	Bohr magneton	Magnetic moment of nucleus	9B
μ_h	$\cdots$	$\cdots$	Huggins polymer–solvent interaction parameter	7C

Symbol	Dimension	Common Unit	Definition	Chapter Where First Defined
μm	L	micrometer	Abbreviation for micrometer	9B
μ_0	MLQ^{-2}	henry/m	Magnetic permeability of vacuum	9B
ν	$\cdots$	$\cdots$	Kinetic chain length	3B
ν	$\cdots$	$\cdots$	Number of items	
ν	T^{-1}	sec^{-1}	Frequency	9B
π	$ML^{-1}T^{-2}$	atm	Osmotic pressure	7C
ρ	$\cdots$	$\cdots$	Ratio of number of functional groups on branch units to total number of such groups	2D
ρ	$\cdots$	$\cdots$	Parameter of distribution functions	3C
ρ	T^{-1}	sec^{-1}	Rate of generation of radicals	6B
ρ	ML^{-3}	g/cm^3	Density	7A
Σ	$\cdots$	$\cdots$	Summation sign	
σ	$\cdots$	$\cdots$	Parameter of distribution functions	3C
σ	$\cdots$	$\cdots$	Parameter of fractionation theory	7D
τ	$\cdots$	$\cdots$	Reduced temperature in scaling theory	7D
τ	L^{-1}	cm^{-1}	Turbidity	8C
τ	T	sec	Relaxation time, retardation time	11C
τ_s	T	sec	Mean lifetime of a free radical	3B
Φ	M^{-1}	$dl/mole\ cm^3$	Universal constant of Flory viscosity theory	8E
ϕ	$\cdots$	$\cdots$	Cross-termination probability constant in copolymerization	5A
χ	$\cdots$	$\cdots$	Polymer–solvent interaction constant	7C

Symbol	Dimension	Common Unit	Definition	Chapter Where First Defined
ψ	$\cdots$	$\cdots$	Thermodynamic constant of Flory–Krigbaum dilute solution theory	7C
ω	$\cdots$	$\cdots$	Denotes last in a series	
ω	T^{-1}	$\sec^{-1}$	Angular velocity	8D
$\propto$	$\cdots$	$\cdots$	Denotes proportionality	
∂	$\cdots$	$\cdots$	Signifies partial derivative	
[ ]	$\cdots$	$\cdots$	Denote concentration of substance within brackets (except in symbol $[\eta]$)	
$\circ$	$\cdots$	$\cdots$	Degree sign	
$\times$	$\cdots$	$\cdots$	Multiplication sign	

APPENDIX TWO

TABLE OF PHYSICAL CONSTANTS

Symbol	Name	Value
Å	Angstrom unit	$1 \text{ Å} = 10^{-10}$ m (not a preferred unit)
e	Base of natural logarithms	2.7183
h	Planck's constant	6.6255×10^{-27} erg-sec
k	Boltzmann's constant	1.3805×10^{-16} erg/deg
N_0	Avogadro's number	6.0226×10^{23} mole^{-1}
R	Gas constant	8.314×10^7 erg/mole-deg 8.314 J/mole-deg 82.1 cm^3 atm/mole-deg
K	Absolute temperature	$0°C = 273.16$ K (Kelvin)
μm	Micrometer	$1 \text{ μm} = 10^{-3}$ mm $= 10^{-6}$ m
nm	Nanometer	$1 \text{ nm} = 10^{-3}$ μm $= 10^{-6}$ mm $= 10^{-9}$ m

APPENDIX THREE

TRADE NAMES AND GENERIC NAMES

This list is not intended to be exhaustive, but to include only a few of the more common names in widespread or recent use. Products are arranged in the order of the chapters in which they are discussed. Full names and addresses of manufacturers should be obtained from current directories, for example, the *Modern Plastics Encyclopedia*.

Name	Probable Composition	Manufacturer
	Chapter 13A. Polyethylene	
Alathon	Polyethylene	Du Pont
Bakelite	Polyethylene	Union Carbide
Chemplex	Polyethylene	Chemplex
Dylan	Polyethylene	Arco
Hi-Fax	Polyethylene	Hercules
Hostalen	Polyethylene	American Hoechst
Marlex	Polyethylene	Phillips Chemicals
Norchem	Polyethylene	Northern Petrochemicals
Poly-Eth	Polyethylene	Gulf Oil
Rexene	Polyethylene	El Paso Polyolefins
Sclair	Polyethylene	Du Pont of Canada
Tenite	Polyethylene	Eastman
Tyvek	Spun-bonded polyolefin	Du Pont

Name	Probable Composition	Manufacturer

Chapter 13B. Polypropylene

Marlex	Polypropylene	Phillips Chemicals
Moplen	Polypropylene	Novamont
Norchem	Polypropylene	Northern Petrochemicals
Profax	Polypropylene	Hercules
Tenite	Polypropylene	Eastman

Chapter 13C. Other Olefin Polymers

Alathon	Ethylene copolymers	Du Pont
Elvax	Ethylene–vinyl acetate copolymers	Du Pont
Eval	Ethylene–vinyl alcohol copolymers	Hercules
Marlex	Ethylene copolymers	Phillips Chemicals
Nucrel	Ethylene–methacrylic acid copolymer	Du Pont
Oxytuf	EPDM–vinyl graft copolymer	Occidental Chemical

Chapter 13D–F. Olefin-Based Elastomers

Ameripol	Polyisoprene (*cis*-1,4)	Goodrich
Chemigum	Butadiene–acrylonitrile copolymer	Goodyear
Hycar	Butadiene copolymers	Goodrich
Hypalon	Chlorosulfonated polyethylene	Du Pont
Krynak	Nitrile and nitrile/butadiene elastomers	Polysar
Natsyn	Polyisoprene (*cis*-1,4)	Goodyear
Neoprene	Polychloroprene	Du Pont
Nordel	EPDM elastomer	Du Pont
Polysar	Butyl, halobutyl, SBR, and EPDM elastomers	Polysar
Taktene	Polybutadiene elastomers	Polysar
TPR	Thermoplastic elastomers	Uniroyal
Vistanex	Polyisobutylene	Enjay

Chapter 14A. Polystyrene and Related Polymers

Bakelite	Polystyrene	Union Carbide
Cadon	Styrene–maleic acid copolymer, impact	Monsanto
Cycolac	ABS	Marbon
Fostarene	Polystyrene	Foster-Grant

Name	Probable Composition	Manufacturer
Kralastic	ABS	U.S. Rubber
Lustran	ABS	Monsanto
Lustrex	Polystyrene	Monsanto
Oxyloy	Alloy of EPDM–vinyl graft copolymer with ABS	Occidental Chemical
Rovel	Styrene-based terpolymer	Uniroyal
Styron	Polystyrene	Dow Chemical

Chapter 14B.   Acrylic Polymers

Acrilan	Acrylonitrile–vinyl acetate copolymer	Chemstrand
Barex	Modified polyacrylonitrile	Hercules
Creslan	Acrylonitrile–vinyl ester copolymers	American Cyanamid
Darvan	Vinylidene cyanide–vinyl acetate copolymer	Celanese
Elvacite	Methyl, ethyl, butyl methacrylate polymers and copolymers	Du Pont
Hycar	Acrylic elastomer	Goodrich
Implex	Poly(methyl methacrylate), impact modified	Rohm and Haas
Lucite	Poly(methyl methacrylate)	Du Pont
Orlon	Polyacrylonitrile	Du Pont
Plexiglas	Poly(methyl methacrylate)	Rohm and Haas

Chapter 14C.   Poly(vinyl esters) and Derived Polymers

Butacite	Poly(vinyl butyral)	Du Pont
Elvacet	Poly(vinyl acetate)	Du Pont
Elvanol	Poly(vinyl alcohol)	Du Pont
Saflex	Poly(vinyl butyral)	Monsanto
Vinylite	Poly(vinyl acetate), poly(vinyl alcohol)	Union Carbide

Chapter 14D.   Chlorine-Containing Polymers

Bakelite	Poly(vinyl chloride) and copolymers	Union Carbide
Dynel	Acrylonitrile–vinyl chloride copolymer	Union Carbide
Geon	Poly(vinyl chloride) and copolymers	Goodrich

Name	Probable Composition	Manufacturer
Saran	Poly(vinylidene chloride) and copolymers	Dow
Tygon	Poly(vinyl chloride) and copolymers	U.S. Stoneware
Vinylite	Poly(vinyl chloride) and copolymers	Union Carbide

Chapter 14E. Fluorine-Containing Polymers

Aclar	Polychlorotrifluoroethylene	Allied
Fluon	Polytetrafluoroethylene	ICI Americas
Halar	Ethylene–chlorotrifluoroethylene copolymer	Allied
Halon	Polytetrafluoroethylene	Allied
Kel-F	Polychlorotrifluoroethylene	3M
Kynar	Poly(vinylidene fluoride)	Pennwalt
Tedlar	Poly(vinyl fluoride)	Du Pont
Teflon	Polytetrafluoroethylene	Du Pont
Teflon FEP	Tetrafluoroethylene– hexafluoropropylene copolymer	Du Pont
Tefzel	Tetrafluoroethylene–ethylene copolymer	Du Pont
Viton	Vinylidene fluoride–hexafluoropropylene copolymer	Du Pont

Chapter 15A. Polyamides

Caprolan	6-Nylon (polycaprolactam)	Allied
Capron	6-Nylon (mineral filled)	Allied
Dartek	66-Nylon (polyhexamethylene adipamide) film	Du Pont of Canada
Grilon	612-Nylon copolymer	Emser Industries
Kevlar	Poly(p-phenylene terephthalate)	Du Pont
Minlon	66-Nylon (mineral reinforced)	Du Pont
Nomex	Poly(m-phenylene isophthalate)	Du Pont
Quiana	Poly(bis-p-aminocyclohexylmethane dodecamide)	Du Pont
Rilsan	Nylons 11 and 12	Rilsan
Tynex	Nylons, undisclosed (monofilament)	Du Pont
Vydane	Nylon (glass and mineral reinforced)	Monsanto

Name	Probable Composition	Manufacturer
Zytel	Nylons 6, 66, 610, and copolymers	Du Pont

Chapter 15B. *Polyesters, Polyethers, and Related Polymers*

Name	Probable Composition	Manufacturer
Adiprene	Polyether-based polyurethane (elastomer)	Du Pont
Ardel	Polyarylate	Union Carbide
Bayblend	Polycarbonate–ABS blend	Mobay
Celcon	Acetal copolymer	Celanese
Chemigum	Polyester-based polyurethane (elastomer)	Goodyear
Dacron	Poly(ethylene terephthalate) (fiber)	Du Pont
Delrin	Acetal polymer (polyoxymethylene)	Du Pont
Estane	Polyester-based polyurethane (elastomer)	Goodrich
Gafite	Poly(butylene terephthalate)	GAF
Isoplast	Thermoplastic polyurethane (impact modified)	Upjohn
Kodapak	Poly(ethylene terephthalate)	Eastman
Kodar	Poly(ethylene terephthalate) (glycol modified)	Eastman
Lexan	Polycarbonate	General Electric
Lycra	Polyurethane (fiber)	Du Pont
Merlon	Polycarbonate	Mobay
Montac	Polyester–amide block copolymer	Monsanto
Mylar	Poly(ethylene terephthalate) (film)	Du Pont
Polathane	Polyurethane (elastomer)	Polaroid
Polyox	Poly(ethylene oxide)	Union Carbide
Radel	Polyaryl sulfone	Union Carbide
Rynite	Poly(ethylene terephthalate) (glass reinforced)	Du Pont
Ryton	Poly(phenylene sulfide)	Phillips
Tenite	Poly(ethylene terephthalate)	Eastman
Thiokol	Polysulfide	Thiokol
Udel	Polysulfone	Union Carbide
Valox	Poly(butylene terephthalate)	General Electric
Vibrathene	Polyurethane prepolymers	Uniroyal
Victrex	Polyether sulfone	ICI Americas

Name	Probable Composition	Manufacturer
Xenoy	Poly(butylene terephthalate)–poly-carbonate blend	General Electric

Chapter 15C. Cellulose-Based Polymers

Acele	Cellulose (di)acetate	Du Pont
Ampol	Cellulose esters and ethers	American Polymers
Arnel	Cellulose triacetate	Celanese
Avisco	Viscose and acetate rayons	American Viscose
Cordura	Viscose rayon, high tenacity	Du Pont
Forticel	Cellulose propionate	Celanese
Fortisan	Saponified cellulose acetate	Celanese
Kodapak	Cellulose acetate, propionate, butyrate	Eastman
Tenite	Various cellulosics	Eastman

Chapter 15D. High-Temperature and Inorganic Polymers

Fortafil	Carbon (fiber)	Great Lakes Carbon
Kapton	Polyimide (film)	Du Pont
Noryl	Poly(phenylene oxide)–polystyrene blend	General Electric
Prevex	Poly(phenylene ether)	Borg Warner
Torlon	Polyamide-imide	Amoco Chemicals
Ultem	Polyether-imide	General Electric

Chapter 16A. Phenolic and Amino Resins

Avisco	Urea–formaldehyde	American Viscose
Bakelite	Phenol–formaldehyde	Union Carbide
Beetle	Urea–formaldehyde	American Cyanamide
Durez	Phenol–formaldehyde	Hooker
Plaskon	Phenol, melamine, and urea–formaldehyde	Allied
Plenco	Phenol–formaldehyde	Plastics Engineering

Chapter 16B. Unsaturated Polyester Resins

Gen-Glaze	Unsaturated polyester	General Tire
Selectron	Unsaturated polyester	PPG Industries

Chapter 16C. Epoxies and Polyurethanes

Araldite	Epoxy	Ciba

Name	Probable Composition	Manufacturer
Bayflex	RIM polyurethane	Mobay
Epon	Epoxy	Shell

Chapter 16D. Silicone Polymers

Rimplast	Silicone-based interpenetrating polymer networks, polyamide and polyurethane components	Petrarch Systems
Silastic	Silicone	Dow

Chapter 16E. Miscellaneous Thermosetting Resins

CR-39	Allyl carbonate	PPG Industries
Durez	Diallyl phthalate	Hooker
Plaskon	Alkyd, diallyl phthalate	Allied

AUTHOR INDEX

SUBJECT INDEX

555